D1619247

FREQUENCY ANALYSIS AND PERIODICITY DETECTION IN HEARING

THE PROCEEDINGS OF THE INTERNATIONAL SYMPOSIUM ON FREQUENCY ANALYSIS AND PERIODICITY DETECTION IN HEARING

Held at Driebergen, The Netherlands
June 23-27, 1969

Sponsored by the Advisory Group on Human Factors
on behalf of the N.A.T.O. Science Committee

Organizing Committee

R. Plomp, Soesterberg—Symposium Director
E. de Boer, Amsterdam
B. L. Cardozo, Eindhoven
R. J. Ritsma, Groningen

FREQUENCY ANALYSIS AND PERIODICITY DETECTION IN HEARING

Edited by

R. PLOMP

and

G. F. SMOORENBURG

Institute for Perception RVO-TNO
Soesterberg, The Netherlands

A. W. SIJTHOFF ★ LEIDEN ★ 1970

ISBN 90 218 9021 6

Copyright © 1971, A. W. Sijthoff, Leiden
All rights reserved. No part of this book may be reproduced by any means without written permission from the Publisher.

PREFACE

The question of to which degree frequency and time periodicity of the sound stimulus are preserved in the auditory system as constituent parameters of our sensation has always been of central significance in hearing theory. This question becomes especially acute in the explanation of the origin of pitch: is the pitch of a periodic sound wave derived from the frequencies of its harmonics or from the periodicity of the waveform? Or should we abandon this alternative and accept the fact that both parameters are relevant for pitch perception?

Von Helmholtz proved already 125 years ago that the cochlea can be thought of as a frequency analyzer, and he considered the fundamental of a complex sound to be the physical correlate of pitch. I think, nowadays nobody will criticize the significant role of frequency analysis in hearing, but it has become apparent that the fundamental is of minor importance in pitch perception. Does this mean that pitch is based upon temporal rather than spectral information or should we reconsider the role of the harmonics in this respect? Is the filtering process essential for pitch perception or, on the contrary, does it interfere? Should the fact that the results of many psychoacoustical experiments can be explained so nicely in terms of a periodicity theory be accepted as proof, or must we ask whether there is enough physiological evidence to arrive at that conclusion?

As one may see there were enough questions unsolved to invite the workers in this field to an international symposium in which recent experimental data could be presented and the theoretical significance of those data discussed. The organizing committee felt it to be undesirable to deal with the pitch problem in too narrow a sense. In reconsidering the origin of pitch as a central question of hearing theory the whole area of the possible roles of frequency analysis and periodicity detection in hearing has to be taken into account. For the same reason it was decided that the theme should be approached both from the psychophysical and physiological sides. Any real progress can be expected only from a fruitful cooperation of these two disciplines.

Since the symposium was organized with the main purpose to discuss the controversy of spectral *versus* temporal origin of pitch, the papers were programmed in such an order that this point was touched on in the very first session — in the introductory papers by Dr. H. Spoendlin and Dr. J. F. Schouten — and would recur many times on the following days. To promote this a sort of cyclic treatment of the symposium's theme was chosen: After the introductory papers the physiology of the cochlea and the auditory pathway are treated; frequency analysis receives much attention in these papers. The next contributions, on single-fibre discharges and combination tones, present a good transition to a series of psychoacoustical papers in which periodicity detection plays an important role. The last section on frequency analysis and masking closes the loop. This order of treatment appeared to work out very well.

This volume contains both the papers and the slightly condensed reports of the discussions. We hope that the time spent in editing these discussion reports will be appreciated by the readers. Perhaps better than the papers they reflect the various points of view and arguments, the certainties and uncertainties of the participants.

The discussion reports also show that no definite conclusions could be drawn, that many question marks had to be maintained or even introduced, that much experimental and theoretical work has to be done before the questions on the relative roles of frequency analysis and periodicity detection are cleared up. If both the symposium itself and these proceedings may stimulate and direct further research, the meeting has fulfilled its purpose.

I wish to express my sincere thanks to all who cooperated in organizing the symposium and preparing the proceedings. Without the valuable intellectual support of my colleagues and friends in the Organizing Committee, the substantial financial support of the N.A.T.O. Science Committee, and the actual all-round support of my associates the symposium could not have been successful. I am grateful that the speakers followed our rules in editing their manuscripts and contributions to the discussions, and that a group of young participants was willing to prepare first drafts of the discussion reports. Drs. W. D. Larkin and F. Ph. van Eyl were of great help in reviewing parts of the text of this volume. My special thanks are directed to my associate Guido F. Smoorenburg in sharing the editor's task and to Sijthoff's Publishing Company for the effective cooperation in publishing these proceedings.

Reinier Plomp

CONTENTS

SECTION 1. INTRODUCTORY PAPERS

H. SPOENDLIN	Structural basis of peripheral frequency analysis	2
	Discussion: Smoorenburg, Spoendlin, Goldstein, Zwislocki, Anderson, Møller, Bosher, Johnstone, Honrubia, Wilson, Kohllöffel	37
J. F. SCHOUTEN	The residue revisited	41
	Discussion: Piazza, Schouten, Terhardt, Zwislocki, Johnstone, Whitfield, Keidel, Plomp, Zwicker, Bosher	54

SECTION 2. THE COCHLEAR FUNCTION

W. D. KEIDEL	Biophysics, mechanics and electrophysiology of the human cochlea — REVIEW PAPER	60
	Discussion: Bosher, Schwartzkopff, Keidel, Schügerl	79
B. M. JOHNSTONE K. TAYLOR	Mechanical aspects of cochlear function	81
	Discussion: Wilson, Johnstone, Kohllöffel, Zwicker, Rose, Whitfield, Terhardt, Plomp	90
V. HONRUBIA	Temporal and spatial distribution of the CM and SP of the cochlea	94
	Discussion: Schwartzkopff, Honrubia, Whitfield	105
L. U. E. KOHLLÖFFEL	Cochlear microphonics distribution and spatial filtering	107
	Discussion: Anderson, Kohllöffel, Dallos, Honrubia	116
J. P. LEGOUIX	Experiments on cochlear analysis for transients in the guinea pig	118
	Discussion: Zwislocki, Legouix, Johnstone, Honrubia	124

R. KUPPERMAN The SP in connection with the movements of the 126
basilar membrane

Discussion: Whitfield, Kupperman, Johnstone, 131
Schwartzkopff, Dallos, de Boer

SECTION 3. THE AUDITORY PATHWAY

I. C. WHITFIELD Central nervous processing in relation to spatio- 136
temporal discrimination of auditory patterns—
REVIEW PAPER

Discussion: Keidel, Whitfield, Boerger, Smoorenburg, 147
Klinke, Zwislocki, Goldstein, de Boer, Schwartzkopff,
Rose, Plomp, Cardozo, Schouten

S. KALLERT et al. Two different neuronal discharge periodicities in the 153
acoustical channel

Discussion: Schwartzkopff, Keidel, Ward, Whitfield, 158
Kallert, Zwicker, Møller

R. KLINKE The influence of the frequency relation in dichotic 161
G. BOERGER stimulation upon the cochlear nucleus activity
J. GRUBER

Discussion: Keidel, Klinke, Rose, Scharf, Johnstone, 165
Zwislocki

A. R. MØLLER Two different types of frequency selective neurons in 168
the cochlear nucleus of the rat

Discussion: Goldstein, Møller, Whitfield, Wilson, 173
Scharf

SECTION 4. PERIPHERAL NERVE-FIBRE DISCHARGES

J. E. ROSE Discharges of single fibers in the mammalian auditory 176
nerve—REVIEW PAPER

Discussion: Smoorenburg, Rose, Ward, Whitfield, 188
Schügerl, Plomp, Zwislocki, Goldstein, Schouten,
de Boer, Honrubia

J. E. HIND *et al.*	Two-tone masking effects in squirrel monkey auditory nerve fibers	193
	Discussion: de Boer, Hind, Smoorenburg, Goldstein, Zwicker, Whitfield, Zwislocki	200
E. DE BOER	Synchrony between acoustic stimuli and nerve-fibre discharges	204
	Discussion: Zwislocki, de Boer, Bosher, Zwicker, Møller, Anderson, Johnstone, Whitfield	212

SECTION 5. ORIGIN OF COMBINATION TONES

P. DALLOS	Combination tones in cochlear microphonic potentials	218
	Discussion: Smoorenburg, Dallos, Goldstein, Johnstone, Plomp, Schwartzkopff, Kohllöffel	226
J. L. GOLDSTEIN	Aural combination tones	230
	Discussion: de Boer, Goldstein, Smoorenburg, Helle	245

SECTION 6. PITCH PERCEPTION

R. J. RITSMA	Periodicity detection — REVIEW PAPER	250
	Discussion: Schouten, Ritsma, Smoorenburg, Piazza, de Boer, Terhardt, Whitfield, Scharf, Wilson	263
G. F. SMOORENBURG	Pitch of two-tone complexes	267
	Discussion: Goldstein, Smoorenburg, Zwislocki, Terhardt, Plomp, Zwicker, Schügerl	275
E. TERHARDT	Frequency analysis and periodicity detection in the sensations of roughness and periodicity pitch	278
	Discussion: Plomp, Terhardt, Zwicker, Smoorenburg, Scharf, Schouten, de Boer, Schügerl	287
F. A. BILSEN	Repetition pitch; its implication for hearing theory and room acoustics	291
	Discussion: Gruber, Schwartzkopff, Bilsen, Goldstein, Carterette, Fourcin	300

J. P. WILSON	An auditory after-image	303
	Discussion: Fourcin, Wilson, Bilsen, Johnstone, Ward, Bosher, Zwicker	316
A. J. FOURCIN	Central pitch and auditory lateralization	319
	Discussion: de Boer, Fourcin, Piazza, Plomp, Wilson, Goldstein	326
I. POLLACK	Jitter detection for repeated auditory pulse patterns	329
	Discussion: Rose, Pollack, Goldstein, Zwicker, Terhardt, Zwislocki	336
B. L. CARDOZO	The perception of jittered pulse trains	339
	Discussion: Zwislocki, Cardozo, Kuyper, Smoorenburg, Schouten, Zwicker	347
G. B. HENNING	A comparison of the effects of signal duration on frequency and amplitude discrimination	350
	Discussion: de Boer, Goldstein, Klinke, Henning, Wilson, Bosher, Zwislocki, Zwicker	360
G. VAN DEN BRINK	Experiments on binaural diplacusis and tone perception	362
	Discussion: Zwislocki, van den Brink, Schügerl, Scharf	373

SECTION 7. FREQUENCY ANALYSIS AND MASKING

E. ZWICKER	Masking and psychological excitation as consequences of the ear's frequency analysis — REVIEW PAPER	376
	Discussion: Hind, Zwicker, Plomp, Cardozo, Ritsma	394
R. PLOMP	Timbre as a multidimensional attribute of complex tones — REVIEW PAPER	397
	Discussion: Schouten, Plomp, Sergeant, Bosher, Goldstein, Scharf	411

K. SCHÜGERL	On the perception of concords	415
	Discussion: Kuyper, Schügerl, Ritsma	425
E. C. CARTERETTE M. P. FRIEDMAN J. D. LOVELL	Mach bands in auditory perception	427
	Discussion: Greenwood, Carterette, de Boer, Hind, Zwicker	436
J. J. ZWISLOCKI	Central masking and auditory frequency selectivity	445
	Discussion: Whitfield, Zwislocki, Terhardt	453
B. SCHARF	Loudness and frequency selectivity at short durations	455
	Discussion: Zwislocki, Scharf, Johnstone, Zwicker, de Boer, Dallos	461
L. C. W. POLS	Perceptual space of vowel-like sounds and its correlation with frequency spectrum	463
	Discussion: Fourcin, Pols, Schouten, Carterette, Cardozo, Lindblom, Plomp	470

NAME INDEX 475

SUBJECT INDEX 479

PARTICIPANTS

International Symposium on FREQUENCY ANALYSIS AND PERIODICITY DETECTION IN HEARING, Driebergen, The Netherlands, June 23-27, 1969.

D. J. ANDERSON, Kresge Hearing Research Institute, University of Michigan, Ann Arbor, Mich., U.S.A.
F. A. BILSEN, Applied Physics Department, Delft University of Technology, Delft, The Netherlands.
G. VON BISMARCK, Institut für Elektroakustik, Technische Hochschule München, München, Germany.
E. DE BOER, Physical Laboratory, ENT-Department, Wilhelmina Hospital, University of Amsterdam, Amsterdam, The Netherlands.
G. BOERGER, Heinrich-Hertz-Institut, Berlin-Charlottenburg, Germany.
S. K. BOSHER, Ferens Institute of Otolaryngology, The Middlesex Hospital, Medical School, London, Great Britain.
M. A. BOUMAN, Laboratory for Medical and Physiological Physics, University of Utrecht, Utrecht, The Netherlands.
G. VAN DEN BRINK, Department of Biological and Medical Physics, Medical Faculty Rotterdam, Rotterdam, The Netherlands.
B. L. CARDOZO, Institute for Perception Research, Eindhoven, The Netherlands.
E. C. CARTERETTE, Department of Psychology, University of California, Los Angeles, Calif., U.S.A.
P. DALLOS, Auditory Research Laboratory, Northwestern University, Evanston, Ill., U.S.A.
H. DUIFHUIS, Institute for Perception Research, Eindhoven, The Netherlands.
A. J. FOURCIN, University College London, London, Great Britain.
M. P. FRIEDMAN, Department of Psychology, University of California, Los Angeles, Calif., U.S.A.
J. L. GOLDSTEIN, Research Laboratory of Electronics, Massachusetts Institute of Technology, Cambridge, Mass., U.S.A.
D. D. GREENWOOD, Department of Psychology, University of British Columbia, Vancouver, Canada.
L. M. GROBBEN, Laboratory of Labyrinthology, ENT-Department, University Hospital, Utrecht, The Netherlands.
J. GRUBER, Heinrich-Hertz-Institut, Berlin-Charlottenburg, Germany.
G. F. HAAS, University of California, Los Angeles, Calif., U.S.A.

R. HELLE, Institut für Elektroakustik, Technische Hochschule München, München, Germany.

G. B. HENNING, Defence Research Establishment Toronto, Downsview, Ontario, Canada.

J. E. HIND, Laboratory of Neurophysiology, University of Wisconsin, Madison, Wisc., U.S.A.

G. HOMBERGEN, ENT-Department, University Hospital, Catholic University, Nijmegen, The Netherlands.

V. HONRUBIA, Department of Surgery/Head and Neck (Otolaryngology), UCLA School of Medicine, Los Angeles, Calif., U.S.A.

T. HOUTGAST, Institute for Perception RVO-TNO, Soesterberg, The Netherlands.

B. M. JOHNSTONE, Department of Physiology, University of Western Australia, Nedlands, Australia.

H. R. DE JONGH, Physical Laboratory, ENT-Department, Wilhelmina Hospital, University of Amsterdam, Amsterdam, The Netherlands.

S. KALLERT, I. Physiologisches Institut der Universität Erlangen-Nürnberg, Erlangen, Germany.

T. S. KAPTEIJN, ENT-Department, University Hospital, Free University, Amsterdam, The Netherlands.

W. D. KEIDEL, I. Physiologisches Institut der Universität Erlangen-Nürnberg, Erlangen, Germany.

R. KLINKE, Physiologisches Institut der Freien Universität Berlin, Berlin, Germany.

L. U. E. KOHLLÖFFEL, Neurocommunications Research Unit, University of Birmingham, Birmingham, Great Britain.

R. KUPPERMAN, Laboratory of Labyrinthology, ENT-Department, University Hospital, Utrecht, The Netherlands.

P. KUYPER, Physical Laboratory, ENT-Department, Wilhelmina Hospital, University of Amsterdam, Amsterdam, The Netherlands.

P. J. J. LAMORÉ, Department of Biological and Medical Physics, Medical Faculty Rotterdam, Rotterdam, The Netherlands.

W. LARKIN, Department of Psychology, University of Maryland, College Park, Md., U.S.A.

J. P. LEGOUIX, Laboratoire de Neurophysiologie Générale, Collège de France, Paris, France.

B. LINDBLOM, Speech Transmission Laboratory, Royal Institute of Technology, Stockholm, Sweden.

A. R. MØLLER, Department of Physiology, Karolinska Institutet, Stockholm, Sweden.

R. PIAZZA, Istituto Elettrotecnico Nazionale Galileo Ferraris, Torino, Italy.

R. PLOMP, Institute for Perception RVO-TNO, Soesterberg, The Netherlands.

I. POLLACK, Mental Health Research Institute, University of Michigan, Ann Arbor, Mich., U.S.A.
L. C. W. POLS, Institute for Perception RVO-TNO, Soesterberg, The Netherlands.
R. J. RITSMA, Institute of Audiology, University Hospital, Groningen, The Netherlands.
M. RODENBURG, Department of Biological and Medical Physics, Medical Faculty Rotterdam, Rotterdam, The Netherlands.
J. E. ROSE, Laboratory of Neurophysiology, University of Wisconsin, Madison, Wisc., U.S.A.
B. SCHARF, Department of Psychology, Northeastern University, Boston, Mass., U.S.A.
J. F. SCHOUTEN, Institute for Perception Research, Eindhoven, The Netherlands.
K. SCHÜGERL, Währinger Strasse 145/15, Wien, Austria.
J. SCHWARTZKOPFF, Institut für allgemeine Zoologie, Ruhr-Universität, Bochum, Germany.
D. SERGEANT, College of Education, Froebel Institute, London, Great Britain.
G. F. SMOORENBURG, Institute for Perception RVO-TNO, Soesterberg, The Netherlands.
H. SPOENDLIN, Otorhinolaryngologische Klinik und Poliklinik der Universität, Kantonsspital Zürich, Zürich, Switzerland.
A. SPOOR, ENT-Department, University Hospital, Leiden, The Netherlands.
E. TERHARDT, Institut für Elektroakustik, Technische Hochschule München, München, Germany.
J. TOLK, ENT-Department, University Hospital, Catholic University, Nijmegen, The Netherlands.
W. D. WARD, Hearing Research Laboratory, University of Minnesota, Minneapolis, Minn., U.S.A.
I. C. WHITFIELD, Neurocommunications Research Unit, University of Birmingham, Birmingham, Great Britain.
R. P. WILLIAMS, Department of Physics, University College of South Wales and Monmouthshire, Cardiff, Wales, Great Britain.
J. P. WILSON, Department of Communication, University of Keele, Keele, Staffordshire, Great Britain.
S. J. WRIGHT, Department of Physics, University College of South Wales and Monmouthshire, Cardiff, Wales, Great Britain.
E. ZWICKER, Institut für Elektroakustik, Technische Hochschule München, München, Germany.
J. J. ZWISLOCKI, Laboratory of Sensory Communication, Syracuse University, Syracuse, N.Y., U.S.A.

MAIN ABBREVIATIONS USED IN THIS VOLUME

μ	micron, 10^{-6} m
μsec	microsecond, 10^{-6} second
Å	Ångström, 10^{-10} m
AM	amplitude modulation
AP	action potential
CF	characteristic frequency
CM	cochlear microphonic(s)
CN	cochlear nucleus
cps	cycles per second
CT	combination tone
dB	decibel
DC	direct current
DL	difference limen
FM	frequency modulation
Hz	Hertz, cycles per second
IPI	interpulse interval
IT	invariant tone
JND	just-noticeable difference
MDC	microphonic distortion component
PP	periodicity pitch
P-P	peak to peak
pps	pulses per second
PST	post-stimulus time
PZT	post positive-going, zero-crossing time
Q	quality of resonance system
rms	root-mean-square
RP	repetition pitch
SAM	sinusoidally amplitude modulated
SL	sensation level
SP	summating potential(s)
SPL	sound pressure level re 2.10^{-5} N/m^2 ($= 2.10^{-4}$ dyne/cm^2)
TTS	temporary threshold shift
VT	variant tone

Section 1

INTRODUCTORY PAPERS

A fruitful discussion on the roles of frequency analysis and periodicity detection in hearing requires an adequate knowledge of the structure of the hearing mechanism and of the phenomena to be accounted for. These subjects are introduced in the following two papers. Dr. H. Spoendlin presents anatomical evidence for frequency analysis and Dr. J. F. Schouten reviews the recurrent question of why complex tones have one distinct pitch even when the fundamental is absent.

STRUCTURAL BASIS OF PERIPHERAL FREQUENCY ANALYSIS

H. SPOENDLIN

*Otorhinolaryngologische Klinik und Poliklinik der Universität
Kantonsspital Zürich
Zürich, Switzerland*

SUPPORTING STRUCTURES OF THE ORGAN OF CORTI

The main components of the cochlea, its supporting structures, the sensory cells, nerve endings, and nerve fibres all participate differently in frequency analysis. After the principle of tonotopic localization was generally accepted as an important factor in cochlear frequency analysis, it also became evident that this phenomenon relies on the mechanical properties of the cochlea, especially of the cochlear partition.

The absolute size of the cochlea does not seem to play an important role for frequency range although it improves mechanical frequency discrimination (von Békésy, 1960). While the length of the basilar membrane varies from 5 mm in the chicken to 60 mm in the elephant, the portion occupied by a travelling wave envelope appears to be a constant fraction of the total distance (Greenwood, 1962). The relative width of the basilar membrane from base to apex and the structure of the cochlear partition determine the tonotopic frequency analysis in the cochlea. A number of structural features change considerably from the base to the apex. The width of the basilar membrane increases by as much as a factor of 5 from the base to the apex (0.1-0.5 mm in man). It is also much thicker at the base. Bony spikes of the osseous spiral lamina and the secondary osseous spiral lamina fix the basilar membrane tightly in the basal turn, where these spikes actually reach the area of the inner hair cells. The outer hair cells become increasingly longer and the reticular membrane more inclined in the upper turns. Whereas Claudius- and Hensen-cells form a large cushion-like mass on the basilar membrane in the basal turn, they are small at the apex. Böttcher-cells are only found in the basal coil.

The basilar membrane itself is doubtless the main structure to carry the travelling wave. It is by no means of uniform appearance. Looking at the basilar membrane from above in phase contrast it appears to be composed of radial fibres, which seems to be compatible with von Helmholtz's resonance theory. In the electron microscope the basilar membrane consists of large numbers of very fine radial filaments connected together by a rather dense ground substance. In the pars tecta under the tunnel the filaments and ground substance form one coherent membrane. In the more lateral pars pectinata the membrane splits into upper and lower fibrous layers with independent fibres mainly in the lower layer. But even these fibres are connected by a loose ground substance, thus providing longitudinal links which are a requirement for the occurrence of travelling waves (Iurato et al., 1967). Whereas the longitudinal connection of the basilar membrane is continuous throughout all turns in pars tecta, it becomes increasingly looser toward the apex in pars pectinata and the membrane gradually becomes thinner (Fig. 1). All these features are responsible for the changing mechanical properties of the cochlear partition from base to apex, as, for instance, a change in stiffness with a ratio of 200:1 (von Békésy, 1960). The shape and localization of the travelling wave are entirely determined by the relative dimensions and the mechanical properties of the cochlear partition, no matter where the driving vibration enters the cochlea (von Békésy, 1960).

The supporting structures of the organ of Corti, mainly pillar cells and Deiter-cells, form with their intracellular struts of tubular fibrils and the reticular membrane a solid framework that holds the sensory cells firmly in a stable connection to the basilar membrane. The pillars are inclined in a radial direction, the Deiter-cell extensions in a longitudinal, basal direction whereas the sensory cells are slightly inclined toward the apex. Such a strutting provides a system which is highly distortion-free in all directions. The movements of the basilar membrane are therefore transmitted to the hair cells without much loss by mechanical distortion, and an effective shearing motion occurs between the receptor pole of the hair cells and the tectorial membrane, which has its fixpoint outside the basilar membrane (Fig. 1).

THE SENSORY CELLS

The sensory cells, firmly fixed within this supporting framework, are considered to be the actual mechano-electric transducers. Their number is relatively small (15,000 in man) compared with other sense organs.

Although outer and inner hair cells present a very different intrinsic organization, their receptor poles exhibit basically the same features. From a structural viewpoint, the receptor pole is certainly the most specialized and specific part of the cell. Each outer hair cell carries more than 100 and each inner hair cell only about 50, somewhat coarser, stereocilia. In the outer hair cells the hairs

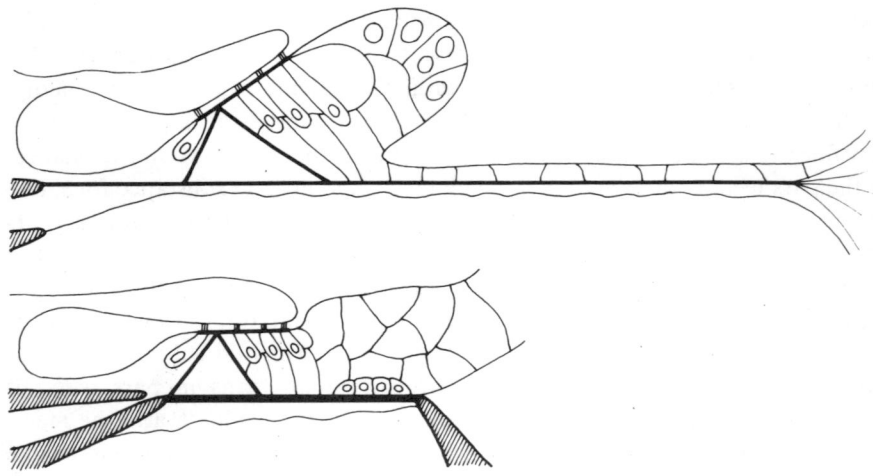

Fig. 1. Schematic representation of the different structural features of the cochlear partition in basal (below) and apical (above) turn.

are arranged in 3 or more rows in the form of a W open toward the modiolus. On top of the inner hair cells there are only 2 rows of full-sized hairs followed by a row of rudimentary hairs, arranged in a very flat W (Fig. 2). The stereocilia have the shape of a club and are 3-5 μ long. The cilia of the outermost row are usually the longest, especially in the inner hair cells. They are exvaginations of the cell surface and are therefore surrounded by the typical three-layered cell unit membrane. The proper substance of the hair is a specific cytoplasmic differentiation. It consists of a dense material with a longitudinal regular fibrillar texture. The cilia are anchored in the cuticular plate with small tubule-like rootlets of very dense material extending upward in the thin neck portion of the hair and deep into the cuticular plate (Fig. 3). The entire stereocilium appears to be a stiff rod even in an unfixed state (Engström et al., 1962).

The upper ends of the hairs are flattened and can be recognized in sections. It has long been debated whether they penetrate into the tectorial membrane or only touch its lower surface. An entire model for frequency analysis in the cochlea has been based on the assumption that the hairs penetrate into the tectorial membrane (Borghesan, 1951). As the electron microscope has demonstrated in all instances so far examined, this is clearly not the case. However, the tips of most hairs not only touch the lower surface of the tectorial membrane but are very slightly embedded in the lower surface of the tectorial membrane (Fig. 4) (Spoendlin, 1966). This connection is very weak since the tectorial membrane can be pulled away easily without damage to the hairs. There is, however, a sufficiently stable connection for the horizontal shearing motions which have been demonstrated to be the effective stimulation movement in all

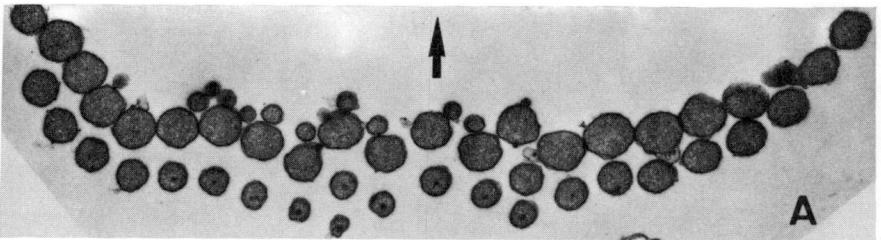
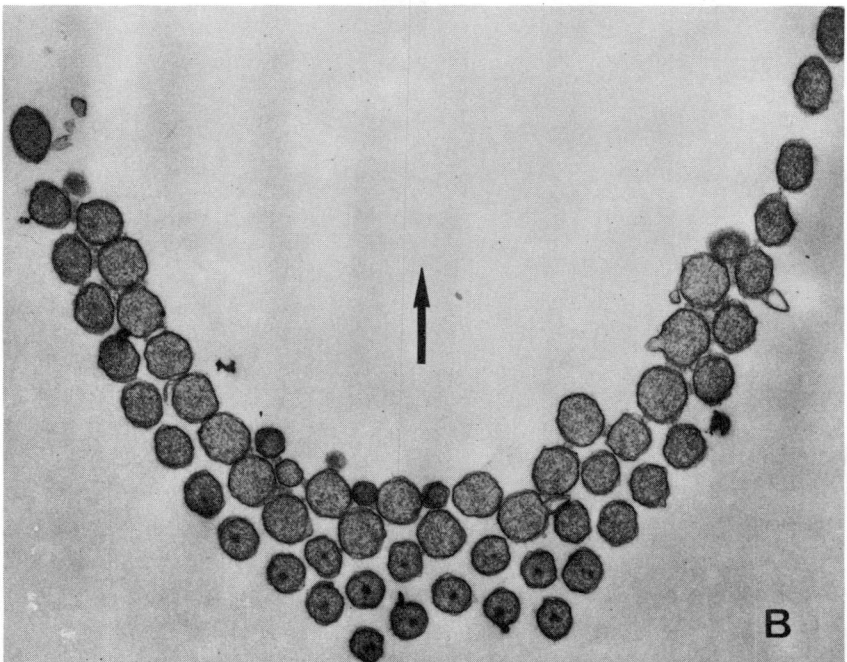

Fig. 2. Horizontal section through the sensory hairs of an inner hair cell (A) and an outer hair cell (B). Arrow points towards modiolus.

inner ear sensory epithelia (von Békésy, 1960; Loewenstein and Wersäll, 1959). The tectorial membrane consists entirely of a spongy extracellular material which appears after fixation as an irregular texture of fine filaments containing mucopolysaccharides. Whereas the stereocilia are rather rigid, the tectorial membrane appears to be more elastic. When it is pulled out from a fresh unfixed cochlea, it behaves like a spring. However, according to von Békésy (1960), the internal friction is high enough that the tectorial membrane behaves as a rigid structure with respect to vibrations.

Even if strict physiological proof may be lacking all the available evidence indicates that the initial mechano-electric transduction takes place at the recep-

Fig. 3. Longitudinal section through one stereocilium. The tubule-like rootlet is anchored in the cuticular plate (C). The cilium is club-like with a narrow neck portion above the cuticular plate and a flattened upper end.

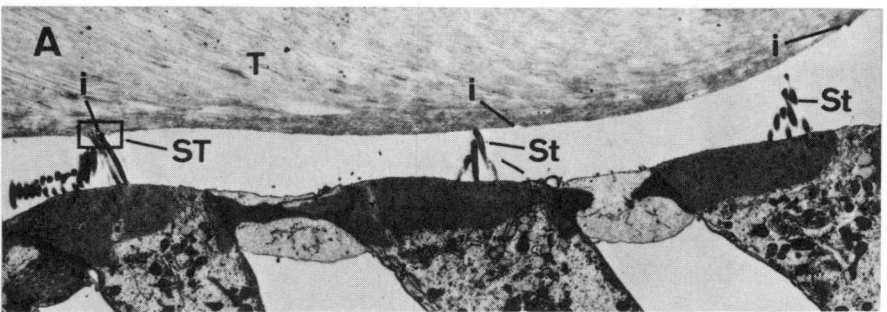

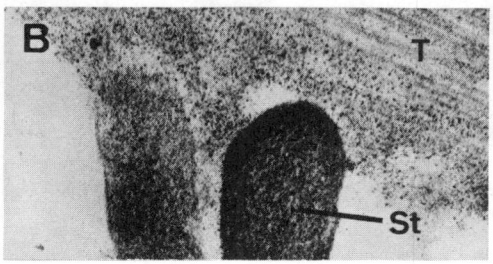

Fig. 4. A: Relation of stereocilia of three outer hair cells with the tectorial membrane (T). The tips of the stereocilia (St) are slightly embedded in the lower surface of the tectorial membrane. During tissue preparation this connection is easily disrupted but the impressions (i) of the tips of the stereocilia in the lower surface of the tectorial membrane are still visible. B: Enlarged detail of figure A with the embedding of the tip of one stereocilium (St) in the lower surface of the tectorial membrane (T).

tor pole of the cell (Fex, 1968). As Engström et al. (1962) proposed, the stereocilia could serve simply as passive levers to transmit shearing forces to the cuticular plate, tilting of which would distort the sensory cell and produce a change in electric resistance and in current. The cuticular plate, however, appears to be tightly fixed all around within the solid framework of the reticular membrane, which would seem to prevent any appreciable tilting (Fig. 6). On the other hand, it is conceivable that the stereocilia with their longitudinally oriented fibrillar ultrastructure have electric properties allowing them to serve as initial mechano-electric transducers. The hypothesis that acid mucopolysaccharides could be primarily responsible for the first step of such mechanisms was mainly suggested by Vilstrup and Jensen (1954) and further developed by Christiansen (1964). Using ruthenium red stain (a specific stain for acid mucopolysaccharides), mucopolysaccharides can be demonstrated between stereocilia (Spoendlin, 1968). Even some structural patterns appear within these layers of mucopolysaccharides, indicating a regular orientation of the molecules (Fig. 5). It is known that acid mucopolysaccharides induce potential changes under mechanical distortion (Vilstrup and Jensen, 1954) and it is therefore conceivable

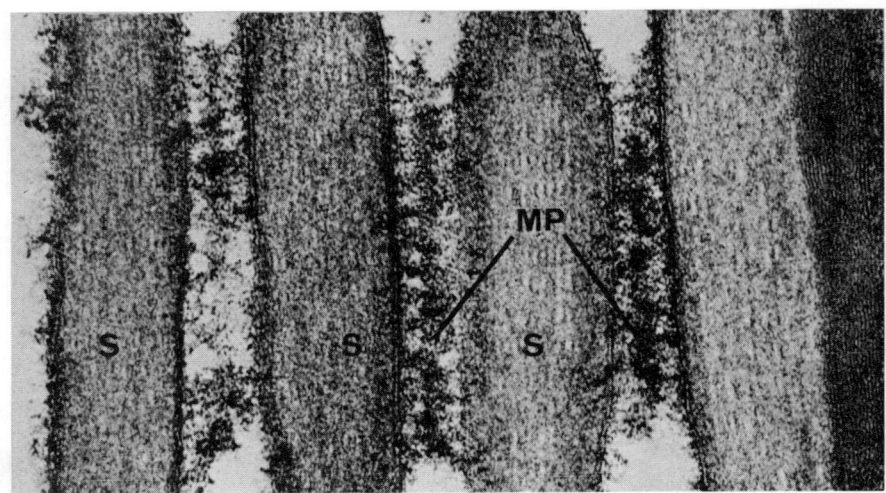

Fig. 5. Portion of some stereocilia (S) stained with ruthenium red. The space between the hairs is filled with acid mucopolysaccharides (MP).

that mucopolysaccharides, which certainly are distorted between the stereocilia under the effect of shearing motion, play a part in the initial transduction. Christiansen (1964) even constructs a model of frequency analysis on the basis of the hyaluronate molecules as regulators, a model which, however, remains entirely hypothetical.

As von Békésy (1960) has shown, the cochlear receptor cells present a specific directional sensitivity. The outer hair cells respond best to radial and the inner hair cells best to longitudinal shearing motion, a selectivity that considerably improves the mechanical frequency analysis in the cochlea (Tonndorf, 1962). Looking for a structural basis of such a directional sensitivity of the sensory cells, a number of features have to be taken into account. In the vestibular sensory epithelia the directional sensitivity corresponds to a morphological polarization of the sensory cells; each of these carries about 100 stereocilia and one single kinocilium at one pole of the cell surface. The position of the kinocilium always points in the direction of the cell's maximum sensitivity to shearing motion. Kolmer (1927) gave an early description of a kinocilium in the cochlear hair cells, but his observations were not confirmed by electron microscopy. Nevertheless there is always an opening in the cuticular plate of the cochlear hair cells located at the distal pole of the cell surface at the base of the stereociliar "W". In some animals, as the guinea pig, a kinociliar basal body is regularly found within this cuticular free area (Flock et al., 1962). On the basis of such findings and in view of the great functional significance of modified kinocilia in other sense organs, the basal bodies were considered by Engström

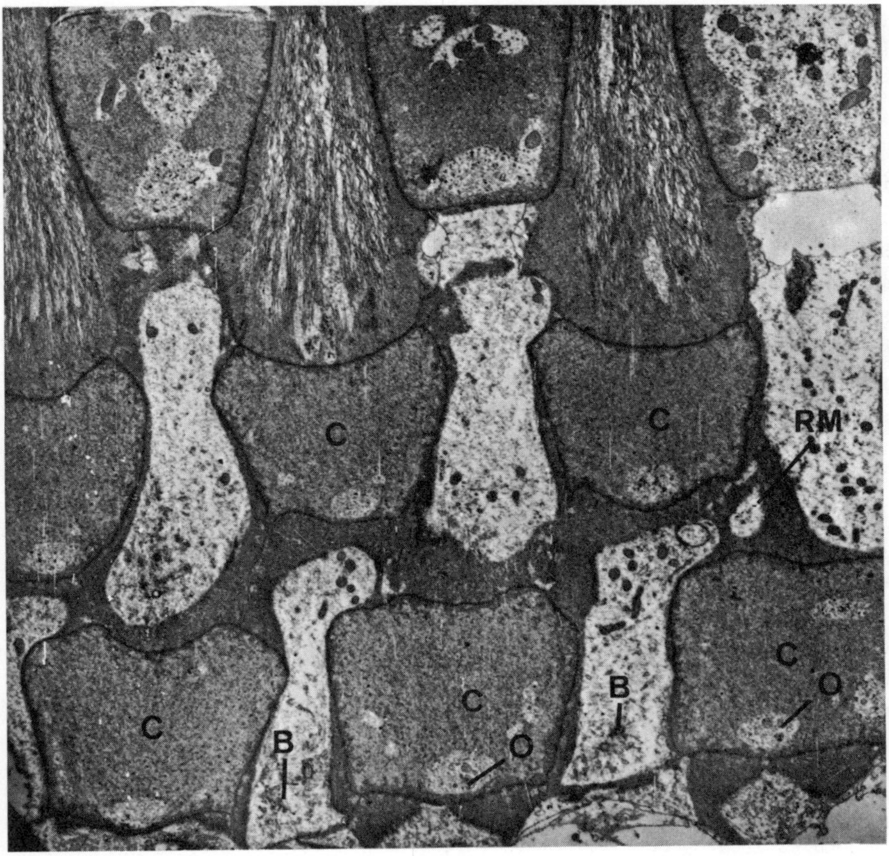

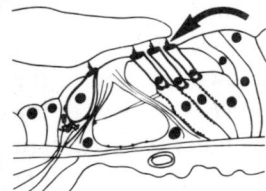

Fig. 6. Horizontal section through cuticular plates (C) of outer hair cells and reticular membrane (RM). Basal bodies (B) are only seen in the supporting cells and not in the openings (O) of the cuticular plates.

et al. (1962) as the essential excitable structure of the hair cells. There are, however, a number of reasons which seem to exclude a direct involvement of the basal body in the mechano-electric transduction in the receptor. The tight fixation of the cuticular plate in the supporting framework (Spoendlin, 1966) prevents its movement; thus, distortion of the sensory cell, including the kinociliar basal body, would not seem to be effective. Moreover in the adult cat, the kinociliar basal body seems to be completely absent (Fig. 6). Only in very young kittens, we were able to find such basal bodies. Thus it is very unlikely that the basal body as such is involved in the receptor mechanism. Its presence in developing animals, its disappearance in adult life, and its definite, constant position seem to indicate an important role in the differentiation of the a-symmetric receptor pole during cell development (Spoendlin, 1968).

The asymmetric organization of the receptor pole as such is probably an important structural basis of its directional sensitivity (Fig. 2). So far we have not been able to find any signs of intrinsic structural polarization in the proper substance of the stereocilia. If the acid mucopolysaccharides between the stereocilia are important for the initial mechano-electric transduction, the asymmetric arrangement of the hairs alone could account for the directional sensitivity of the sensory cells. In the outer hair cells there is a clear tendency of radial, and in the inner hair cells of longitudinal arrangement of the rows of stereocilia. Mechanical distortion of the links between the hairs is maximum if the rows are oriented in the direction of the bending of the hair, and minimum if they are oriented perpendicular to it. Since the stereociliar rows in the inner hair cells have a predominantly longitudinal and those in the outer hair cells a predominantly radial orientation, the respective longitudinal and radial directional sensitivity of inner and outer hair cells could be explained.

The cell body of the hair cells, with its cytoplasmic components such as mitochondria, membranes and ribosomes, probably has a primarily metabolic function to maintain the negative intracellular potential and to transmit the receptor potential (cochlear microphonics) for the excitation of the afferent nerve endings. It is not directly involved in frequency detection in the cochlea.

THE INNERVATION OF THE ORGAN OF CORTI

The processing of the receptor potential to the action potential in the cochlear nerve fibres, which includes the coding of the acoustic message to the brain, is the most elaborate and complicated, but also the least known function of the cochlea. Direct electrophysiological data on the initial nervous activity in the organ of Corti are very scarce. However, the study of the innervation pattern of the cochlear hair cells might give us some cues to understand this important step in hearing.

For the following reasons our studies have been carried out mainly in the cat: The acoustic system of this animal is best known from an electrophysiological viewpoint and its organ of Corti happens to be most favourable for investigations of the innervation pattern mainly because efferent and afferent nerve fibres show clear-cut differences in their distribution pattern and synaptic connections. In other animals such as the guinea pig, these differences appear to be less systematic, which makes a differentiation of afferent and efferent fibres more difficult and explains some divergent results published by other investigators (Smith and Rasmussen, 1963, 1965; Engström, 1968). Our results to be presented are based on consistent findings in 59 cats and 20 guinea pigs the ears of which have been extensively studied mostly by electron microscopy under normal conditions and after different nerve lesions (Fig. 8).

All fibres lose their myelin sheaths before they enter the organ of Corti, where they expand either in a spiral (spironeurons) or a radial direction (orthoneurons). According to these principles, the following groups of fibres are distinguished (Lorente de Nó, 1937): the radial fibres to the inner hair cells and the tunnel crossing radial fibres to the outer hair cells; the internal spiral bundle, the tunnel spiral fibres and the outer spiral fibres (Fig. 7).

Afferent and efferent nerve fibres participate in the innervation of the organ of Corti. Rasmussen (1946) showed that the olivo-cochlear bundle brings an efferent nerve supply to the cochlea from the contralateral and homolateral superior olives, reaching the cochlea with the vestibular nerve and through the anastomosis of Oort (Fig. 8).

The efferent terminals in the organ of Corti are characterized by a relatively large size, by a great number of synaptic vesicles and by a postsynaptic cisterna along the postsynaptic membrane in the outer hair cells. At the base of the outer hair cells the efferent endings contact directly the sensory cells, whereas a direct contact with the inner hair cells is rare (Figs. 9, 20). After selective

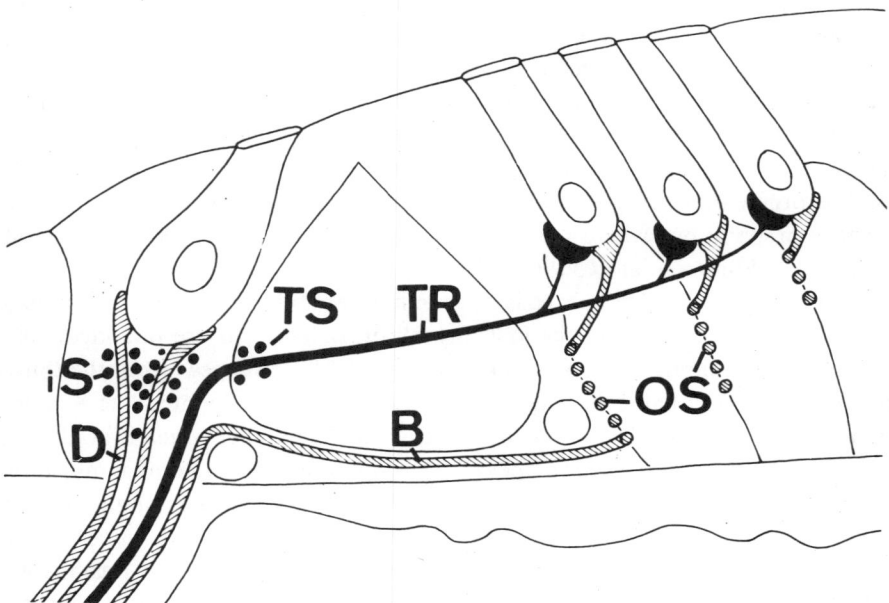

Fig. 7. Schematic representation of different groups of nerve fibres in the organ of Corti. The efferents are drawn in black. D: radial fibres to inner hair cells; iS: internal spiral fibres; TS: tunnel spiral fibres; TR: tunnel radial fibres; B: basilar fibres; OS: outer spiral fibres.

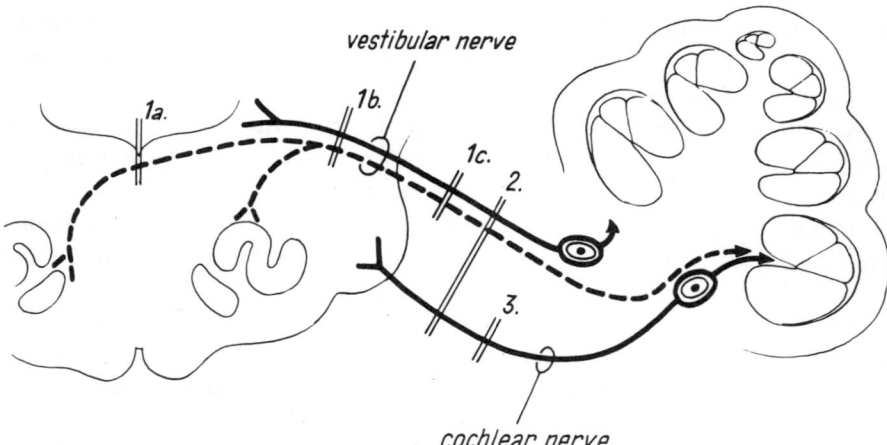

Fig. 8. Outline of experimental lesions, which have been carried out on cats. 1a-1c show the possible sites of interruption of the olivo-cochlear fibres; 2 the total transsection of the eighth nerve, and 3 a selective lesion of the cochlear nerve.

transsection of the entire olivo-cochlear bundle in the vestibular root (Iurato, 1962; Spoendlin and Gacek, 1963; Smith and Rasmussen, 1963) or of the contralateral bundle at the floor of the IVth ventricle (Kimura and Wersäll, 1962), it could be demonstrated in different animals that many of these large vesiculated nerve endings degenerated (Fig. 10). In the cat we were able to show that practically all vesiculated endings at the outer hair cells as well as all inner and tunnel spiral fibres disappeared after complete transsection of the olivo-cochlear fibres in the vestibular nerve (Spoendlin, 1966) (Figs. 10-12). This brings up the surprising fact that there is an enormous efferent nerve supply of the organ of Corti. This also can be demonstrated in light-microscopic surface preparations where the efferents are selectively stained by the method of Maillet (Engström et al., 1966) (Fig. 13).

The first row of outer hair cells has the most abundant efferent nerve supply throughout the cochlea. In the first turn all outer hair cells are provided with efferent nerve endings. In upper turns they gradually disappear from the third and second row of hair cells (Fig. 13). In serial sections we counted 6 to 8 nerve endings per outer hair cell in the lower basal turn, a number which is gradually reduced towards the cochlear apex. The roughly estimated total number of efferent nerve endings at the outer hair cells in the cochlea is about 40,000 (Spoendlin, 1966). All these endings together with several hundred inner spiral fibres originate from the olivo-cochlear bundle which, according to Rasmussen (1960), consists of only about 500 neurons. This necessitates an extensive ramification of the efferent nerve fibres, which has been demonstrated in the osseous spiral lamina (Nomura and Schuknecht, 1965), in the inner spiral plexus and below the outer hair cells (Fig. 9A) (Spoendlin, 1966). At the level

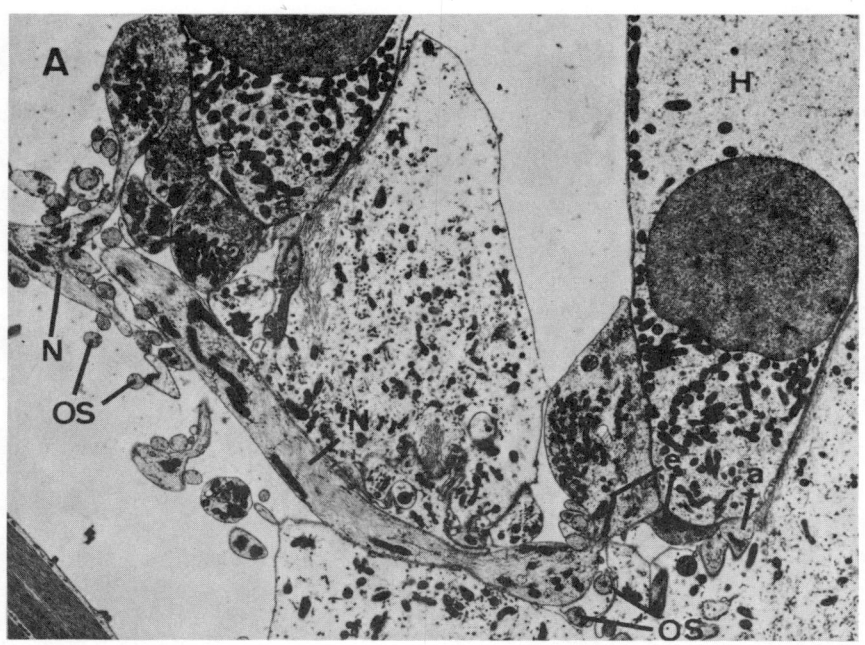

Fig. 9. A: Bases of 2 outer hair cells (H) with coarse efferent nerve fibres (N), efferent nerve endings (e) and a few afferent nerve terminals (a). OS: outer spiral fibres. B: Detail of receptor neural junction of an outer hair cell with efferent (e) and afferent (a) nerve endings and synaptic membrane differentiations. Ci: postsynaptic cisterna.

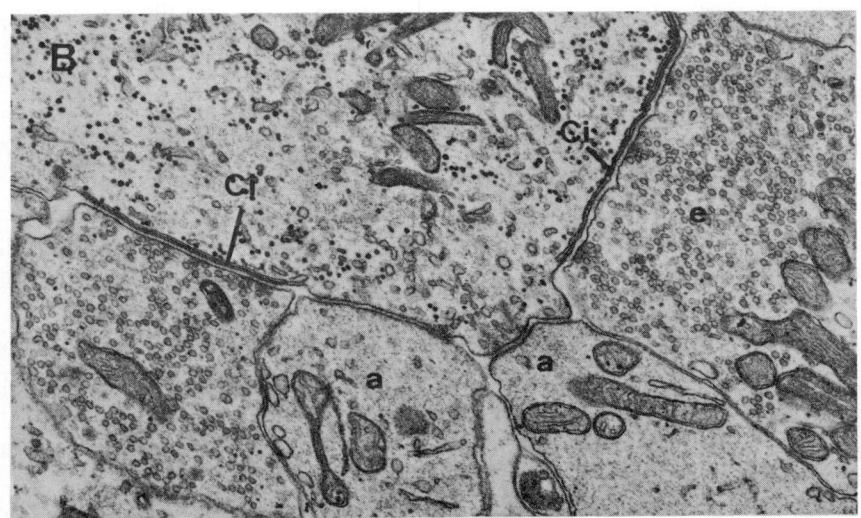

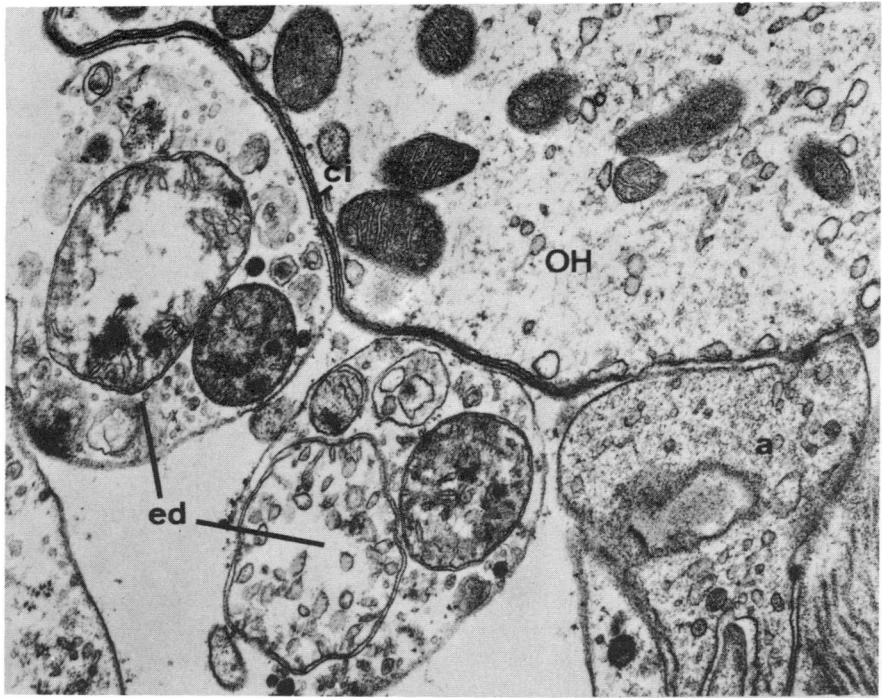

Fig. 10. Base of an outer hair cell (OH) four days after transsection of the olivo-cochlear fibres in the vestibular nerve. Two efferent nerve endings (ed) are in obvious degeneration whereas the afferent ending (a) looks still normal. The postsynaptic cisterna (ci) at the junction of efferent endings with the hair cell are still present and of normal appearence.

of habenula perforata only a very small percentage of all nerve fibres is efferent. The number of all nerve fibres which run through each habenular opening ranges in normal animals from less than 10 to over 60. The average is between 20 and 30 (Fig. 30b). This number is only slightly smaller after elimination of the efferent nerve supply.

The nerve fibres destined for the outer hair cells cross the tunnel either at a middle level, entirely free of any sheath, in fascicles of 2 to 6 fibres of diameters from 0.3 to 1.5 μ, or at the bottom as basilar fibres (Fig. 14). It appears that all upper tunnel radial fibres in the cat belong to the efferent system since they all disappear after selective transsection of the olivo-cochlear bundle. After elimination of the efferent fibres there is, however, no appreciable reduction of the number of outer spiral fibres, which allows the conclusion that the distribution of the efferent fibres for the outer hair cells is predominantly radial, whereas the efferents in the internal spiral plexus expand in a spiral direction (Figs. 32, 33).

Three-fourths of the olivo-cochlear fibres originate from the contralateral and only 1/4 from the homolateral superior olive (Rasmussen, 1960). According to Iurato (1964) the contralateral fibres lead in the rat almost exclusively to the outer hair cells whereas the homolateral bundle leads to the inner spiral fibres as well as to the outer hair cells. We studied the distribution of the contralateral fibres in five cats 2 to 21 days after midline lesions. Already 12 hours after the

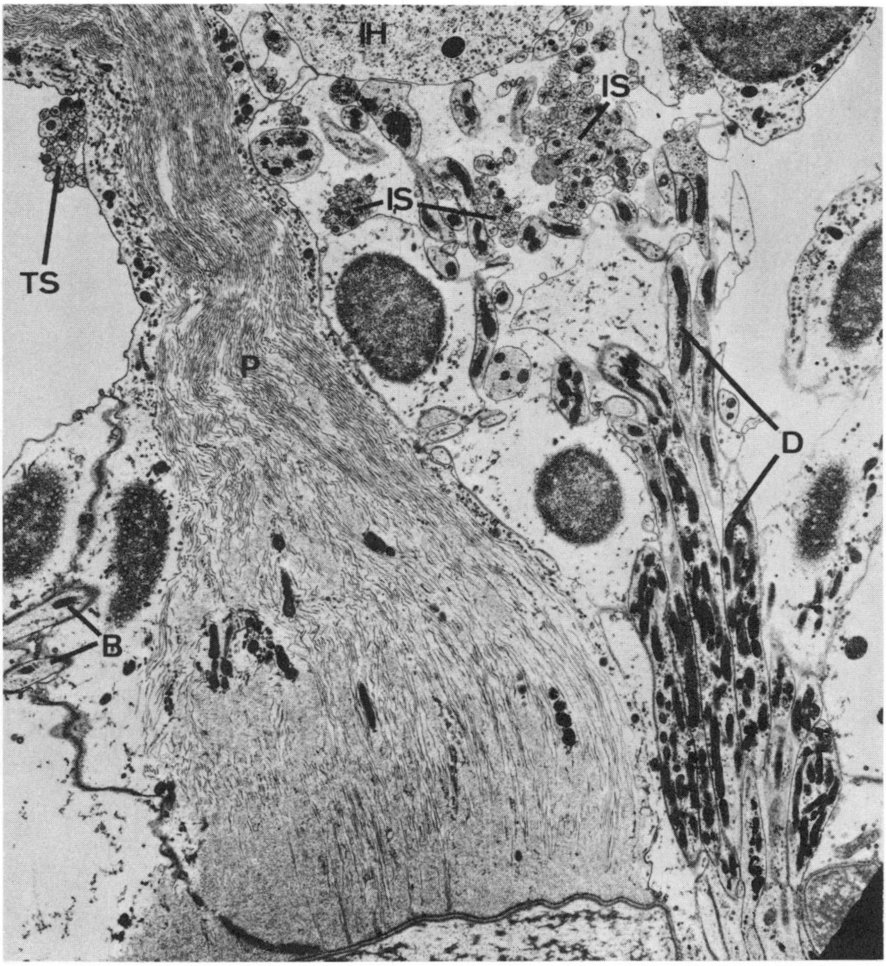

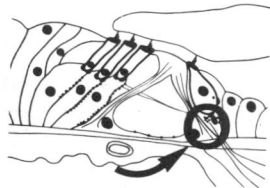

Fig. 11. Area below inner hair cell (IH) of normal cat. Afferent fibres (D) to inner hair cells and basilar fibres (B) between inner pillar cells (P). Tunnel spiral fibres (TS) and inner spiral fibres (IS) disappear after degeneration of the olivo-cochlear fibres.

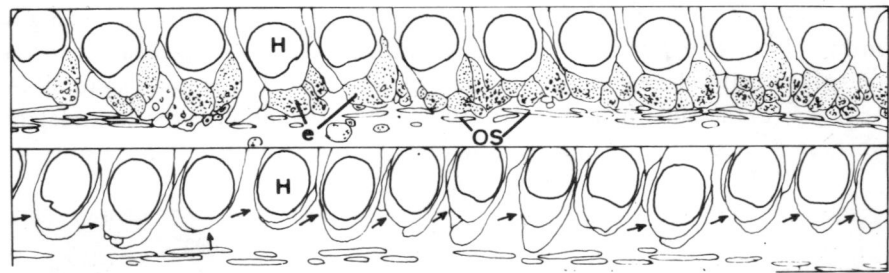

Fig. 12. Longitudinal section through the first row of outer hair cells (H).
Above: normal, large number of efferent nerve endings (e). OS: outer spiral fibres.
Below: 3 weeks after transsection of the olivo-cochlear fibres. All efferent endings are gone (arrows).

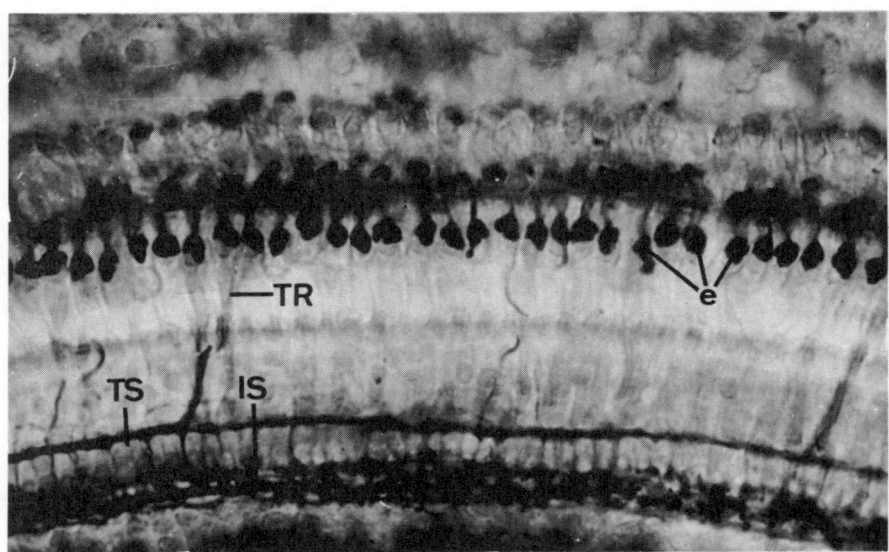

Fig. 13. Surface preparation of the organ of Corti of a cat with maillet stain (upper basal turn). All efferent nerve fibres and nerve endings (e) are black. IS: inner spiral fibres; TS: tunnel spiral fibres; TR: tunnel radial fibres.

lesion obvious signs of initial degeneration appeared in some nerve endings at the outer hair cells. After 4 days desintegration of the nerve endings started and continued for 2 weeks (Fig. 15). The majority but not all nerve endings at the outer hair cells were affected (Fig. 29). If there were, for instance, an average of 60 efferent nerve endings in a row of 8 outer hair cells, only 2 to 5 remained after degeneration of the contralateral olivo-cochlear fibres. This means that less than 10% of the efferent nerve endings at the outer hair cells originate from the homolateral olivo-cochlear bundle. Iurato (1968) found no reduction in

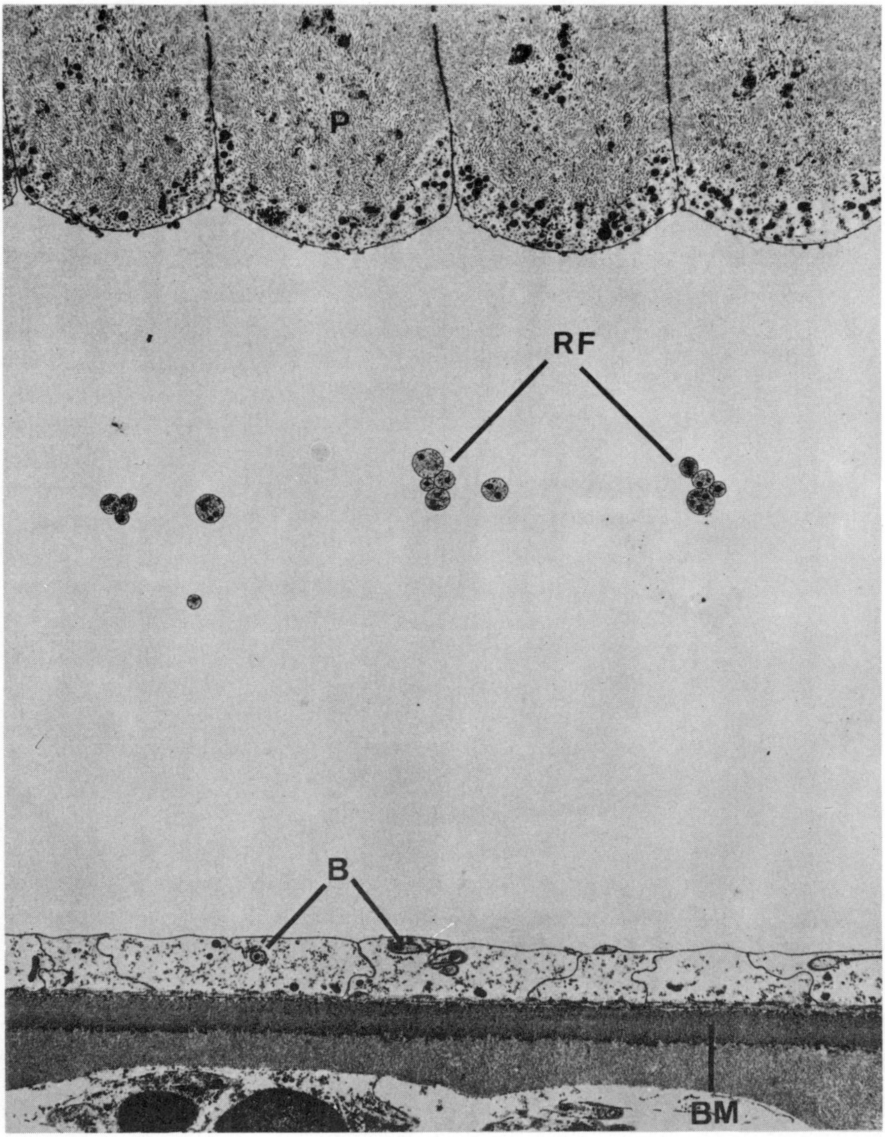

STRUCTURAL BASIS OF PERIPHERAL FREQUENCY ANALYSIS 17

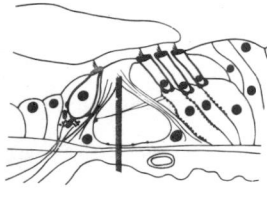

Fig. 14. Tangential section through the tunnel showing the fascicles of upper tunnel radial fibres (RF) which run entirely free in small groups through the tunnel. At the base of the tunnel embedded in the cytoplasm of the pillar cells the basilar fibres (B) are seen. Basilar membrane (BM) and pillar-heads (P).

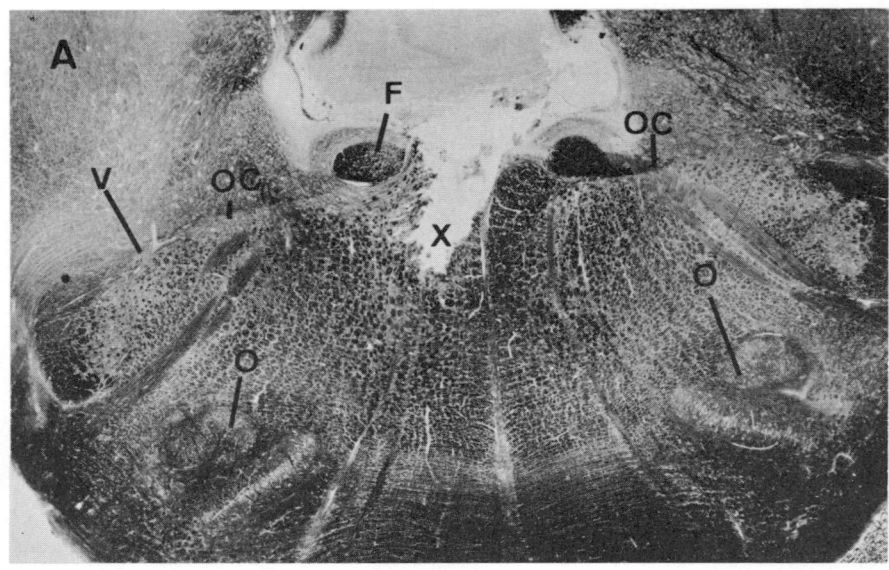

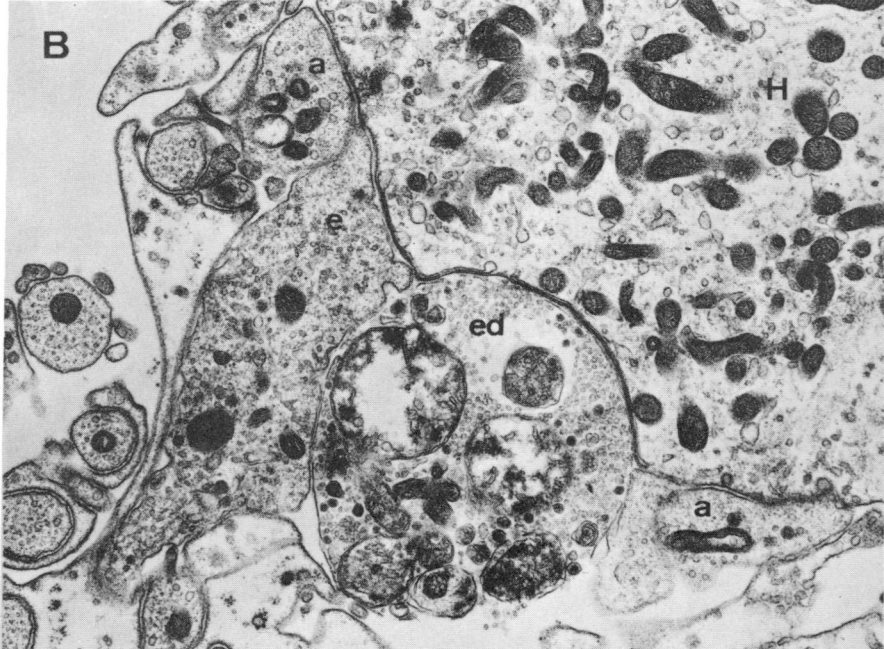

Fig. 15. Cat 4 days after midline lesion.
A: Brainstem showing the lesion (X) between the facial colliculi (F), the superior olive (O), the olivo-cochlear fibres (OC) and the vestibular root (V).
B: Base of an outer hair cell (H) with one intact (e) and one degenerating (ed) efferent ending. a: afferent terminals.

number of the internal spiral fibres after midline lesions in the chinchilla. In our cats we observed, however, a significant reduction of the inner spiral fibres to about half the normal average number (Figs. 16-18). The degree of reduction, however, varied widely from animal to animal. Although the number was clearly reduced, we rarely found signs of degeneration and there were no fibres

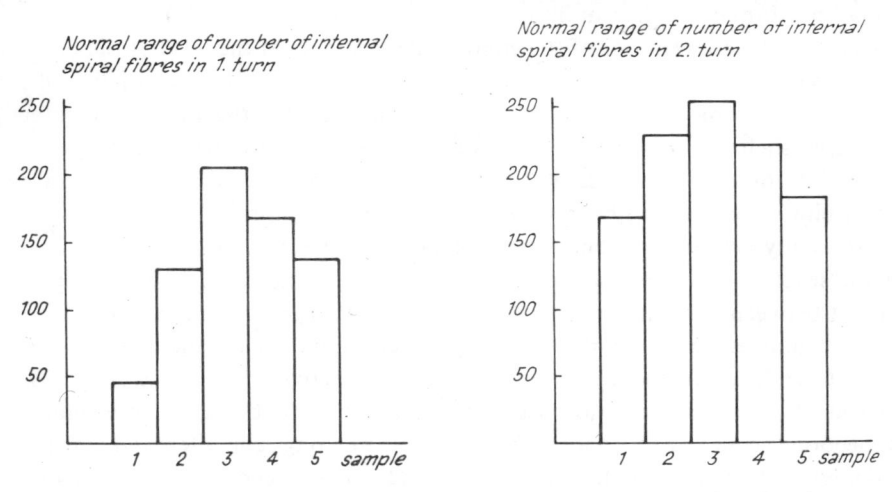

Fig. 16. The internal spiral fibres have been counted in five samples of the basal turn and five samples of the second turn in normal cats. The number of internal spiral fibres varies greatly in the basal turn and is more regular in the second turn.

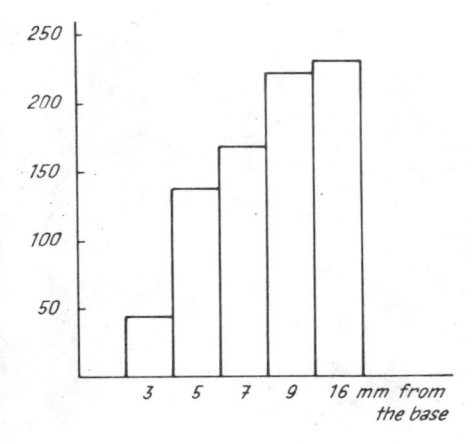

Fig. 17. Number of internal spiral fibres in different parts of the cochlea in the cat. There is a marked increase in numbers of the internal spiral fibres from the lower basal to the second turn of the cochlea.

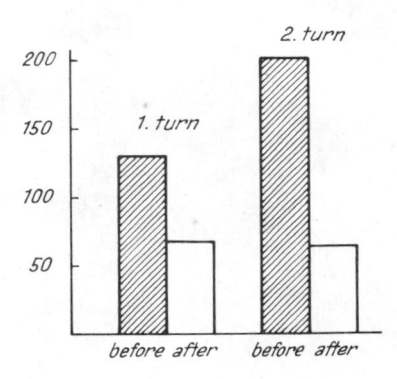

Fig. 18. Average number of internal spiral fibres before and after transsection of the crossed olivo-cochlear fibres (average of 6 ears). There is a significant reduction of the number of internal spiral fibres after transsection of the olivo-cochlear fibres.

in an actual state of degeneration. It might be that, after degeneration has started, the degenerating fibre disappears very fast and escapes observation. Thus, in the cat, the contralateral olivo-cochlear fibres provide most of the efferent innervation for the outer hair cells and about half of the inner spiral fibres, whereas the few homolateral fibres contribute the other half of inner spiral fibres and only a small portion of the efferent fibres to the outer hair cells (Fig. 33).

The synaptic connections of the efferent nerve supply at the level of the outer hair cells differ basically from the connections at the level of the inner hair cells. In the cat morphological evidence for synaptic activities at the level of the outer hair cells, such as agglomeration of synaptic vesicles at the presynaptic membrane, membrane thickenings and postsynaptic membrane differentiations, are found almost exclusively between nerve endings and hair cells and only very exceptionally between efferent nerve endings and afferent endings or dendrites. At the level of the inner hair cells the opposite is the case: direct synaptic contact between efferent elements and hair cells is a rare exception whereas an intimate and extensive contact of the efferent fibre enlargements with the afferent dendrites, including evidence for synaptic activity, is the rule (Fig. 19). (In rodents this differentiation seems to be less clear-cut insofar as synaptic connections occur more frequently between nerve endings and afferent dendrites at the level of the outer hair cells on one hand and between efferent fibres and

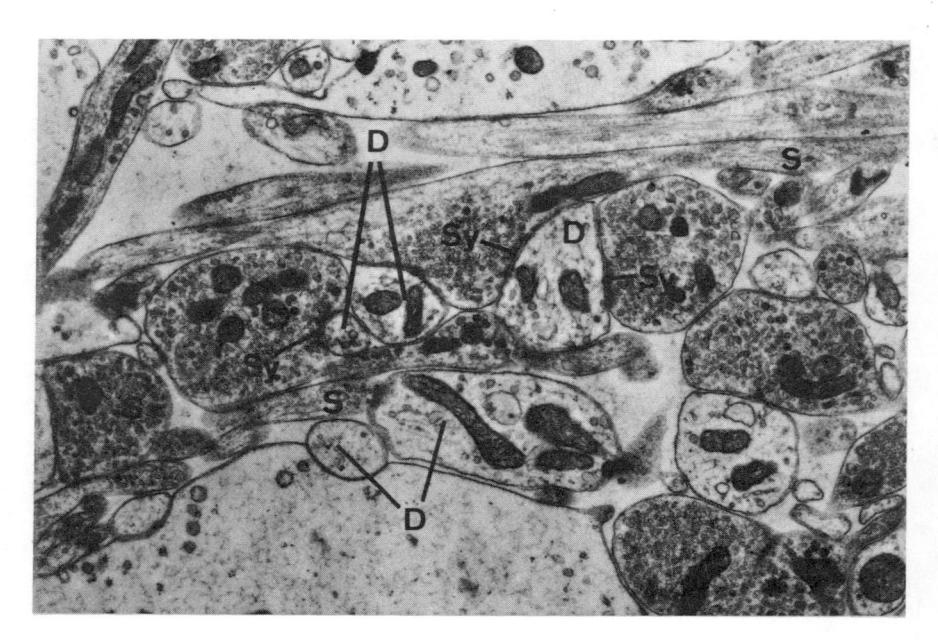

Fig. 19. Horizontal section through a portion of the inner spiral plexus. There are synaptic vesicle agglomerations (Sy) at sites of contact between efferent internal spiral fibres (S) and afferent dendrites (D).

inner hair cells on the other hand.) Since inhibition is so far the main demonstrated effect of the olivo-cochlear fibres (Galambos, 1956; Desmedt, 1962; Fex, 1962), these anatomical relationships indicate that we are dealing with presynaptic inhibition in the outer hair cells and postsynaptic inhibition at the dendrites to the inner hair cells (Fig. 20).

The distribution of the afferent nerve supply to the organ of Corti is best studied after elimination of the efferent innervation, which is relatively easily achieved by cutting all efferent fibres, together with the vestibular nerve, in the internal acoustic meatus without damaging the cochlear nerve and the blood supply to the cochlea. Strangely, it appears to be impossible to induce degeneration of the afferent dendrites to the outer hair cells by transsection of the cochlear nerve even though the spiral ganglion cells degenerate promptly (Spoendlin and Gacek, 1965). After efferent innervation is eliminated the following fibres with their endings at the hair cells remain: radial fibres to the inner hair cells, basilar fibres, and outer spiral fibres along with their terminal collaterals to the outer hair cells (Fig. 27). However, as mentioned above, all upper tunnel radial fibres are gone completely (Fig. 27). The only remaining afferent dendrites leading to the outer hair cells are the basilar fibres which cross the tunnel at the bottom in a slightly spiral direction, penetrate between the outer pillar bases, turn basalward and approach gradually, as outer spiral fibres, the base of the outer hair cells. These afferent dendrites can be counted in tangential sections as they penetrate between the outer pillars. Between every two pillar cells one finds 1-3 nerve fibres or at some places no fibres at all (Fig. 21). From a number of cats, tangential sections of the first and second turns covering a total distance of 200 pillars have been carefully examined. We found 190 fibres per 200 pillar cells or approximately 1 nerve fibre per pillar. If these data can be extrapolated to the entire cochlea, there are only 3000 to 4000 such fibres; an astonishingly small number of afferent dendrites destined for the outer hair cells. This would

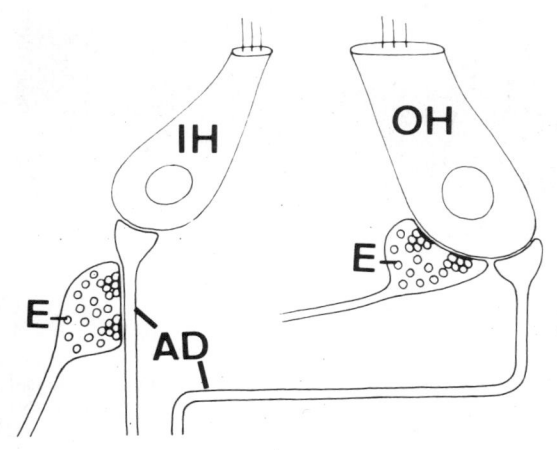

Fig. 20. Schematic representation of efferent synaptic connections in the organ of Corti of the cat. At the outer hair cells (OH) synaptic contacts are almost exclusively with the sensory cell and at the inner hair cells (IH) only with the afferent dendrites (AD). E: efferent ending.

imply that the majority of the 50,000 cochlear neurons lead to inner hair cells. Several doubts have been expressed about the reliability of these data (Eldredge, 1967; Fex, 1968).

The crucial question is whether all upper tunnel radial fibres are efferent and we took considerable care to prove this point (Spoendlin, 1969a). In 10 animals

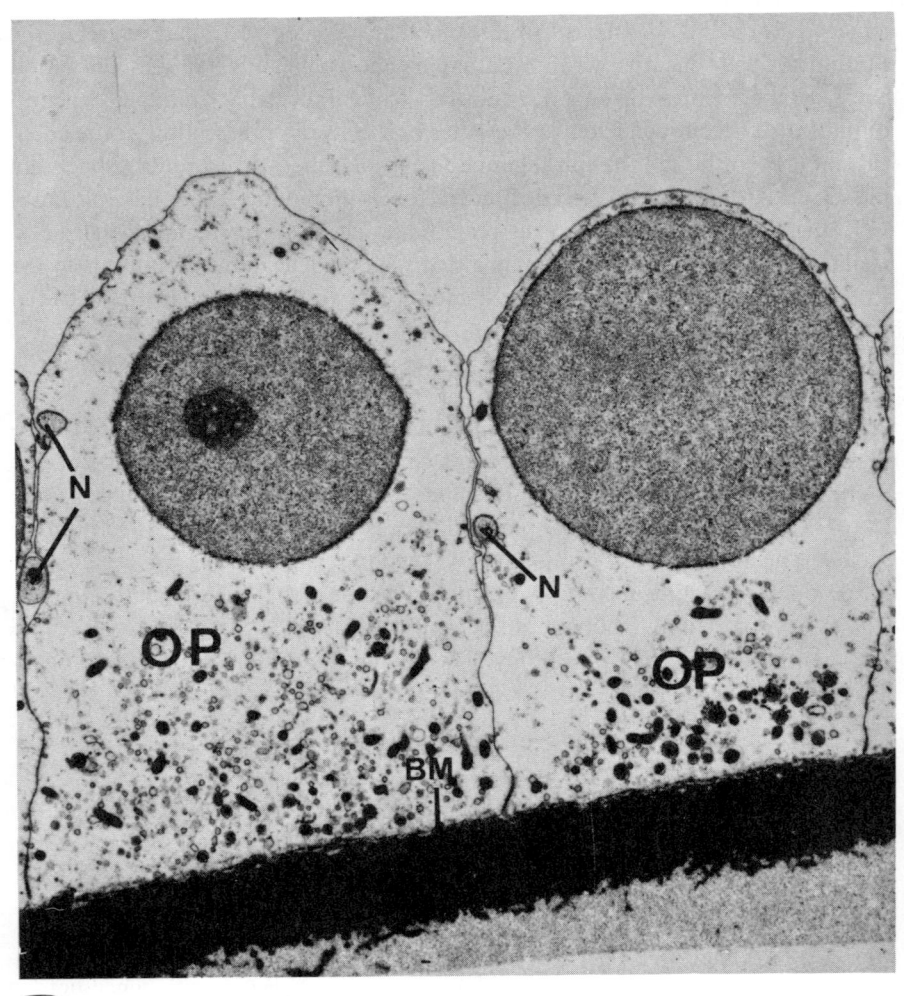

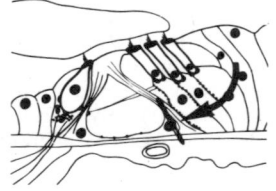

Fig. 21. Longitudinal tangential section through the basal portion of outer pillars (OP) with basilar nerve fibres (N) penetrating between them. BM: basilar membrane.

the olivo-cochlear fibres were sectioned in the vestibular nerve of one side. The efficiency and selectivity of the lesion were checked by examination of the spiral ganglion cells and the intraganglionic spiral bundle. The best way to check the presence or absence of upper tunnel radial fibres is by examination of the organ of Corti in tangential sections through the tunnel, where the fascicles of radial fibres crossing the tunnel are clearly seen in cross section (Figs. 14, 23). Even in phase contrast preparations they are easily recognized as black spots in the middle of the tunnel. In this way 6 cochleas have been evaluated, carefully taking several sections of each turn covering about 1/3 of the total cochlear length (Fig. 22). In the remaining 4 animals the absence of upper tunnel radial fibres was noticed in the course of routine examination. We found no remaining upper tunnel radial fibres after degeneration of the olivo-cochlear fibres. In the cat, therefore, it seems to be a fact that the totality of the upper tunnel radial fibres are efferent and that the basilar fibres are the only afferent fibres for the outer hair cells.

Further evidence for such a predominant nerve supply to the inner hair cells is provided by reconstructions of the area between the habenula perforata and the inner hair cells in animals where only the afferent dendrites are present; the efferents having been eliminated by transsection and degeneration of the olivo-cochlear fibres. In a series of several hundred sections, an individual

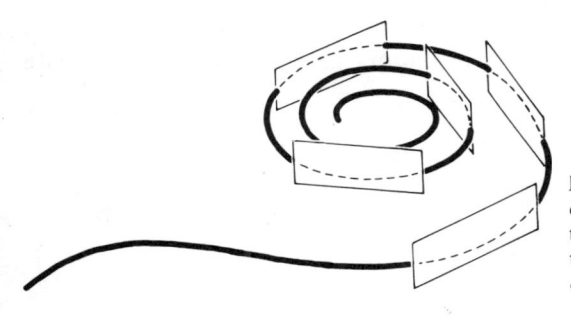

Fig. 22. Schematic representation of several tangential sections through the organ of Corti, taken to evaluate the presence or absence of tunnel radial fibres.

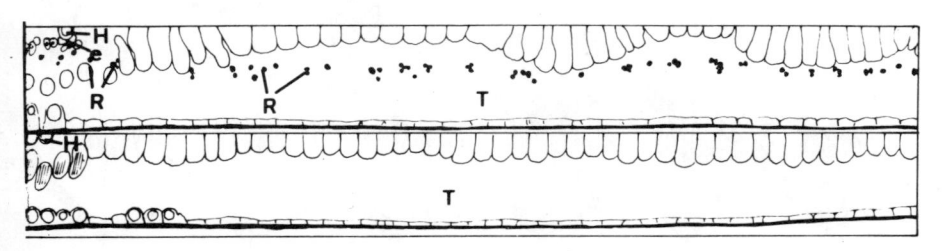

Fig. 23. Tangential section through the tunnel (T).
Above: Normal state; the upper tunnel radial fibres (R) are clearly seen in cross section. In the left corner an outer hair cell (H) with efferent endings (e) are noticed.
Below: 3 weeks after transsection of the olivo-cochlear bundle; all tunnel radial fibres are gone.

fibre can be followed from the habenula to its ending at the inner hair cell or to its passage between the inner pillars. For better demonstration we made a graphic construction of the area of the inner hair cells on the basis of such serial sections. Six representative sections of these series were chosen to reconstruct the area in a drawing. The sectional plans used to make this reconstruction are exact copies of the original electron micrographs from the series. The fibres enter the organ of Corti in well-packed bundles of about 20 fibres per habenular opening. These nerve bundles are maintained for a short distance, after which the fibres disperse between the supporting cells. The majority continue in an essentially radial direction to the closest inner hair cell (Fig. 24). Only very few fibres take a short spiral course before they penetrate between the inner pillars to continue as basilar fibres or to end at one of the neighbouring inner hair cells. In reconstructions covering the area of 5 inner hair cells we found of 20 afferent fibres entering the organ of Corti, only one or two turned toward the outer hair cells. These numbers are consistent with the number of basilar fibres that pass between the outer pillars. The great majority of all fibres end unbranched with a single nerve ending at the inner hair cells (Fig. 25). Each ending presents a typical synaptic complex characterized by a marked thickening of the postsynaptic membrane, a minor accentuation of the presynaptic membrane, and a pronounced synaptic bar surrounded by vesicles within the sensory cell cytoplasm (Fig. 25) (Smith and Sjöstrand, 1961).

All these observations give, I think, convincing evidence that, contrary to the prevalent opinion, the great majority of sensory neurons are associated with

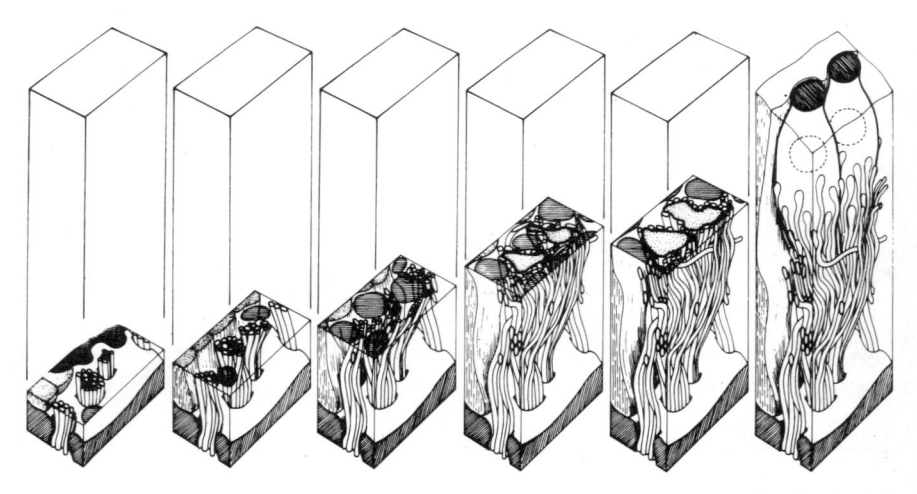

Fig. 24. Graphic reconstruction of the area of two internal hair cells of an animal in which the olivo-cochlear fibres have been transsected 4 weeks previously. It is clearly seen that the great majority of afferent dendrites go to the internal hair cells and end there unbranched with a single terminal.

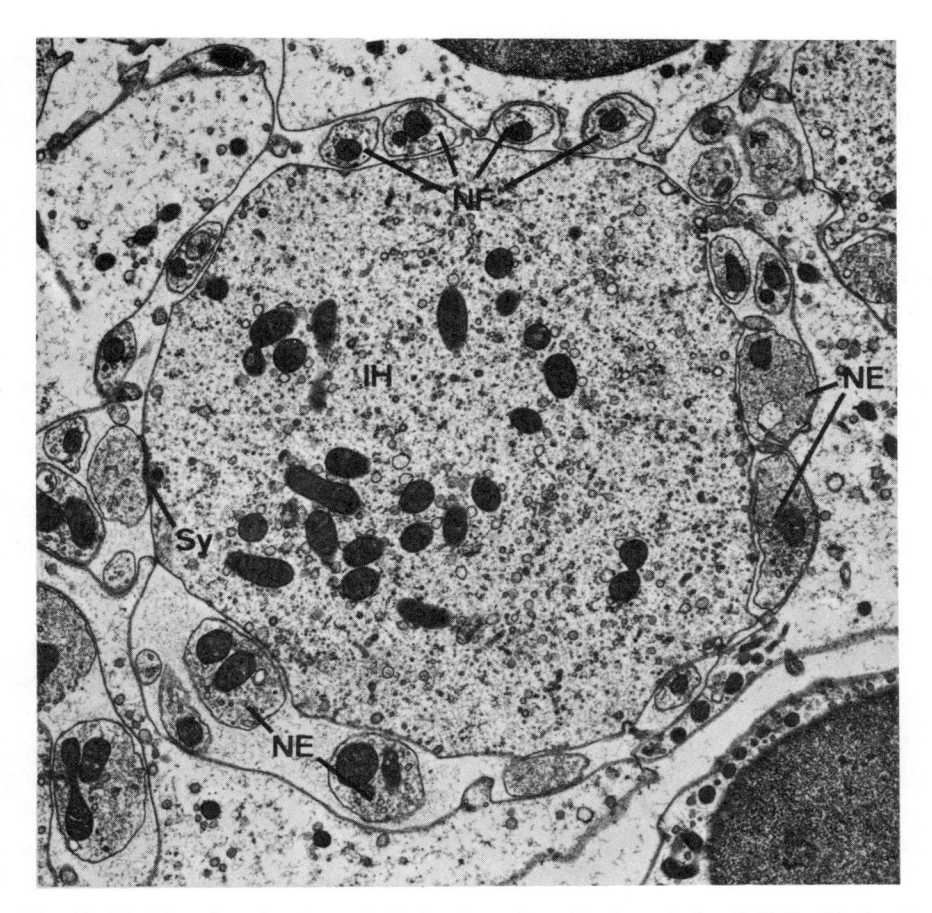

Fig. 25. Horizontal section through the basal portion of an inner hair cell (IH) with about 20 nerve endings (NE) and nerve fibres (NF) which all end at this cell. A typical synaptic complex is seen at the left (Sy).

the inner hair cells and only a minority with outer hair cells.

In contrast with the dendrites to the inner hair cells, the afferent fibres to the outer hair cells take a spiral, basalward course of considerable length before they end at the outer hair cells. Because of the long course of these fibres it is impossible to follow them in serial sections. However, some conclusions can be reached from available data. Since we know that practically all outer spiral fibres are the continuation of the basilar fibres, their relative numbers allow rough estimates of their average spiral extension (Spoendlin, 1968). There are from 70 to 100 outer spiral fibres at any given place in the basal turn. (The number increases slightly in upper turns.) Each fibre is the continuation of one basilar fibre. If only one basilar fibre penetrates between two outer pillars we must assume that each fibre runs in a spiral direction over a distance of at least

70-100 pillars (0.7-1.0 mm) in order to account for the number of outer spiral fibres found at any one place (Fig. 26). However, this estimate is applicable only under the assumption that all outer spiral fibres are of approximately equal length.

There is another way to estimate the spiral extension of the outer spiral fibres in the basal turn. These fibres start their spiral course close to the basilar membrane and ascend slowly in a remarkably straight line toward the hair cell base (Fig. 29). By measuring the angle of ascent, the distance needed for one fibre to reach the hair cell base can be estimated; it is in the order of 0.7 mm.

The outer spiral fibres run over a considerable distance unbranched before they send off small terminal collaterals which end with small bud-like endings at the hair cells. Only when the main fibre has reached the level of the Deiter-cell nuclei, the first collaterals are given off. Every sensory cell is innervated by several different neurons and each neuron sends its terminal branches to a number of different hair cells according to the principle of multiple innervation. One hair cell might receive very early collateral branches of one dendrite as well as terminal collaterals of another dendrite. Even in the terminal branching portion one dendrite does not send collaterals to all hair cells along its course. There may be gaps of several hair cells between the individual collaterals. It happens, but it is rather exceptional, that one dendrite sends several collaterals to one sensory cell. The range over which one dendrite sends its terminal branches to the sensory cells probably does not much exceed 200 μ (Fig. 33). By comparing the number of afferent nerve endings with the number of afferent neurons, the average number of outer hair cells innervated by the collaterals of one single neuron can be estimated. In the basal turn we found in serial sections in 9 outer hair cells an average of 4 afferent nerve endings per cell and there is an average of one afferent neuron per 3 outer hair cells (Fig. 26). If every dendrite

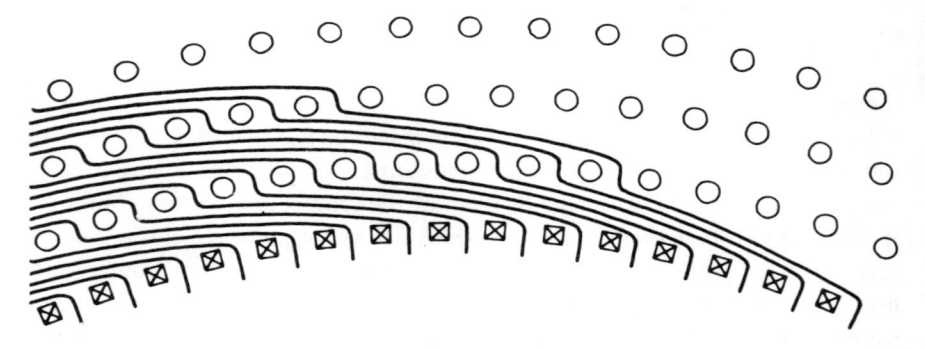

Fig. 26. Schema showing the evaluation of the average spiral extension of outer spiral fibres. The outer pillars are represented by crossed squares and the outer hair cells by circles. Between every two pillars one basilar fibre penetrates to continue as an outer spiral fibre. The number of outer spiral fibres corresponds to the number of pillars over which the nerve runs in a spiral direction.

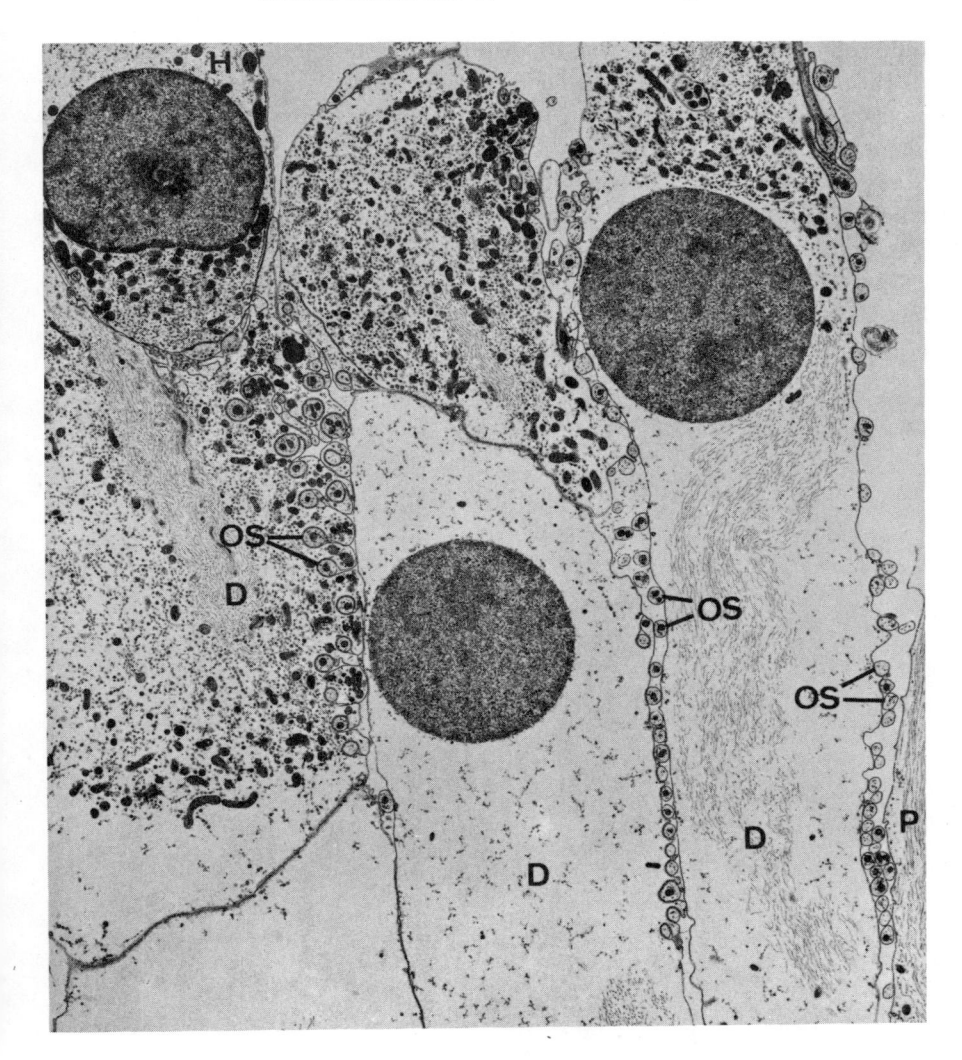

Fig. 27. Area below the outer hair cells (H) of cat after elimination of the efferents, with outer spiral fibres (OS) between the Deiter-cells (D). P: outer pillar.

gives only one collateral to one hair cell, we can estimate that about 10 outer hair cells are innervated from one neuron. These numbers are probably smaller at the lower end of the cochlea and larger in upper turns where the afferent terminals at the outer hair cells are more numerous.

The dendrites to the inner and outer hair cells are not only different in their distribution but also in their behaviour. The fibres to the inner hair cells are more susceptible to anoxia and acoustic trauma (Spoendlin, 1970). After transsection of the cochlear nerve, these fibres degenerate, but the fibres to the outer hair cells are not affected (Spoendlin, 1966).

The caliber of the afferent nerve fibres varies and the individual fibres change considerably in diameter during their course through the organ of Corti. The dendrites to the inner hair cells are 0.5 to 1.5 μ thick. The diameters of all fibres shrink by about half when they pass the habenula, and they again become much larger at the beginning of the myelin sheath below the habenula (Fig. 30). The outer spiral fibres have a fairly constant diameter of about 0.4 to 0.6 μ except the terminal portion of the fibre which is smaller. All collaterals to the hair cells are much smaller with about 0.2 μ diameters (Fig. 28). At the bottom of the tunnel the main dendrites are somewhat smaller and especially in the

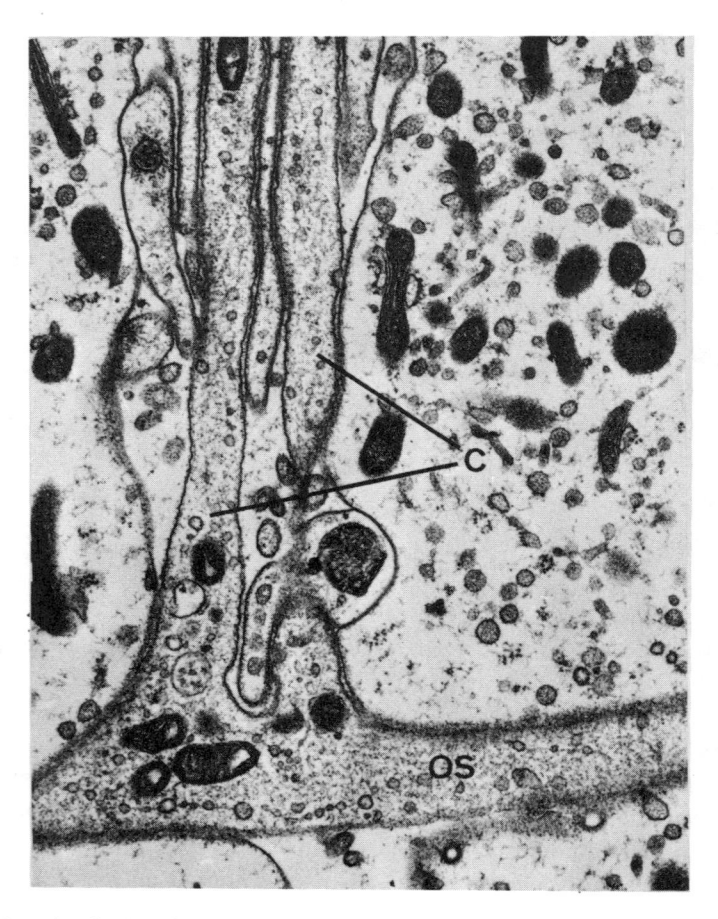

Fig. 28. Longitudinal section through an outer spiral fibre (OS) which gives off a terminal collateral (C) to a hair cell. There is no striking change in the axoplasm at the branching point.

passage between the outer and inner pillars their diameter is reduced to about 0.3 μ (Fig. 31).

The caliber of nerve fibres is probably of great functional importance. It influences the speed of propagation of nervous activity and might determine the initial segment of a neuron. The important question at which point the initial segment of the afferent dendrites, the site at which the action potential is initiated, is located has frequently been discussed (Spoendlin and Lichtensteiger, 1966; Bosher, 1968). Unequivocal morphological characteristics of the initial segment are not known. There are reasons to assume that it is at the place where myelin sheaths begin under the habenula perforata (Spoendlin, 1968; de Boer and Jongkees, 1968). According to the electrophysiologists, the initial segment is characterized by a change of the capacity of the axon membrane, which might occur at places where the diameter of the nerve fibre changes abruptly. Such a considerable reduction of the dendrite's diameter is found in the habenula perforata, which therefore might be the place of the initial segment, at least for the dendrites coming from the inner hair cells. The question is more complex for the long dendrites coming from the outer hair cells. If we assume conduction by electrotonic propagation of the graded postsynaptic potential along the entire distance of these dendrites, a large decrement would take place. So it can not be excluded that in these dendrites the site of the initial segment is more peripheral than at the habenula, although we have no hints for such an assumption. In any case, the initial segment is most likely proximal to the terminal collaterals where, presumably, spatial summation takes place, which would not be possible with all-or-none responses (Davis, 1961). The length of the dendrite alone probably does not exclude electrotonic spread of the excitation. We have other examples of unmyelinated dendrites of considerable length with ramifications along their entire course, such as the Purkinje cells of the cerebellum. The presence of an adrenergic terminal plexus independent of blood vessels below the habenula (Spoendlin and Lichtensteiger, 1966) might also suggest that at least some fibres have their initial segment at this site, since autonomic influence on nervous activity probably can take place only where the activity consists of graded potentials and not of spikes.

CONCLUSIONS

The nerve fibre distribution in the organ of Corti, the synaptic connections with the hair cells, and the interrelations between afferent and efferent nerve fibres may be important factors for frequency discrimination in the cochlea. The tuning curve of a unit in the cochlear nerve (Kiang, 1965) is considerably sharper than the resonance curve of a corresponding point in the cochlear partition (von Békésy, 1960). This sharpening process must take place somewhere in the cochlear receptor. It might be due to mechanical phenomena in

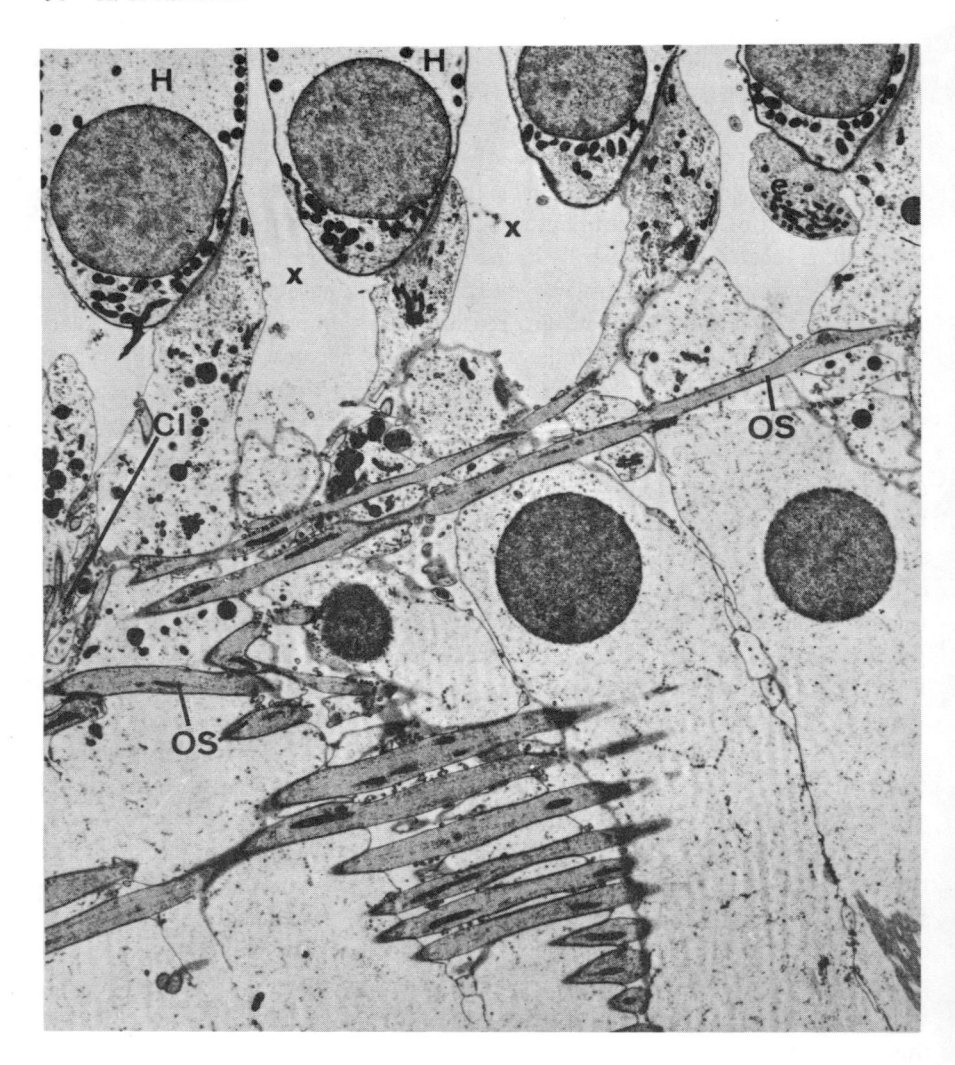

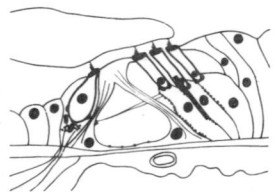

Fig. 29. Preparation from a cat 4 days after midline lesion. Tangential section through the area below outer hair cells (H) with outer spiral fibres (OS) and their collaterals (CI) to the outer hair cells. The outer spiral fibres climb in a straight line gradually up to the base of the outer hair cells. The majority of efferent nerve endings have disappeared (x), only a few remain (e).

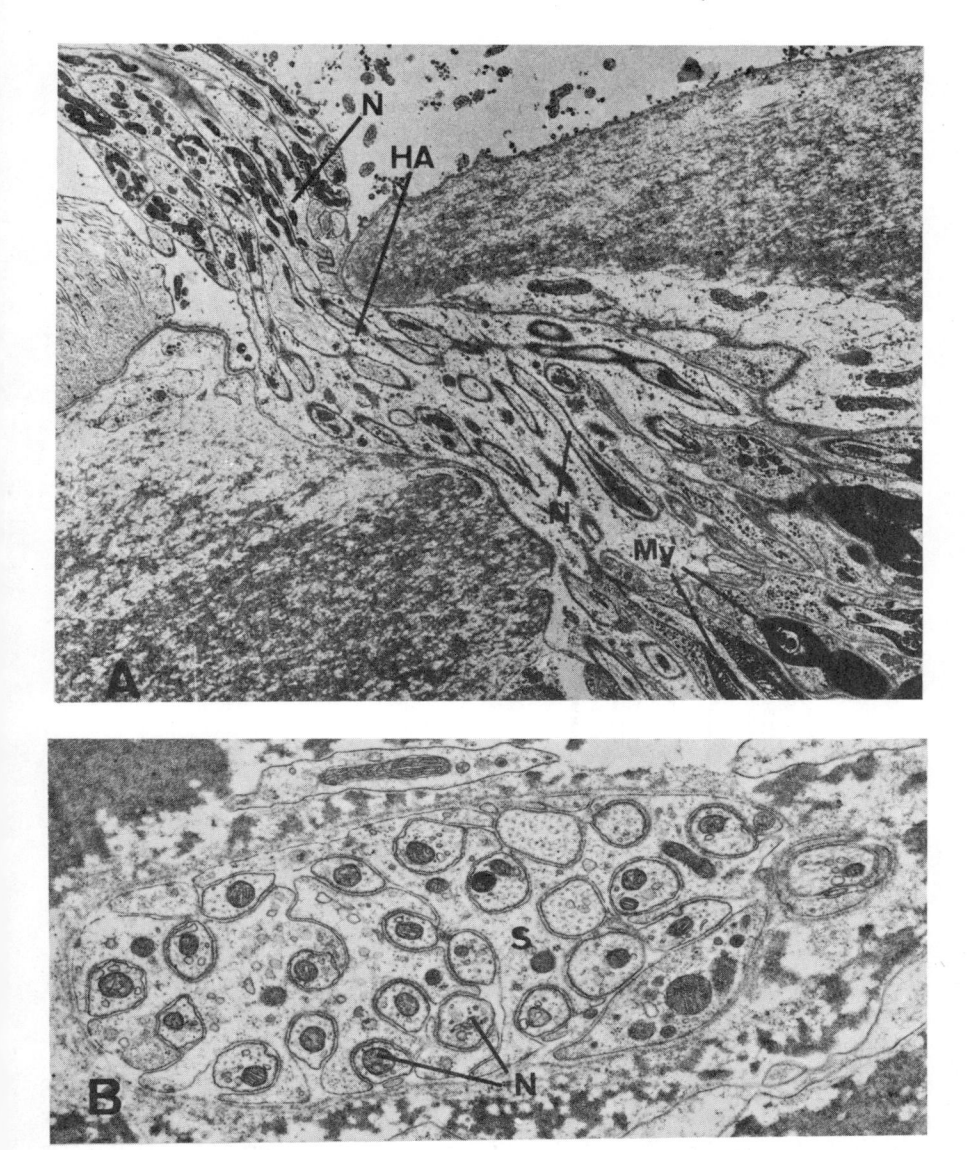

Fig. 30. The habenula.
A: Radial section through a habenular opening (HA) with penetrating nerve fibres (N) which have lost their myelin sheaths (My), shown below. Within the habenula the fibres are thin and contain very few mitochondria.
B: Transverse section through one habenular opening with the nerve fibres (N) individually surrounded by special satellite cells (S).

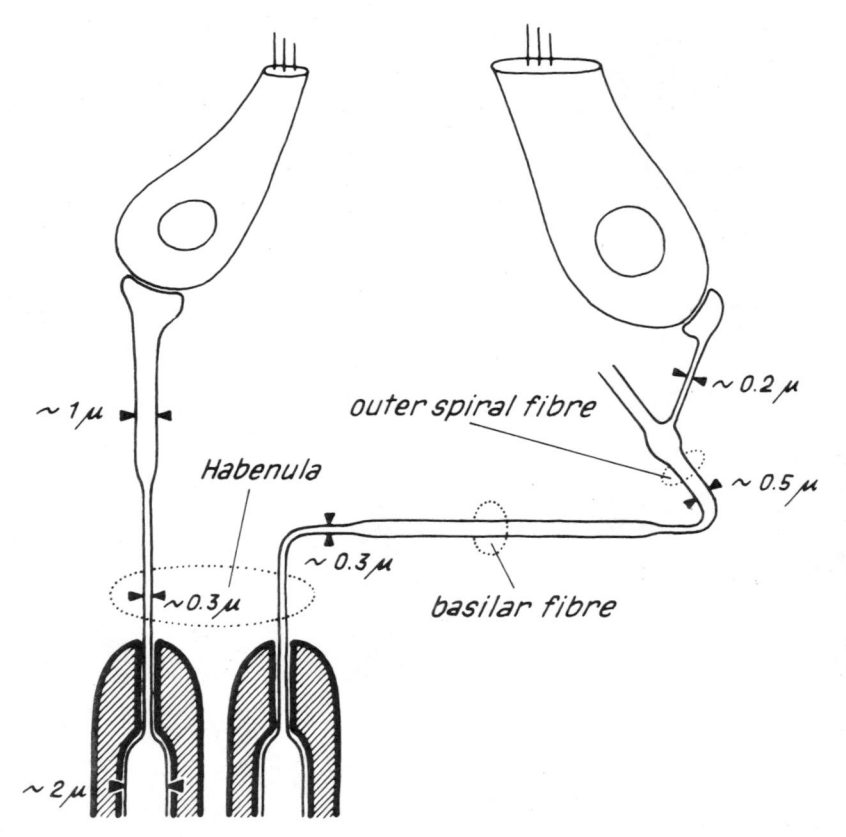

Fig. 31. Diagram showing the relative average diameter of afferent nerve fibres in the organ of Corti.

the cochlea (Johnstone, 1968), or to neural processing in unmyelinated portions of the nerve fibres in the organ of Corti (Spoendlin, 1966; de Boer and Jongkees, 1968).

There is morphological evidence that the synaptic transmission from hair cells to afferent nerve endings is a chemical process dependent on a transmitter substance. Under the effect of such synaptic activities a graded postsynaptic potential arises within the nerve ending. This graded potential propagates by electrotonic spread and has no threshold. At what point it is transformed into the all-or-none response of the action potential depends on the location of the initial segment. The place of the initial segment is presumably central to dendritic ramifications and synaptic connections with efferent fibres. Such graded generator potentials provide the possibility of temporal summation (Davis, 1961) in all dendrites and spatial summation in the branched dendrites to the outer hair cells only, in such a way that all electrotonically propagated potentials

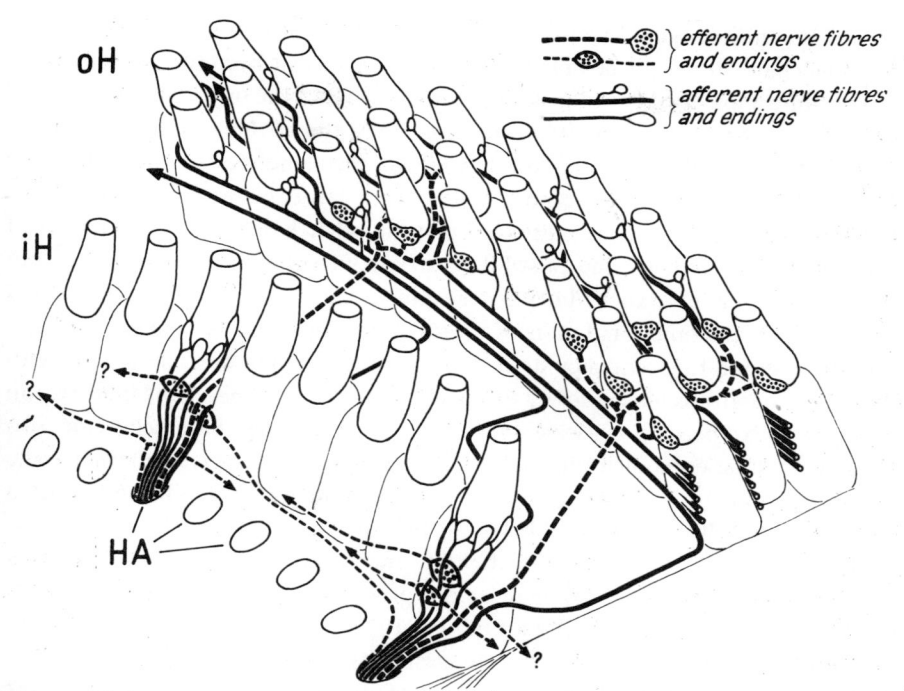

STRUCTURAL BASIS OF PERIPHERAL FREQUENCY ANALYSIS 33

Fig. 32. Schema of the general innervation pattern of the organ of Corti. oH: outer hair cells; iH: inner hair cells; HA: habenular openings.

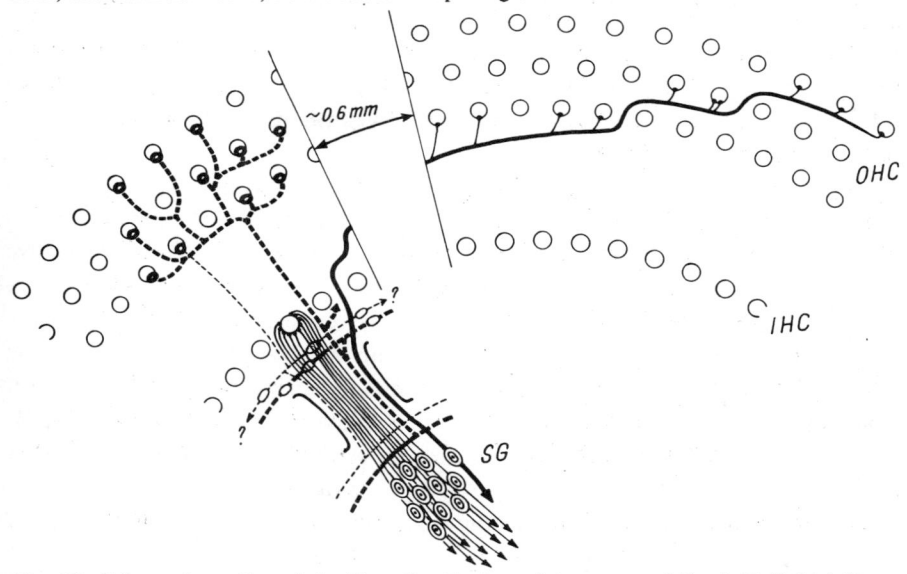

Fig. 33. Schematic outline of the fibre distribution of the organ of Corti. Full thick lines: afferent fibres to the outer hair cells. Full thin lines: afferent fibres to the inner hair cells. Interrupted thick lines: efferent nerve fibres from the contralateral olivo-cochlear bundle. Thin interrupted lines: efferent nerve fibres from the homolateral olivo-cochlear bundle.

add at the initial segment, and if the threshold is reached, a nerve spike occurs. Such mechanisms might be able to increase the contrast by acting as a filter.

The almost exclusively radial afferent innervation of the inner hair cells, where each nerve fibre represents only one spot along the cochlea, tends to provide maximum tonotopic discrimination, whereas the dendrites to the outer hair cells innervate a group of sensory cells over a larger area; a less favourable situation for tonotopic discrimination. On the other hand, the ramifications of these dendrites provide the possibility of spatial summation, which can not take place in the unbranched dendrites to the inner hair cells. Thus, one would conclude that the functional units of the inner hair cells and associated dendrites are better suited for frequency discrimination whereas the outer hair cells with associated dendrites have a lower threshold (Spoendlin, 1968). This innervation system with a specially massive innervation of the inner hair cells can also explain the great dynamic range of the cochlear neurons at a given characteristic frequency which is as much as 60 dB whereas a single nerve fibre has only a range of about 30 dB (Davis, 1968).

The question of how much the efferent innervation is involved in frequency discrimination, especially in sharpening processes, is not yet settled. It is certainly tempting to relate efferent inhibitory effects to an increase of contrast by means of lateral inhibition as first suggested by von Békésy (1960). Capps and Ades (1968) reported on deterioration of frequency discrimination after transsection of the olivo-cochlear fibres in monkeys. Klinke et al. (1969; see also pp. 161-164 of this volume) reported an improvement in pitch discrimination in cochlear neurons by simultaneous contralateral stimulation, which they attributed to the efferent innervation. On the other hand, doubts have been expressed about an important effect of the efferents on frequency discrimination (Johnstone, 1968; Fex, 1968).

Any efferent effect coming from the central nervous system is delayed by a long latency (Fex, 1962) and is presumably of a very general character because of the small number of olivo-cochlear fibres (approximately 500) with extensive ramifications. The question arises whether the olivo-cochlear efferent innervation is only activated in the central nervous system in the sense of a feed-back loop as proposed by Fex (1962) or whether it can also be activated directly in the periphery by afferent nerve potentials. Presumably such a peripheral activation would be, however, of an electric and not a chemical nature since morphological evidence for synaptic connections leading from afferent dendrites to efferent fibres is lacking.

The size of the efferent endings, containing a great number of mitochondria, certainly suggests a long-lasting energy-consuming function. On the basis of such reasons other functions of the efferent innervation have been proposed such as stabilization of the threshold (Johnstone, 1968) or prevention of wastage of chemical mediator in the outer hair cells (Davis, 1968). An effect on adaptation was proposed by Leibbrandt (1965) and a reduction of masking effect of noise by Trahiotis and Elliott (1970).

Whatever the final answer about the functional significance of the efferent innervation will be, the enormous anatomical representation of efferent endings in the cochlea allows the conclusion that it must fulfill an important role in the cochlear receptor, especially in the basal turn where the efferent nerve supply is much greater than at the apex.

REFERENCES

Békésy, G. von (1960): Experiments in Hearing (McGraw-Hill, New York).
Boer, E. de, and Jongkees, L. B. W. (1968): On cochlear sharpening and cross-correlation methods, Acta Oto-Laryngol. 65, 97-104.
Borghesan, E. (1951): Tectorial membrane and organ of Corti considered as a unique anatomical and functional entity, Acta Oto-Laryngol. 42, 473-486.
Bosher, S. K. (1968): Contribution to discussion, in: Hearing Mechanisms in Vertebrates, A. V. S. de Reuck and J. Knight, Eds. (Churchill, London), pp. 124-125.
Capps, M. J., and Ades, H. W. (1968): Auditory frequency discrimination after transsection of the olivo-cochlear bundle in squirrel monkeys, Exper. Neurol. 21, 147-158.
Christiansen, J. A. (1964): On hyaluronate molecules in the labyrinth as mechano-electrical transducers, and as molecular motors acting as resonators, Acta Oto-Laryngol. 57, 33-49.
Davis, H. (1961): Some principles of sensory receptor action, Physiol. Rev. 41, 391-416.
Davis, H. (1968): Contribution to discussion, in: Hearing Mechanisms in Vertebrates, A. V. S. de Reuck and J. Knight, Eds. (Churchill, London), pp. 119, 305.
Desmedt, J. E. (1962): Auditory-evoked potentials from cochlea to cortex as influenced by activation of the efferent olivo-cochlear bundle, J. Acoust. Soc. Amer. 34, 1478-1496.
Engström, H. (1968): Contribution to discussion, in: Hearing Mechanisms in Vertebrates, A. V. S. de Reuck and J. Knight, Eds. (Churchill, London), p. 124.
Engström, H., Ades, H. W., and Hawkins, J. E. (1962): Structure and functions of the sensory hairs of the inner ear, J. Acoust. Soc. Amer. 34, 1356-1363.
Engström, H., Ades, H. W., and Andersson, A. (1966): Structural Pattern of the Organ of Corti (Almqvist & Wiksell, Stockholm).
Eldredge, D. H. (1967): Book review on H. Spoendlin: The Organization of the Cochlear Receptor, J. Acoust. Soc. Amer. 41, 1386-1388.
Fex, J. (1962): Auditory activity in centrifugal and centripetal cochlear fibres in cat, Acta Physiol. Scand. 55, Suppl. 189, 5-68.
Fex, J. (1968): Efferent inhibition in the cochlea by the olivo-cochlear bundle, in: Hearing Mechanisms in Vertebrates, A. V. S. de Reuck and J. Knight, Eds. (Churchill, London), pp. 169-181.
Flock, A., Kimura, R., Lundquist, P.-G., and Wersäll, J. (1962): Morphological basis of directional sensitivity of the outer hair cells in the organ of Corti, J. Acoust. Soc. Amer. 34, 1351-1355.
Galambos, R. (1956): Suppression of auditory nerve activity by stimulation of the efferent fibers to cochlea, J. Neurophysiol. 19, 424-437.
Greenwood, D. D. (1962): Approximate calculation of the dimensions of traveling-wave envelopes in four species, J. Acoust. Soc. Amer. 34, 1364-1369.
Iurato, S. (1962): Efferent fibres to the sensory cells of Corti's organ, Exp. Cell Res. 27, 162.
Iurato, S. (1964): Atti. Soc. Ital. Anat. 72, 60.
Iurato, S. (1968): personal communication.
Iurato, S., et al. (1967): Submicroscopic Structure of the Inner Ear (Pergamon Press, Oxford), pp. 61-76.

Johnstone, B. M. (1968): Contribution to discussion, in: Hearing Mechanisms in Vertebrates, A. V. S. de Reuck and J. Knight, Eds. (Churchill, London), pp. 299-300, 303-305.
Kiang, N. Y.-S. (1965): Discharge Patterns of Single Fibers in the Cat's Auditory Nerve (Research Monograph No. 35, M.I.T. Press, Cambridge, Mass.), pp. 84-92.
Kimura, R., and Wersäll, J. (1962): Termination of the olivo-cochlear bundle in relation to the outer hair cells of the organ of Corti in guinea pig, Acta Oto-Laryngol. 55, 11-32.
Klinke, R., Boerger, G., and Gruber, J. (1969): Studies on the functional significance of efferent innervation in the auditory system, Pflügers Arch. ges. Physiol. 306, 165.
Kolmer, W. (1927): Gehörorgan, in: Handbuch der mikroskopische Anatomie des Menschen, A. von Möllendorf, Ed. (J. Springer, Berlin), Vol. 3, Part 1, pp. 250-478.
Leibbrandt, C. C. (1965): The significance of the olivo-cochlear bundle for the adaptation mechanism of the inner ear, Acta Oto-Laryngol. 59, 124-132.
Loewenstein, R., and Wersäll, J. (1959): A functional interpretation of the electron microscope structure of the sensory hairs in the cristae of the Elasmo-branch Raja clavata in terms of directional sensitivity, Nature 184, 1807.
Lorente de Nó, R. (1937): The sensory endings in the cochlea, Laryngoscope 47, 373-377.
Nomura, Y., and Schuknecht, H. F. (1965): The efferent fibers in the cochlea, Ann. Otol. Rhinol. Laryngol. 74, 289-302.
Rasmussen, G. L. (1946): The olivary peduncle and other fiber projections of the superior olivary complex, J. Comp. Neurol. 84, 141-219.
Rasmussen, G. L. (1960): Efferent fibers of the cochlear nerve and cochlear nucleus, in: Neural Mechanisms of the Auditory and Vestibular Systems, G. L. Rasmussen and W. F. Windle, Eds. (Charles C. Thomas, Springfield, Ill.), pp. 105-115.
Smith, C. A., and Rasmussen, G. L. (1963): Recent observations on the olivo-cochlear bundle, Ann. Otol. Rhinol. Laryngol. 72, 489.
Smith, C. A., and Rasmussen, G. L. (1965): Degeneration in the efferent nerve endings in the cochlea after axonal section, J. Cell Biol. 26, 63-77.
Smith, C. A., and Sjöstrand, F. (1961): A synaptic structure in the hair cell of the guinea pig cochlea, J. Ultrastruct. Res. 5, 184.
Spoendlin, H. (1966): The Organization of the Cochlear Receptor (Karger, Basel-New York).
Spoendlin, H. (1967): The innervation of the organ of Corti, J. Laryngol. Otol. 81, 717-738.
Spoendlin, H. (1968): Ultrastructure and peripheral innervation pattern of the receptor in relation to the first coding of the acoustic message, in: Hearing Mechanisms in Vertebrates, A. V. S. de Reuck and J. Knight, Eds. (Churchill, London), pp. 89-119.
Spoendlin, H. (1969a): Innervation patterns in the organ of Corti of the cat, Acta Oto-Laryngol. 67, 239-254.
Spoendlin, H. (1969b): Das ischämische Syndrom des Innenohres, Pract. Oto-Rhino-Laryngol. 31, 257-268.
Spoendlin, H. (1970): in: Ultrastructure of the Peripheral Nervous System and the Sensory Organs, Babel, Bischoff and Spoendlin, (Thieme, Stuttgart), in press.
Spoendlin, H., and Gacek, R. R. (1963): Electron microscopic studies on the efferent and afferent innervation of the organ of Corti in the cat, Ann. Otol. Rhinol. Laryngol. 72, 1-27.
Spoendlin, H., and Gacek, R. R. (1965): Survival of the peripheral dendrites after section of the cochlear nerve, in: Proceedings of the Vth International Congress of Neuropathology, F. Lüthi and A. Bischoff, Eds. (Excerpta Medica Foundation, Amsterdam), pp. 926-934.
Spoendlin, H., and Lichtensteiger, W. (1966): The adrenergic innervation of the labyrinth, Acta Oto-Laryngol. 61, 423-434.
Tonndorf, J. (1962): Time/frequency analysis along the partition of cochlear models: a modified place concept, J. Acoust. Soc. Amer. 34, 1337-1350.
Trahiotis, C., and Elliot, D. N. (1970): Behavioral investigation of some possible effects of sectioning the crossed olivocochlear bundle, J. Acoust. Soc. Amer. 47, 592-596.
Vilstrup, Th., and Jensen, C. E. (1954): Three reports on the chemical composition of the fluids of the labyrinth, Ann. Otol. Rhinol. Laryngol. 63, 151-163.

DISCUSSION

Smoorenburg: Are you able to correlate the different types of nerve fibres you found with the two groups Kiang (1965) found with respect to the spontaneous firing rate?

Spoendlin: No, I am not able to do so.
I also looked to see whether there was evidence for two different threshold groups in Kiang's (1968) data; there was none.

Smoorenburg: Although you are not able to come to conclusions on behalf of the type of conduction in the dendrites in the cochlear duct—whether it is electrotonic spread or action potentials—, can you give some estimates of the propagation velocities? I think this is especially important with respect to the time delays in the outer spiral afferent dendrites.

Spoendlin: The speed of electrotonic spread is not known. The conduction time depends on the place of the initial segment and we are not sure about that either.

Goldstein: Kiang told me that he has no data on the propagation velocities in the unmyelinated dendrites.

Zwislocki: The threshold depends very much on the way it is defined. Another definition may result in a greater spread of threshold values.

Anderson: Defining the threshold in terms of synchronization to the stimulus cycle may lower some threshold determinations by as much as 20 dB.

Møller: Kiang (1968) recently showed that degeneration of the efferents had no influence on the tuning curve or on the two-tone inhibitory areas.

Spoendlin: How long after the lesion did he record?

Møller: He waited long enough to be sure that the efferents had degenerated.

Spoendlin: I am very interested in his results, but, in contrast, Capps and Ades (1968) found a marked deterioration of frequency discrimination in monkeys after transsection of the olivo-cochlear bundle. Recently also Nieder and Nieder (1969), measuring the modulating effect of a low frequency on a high frequency, describe a modulation pattern which disappears after transsection of the olivo-cochlear fibres.

Bosher: It seems to me that the efferent fibres to the outer hair cells do not synapse substantially on the afferent fibres from the inner hair cells en route, and that the efferent fibres of the inner spiral plexus synapse almost entirely on the inner hair cell dendrites. Is it not possible, therefore, that we are dealing with two efferent systems; one system destined for the outer hair cells with large nerve endings which are well designed to either stabilize the threshold as suggested by Johnstone (1968) or prevent the excessive leakage of transmitter substances as suggested by Davis (1968). As the basal portion is activated by every travelling wave, this would explain the predominant efferent innervation there. The first row of outer hair cells is also predominantly innervated, suggesting these require greater stabilization than the other two rows which will be the case, for example, if the initial segment is near the first row of outer hair cells.

These concepts cannot apply to the efferent-afferent fibre synapses and so these then might well constitute a second system possibly producing the frequency selectivity effects described by Capps and Ades (1968).

Spoendlin: There might be two different efferent systems. There is even morphological evidence for that: the efferent fibres for the outer hair cells have a large average diameter (1.0-1.5 μ) and the efferent fibres of the inner spiral bundle are very thin (0.1-0.2 μ). The concept that the efferents could have some stabilizing or protective function is appealing. Judging from the great number of mitochondria, one must conclude that the efferent endings fulfill a highly consuming task. For instance, we find a tremendous increase of synaptic vesicles in the endings after exposure to 120 dB SPL white noise during one hour.

The question about the location of the initial segment is discussed on pp. 29-34.

Johnstone: The idea presented in the paper that the mucopolysaccharides play an important part in the mechano-electric transduction has its attraction, but I can not see how it works. I would propose that a modulation of the area of the cuticular membrane pore plays an important part. This locates the microphonic generator at the site of the cuticular plate. The presence of a kinocilium may not be so important then, provided a force can still modulate the area.

Spoendlin: The kinocilium can not play an important role because it is not there in all species. I agree with you that the cuticular plate itself is elastic. Still, I can not understand the mechanism of the modulation. Mechanical distortion, which in one way or another must be the crucial event in mechano-electric transduction, is most likely to occur much easier and more efficiently at the level of the elongated, thin, stereocilia than at the level of the massive cuticular plate, which is tightly hold all around by the reticular membrane (see Fig. 6). A significant distortion of the cuticular pore area by the extremely small amplitude of movement involved seems most unlikely.

Honrubia: The question of the influence of the number of synapses on the threshold can be answered if we look to what is happening in the entire hair cell. If the CM is produced by a change in the resistance across the cuticular lamina, the number of afferent synapses can make a difference in the triggering facility, but from this still is hard to understand threshold differences of 20 dB in the primary fibres.

Are you, Dr. Spoendlin, able to fit your ideas about the role of the mucopolysaccharides with the well-known properties of the CM, such as the change in polarity between the scala tympani and the scala vestibuli, and the influence of the efferent system on the size of the CM? If the production of the CM is not at the boundary of the cuticular lamina, but somewhere floating in the scala media, how are you able to close the current loops?

Spoendlin: We do not know precisely the place where CM is generated. The place of production of the cochlear microphonics is certainly not located somewhere floating in the scala media. If mucopolysaccharides play a role in the initial step of mechano-electric transduction, it would only be at the dense coat of mucopolysaccharides closely attached to the stereocilia (see Fig. 4). This would not be in contradiction to the change of polarity of the cochlear microphonics at the boundary between organ of Corti and scala media. The efferent system may influence the cochlear microphonics by changing the intracellular potential in the outer hair cells.

Wilson: I am not convinced by your Fig. 2 that the arrangement of the stereocilia is really closer in the longitudinal direction for the inner hair cells and closer in the radial direction for the outer hair cells. Do you have other pictures which show this difference more clearly? If not, is there other histological evidence — like a structural difference in the spaces between the stereocilia — which would lead one to believe that the hair cells could have this different directional sensitivity?

Spoendlin: I am aware of the problem. The directional sensitivity is not an absolute matter. Von Békésy (1960) spoke of hair cells responding *best* to some direction. So far I could not detect structural difference in the spaces between the stereocilia.

Kohllöffel: Dr. Spoendlin, what do you mean by "morphological influence" of the basal body in the young kitten? Do you mean, for instance, that each hair cell gets a particular collection of mucopolysaccharides which makes it frequency-selective or which makes the hairs grow in a particular way?

Spoendlin: The latter is meant. The basal bodies are morphologically the same as centrioles, which organizes the cell during its development. The topographical arrangement of the structural elements of the cell depends on the centriole. The centriole therefore is most probably responsible for the asymmetric arrangement of the receptor pole of the hair cells. The fact that the centriole induces structural differentiation at certain places in the cell can be illustrated by some observations we made in vestibular sensory cells. In some instance we found the centriole (or basal body) mislocated at the base of the sensory cell and the kinocilium as well as the cuticular plate were also found in its neighbourhood at the base of the cell instead, as normally, at the apical surface of the cell.

REFERENCES

Békésy, G. von (1960): Experiments in Hearing (McGraw-Hill, New York).
Capps, M. J., and Ades, H. W. (1968): Auditory frequency discrimination after transsection of the olivo-cochlear bundle in squirrel monkeys, Exp. Neurol. *21*, 147.
Davis, H. (1968): Contribution to discussion in: Hearing Mechanisms in Vertebrates, A. V. S. de Reuck and J. Knight, Eds. (Churchill, London), p. 305.

Johnstone, B. M. (1968): Contribution to discussion in: Hearing Mechanisms in Vertebrates, A. V. S. de Reuck and J. Knight, Eds. (Churchill, London), p. 303.

Kiang, N. Y.-S. (1965): Discharge Patterns of Single Fibers in the Cat's Auditory Nerve (Research Monograph No. 35, M.I.T. Press, Cambridge, Mass.).

Kiang, N. Y.-S. (1968): A survey of recent developments in the study of auditory physiology, Ann. Otol. Rhinol. Laryngol. 77, 656-676.

Nieder, P., and Nieder, I. (1969): Cochlear responses to bitonal stimuli and function of efferents, J. Acoust. Soc. Amer. 45, 789-790.

THE RESIDUE REVISITED

J. F. SCHOUTEN

Institute for Perception Research
Eindhoven, The Netherlands

> Wodurch kann über die Frage, was zu
> einem Tone gehöre, entschieden werden,
> als eben durch das Ohr?*
> August Seebeck (1844a)

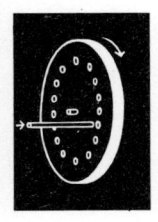

In the zigzagging road towards wisdom about the human auditory system we collect knowledge from two entirely different sources of experimental information. First, from anatomy and physiology, as presented in the one foursome of reviews in this symposium. Secondly, from perception and psychoacoustics, as to be presented in another foursome. Our ever-wondering mind tries to combine and to explain these findings in terms of some model, law, hypothesis or theory.

This pattern represents the eternal triangle redolent of mutual inspiration or contempt, agreement or controversy, typical of the historical breakthroughs or stumbling blocks on our road towards some wisdom in the field of human hearing as well as in all fields of human perception.

It is the aim of this paper to retrace the older history of hearing research in terms of the contributions from the three corners of this triangle and of the interactions along its sides.

We hope to show that many of the fateful misunderstandings can be attributed either to a scientist, well entrenched in his own corner, ignoring the evidence from other corners, or to scientists being insufficiently aware of their own peregrinations along the triangle (Fig. 1).

References in this paper are fragmentary. For more complete information readers are referred to Plomp's (1966) excellent and balanced survey.

---

* How else can the question as to what makes out a tone be decided but by the ear?

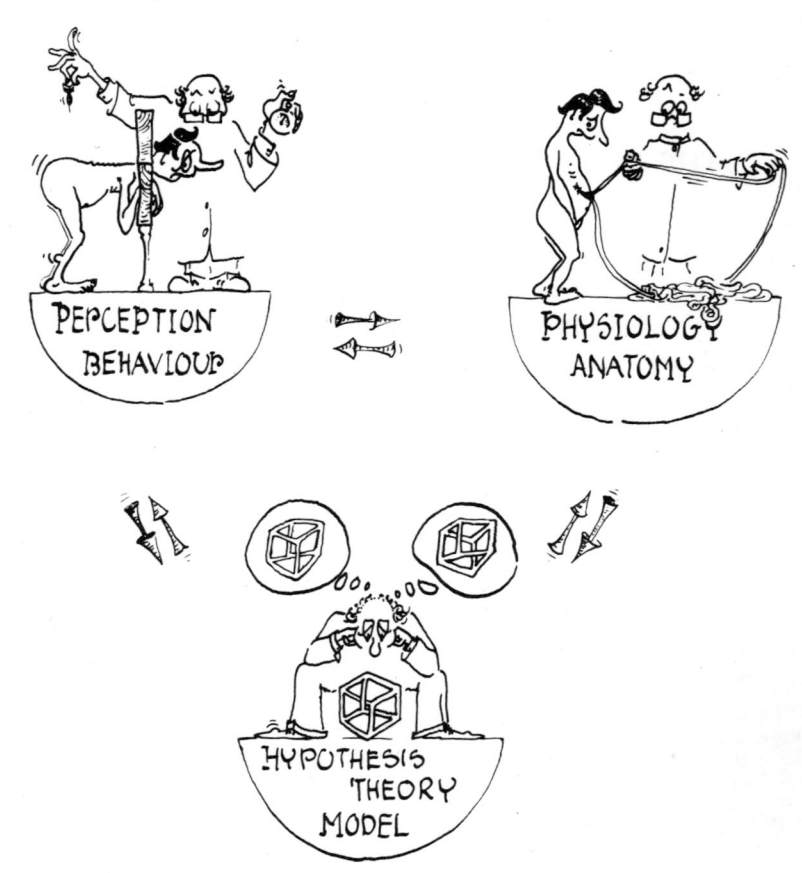

Fig. 1. The eternal triangle in the research on human perception.
Top left corner: Experimental evidence from perception and behaviour.
Top right corner: Experimental evidence from anatomy and physiology.
Bottom corner: Modelmaking, hypothesis and theory.
Along the sides: Mutual interaction.

The story may be taken to begin with Seebeck's (1841) ingenious observations with his acoustic siren. Blowing air through one or more pipes placed against various regular concentric arrangements of holes in a turning disc, he observed tones, shifts of an octave, etc. A case of pure perception and data collection.

Ohm (1843) suggested applying Fourier's theorem and assuming that each tone of different pitch in a complex sound originates from the objective existence of that particular frequency in the Fourier analysis of the acoustic wave pattern (Ohm's acoustical law). In order to explain Seebeck's observations, Ohm had to assume that the siren produced not only the fundamental frequency but also some higher harmonics like the octave, the duodecime, etc. A case of pure mathematical modelmaking and of inspiration towards further observations.

Seebeck (1843) admitted to having heard overtones and to having ascribed them to an artefact. Then, whilst agreeing upon the faint audibility of some higher harmonics, he had the cheek (as we might judge easily in our era of prolific electroacoustical engineering) to claim that Ohm's theory, though qualitatively acceptable, fails completely to account for the quantitative aspects.

As an example we quote one of his earliest experiments* (Fig. 2), of which he provides a quantitative analysis in his second paper. In general Seebeck's

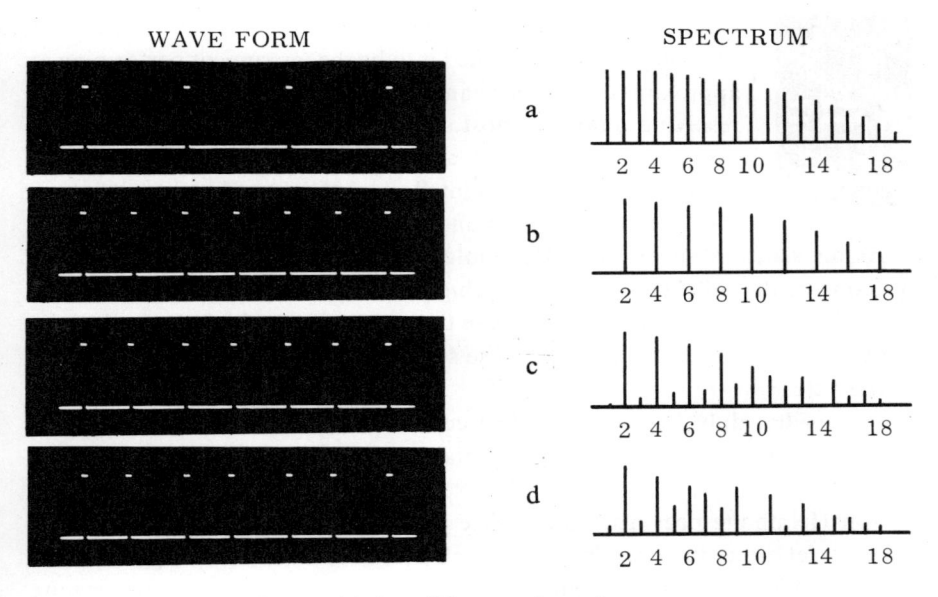

Fig. 2. Seebeck's experiment with four different pulse series.
The amplitudes in the spectra are those derived from the Fourier analysis of the four waveforms. The harmonics are indicated with their respective numbers. The indications of strength given below are in terms of perceived loudness.
(a) Spacing of holes 20°. Fundamental strong, harmonics weak.
(b) Spacing of holes 10°. The octave (now the fundamental) strong, higher even harmonics weak.
(c) Spacing alternately 9°.5 and 10°.5. The octave is stronger than the fundamental even though the latter is negligible in the Fourier spectrum.
(d) Spacing alternately 9° and 11°. The fundamental is somewhat louder than the octave, although still very small in the Fourier spectrum.

---

* We state with regret that von Békésy's (1964) account of this story and its further developments contains several misquotations. For instance, his numbering of Seebeck's experiments is not Seebeck's but the present author's (Schouten, 1940a) for examples taken from Seebeck's manifold experiments.

point is that in all cases observed, the fundamental tone is far louder as compared to the higher harmonics than the calculated Fourier spectrum accounts for. Perception and profound logical deduction. Seebeck, modelmaking, suggested that in some way the higher harmonics contribute to the loudness of the fundamental tone.

Ohm (1844), far from pleased, turned his modelmaking back towards the perceptual evidence (Fig. 1) and finally begged the question by proclaiming Seebeck's observations to be the result of an illusion.

The story ends with Seebeck (1844a; 1844b). As a modelmaker, he returned to his "wider interpretation" of Ohm's law. With respect to the imputed illusion, he provided us with the motto of this paper.

Helmholtz (1863), as a physiologist, investigated the cochlea and, as a modelmaker, assumed it to be the instrument performing an approximate Fourier analysis, characterized by a limited resolving power. Almost as a matter of logical consequence he assumed that each resonator leads to its own characteristic sensation of pitch. This intellectual sleight of hand fooled many investigators, except Wundt (1880), for scores of years to come.

As an astute observer, von Helmholtz rendered a remarkably well written account of the difficulty of observing higher harmonics, of the preponderant role of attention, and of the appropriate tricks to perform the feat of subjective sound analysis: an unparalleled treatise still worth reading by anyone engaged in this field.

The reader, duly impressed, is then confronted with Seebeck's criticism of Ohm's hypothesis. In this presentation the wisdom of hearing suffered a fateful setback.

Von Helmholtz provided no evidence that he had repeated Seebeck's experiments, let alone the *experimentum crucis* described above. He suggested that Seebeck failed to hear the higher harmonics by not using the proper means. He ignored the fact that Seebeck heard the fundamental tone even when it was all but absent objectively. Quite unforgivably, he presented the case against Seebeck: "although an investigator admirably versed in acoustical experiments and observations" with undertones casting doubt on the reliability of Seebeck's observations. No wonder that Seebeck's work remained almost forgotten for a long time to come, a fact for which von Helmholtz's authoritative statements are largely to blame.

Let us not put all the blame on von Helmholtz. The adherence to theories is and remains a matter of taste as long as definite proof or disproof is outstanding. Von Helmholtz's great asset was that his model was a well defined measuring instrument. Critical remarks, like those of Seebeck and, for instance, Hermann (1890, 1912), left the reader with a decided uncertainty with respect to better models that might be substituted, and thus with little taste for unspecified assumptions.

We skip the next half century (see *e.g.* Wever, 1949; Plomp, 1968), and pass on to the era of acoustical and auditory revival, initiated by its patriarchs von Békésy and Fletcher in the twenties.

Von Békésy (1928), as a physiologist (though a telephone engineer by profession), brought proof of localized stimulation by different frequencies in the cochlea, as anticipated by von Helmholtz. The resolving power, though, was poorer than expected.

Fletcher, equally a telephone engineer, stands for a new school employing electronically generated and measured sounds. It is of interest to remark in passing that equipment for vision research remained in the mechanical disk-and-shutter stage until recently.

The new devices, both in generation and in measuring, brought the "case of the missing fundamental" home with increasingly convincing evidence. It is a pity that this departure from Ohm's acoustical law was put forward so onesidedly in the prosecution. The full point is not only that the fundamental tone (or, as we shall see later, another subjective component of equal pitch) is heard whilst the first harmonic is absent in the objective sound, but also, and concurrently, that a host of higher Fourier components fail to make their appearance in subjective sound analysis. This "case of the profuse harmonics" scarcely drew any attention.

The case of the missing fundamental was solved for the time being by a modelmaker's brainwave that the nonlinearity of the ear could produce the missing fundamental as a difference tone between the higher harmonics. Both Fletcher (1929) and von Békésy (1934), now as observers, proved this point by the method of best beats. Sadly enough, the method of best beats is unreliable unless one knows "who beats whom".

By a compensation method the author (Schouten, 1938) found the difference tone to be negligible for moderate loudnesses, a finding later corroborated, for instance, by Plomp (1966) and Goldstein (1967). The author furthermore observed (Schouten, 1940a):

(1) The ear follows Ohm's law for frequencies wider apart than, say, a full tone (12%). In a harmonic series some 8 to 10 lower harmonics can be perceived by the unaided ear.
(2) Higher harmonics are heard collectively as *one* subjective component (one percept) called the residue.
(3) The residue has a sharp timbre.
(4) The residue has a pitch equal to that of the fundamental tone. If both are present in one sound they can be distinguished by their timbre.

Modelmaking, point 2 is evident from the limited resolving power of the ear's quasi-Fourier analysis. Point 3 is found to correlate with the location of the frequencies contributing to the residue. Our acuity of discrimination between these locations is rather good, as illustrated by the vowel formants, which are typical examples of residues. Point 4 implies that one region of the basilar membrane may give rise to sensations of widely different pitch. So von Helmholtz's first hypothesis regarding subjective frequency analysis is fully confirmed. His second hypothesis, regarding "one place, one pitch" is proved wrong as already anticipated by Wundt (1880).

In a way the discovery of the residue confirmed Seebeck's "wider interpretation" that the high harmonics might contribute somehow to the fundamental tone, except that it is not the fundamental tone itself which is enhanced but a different subjective component: the residue.

As a modelmaker, the author ascribed the low pitch of the residue to the *period* in the time pattern of the joint harmonics striking a particular area of the basilar membrane (Schouten, 1940b). This amounted to a revival of many older assumptions, except that the initial quasi-Fourier analysis was left standing in full glory.

If a set of harmonics is shifted collectively over a small distance in the frequency scale, the pitch of the residue shifts *in proportion* to the constituent frequencies (Schouten, 1940c). This crucial experiment (now called the first effect of pitch shift) kills two birds with one stone. It implies that pitch is determined neither by the *spacing* of the harmonics, nor by the *time envelope* of the wave pattern since both remain invariant. Hence, in the realm of modelmaking, it must be another form of periodicity detection in the time domain. This notion goes back to some extent to Seebeck's (1844a) idea of "the periodic recurrence of an equal or similar state of movement" and to Hermann's (1912) "intermittence tones".

These results, both in the field of perception and of modelmaking, drew little attention, until Hoogland (1953) found no evidence whatsoever of the residue and turned the clock back to Fletcher and von Békésy. Licklider (1954) heard the residue and provided additional evidence for the absence of mutual masking between a residue and a pure tone of equal pitch.

Spurred by these controversies de Boer (1956) took up the matter anew. He corroborated the existence of the residue, introduced the concept of a tonal and an atonal residue, performed a systematic investigation of the first effect of pitch shift and discovered the second effect of pitch shift. If a residue is obtained by modulating a carrier frequency f of say 2000 Hz with a modulating frequency g of say 200 Hz, the pitch p of the residue corresponds to that of g. The first effect consists in a pitch shift Δp proportional to the change Δf in the carrier frequency. The second effect consists in a slight but systematic deviation of the constant of proportionality. It also consists in a *downward* shift in pitch when the modulating frequency g is *raised*. The latter effect was met with doubt up to incredulity.

On the basis of modelmaking, de Boer ascribed the first effect to a shift in the fine structure of the modulated sound from one period of the modulating frequency g to the next. As to the second effect, he suggested that somehow the lower Fourier components might carry more weight in the formation of the residue than the higher ones.

In the Institute for Perception Research, founded in 1957, the controversial problem was tackled in full force by Cardozo and Ritsma. De Boer's experimental findings were fully confirmed, including the second effect. The range of the first effect was found to extend further than de Boer's sawtooth graph, in fact up to the next harmonic situation and even beyond. The phenomenon of ambiguity of pitch, anticipated earlier (Schouten, 1940b), was verified experimentally (Schouten et al., 1962). In terms of modelmaking this finding was in line with de Boer's peak-to-peak idea in the fine structure of the oscillogram (Fig. 3).

To Ritsma (1962, 1963) we owe the systematic exploration of the existence diagram of the residue (Fig. 4), corroborating de Boer's distinction between tonal and atonal residues. Roughly speaking the existence diagram is limited to values of $f = 5000$ Hz, $g = 60$ Hz and $n = 20$ ($f = ng$).

It turned out to have been Hoogland's bad luck to have chosen values of f and g outside the existence diagram.

Let us sum up the theoretical consequences of the findings so far. The Fourier spectrum of a complex sound is represented on the basilar membrane and possibly on some higher operationally relevant representation as a quasi-Fourier spectrum with a limited resolving power and its respective time constants (Fig. 4).

The existence of the residue and its properties in terms of

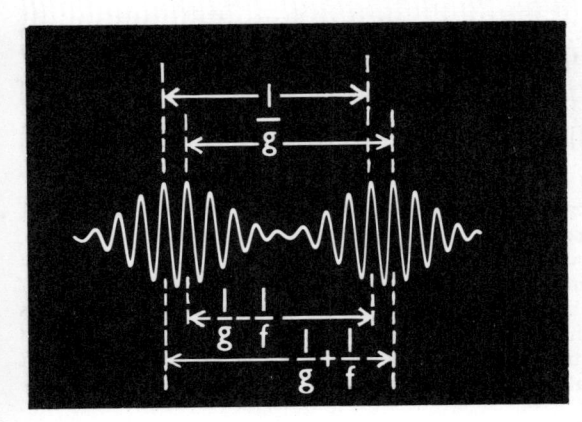

Fig. 3. Ambiguity of pitch. A modulated waveform consisting of the eighth, ninth and tenth harmonics ($f = 9g$, modulation depth 80%). The main interval is $\frac{1}{g}$. Ambiguous intervals are e.g. $\frac{1}{g} + \frac{1}{f}$ and $\frac{1}{g} - \frac{1}{f}$.

48 J. F. SCHOUTEN

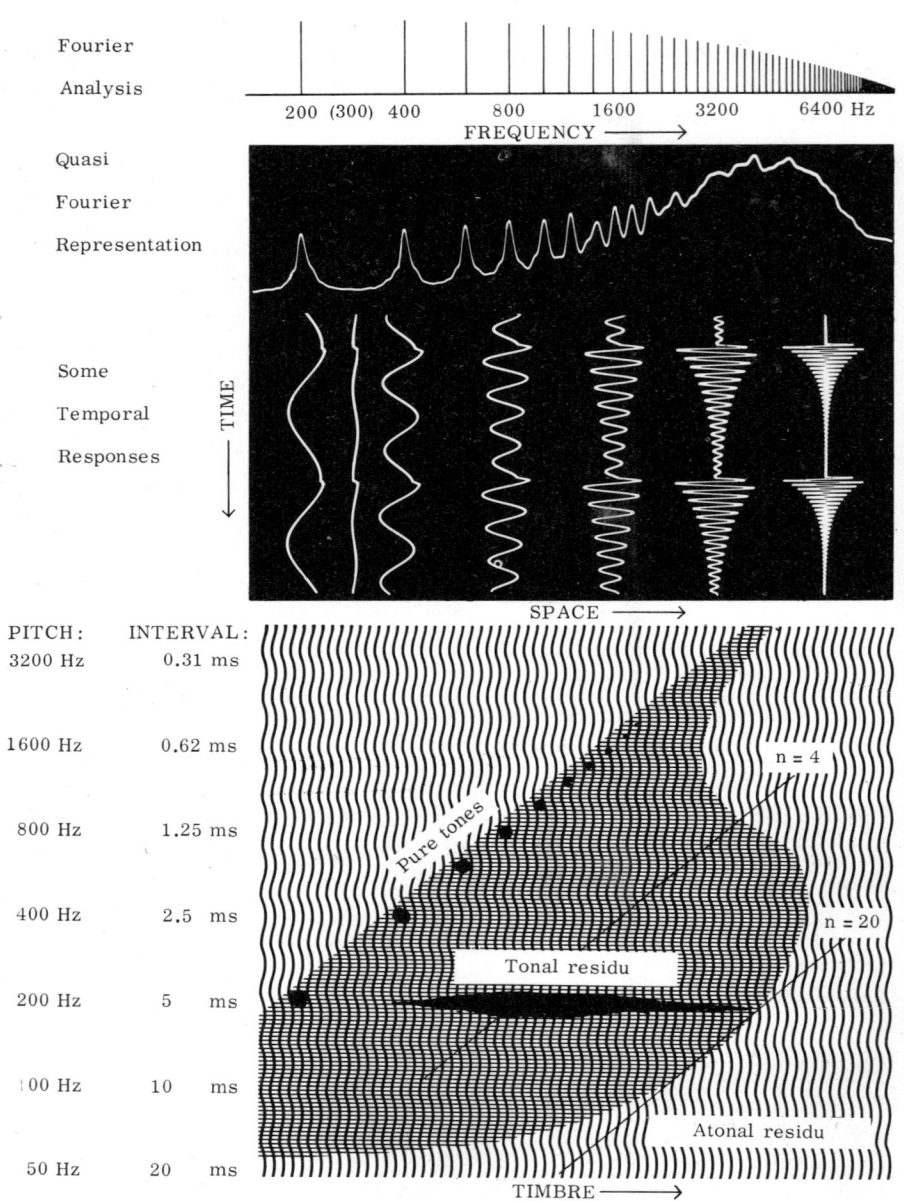

Fig. 4. A model of the genesis of the residue.
A sharp periodic pulse of 200 Hz is taken as an input. The Fourier analysis contains all harmonics in gradually decreasing amplitude. At some relevant representation of the basilar membrane a quasi-Fourier spectrum (excitation pattern) is formed. The waveform of the responses at the place corresponding to 200, 300, 400, 800, 1600, 3200 and 6400 Hz is shown. In the pitch extractors (below) the horizontal lines indicate the measurement of time intervals perceived as pitch. The vertical lines indicate a further projection of frequencies, perceived as timbre (coloration). For pure tones pitch and timbre have a fixed relation. This is not the case for the residue. The tonal residue is obtained within the existence region.

pitch perception forcibly lead to the assumption of a set of parallel pitch extractors connected to the relevant representation of the basilar membrane. Each extractor is assumed to respond to a limited area of frequencies and to determine pitch by the measurement of the *time interval* between recurrent time patterns in the signal presented to it. Thus a particular pitch, say of 200 Hz, may be reported by a number of pitch extractors simultaneously.

The existence diagram provides the modelmaker with two important clues. First, the limitation to $f = 5000$ Hz, to which we shall return later. Secondly, the limitation to a *minimum* value of g dependent upon the region. This is a highly important clue. Let us remember that, due to the limited resolving power, there will be overlap in excitation curves: one area of the basilar membrane (or of its relevant representation) may respond to several Fourier components of the presented waveform. This overlap ties in with the concept of critical bandwidth: the average area over which loudness summation takes place (Zwicker *et al.*, 1957).

Moreover the overlap acts as a *source* for three phenomena: beats, nonlinear distortion, and the residue. In fact the residue may be interpreted as the tonal sensation of beats between higher harmonics. As far as this overlap is concerned one might expect the residue to become the more pronounced the closer the harmonics are. Experimentally, almost the contrary is the case.

According to Ritsma's (1967) principle of dominance the tonal residue is most pronounced for $f = 4g$ and, keeping f constant, it becomes less pronounced as we lower g until it loses its tonality at the border of the existence diagram.

As a modelmaker it seems feasible to look at the pitch extractors for an explanation of this paradoxical effect and to assume that each pitch extractor has a limited range of measurable time intervals. The lower limit of the existence diagram then would be determined by the largest time interval measurable by each pitch extractor as a function of its area f.

Knowledge of the nonlinear behaviour of the human ear was increased considerably by the measurements of Zwicker (1955), Plomp (1965) and Goldstein (1967) on the combination tone $2f_1 - f_2$ (f_1 being the lower and f_2 the higher frequency presented). This combination tone is surprisingly strong as compared, for instance, to the notorious difference tone $f_2 - f_1$.

Goldstein investigated the combination tones $(n+1)f_1 - nf_2$ thoroughly and found no evidence of the corresponding tones $(n+1)f_2 - nf_1$ which flank the mother frequencies at the high-frequency side. Plomp (1966) offers masking as an explanation. We are tempted, however, to embroider upon Goldstein's (1967) suggestions regarding "a frequency selectivity at the site of the nonlinearity".

As a simple example, let us take the experimental fact that, for sufficient loudness, the difference tone $f_2 - f_1$ is heard distinctly but not the summation tone $f_1 + f_2$. It is proved that the site or source of the nonlinearity is located at the overlap of the excitation curves of f_1 and f_2. If we assume that this site, due to the asymmetry of its response curve (or tuning curve), can respond to frequencies lower than its critical frequency but *not* to higher ones, we obtain the model of a nonlinear source constrained by a frequency-dependent network which forbids higher frequencies to be formed. The source has to adjust itself to a working point fulfilling these restrictions. In the case of quadratic nonlinearity the solution is obvious and entails the suppression of other high-frequency products like the octave $2f_1$ and $2f_2$ (Fig. 5).

Whether the same argument holds for the much closer spacing of the higher order combination tones still remains to be investigated.

The asymmetric appearance of the combination tones $(n+1)f_1 - nf_2$ may have a great bearing on our modelmaking to account for the second effect of pitch shift. If the signal presented contains several harmonics, the nonlinearity will produce diverse additional *lower* harmonics. In addition the lower mother frequencies will be biased with respect to the higher ones. Thus the effective

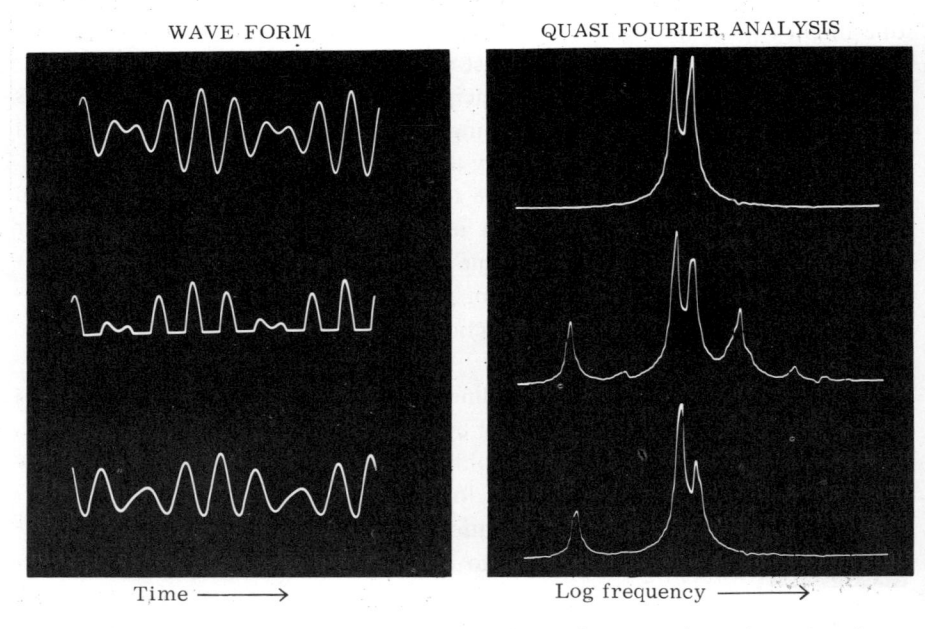

Fig. 5. Quadratic nonlinear distortion constrained by a frequency dependent network.
Above: Two frequencies, $f_1 = 800$ and $f_2 = 1000$ Hz, of equal amplitude.
Middle: Effect of onesided limiting. Note the difference tone f_2-f_1 and *e.g.* the cluster formed by $2f_1$, f_1+f_2 and $2f_2$.
Below: The effect of the limiter is constrained by a network forbidding the formation of higher components. Note the disappearance of the cluster and the bias of f_1 with respect to f_2.

centre of gravity will shift even more towards a lower frequency than de Boer envisaged.

Let us finish this section by noting a remarkable success along the upper side of our triangle obtained by Goldstein and Kiang (1968) who found the neurophysiological correlate of the aural combination tone $2f_1 - f_2$.

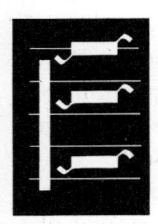

Even though it seems certain by now that the pitch of the residue is determined by a pitch extractor measuring time intervals, it should be noted that no direct proof exists that the pitch of pure tones is equally determined by a time measuring instrument. The available evidence is circumstantial rather than direct. An example is Cardozo's (1962) finding that the acuity of pitch discrimination for short tone pips beats *any* measuring system based on spectral resolution, independent of resolving power.

The experimental fact that the upper limit of 5000 Hz of the existence diagram corresponds roughly to the upper limit of our sense of musical interval also provides circumstantial rather than direct evidence.

Yet, our very remarkable sense of musical interval and the well-known ambiguity in judging octaves, fifths, etc., provide us with another clue, be it circumstantial again, that pitch is determined by one and the same mechanism. Our musical scale is *not* one-dimensional but repeats itself every octave. This applies both to pure tones and to residues.

Spiral models with polar coordinates have been proposed. The essence of any polar coordinate system, though, is that not only an angle of 360° (an octave) but also an angle of 180° plays an essential role. The latter corresponds to the interval of a tritonus which can hardly be called essential as far as musical intervals are concerned.

An interval measuring pitch extractor could easily misfire by skipping one or more interval endings and thus lead to ambiguities of an octave, a fifth, etc.

> The author is aware of the fact that the weight of this argument boils down to the question whether our sense of musical interval is innate or acquired. If it were acquired, how could we possibly explain, not that we judge an octave so well, but rather that we err systematically? The mathematically pure octave, for instance, is heard definitely, alarmingly and persistently flat. This, to the author's opinion points rather towards an innate equipment measuring musical intervals with systematic errors (be it, as far as this argument is concerned, of the time measuring *or* the frequency measuring type) rather than towards an acquired faculty which never quite reaches the original task setting.

Three examples of circumstantial evidence do not provide a direct proof. Yet, the author would be immensely surprised if the pitch of a pure tone were *not* assessed by a time interval measurement.

ringing to mind again the three angles from which we viewed the progress through knowledge towards wisdom about the auditory system we should like to make some closing remarks.

With respect to physiology and anatomy a paramount problem is to find the neurophysiological correlate of the residue. Auditory physiology is making great progress (see *e.g.* Kiang, 1968), but it should be remarked that if the variety of acoustic stimuli used remains restricted to pure tones, noise, and pulses, the results will be of little avail with respect to the problem raised. The residue phenomenon and the location of pitch extraction may slip through the physiologist's fingers unless specially tailored signals, like modulated sine waves, are used.

With respect to perception, the old adage that listening should be performed with one's ears rather than with one's pen is still valid. New perceptual phenomena are bound to be observed and may turn out to be of crucial importance in deciding between alternative hypotheses.

With respect to theoretical insight, irreverently called modelmaking throughout this paper, we have to find out whether the observed phenomena can be attributed either to the basilar membrane (or its representation) or to the pitch extractors or to both. The distinction in terms of "frequency domain" and "time domain" respectively may easily lead into logical pitfalls. The basilar membrane, in *not* performing an ideal Fourier analysis, operates *both* in the frequency domain *and* in the time domain. Any onset of or any change in a regular vibration of the basilar membrane leads to transients in its response which, in principle, can be detected on the spot.

Finally, with respect to the three sides of our triangle, the importance of cooperation and mutual inspiration cannot be sufficiently stressed.

The author is greatly indebted to Messrs. Cardozo, Ritsma and Cohen for inspiring discussions and to Messrs. Admiraal, Alewijnse and de Jong for preparing the demonstrations.

DEMONSTRATIONS

The lecture was given in the form of a set of demonstrations. These consisted of various sources of sound which were displayed both as a function of time on an oscilloscope and as a function of frequency on an acoustical spectroscope with a resolving power of half a tone (see "quasi-Fourier representation" in Fig. 4 as also in Fig. 5).

The listener could compare his perception of these sounds with the objective displays of both the waveform and the spectrum.

All special apparatus was developed in our laboratory.

(1) *One single and variable freuqency*. A pure tone is heard, the pitch of which goes up with smaller period and larger frequency.

(2) *Masking of a single frequency by a narrow band of noise.* The pure tone is heard when it is visible on the spectroscope either below or above the frequency region of the noise band. It becomes inaudible when it is seen to merge with the noise band on the spectroscope.

(3) *Demonstration of Seebeck's observation of pulses with alternating time intervals* (see Fig. 2). In one demonstration the pulses were produced by a turning disc with holes along four concentric circles, spaced equally to Seebeck's experiment. The sound was produced by processing the response of a photo-electric cell to a constant light source placed behind the disc. In another demonstration the pulses of alternating spacing were produced by a modular time source.

(4) *Periodic pulse with suppression of lower harmonics.* If the lower harmonics are suppressed by a high-pass filter of variable cut-off frequency these harmonics are seen and heard to disappear. The sharp timbre and the pitch equal to that of the fundamental frequency remain unaffected (the residue).

(5) *Audibility of the lower harmonics.* With the "Pan's flute" the lowest ten harmonics can be generated with adjustable amplitude and phase. The disappearance and reappearance of the lowest five harmonics, both objectively and perceptually, was demonstrated.

(6) *The residue obtained by modulation of a carrier frequency.* A residue with a pitch corresponding to that of a pure tone of 250 Hz is clearly heard when a carrier frequency of 2000 Hz is modulated in amplitude with a frequency of 250 Hz. The first and second effect of pitch shift were demonstrated.

(7) *Masking of pure tones and residues with noise.* The demonstration is a variant of one originally given by Licklider. Each tone of the Westminster chime is given twice: first by a pure tone in the frequency region around 200 Hz, second by a residue of the same pitch but consisting of harmonic frequencies in the region of 2000 Hz. If low-frequency noise is added the pure tones are masked and one hears the melody of the residues. If high-frequency noise is added the residues are masked and one hears the melody of the pure tones.

REFERENCES

Békésy, G. von (1928): Zur Theorie des Hörens; die Schwingungsform der Basilarmembran, Phys. Z. *29*, 793-810.
Békésy, G. von (1934): Ueber die nichtlinearen Verzerrungen des Ohres, Ann. Phys. *20*, 809-827.
Békésy, G. von (1963): Hearing theories and complex sounds, J. Acoust. Soc. Amer. *35*, 588-601.
Boer, E. de (1956): On the Residue in Hearing, Doctoral Dissertation, University of Amsterdam.
Cardozo, B. L. (1962): Frequency discrimination of the human ear, in: Reports 4th International Congress on Acoustics, Copenhagen, Vol. I, H 51.
Fletcher, H. (1929): A space-time pattern theory of hearing, J. Acoust. Soc. Amer. *1*, 311-343.
Goldstein, J. L. (1967): Auditory nonlinearity, J. Acoust. Soc. Amer. *41*, 676-689.
Goldstein, J. L., and Kiang, N. Y.-S. (1968): Neural correlates of the aural combination tone $2f_1-f_2$, Proc. IEEE *56*, 981-992.
Helmholtz, H. L. F. von (1863): Die Lehre von den Tonempfindungen als physiologische Grundlage für die Theorie der Musik (F. Vieweg & Sohn, Braunschweig), 1st. Ed.
Hermann, L. (1890): Phonophotographische Untersuchungen III, Arch. ges. Physiol. *47*, 347-391.
Hermann, L. (1912): Neue Versuche zur Frage der Unterbrechungstöne, Arch. ges. Physiol. *146*, 249-294.
Hoogland, G. A. (1953): The Missing Fundamental, Doctoral Dissertation, University of Utrecht.
Kiang, N. Y.-S. (1968): A survey of recent developments in the study of auditory physiology, Ann. Otol. Rhinol. Laryngol. *77*, 656-675.
Licklider, J. C. R. (1954): 'Periodicity' pitch and 'place' pitch, J. Acoust. Soc. Amer. *26*, 945 (A).

Ohm, G. S. (1843): Ueber die Definition des Tones, nebst daran geknüpfter Theorie der Sirene und ähnlicher tonbildenden Vorrichtungen, Ann. Phys. Chem. *59*, 513-565.
Ohm, G. S. (1844): Noch ein Paar Worte über die Definition des Tones, Ann. Phys. Chem. *62*, 1-18.
Plomp, R. (1965): Detectability threshold for combination tones, J. Acoust. Soc. Amer. *37*, 1110-1123.
Plomp, R. (1966): Experiments on Tone Perception, Doctoral Dissertation, University of Utrecht.
Plomp, R. (1968): Pitch, timbre and hearing theory, Internat. Audiol. *7*, 322-344.
Ritsma, R. J. (1962): Existence region of the tonal residue. I, J. Acoust. Soc. Amer. *34*, 1224-1229.
Ritsma, R. J. (1963): Existence region of the tonal residue. II, J. Acoust. Soc. Amer. *35*, 1241-1245.
Ritsma, R. J. (1967): Frequencies dominant in the perception of the pitch of complex sounds, J. Acoust. Soc. Amer. *42*, 191-199.
Schouten, J. F. (1938): The perception of subjective tones, Proc. Kon. Nederl. Akad. Wetensch. *41*. 1086-1093.
Schouten, J. F. (1940a): The residue, a new component in subjective sound analysis, Proc. Kon. Nederl. Akad. Wetensch. *43*, 356-365.
Schouten, J. F. (1940b): The residue and the mechanism of hearing, Proc. Kon. Nederl. Akad. Wetensch. *43*, 991-999.
Schouten, J. F. (1940c): The perception of pitch, Philips Techn. Rev. *5*, 286-294.
Schouten, J. F. (1960): Five Articles on the Perception of Sound (1938-1940) (Instituut voor Perceptie Onderzoek, Eindhoven, The Netherlands).
Schouten, J. F., Ritsma, R. J., and Cardozo, B. L. (1962): Pitch of the residue, J. Acoust. Soc. Amer. *34*, 1418-1424.
Seebeck, A. (1841): Beobachtungen über einige Bedingungen der Entstehung von Tönen Ann. Phys. Chem. *53*, 417-436.
Seebeck, A. (1843): Ueber die Sirene, Ann. Phys. Chem. *60*, 449-481.
Seebeck, A. (1844a): Ueber die Definition des Tones, Ann. Phys. Chem. *63*, 353-368.
Seebeck, A. (1844b): Ueber die Erzeugung von Tönen durch getrennte Eindrücke, mit Beziehung auf die Definition des Tones, Ann. Phys. Chem. *63*, 368-380.
Wever, E. G. (1949): Theory of Hearing (John Wiley & Sons, New York).
Wundt, W. (1880): Grundzüge der physiologischen Psychologie, Vol. 1 (Verlag W. Engelmann, Leipzig), 2nd Ed., pp. 315-317.
Zwicker, E. (1955): Der ungewöhnliche Amplitudengang der nichtlinearen Verzerrungen des Ohres, Acustica *5*, 67-74.
Zwicker, E., Flottorp, G., and Stevens, S. S. (1957): Critical band width in loudness summation, J. Acoust. Soc. Amer. *29*, 548-557.

DISCUSSION

Piazza: I did some experiments in which the pitch shifts of pure and complex tones as a function of SPL were compared (Piazza, 1967). The pitch of the pure tone shifted in accordance with the data of other investigators, but for the complex tone, consisting of the first and second harmonics, no significant pitch shift was found. Has Dr. Schouten any explanation of this result?

Schouten: May I give an example? If you take a pure tone of 200 Hz and you raise the loudness, then you get a considerable drop in pitch. If you take the residue of 200 Hz, however, you don't get such a shift. In the place theory, one could relate the drop in pitch to a shift of the maximum of stimulation along the basilar membrane. If one has grounds to believe that pitch is determined by the periodicity of the stimulus, irrespective of the location along the basilar membrane, rather than by the maximum of stimulation, then one has to explain the pitch shift in terms of the pitch extractors. We have no feasible explanation of how these pitch extractors react differently to signals of different loudness.

Terhardt: Walliser's experiments have shown clearly, I think, that the pitch of pure tones and periodicity pitch have very much to do with each other (see pp. 281-286 of this volume). This does not decide whether the pitch of a pure tone derives from time analysis or place analysis. In my opinion, there is some evidence for saying that the residue pitch is quite related to the pitches of the harmonics that are actually presented. I will go further into this question in my paper.

Zwislocki: May I ask Dr. Schouten whether I understood him correctly: is your point that the mechanism for pitch perception is only in terms of the time domain or do you accept also place as a basis for pitch?

Schouten: First of all, we have several grounds for assuming that time measurement holds only for frequencies lower than about 5000 Hz. Our sense of musical intervals breaks down there, but still we can hear a sound going higher up in pitch, so there must be some mechanism for that. We don't know whether just time alone, or also location along the basilar membrane, determines pitch perception but, personally, I would be extremely surprised if pitch sensation would come from two entirely different mechanisms. One of the most curious of all phenomena in pitch perception is our sense of musical scale; our sense of octave, fifth and fourth intervals. For me, no theory is quite finished unless it accounts for that sense. But, as I said in my paper, we have no direct proof that the pitch of pure tones is indeed determined only by time measuring.

Zwislocki: In physics, we are used to measuring frequency either in terms of the time domain or by using frequency-selective filters. But this does not have to be done that way necessarily in the auditory system. Why should we exclude any one of these analyses? I think, it would be quite reasonable to assume that in the biological system as much information is being used as is available. Looking back at all the experiments I have seen done in the past, it seems to me that some experiments are very difficult to explain in terms of the place theory, on the other hand, other experiments are very difficult to explain in terms of periodicity pitch. Many experiments on pitch are done by means of pure tones; in this case, you cannot separate frequency and periodicity. When you do experiments with complex sounds, some show that place is important, others that the time domain is important. I begin to believe that both are involved. Although I am of the opinion that the place mechanism is the more important one, I do not believe that periodicity is irrelevant. There have been many experiments indicating that one can easily differentiate vibrations on the skin up to about 500 Hz, although no place analysis intervenes.

Schouten: Yes, but I would not like to insinuate that place does not play any role.

Johnstone: This is the first time that I have ever heard what Dr. Schouten calls a residue and I must say I am mightily impressed. It is obvious that the residue pitch does not arise through the nonlinearity of the basilar membrane because otherwise it could be masked at the fundamental frequency. But it is not so obvious that this pitch does not arise through nonlinearities at the same place of the basilar membrane that is simulated by the harmonics. One possible source is the hair cell bending distortion which several people, including myself, have analyzed, and the other one is the nonlinearity of the electric transducer system, discussed by Dr. Dallos in his paper (see p. 218). Both nonlinearities might give a patterning in the nerve fibre in which the fundamental frequency is present, though I don't know how.

Whitfield: In relation to what Dr. Johnstone said, I may remark that the hair cell transducer itself is highly nonlinear. It fires off a pulse when the hair is moved in one direction, so that you are in fact extracting the periodicity of the stimulus from this nonlinearity. I thought that this was just the argument used in the periodicity theory. What strikes me, however, about the residue is how different it sounds from a pure tone of the same pitch; you have no doubt about which is which. This makes me wonder whether one ought to draw the conclusion that if periodicity is responsible for the pitch of the residue, it must also be responsible for the pitch of a pure tone.

Schouten: Concerning the role of nonlinearity, I may say that, whatever nonlinearities one would like to contemplate, one has to bear in mind that the low pitch observed correlates with the location of the originating frequencies and not with that of a pure tone of equal pitch. However, they can be matched very well, provided one mentally discards the difference in timbre.

Keidel: If we look at the whole problem in the light of biology, we should keep in mind that all tactile receptors are able to follow periodicity up to about 500 Hz. Now, what is the point of evolution in the human ear? Obviously, there are two tricks used for improving pitch discrimination. The first is the volley theory, and in this way pitch perception up to perhaps 5000 Hz can be explained. Then, you need a second trick which implies the distribution of frequencies along a given scale, and in this, I would like to sequel what was mentioned by Dr. Zwislocki. Guelke and Huyssen (1959) in Cape Town tried to use the mechanical receptors of the fingers to receive information of speech. They failed for consonants, but succeeded for the fundamentals of vowels. This means that, since the skin can distinguish frequencies over a limited frequency range up to about 500 Hz, as is also demonstrated by other experiments (Keidel, 1958; Biber, 1961), it is obvious that the same thing can occur at any point along the basilar membrane. Now, if we take the next step in the direction nature obviously did, and use a dispersion model of the form as we have tried in our laboratory (Keidel, 1958; Biber, 1961), it is easy to receive even the speech formants. So, this dispersion mechanism provides an increased frequency range. In my opinion, volley principle and frequency dispersion can be considered as tricks of nature to improve human hearing. The latter one is represented by the frequency-place distribution along the basilar membrane and, at any place, there is the general capacity of the mechanical receptor to process periodicity information up to about 500 Hz.

Schouten: The whole problem is indeed manifest in the perception of speech formants. If we have a vowel with a formant of say, 2000 Hz, then we have to account for the fact that this formant can be heard with a constant timbre corresponding to its frequency area irrespective of the periodicity of the sound, which may correspond to 100, 200, 300 or 400 Hz. So, we have in all these cases the same spectral envelope, which is constant along the basilar membrane, and yet we can hear it with different pitches. Therefore, there must be some pitch extractor.

Plomp: In relation to this and to what is said about the role of nonlinearities in pitch perception, we should bear in mind that the problem is not what exactly happens at the location corresponding to the formant frequency of, taking the same example, 2000 Hz, but the way in which the low pitch is coded in the auditory nerve. Since we may exclude place as a pitch code for signals consisting of only a group of higher harmonics, it seems to be inescapable that this code is the time pattern of the nerve impulses. We may not be able at the moment to say how significant the various sources of nonlinearity are, but this uncertainty does not suspend the role of the time pattern in pitch perception.

Schouten: By introducing the word coding, we have to bear in mind also that, at the end of the line, the coded message is received and measured. We have one measuring apparatus in the form of frequency analysis, but we need another for the periodicity of the signal. Whether the last one is exclusive for pitch perception or not, let us leave that alone for the moment.

Zwicker: The demonstrations by Dr. Schouten have convinced me again that there are two different kinds of pitches, if you like to say so. One is represented in the sinusoidal tone and the other one in the complex signals we have heard. Furthermore, I would like to point out that in the comparison of the pitch difference between the Seebeck siren discs with 10° spacing and with alternate 9.5° and 10.5° spacing, a big part of the sound actually was the same in the two cases, and only a very small part was different. In the sounds demonstrated, the pitch relation suggested by Dr. Schouten is quite faint compared to the similarity of the sound impression as a whole. This large similarity, I would say, is correlated with place, and the faint pitch shift is correlated with the time pattern. I still think, and I agree in this with Dr. Zwislocki, that we need both place theory and time pattern to understand what is going on, and I am very sure that each sensory organ tries to pick up as much information as possible.

Schouten: We are able to pronounce vowels with completely different formants on the same pitch or on different pitches, and in all cases we can distinguish them. In this respect, we come into the realm of the differences in timbre, which will be also discussed in this symposium. Of course, for pure tones one cannot make a distinction between the pitch and the timbre because they just go hand in hand up and down the scale. So, perhaps a part of the argument of Dr. Zwicker is a matter of definition, what is to be called pitch and what timbre.

Bosher: I would like to make only a brief comment to supplement what Dr. Zwicker has just said. The quotation of Seebeck that only the ear can decide what makes out a tone is, in my opinion, quite wrong as, of course, it is in the brain that pitch is decided. In the field of vision, a lot of work has been done on pattern recognition. If you present the eye with a series of circle-like shapes and ask the subject to say whether they are circles or squares, he will go on calling them circles until they are almost square. If you present him with square-like shapes initially, he will call them squares until they are almost circles. So, the same shape is called a circle in one context and a square in another. The same thing may hold for the sounds to which Dr. Zwicker referred. Decisions on their similarity and dissimilarity depend upon the question asked and what one's mind is trying to do. Again, if you present random visual objects, such as ink thrown on a piece of paper, the subjects will try to find patterns. Indeed, they will report patterns when in fact no patterns exist. So, it seems that in man the visual system is always trying to find patterns, and presumably this holds also for the auditory system. This makes the interpretation of the demonstrated sounds very complex, as Dr. Zwicker has said.

Schouten: I quite agree that our perceptions depend upon our cumulative perceptual experience. Yet, let us not forget that our cognitive mind operates upon the preprocessed perceptual information. Therefore, the preprocessing by the preceding receptors, detectors, extractors, etc. does enter into ultimate cognition.

REFERENCES

Biber, K. W. (1961): Ein neues Verfahren zur Sprach-Kommunikation über die menschliche Haut, Doctoral Dissertation, University of Erlangen, Germany.
Guelke, R. W., and Huyssen, R. M. J. (1959): Development of apparatus for the analysis of sound by the sense of touch, J. Acoust. Soc. Amer. *31*, 799-809.
Keidel, W. D. (1958): Note on a new system for vibratory communication, Perceptual Motor Skills *8*, 250.
Piazza, R. (1967): Effetto della struttura armonica su alcuni fattori che influenzano il giudizio d'altezza, Giornale dell'Accademia di Medicina di Torino *130*, Fasc. 7-12, 325-335.
Walliser, K. (1969): Zusammenhänge zwischen dem Schallreiz und der Periodentonhöhe, Acustica *21*, 319-329.

Section 2

THE COCHLEAR FUNCTION

Since the time of von Helmholtz, frequency analysis has been accepted generally as the major function of the cochlea. In this section research is reported which deals with the underlying filtering process directly through studies of the movements of the basilar membrane and indirectly by measurements of such related electrophysiological activities as cochlear microphonics and summating potentials. The selectivity of this filtering is the central subject of the following papers.

BIOPHYSICS, MECHANICS AND ELECTROPHYSIOLOGY OF THE HUMAN COCHLEA

WOLF D. KEIDEL

*I. Physiologisches Institut der Universität Erlangen-Nürnberg
Erlangen, Germany*

Dealing with the topic mentioned in the title it seems obvious, that this topic should be divided into at least two (better than into three) parts: namely biophysics from the standpoint of mechanics and hydrodynamics on one side and the electrophysiological events within the cochlea on the other hand.

MECHANICS OF THE COCHLEA

Up to very recent times it looked like that the first part of this review article would be settled in a nearly classical sense. This means the so-called *travelling wave theory* (von Békésy, 1960; Ranke, 1931, 1950a, 1950b; Zwislocki, 1948; Tonndorf, 1958; Fletcher, 1951; Siebert et al., 1962) was so well established by so many exactly fitting experiments of very different types that nobody could imagine that not later than in March 1969 Andrew Huxley, one of the Nobel-price winners, would publish an article entitled "Is resonance possible in the cochlea after all?". This publication certainly emphasizes the need for a clearcut and critical discussion, how well the travelling wave theory would be established nowadays in fact and what evidence we would actually have for and against it on one side and for and against the resonance theory on the other side. I certainly would not like to repeat all the facts brought together so beautifully in G. von Békésy's "Experiments in Hearing" (1960), a book which actually describes the results of a life filled with so many ingenious experiments as even possible for the merit of another Nobel-price and establishing him as the expert on this field in the world.

It is absolutely convincing for everybody that a system vibrating in the mode of pure resonance (1) has to have a radial tension what was experimentally disproved by von Békésy's (1960) longitudinal cut into the cochlear partition revealing no radial displacement after the cut at all. The only remaining objection might be, that this experiment was done on a human cadaver and not on a living ear as mentioned by Huxley (1969). (2) The maximum possible phase at any place behind the site of resonance relative to the moving force at the stapes can only be one π and even only half π at the resonant site of the cochlear partition. The measured size of this phase angle, however, was clearly much higher, namely in the order of 3π. (3) It was as early as in 1905 that Wien stated his classical objection against the resonance theory of hearing: either the damping factor of a resonating system is high, then this might be in accordance with the quick transients observed in hearing, when perceiving for instance trill frequencies in the order of 20 Hz, or the damping factor is low, then this could explain the fine difference limen for frequency in the order of 0.2% for optimum frequency. Since *both* observations, however, can be made simultaneously at the same analysing system, namely the human ear, obviously something must be wrong with any trial to explain the ear's hydrodynamics on the basis of a pure resonance theory in the sense of von Helmholtz (1863). Nevertheless, there have been published quite a few purely speculative theories as, for instance, the "whip-wave theory" (Kietz, 1959, 1960) and others which all neglect the fact that the damping factor of the inner ear has been measured already by von Békésy and has been found to be relatively high, so it would be inconsistent with the quick transients (necessary for speech recognition) and also explaining the broad band of masking high sinusoidal tones by low ones as well as a relatively long distance upon which long lasting fatigue on *one* ear by a sinusoidal tone influences the pitch shift relative to the other ear which was not exposed to the sound (von Békésy, 1929).

So, what still has to be explained is not that the envelope of the partition's displacement should be of the order to correspond to the difference limen for frequency of 0.2%, for this is not possible hydrodynamically because of the high damping factor measured. What has to be explained therefore rather should be the question of how is it managed by the human ear that although the envelope of the cochlear partition's displacements is relatively flat (in accordance with the high damping factor and with the travelling wave theory), the auditory information is handled in such a way that the extremely low difference limen for frequency in human hearing results *in spite of* the high damping factor. This problem becomes even more complicated since tuning curves at all levels of the auditory pathway now are available and partially (at least in cat for a few of all fibres) show much steeper flanks than the envelope of the cochlear partition's displacements. This should be shown by some typical figures: Fig. 1 shows Johnstone's (1967; p. 85 of this volume) direct measurements of the cochlear partition's displacements of the cat by means of the Mössbauer effect. These curves are in excellent agreement with the measurements of von Békésy (1960) on the human ear near the helicotrema. Since Johnstone's measurements are made near the stapes the entire length of the

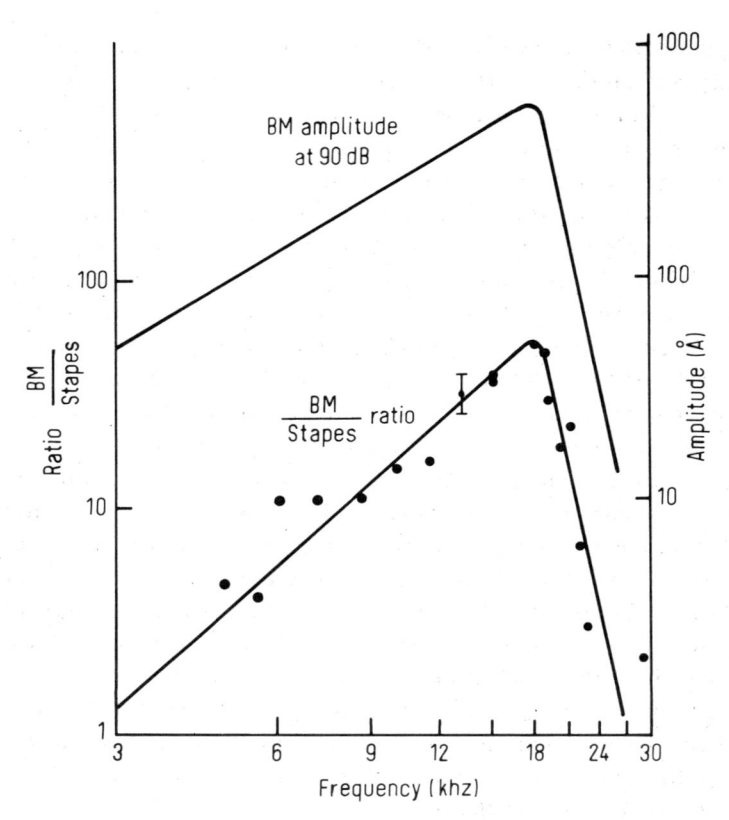

Fig. 1. Tunning curve of the basilar membrane at 1.4 mm from stapes. The lower curve is referred to a constant stapes amplitude. The upper curve is for a constant sound pressure level of 90 dB SPL; the amplitudes are peak ($\sqrt{2}$ times the root-mean-square) values (Johnstone and Boyle, 1967; Copyright 1967 by the American Association for the Advancement of Science).

cochlear partition obviously vibrates in a very similar way and with a relatively flat envelope for the displacements of the cochlear partition. There are on the other hand the old measurements of Fernandez (1951) about the dimensions of the cochlea of the guinea pig. He compared the width of the basilar membrane for guinea pigs with that of the human ear and showed somehow clearly that along the basilar membrane this width of the cochlear partition increases toward the helicotrema about logarithmically. According to that finding in the model of Siebert (1962) for the dynamic behaviour of the cochlear partition, it is assumed that the ratio of the volume displacement at stapes divided through the maximum amplitude at cochlear partition at any point drops logarithmically up to about 1 kHz. In addition he programs his model so that from thereon for the last 10 mm to the stapes, with other words for frequencies between 1 and 18 kHz, this ratio would be constant. This again is in agreement with measurements for the input impedance of the inner ear of the cat made by Tonndorf et al. (1966). They found that the input impedance of the inner ear drops with a steepness of 60 dB/octave from the low frequencies, say 20 Hz to 1 kHz, and then is nearly constant up to the highest perceived frequencies. Putting all those data together it is well argued if one in a model of the vibrations of the cochlear partitions assumes this amplitude displacement behaviour along the basilar membrane. This was done as well in Siebert's (1962) model as in a model calculated by Miss Breuing (1969) in our laboratory. She used a slightly modified formula of Ranke's (1931, 1950a, 1950b) type and fed it into a CDC 3300 computer using the correct damping factor. By this means she was able to display the vibrations of the cochlear partition completely in accordance with the travelling wave theory of von Békésy. One example is shown in Fig. 2 for a given frequency. It might be of interest in this connection, that with Breuing's model it is possible to compare the displacement of the cochlear partition at any given point, with other words: the actual travelling wave, with its envelope for a situation where two sinusoidal tones are mixed together. The result of this special case is shown in Fig. 3 for the envelope, the travelling wave itself and the energy of the wave. In Fig. 3a-c it can be seen clearly that for two sinusoidal tones which differ in frequency by a factor of two the envelope of the displacement shows nearly no separation between the two tones while the travelling wave itself, with other words its periodicity, its time function, differs clearly for the two frequencies. The energy curve lies somewhere between. Finally another model was fed into a LINC 8 computer by Finkenzeller at our laboratory. In this model the input impedance according to Tonndorf's measurements was corrected, the intensity, frequency and phase parameters could be computed continuously and displayed on the screen of a cathode ray tube. For this model a special program was developed which was shown in the symposium by means of a movie which was a photograph from the screen of the CRT of the LINC 8 computer when delivering Finkenzeller's program (Fig. 4).

Up to now we discussed only the perpendicular movements of the cochlear partition from the point of view of the hydrodynamics and mechanics of the inner ear as a complex system. Now if one proceeds to the crucial question how this perpendicular displacement is translated into the radial bending of the hairs of the receptor cells of the organ of Corti some new and very difficult problems arise. It could be shown clearly in the past that the adequate stimulus is not the first derivative of the perpendicular displacement, but is linearly proportional to the vibrations of the cochlear partition (v. Békésy, 1960; Davis, 1957; Ranke et al., 1952; etc.).

This means that the theory of Neubert according to which the endolymph is accelerated by the cochlear partition's displacements like a jet can no longer be accepted (Neubert and Wüstenfeld, 1955; Neubert, 1960). For in this case there had to be the adequate stimulus the first derivative of the displacement

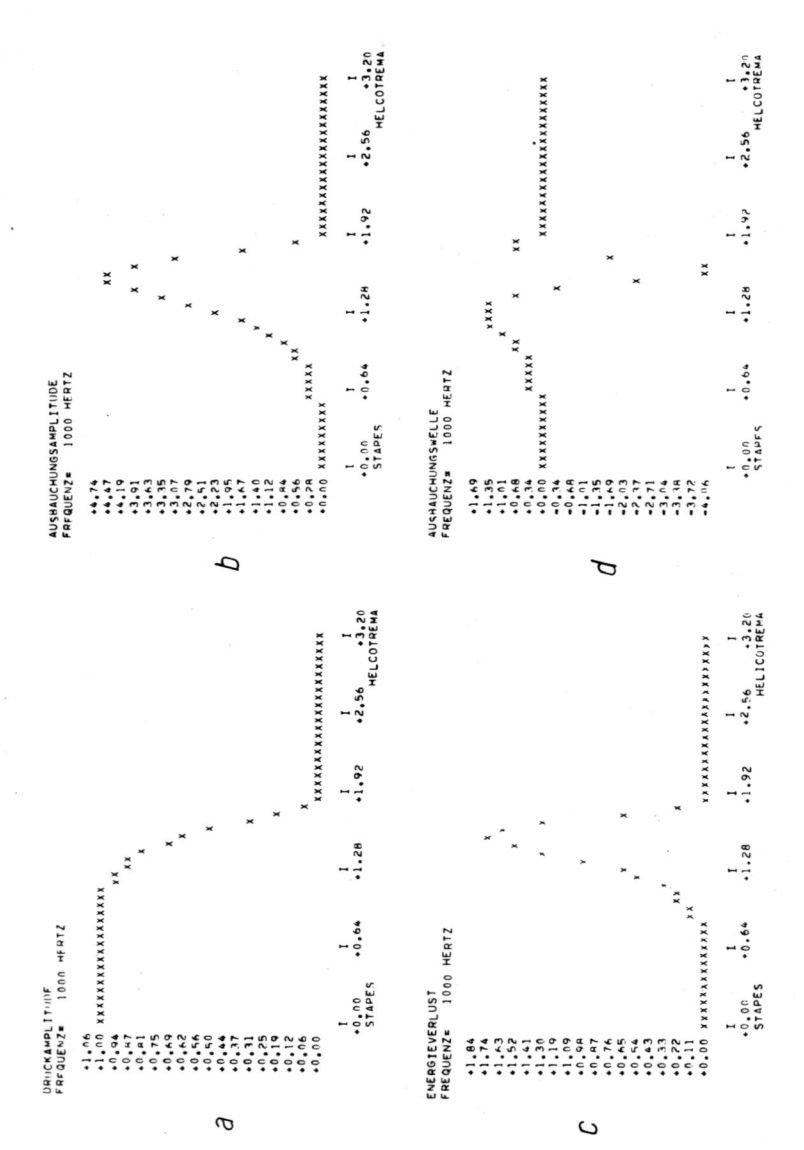

Fig. 2. Sinusoid of 1 kHz. (a) envelope of pressure of the cochlear partition; (b) envelope of displacement of the cochlear partition; (c) energy loss along the cochlear partition; (d) momentary picture of travelling wave of displacement of the cochlear partition.

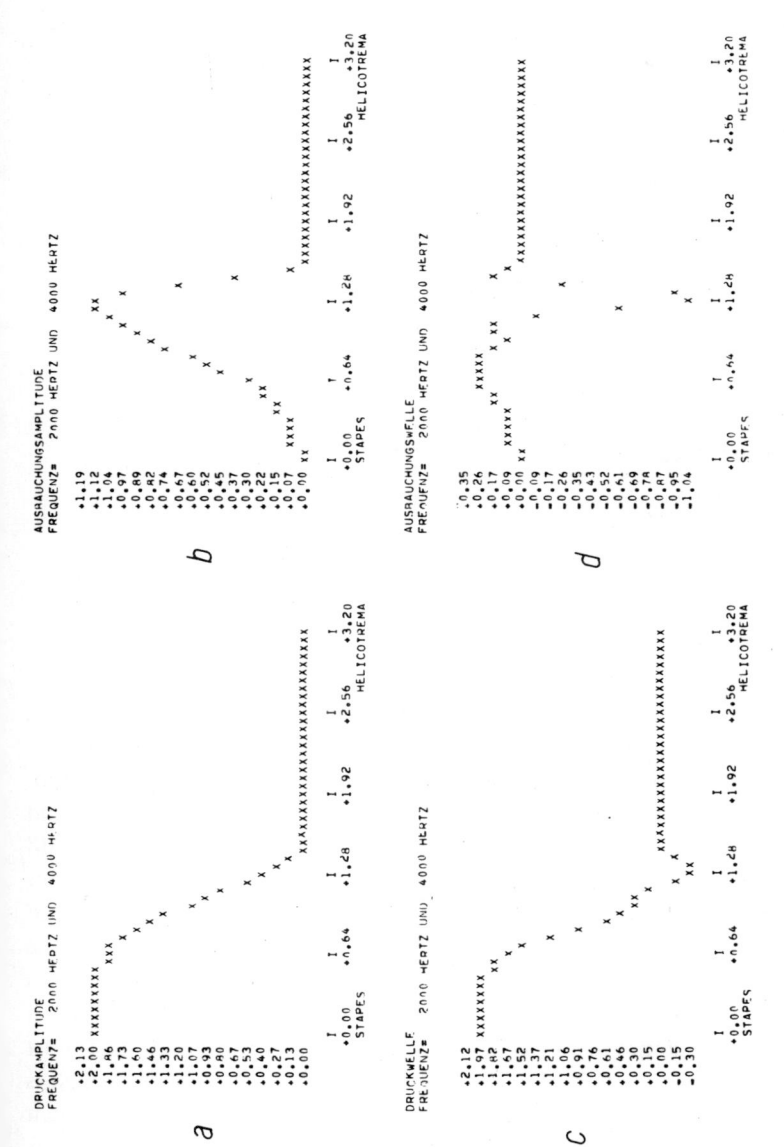

Fig. 3. Sinusoids of 2 kHz plus 4 kHz simultaneously. (a) envelope of pressure of the cochlear partition; (b) envelope of displacement of the cochlear partition; sinusoids of 2 kHz plus 4 kHz simultaneously; (c) energy loss along the cochlear partition; (d) momentary picture of travelling wave of displacement of the cochlear partition.

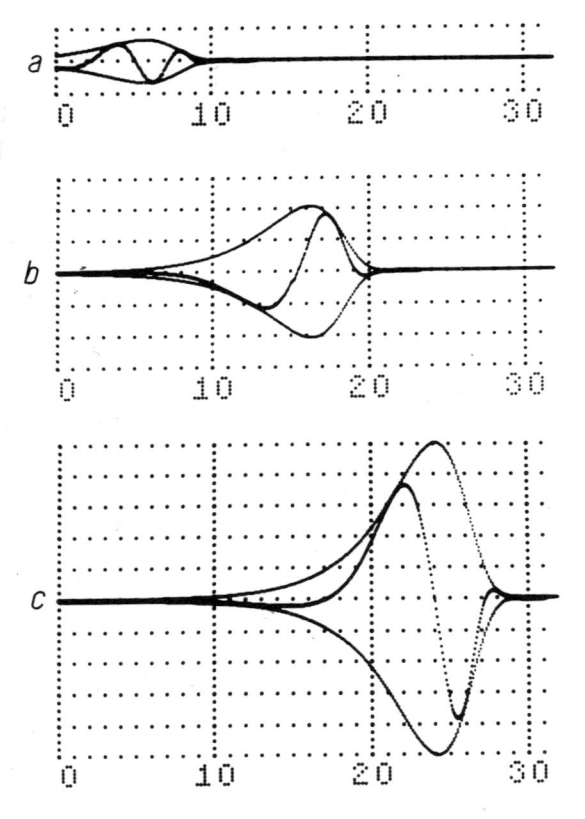

Fig. 4. Program Finkenzeller, LINC-8, LAP 4. Displacement (envelope) plus displacement of travelling wave at t_o. Three different frequencies decreasing from top to bottom trace. Abscissa: distance from stapes in mm.

itself what is not the case. Therefore von Békésy developed a model according to which the perpendicular displacement of the basilar membrane bends like a lever in a fixed position the hairs of the receptor cells of the organ of Corti. This again is in agreement with new findings of the anatomists that there would be no gap between the tectorial membrane and the hair cells but they would stick together closely and in a somehow fixed manner. There exists another additional observation of von Békésy according to which the transformation of the perpendicular movement of the cochlear partition into the bending of the hairs takes place in such a way that first the cochlear partition moves in a longitudinal direction, then purely perpendicular and then in a radial direction if one goes from the stapes towards the helicotrema. This means that there must be a mechanical process involved in the hydrodynamics of the cochlea which is responsible for some sort of a sharpening of the width of the displacement relative to the envelope of the displacement itself. If, for instance, what is very probable, only the radial movements are able to bend the hair cells in the way to trigger the excitation processes within the first neuron, only this small part of the vibrating partition would be responsible for the adequate stimulation.

This type of sharpening of the endolymph movement compared with the perpendicular displacement of the cochlear partition may be called the mechanical theory of the adequate stimulation within the human cochlea.

NEURONAL TUNING POSSIBILITIES

Further experiments will be necessary to clear this problem completely. Besides any mechanical explanation for the difference of the sharpness of the tuning curves of the first neuron and the flatness of the envelope of the displacement of the cochlear partition there are, however, neuronal possibilities which have to do with the special type of neuronal network which is realized within the layer of receptor cells of the human cochlea. Since the classical observations of Held (1926) and later on of Lorento de Nó (1937) we are well aware of the fact that the outer hair cells at least have interneural connections for quite a few sensory cells. According to Held one fiber might innervate a distance as long as one third of a turn although the mean value might be in the order of a few hair cells being connected by one single nonmyelinated fibre. We have more evidence for this concept by the work done by Engström and Wersäll (1953a, 1953b). Especially in this symposium Spoendlin (p. 26 of this volume; Spoendlin, 1968) showed very clearly and beautifully that this concept is true. The first who reflected about the functional consequences of such a network was Ranke 1950c, 1953). He considered this network as a possible principle for a tuning and sharpening process taking place at the ramifications of the first neuron. Later on this concept was related to the general principle of the lateral inhibition which was found by Hartline, McNichol and Ratliff (McNichol and Hartline, 1948; Hartline, 1949; Ratliff and Hartline, 1959) at the limulus eye so beautifully and was treated mathematically so clearly by Reichardt (Reichardt and Mc Ginitie, 1962). It was a long quarrel during the last years whether there would be a real lateral inhibition on the first neuron level also in the cochlea. It was first neglected, then accepted, then criticized again. There is, for instance, a very nice discussion in the monograph of Kiang (1965). There is, however, no question that this sort of neuronal interconnections should influence the information handling on this level although we do not know very clearly in what detailed manner. We would like to label this theory for the adequate stimulus *the neuronal theory*.

Interestingly enough there is a third aspect to be considered when speculating about the sharpening process taking place within the cochlea. Based upon measurements of Desmedt (1963), Fex (1962), and Pfalz (1962a, 1962b) it could be proved that at least the second order neuron, the cochlear nucleus, is influenced in its functions by a descending fibre system. This is in agreement with the morphological studies of the Rasmussen bundle (Rasmussen, 1942). According to those findings the descending system is able to inhibit as well as to facilitate the information handling at very low levels of the auditory pathway. Whether this influence also affects the first order neuron and the hair cells themselves is still not absolutely clear. We saw very nice evidence in Dr. Spoendlin's demonstrations about the double innervation of the hair cells and this again is in agreement with earlier findings of Engström and Wersäll (1953a, 1953b). Especially electron microscopy showed clearly the existence of two different types of synaptic connections between the hair cells themselves and the connecting fibre systems. It is certainly very tempting to assume that one type of this connections is due to the afferent information transfer and the other one to the efferent. It was also Pfalz in our laboratory who first claimed for this effect (1962b). It is shown in another publication that this effect is very sophisticated in so far as at least the inhibition by descending fibre systems is highly frequency sensitive and is related only to a relatively small frequency band which is inhibited while the vicinity is not. In the meantime this effect is confirmed by Klinke (Klinke et al., 1966) who will speak about his findings in this symposium too (p. 161 of this volume). It should be emphasized that the latency for this effect is in the order somewhere between 30 and 90 msec what claims strongly for a central effect and not for a direct inhibition within the ear or from one ear to the other one. So this effect must be separated clearly from the neuronal theory and therefore might be labeled the efferent fibre theory.

Closely connected as well with the question of lateral inhibition in hearing as with the problem of the adequate stimulus in general is the so-called doublicity theory of hearing. Long ago it was found in vision that the receptor cells of the eye consist of two sets, namely the cones and the rods. It was inaugurated by Meyer zum Gottesberge (1948) relatively early, in analogy to this situation in vision, also to separate the function of the inner and outer hair cells in audition. Katsuki (1961) was able to measure different tuning curves for the inner and outer hair cells and although his results are not uncriticized, there are still new facts which prove at least the general concept. One proof is of morphological structure. Neubert and Wüstenfeld (1955) were able to show that for long lasting stimulation with intense sound there is a destruction of the nuclei of the outer hair cells in the sense of a swallowing if continuous sound is delivered to the animals and of inner hair cells if the sound is delivered intermittent. His original data could be confirmed by Beck and his coworkers in about the same sense (Beck, 1959; Beck and Beickert, 1958; Beck and Michler, 1960). On the other hand it is known since about twenty years that the two sets of hair cells can be poisoned separately: the inner hair cells by quinine, the outer ones by the basic antibiotics. A review about those facts is given by Davis (1957; Davis et al., 1958). We continued those experiments in our laboratory (Stange et al., 1962; Plattig et al., 1967; Theissing and Keidel, 1962) and also could show that it is possible to damage nearly exclusively the outer hair cells and leave the inner ones undisturbed. One example for these results which were confirmed by the work of Lawrence (personal communication) on this topic is that the notch in the intensity function of the compound action potential of the auditory nerve which was described first by Frishkopf (1956) will be vanished if, for instance, kanamycine is injected in high dose for a long time in cats (Keidel, 1965). In this case the intensity function becomes monotone without a step in its steepness related to the two different populations of outer and inner hair cells and besides that as well the threshold is considerably enlarged as the adaptation functions are altered what will be discussed later on in more detail. Since we have published these data elsewhere (Keidel, 1965) I would like to restrict myself here to a model of the function of the inner and outer single hair cell which is shown in Fig. 5. In this scheme one ordinate is the frequency and shows the best frequency of the unit as well as the tuning curve as such; the second ordinate is the intensity and the third one is the degree of excitation. It can be seen clearly in this model that the outer hair cells seem to be specialized in evolution to a very high sensitivity, therefore being more adaptable and with a greater dynamic range, while the inner hair cells are more sharply tuned, have a steeper intensity function and are less adaptable. So they are more fitted for

Fig. 5. Model for the excitation of the hair cells and their tuning to central nervous processing (outer hair cells—black—with flat intensity function; inner hair cells—white— with steep intensity function). Horizontal ordinate from left to right: stimulus frequency; from front to back: sound intensity. Vertical ordinate: excitation level of the cell.

conveying information about transients in speech while the outer hair cells seem to be more related to perceive steady-state conditions of sound stimuli. The greater sensibility of the inner hair cells for separation of different frequencies seems to be established morphologically since they are innervated each one by a single fibre while the outer hair cells are innervated in such a way that many receptors hang only on one nerve fibre. This again is somehow analogous to what we know about the innervation within the eye: the cones are innervated separately, the rods are hanging in a neuronal network and therefore switched together as a group of cells. This certainly improves their excitability besides the chemical sensitization which occurs in the rods in vision. It is very likely that the higher sensitivity of the outer hair cells to poisoning by basic antibiotics is related to their chemical sensitization too. But this is not clearly proved up to now.

Another aspect of the transformation of the hydrodynamics to the adequate stimulus and the excitation of the higher levels within the auditory pathway is the so-called *funneling*. We know for certain that the tuning curves are funneled what means they become steeper as higher they are measured within the central pathway in audition. There are many experiments of Katsuki (1961, 1966) which prove this statement, although the newer trends go into that direction that even the transformation from the excitation of the hair cells which can be measured by the tuning curves of the cochlear microphonics and the tuning curves of the first order neuron, with other words of the auditory nerve, includes the main funneling process and all later funneling is only a secondary additional effect (Kiang, 1965). Also on this topic further experimentation has to be done to clear up the question completely. More detail can be read in the handbook articles mentioned above (Keidel, 1965).

ADAPTATION AND THRESHOLD SHIFT

In the context with the consequences of exposition to intense noise the problems of adaptation and related features of the ear became not only more important, there are also more experimental data available. Especially the separation of adaptational processes, temporary threshold shift and fatigue was a big step forward. Here the work of Ward (1967, 1968) has to be mentioned with emphasis. But let us start with the adaptation itself. While in vision the intensity range for adaptational processes is in the order of 10^5, in audition adaptation usually makes no more shift in the intensity threshold than a factor of 10^2 maximally. We know that adaptation does not only mean a shift of the intensity threshold but also affects the difference limen for intensity in the sense that in the adaptive state the difference limen for intensity is smaller. This makes a separation towards fatigue possible since in the case of fatigue also the difference limen for intensity is increased, not only the threshold for intensity

(Ranke, 1952; Keidel, 1961; Keidel et al., 1961). The mechanisms of adaptation certainly are mainly chemical. This was proved in vision especially by Hecht (1931) and Ranke (1953; Keidel, 1961). The latter developed an adaptation theory which is based upon the hypothesis that the velocities for the chemical processes in the sense of the building of the excitation substance and that of the rebuilding of the original chemical substratum would differ in that, that to a step in intensity first an increase in velocity for the set up of the excitations substance is observed (initial overshoot) and, finally, during the process of adaptation a steady state between the two processes the building and the rebuilding will be reached (a steady state of adaptation). In audition the main question was not so much the chemical substance for excitation as which, for instance, acetylcholin was suggested (Katsuki, 1961, 1966); the question was rather at which place adaptation occurs in audition, at periphery or at higher levels of the central pathway. The classical experiments of Luescher (1941) and others are well known and have not to be repeated here. In our laboratory one experiment was performed in the following way: In the cat simultaneously the cochlear microphonics and the compound action potentials were recorded when the intensity of a 18 dB noise was decreased to an intensity level of 15 dB. In this case immediately after the downward step of the stimulus intensity the cochlear microphonics will be at the normal level of the steady state while the action potential needs a time of about 3.2 msec after some sort of a silent period to recover to the new steady-state value. This proves clearly that, at least partly, the adaptation process takes place between the hair cells and the first neuron and not within the hair cells alone. Furthermore it proves that adaptation might not be a mechanical process as inaugurated by Kietz (1960) according to a pendulum theory of the tectorial membrane, since if this would be true the cochlear microphonics should also reveal a silent period as does the action potential. On the other hand ten years before Keidel et al. (1960) could show by using the strichnine technique at the cortical level of the cat that the time course of adaptation can be changed chemically also at the cortical level. This suggests that adaptation takes place mainly at the link between the hair cells and the action potential of the auditory nerve, but additional adaptational processes may occur at any level of the central auditory pathway including the cortex itself. A summary of this idea might be shown in Fig. 6. Here the short time adaptation at the periphery is drawn in the scheme and as higher the level in the auditory pathway will be as longer is the process of adaptation as it is described in the figure in detail. In accordance to the fact that in vision local adaptation takes place, the physiological type of adaptation at all, also in audition adaptation usually does not mean a threshold shift for all frequencies simultaneously but rather in a very sophisticated type with local limitation along the organ of Corti. This means in audition restriction of threshold shift to a special frequency band and this is what Ward (1967, 1968) especially could show clearly in his

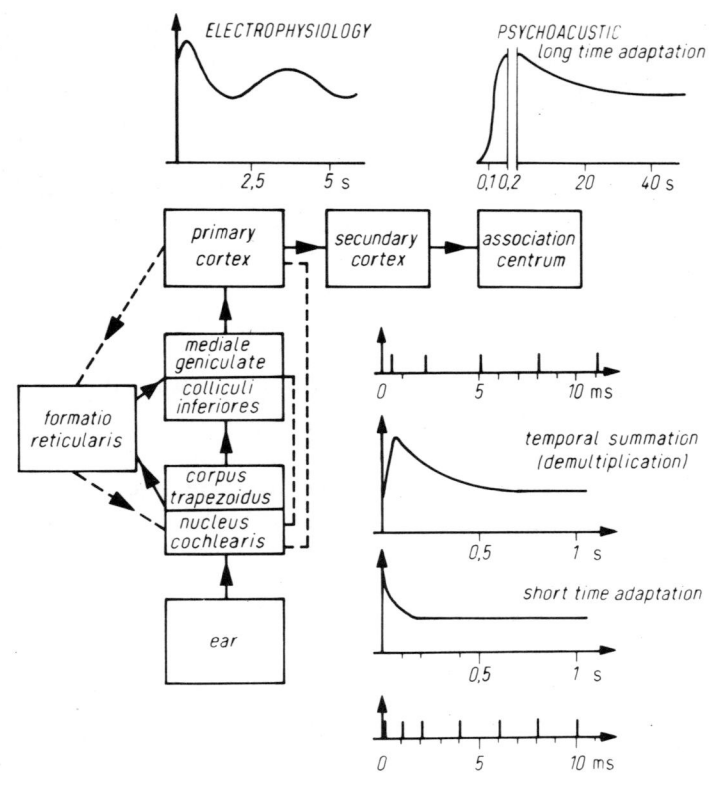

Fig. 6. Transition of the short-time adaptation of single receptors into psychophysiologically measurable long-time adaptation in the auditory pathway. The increasing time constant in transition function with ascending neuronal level within the central nervous system can be seen.

temporary threshold shift. The chemical side of the problem of adaptation could be shown especially by poisoning the ears with basic antibiotics. Here with increasing dosis the steepness of intensity functions as well for the initial overshoot as for the steady-state condition of adaptation was changed in a characteristic manner: first the steepness dropped to a certain level at a given intensity level and then, after long lasting high dosis, this steepness after having passed a minimum was increased again. This is shown in Fig. 7. This clearly reproducible effect cannot be explained otherwise than that first the chemical process setting up to excitable substance is altered, thereby diminishing the intensity functions, and then also the rebuilding process is poisoned what obviously results in an increased amount of excitable chemical substance and therefore an increase in steepness of intensity functions for higher dosis (Stange et al., 1962; Theissing, 1962; Theissing and Keidel, 1962; Keidel, 1965; Plattig et al., 1967). As excitable substance with a high degree of probability acetylcholi-

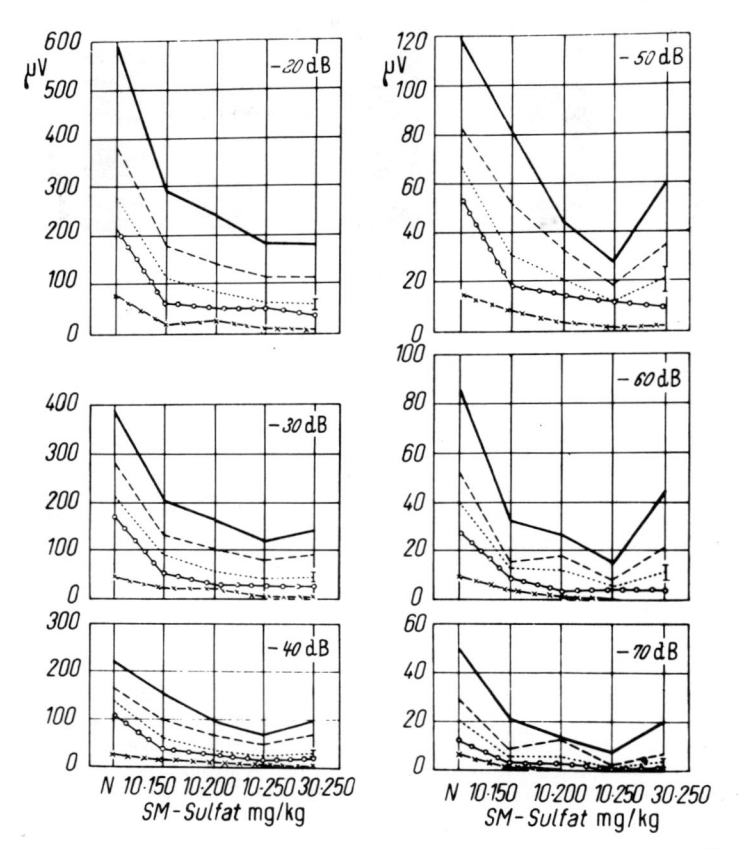

Fig. 7. Behaviour of adaptation curves with increasing streptomycin-sulfate dosis. Ordinate: amplitude of the potentials. Abscissa: increasing dosis of streptomycin-sulfate. Parameter is the repetition rate. It can be seen that especially in the range of speech the curves meet closely for a dosis of 10 times 250 mg/kg streptomycin-sulfate and spread again for maximal dosis.

ne should be taken into account as it was proved very nicely recently by Katsuki (1966). As Fig. 8 demonstrates the initiation of the action potentials is obviously of chemical nature. Otherwise the effect of acetylcholine upon the size of action potential as shown in Fig. 8 could hardly be explained by a purely physical mode of excitation.

CONCLUSIONS

In this context it certainly is not necessary to repeat all facts known about electrophysiology of the human cochlea. I just would like to summarize in a table the different types of electrical events after auditory stimulation. It might be mentioned that the temperature coefficient of the cochlear potentials in

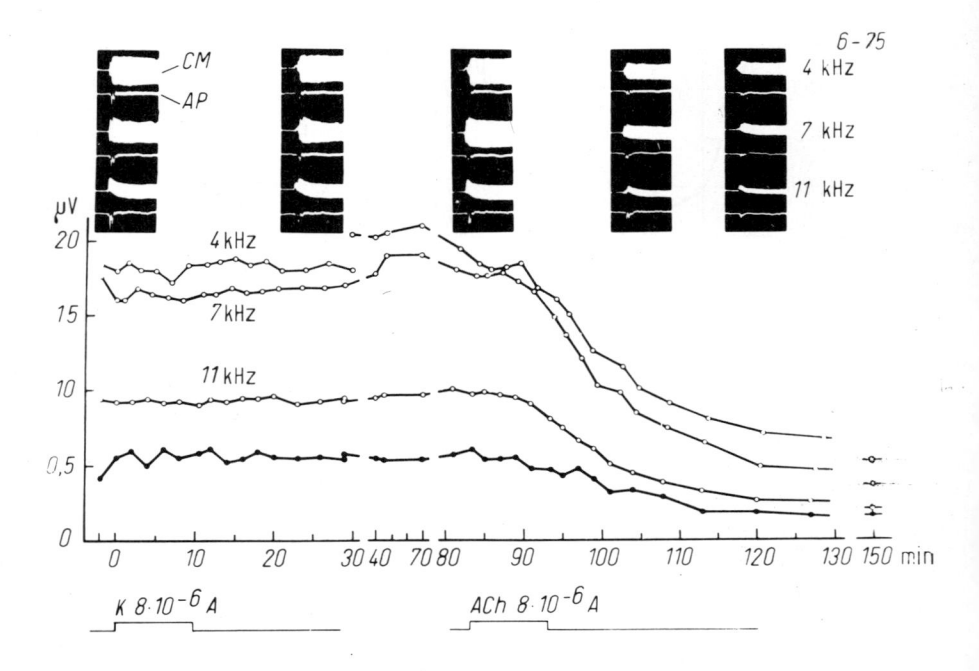

Fig. 8. Effect of combined application of potassium (left) and acetylcholine (right) on cochlear and nerve responses in guinea pig. The cochlear microphonics (CM) and neural responses N_1 (AP) were recorded from the basal turn of the cochlea of the guinea pig. Potassium ions were applied with a current of 8×10^{-6} A from 0 to 10 min, and then acetylcholine from 83 to 93 min. Abscissa: time in min. Ordinate: amplitude of CM (open circles show results obtained with sound at 4.7 and 11 kHz) and the N_1 response (closed circles) in mV. Records above are CM (upper beam) and N_1 (lower beam) responses obtained at different times on the abscissa (Katsuki, 1966).

guinea pig is of the order of more than two (Butler et al., 1960) what again speaks for a chemical process of the initiation, at least of N_1, in the action potential. It might be worth to add that Crowley et al. (1965) could prove that the cochlear potential in the albino rat shows in a considerably great range a log-log intensity function even at the first neuron level. The same result was obtained in 1969 by Plattig in our laboratory who could prove the same in the cat. This fact is important for the question at what place the programming of information handling in audition takes place since we know that the cortical evoked potentials obey a log-log intensity function too. In the same connection it is of interest that the critical bandwidth can be demonstrated physiologically as low as in the cochlear microphonics, with other words in the hair cells and their neuronal network (Spreng, 1967). And finally it might be of interest that at higher levels a frequency following microphonic-like neural response evoked by sound could be recorded by Worden and Marsh (1968).

So far the classical concept of the electrophysiology of the human cochlea.

This chapter, however, should not be concluded without mentioning some new results found by Lawrence (1967) about the electric polarization of the tectorial membrane in animals. As Fig. 9 shows he claimed recently that the difference limen potential change, which one can record when penetrating the organ of Corti with micropipettes at the tectorial membrane, would be relative to the polarization of the perilymph and not equal to the high positive potential of the endolymph as it is considered according to the classical theory. If those results can be confirmed this would mean that the tectorial membrane itself would be polarized at the lower end toward the basilar membrane with -80 mV and at the upper end towards the endolymphatic fluid with $+80$ mV as it is shown in the scheme of Fig. 9. If this is true the tectorial membrane at least could act as a polarized crystal and as a mechanical-electrical transducer. This again would be a hint towards the hypothesis proposed by Naftalin in 1965 who reasons a detailed theory which involves a polarized tectorial membrane.

I would not like to overemphasize those last findings but I consider it as the duty of a reviewer to speak about them. It certainly needs confirmation and a lot of further experimental work to solve the main problem of the physiology of the human cochlea, namely to solve the question of how does the cochlea sharpen the flat envelope of the displacements of the cochlear partition to the sharply tuned tuning curves of the first order neurons. Summarizing what was said to this question we can say, there are three theories about possible modes,

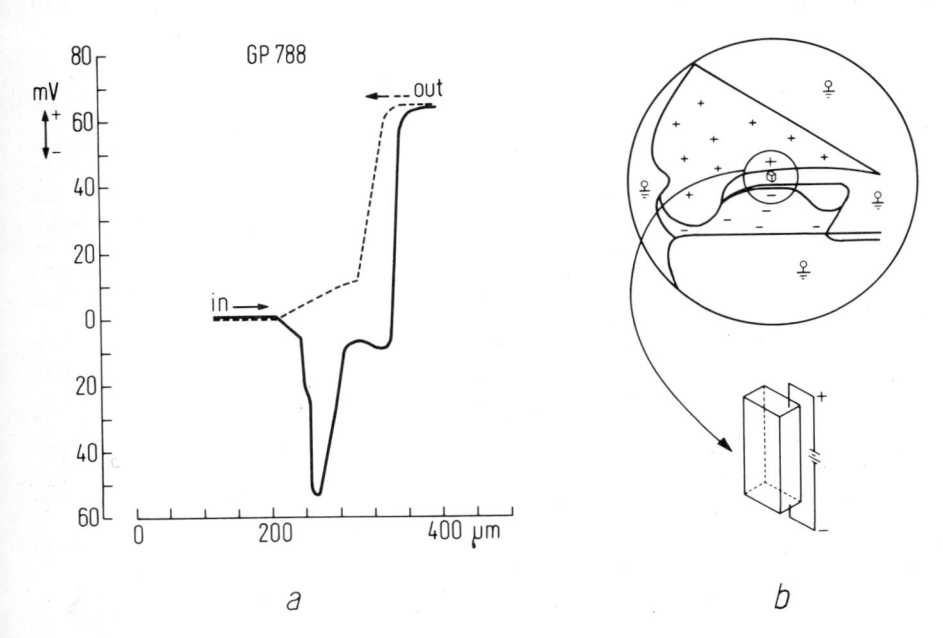

Fig. 9. (a) The DC electrical potential change as the electrode is advanced through the arch of Corti. (b) The polarization of the tectorial membrane (Lawrence, 1967).

namely first the mechanical theory of the transformation of the perpendicular displacements into the shearing forces for the bending of the hair cells, secondly there is a neuronal theory based upon some lateral inhibition mechanism within the neuronal network at least for the outer hair cells, and thirdly there is the descending fibre theory according to which the double innervation of the hair cells may be interpreted in that way that one synaptic connection is an efferent one.

I would like to conclude this review with the statement that we do not have a real argument against the travelling wave theory which according to my feeling without any doubt is absolutely valid and correct in describing the hydrodynamics of the inner ear, and secondly we have no clear concept up to now how the funneling process from the hydrodynamics of the ear to the electro- and neurophysiology of neuronal auditory pathway takes place. Those two statements include finally the fact that, although the resonance theory is disproved and the travelling wave theory is well established, Wien's (1905) objection is still valid but in a higher sophisticated way, in so far that we now know that the performance of hearing is not done by the hydrodynamics of the ear solely but certainly needs an additional neuronal information handling. This again includes the consequence that we need both: a frequency dispersion along the basilar membrane, with other words a place theory of hearing, and we need the neuronal use of the periodicity of the auditory stimuli, with other words a periodicity theory. The question can be only how do they fit together in detail.

REFERENCES

Beck, C. (1959): Feinere Stoffwechselreaktionen an den Sinneszellen des Cortischen Organs nach Reintonbeschallung, Arch. Ohr.-, Nas.-, Kehlk.-Heilk. *175*, 374-378.
Beck, C., and Beickert, P. (1958): Morphologische Veränderungen an der Schnecke des Meerschweinchens bei Sauerstoffmangel und Lärmbelastung, Arch. Ohr.-, Nas.-, Kehlk.-Heilk. *172*, 238-245.
Beck, C., and Michler, H. (1960): Feinstrukturelle und histochemische Veränderungen an den Strukturen der Cochlea beim Meerschweinchen nach dosierter Reintonbeschallung, Arch. Ohr.-, Nas.-, Kehlk.-Heilk. *174*, 496-567.
Békésy, G. von (1929): Zur Theorie des Hörens; Ueber die Bestimmung des einem reinen Tonempfinden entsprechenden Erregungsgebietes der Basilarmembran vermittels Ermüdungserscheinungen, Physikal. Z. *30*, 115-125.
Békésy, G. von (1960): Experiments in Hearing (McGraw-Hill, New York).
Breuing, G. (1969): Modellvorstellung zum Periodizitätshören, Pflügers Arch. ges. Physiol. *312*, R 131.
Butler, R. A., Konishi, T., and Fernandez, C. (1960): Temperature coefficients of cochlear potentials, Amer. J. Physiol. *199*, 688-692.
Crowley, D. E., Hepp-Reymond, M. C., Tabowitz, D., and Palin, J. (1965): Cochlear potentials in the albino rat, J. Audit. Res. *5*, 307-316.
Davis, H. (1957): Biophysics and physiology of the inner ear, Physiol. Rev. *37*, 1-49.
Davis, H., Deatherage, B. H. Rosenblut, B., Fernandez, C., Kimura, K., and Smith, C. A. (1958): Modification of cochlear potentials produced by streptomycin poisoning and by extensive venous obstruction, Laryngoscope *68*, 596-627.

Desmedt, J. E. (1963): Efferent olivo-cochlear gating of acoustic input and the resultant changes in auditory cortex potentials, J. Physiol. *165*, 33P-34P.
Engström, H., and Wersäll, J. (1953a): Structure of the organ of Corti I. Outer hair cells, Acta Oto-Laryngol. *43*, 1-10.
Engström, H., and Wersäll, J. (1953b): Structure of the organ of Corti II. Supporting structures and their relations to sensory cells and nerve endings, Acta Oto-Laryngol. *43*, 323-334.
Fernandez, C. (1951): The innervation of the cochlea (guinea pig), Laryngoscope *61*, 1152-1172.
Fex, J. (1962): Centrifugal activity in the crossed olivo-cochlear bundle: a single unit analysis from a feedback loop, in: XXIInd International Congress of Physiological Sciences (Internat. Congr. Ser. 48, Excerpta Medica Foundation, Amsterdam), p. 1013.
Fletcher, H. (1951): On the dynamics of the cochlea, J. Acoust. Soc. Amer. *23*, 637-645.
Frishkopf, L. S. (1956): A probability approach to certain neuroelectric phenomena, M.I.T. Research Laboratory of Electronics, Technical Report No. 307, Cambridge, Mass.
Hartline, H. K. (1949): Inhibition of activity of visual receptors by illuminating nearby retinal elements in the Limulus eye, Federation Proc. *8*, 69.
Hecht, S. (1931): Die physikalische Chemie und die Physiologie des Sehaktes, Ergebnisse Physiol. *32*, 243-390.
Held, H. (1926): Die Cochlea der Säuger und der Vögel, ihre Entwicklung und ihr Bau, in: Handbuch der normalen und pathologischen Physiologie, A. Bethe, G. v. Bergmann, G. Embden, A. Ellinger, Eds. (Springer, Berlin), Vol. 11, p. 467-534.
Helmholtz, H. L. F. von (1863): Die Lehre von den Tonempfindungen als physiologische Grundlage für die Theorie der Musik (F. Vieweg & Sohn, Braunschweig).
Huxley, A. (1969): Is resonance possible in the cochlea after all?, Nature *221*, 935-940.
Johnstone, B. M., and Boyle, A. J. F. (1967): Basilar membrane vibration examined with the Mössbauer technique, Science *158*, 389-390.
Katsuki, Y. (1961): Neural mechanism of auditory sensation in cats, in: Sensory Communication, W. A. Rosenblith, Ed. (Wiley, New York), pp. 561-583.
Katsuki, Y. (1966): Neural mechanisms of hearing in cats and monkeys, Progress in Brain Res. *21A*, 71-97 (Elsevier, Amsterdam).
Katsuki, Y., Watanabe, T., and Maruyama, N. (1959): Activity of auditory neurons in upper levels of brain of cat, J. Neurophysiol. *22*, 343-359.
Keidel, W. D. (1961): Rankes Adaptationstheorie, Z. Biol. *112*, 411-425.
Keidel, W. D. (1964): Kybernetische Leistungen des menschlichen Organismus, Elektrotechn. Z. *24*, 769-808.
Keildel, W. D. (1965): Physiologie des Innenohres, in: Hals-Nasen-Ohren-Heilkunde, Berendes-Link-Zöllner, Eds. (Thieme, Stuttgart), Vol. III/1, pp. 235-310.
Keidel, W. D., and Spreng, M. (1963): Elektronisch gemittelte langsame Rindenpotentiale des Menschen bei akustischer Reizung, Acta Oto-Laryngol. *56*, 318-328.
Keidel, W. D., Keidel, U. O., and Kiang, N. Y.-S. (1960): Cortical and peripheral responses to vibratory stimulation of the cat's vibrissae, Arch. Internat. Phys. Biochim. *68*, 241-262.
Keidel, W. D., Keidel, U. O., and Wigand, M. E. (1961): Adaptation: loss or gain of sensory information?, in: Sensory Communication, W. A. Rosenblith, Ed. (Wiley, New York), pp. 319-338.
Kiang, N. Y.-S. (1965): Discharge Patterns of Single Fibers in the Cat's Auditory Nerve (Research Monograph No. 35, M.I.T. Press, Cambridge, Mass.).
Kietz, H. (1959): Ein Einwand gegen die Wanderwellen im Innenohr und eine Abwandlung dieser Theorie, Z. Laryngol. Rhinol. *38*, 295-300.
Kietz, H. (1960): Physik des Hörens (Schwiefert, Bremen).
Klinke, R., Boerger, G., and Gruber, J. (1966): Studies on the functional significance of efferent innervation in the auditory system, Pflügers Arch. ges. Physiol. *306*, 752-763.
Lawrence, M. (1967): Electric polarization of the tectorial membrane, Ann. Otol. Rhinol. Laryngol. *76*, 289-312.

Lawrence, M., and Wever, E. G. (1952): Effects of oxygen deprivation upon the structure of the organ of Corti, Arch. Otolaryngol. 55, 31-37.
Lawrence, M., and Yantis, P. A. (1957): Individual differences in functional recovery and structural repair following overstimulation of the guinea pig ear, Ann. Otol. Rhinol. Laryngol. 66. 595-621.
Lorente de Nó, R. (1937): The neural mechanism of hearing. I. Anatomy and Physiology. (b) The sensory endings in the cochlea, Laryngoscope 47, 373-377.
Lüscher, E. (1941): Ueber Regulationsmechanismen des Gehörorgans, Schweiz. Med. Wschr. 1, 430-432.
Mac Nicol, E. F., and Hartline, H. K. (1948): Responses to small changes of light intensity by light-adapted photoreceptor, Federation Proc. 7, 76.
Meyer zum Gottesberge, A. (1948): Zur Physiologie der Haarzellen, Arch. Ohrenheilk. 155, 308-314.
Naftalin, L. (1965): Some new proposals regarding acoustic transmission and transduction, Sensory Receptors 30, 169-180.
Neubert, K. (1960): Innere Haarzellen des Cortischen Organs und Schallanalyse, Naturwissenschaften 47, 526-527.
Neubert, K., and Wüstenfeld, E. (1955): Nachweis der zellulären Ansprechgebiete im Innenohr, Naturwissenschaften 42, 350-351.
Pfalz, R. (1962a): Einfluss schallgereizter efferenter Hörbahnteile auf den de-afferentierten Nucleus cochlearis (Meerschweinchen), Pflügers Arch. ges. Physiol. 274, 533-552.
Pfalz, R. (1962b): Erregung efferenter Einzelneurone innerhalb der Cochlea durch kontralaterale Beschallung mit Klickfolgen (Meerschwein), Z. Biol. 113, 215-221.
Plattig, K. H. (1969): Periphere Codierungsfunktionen in der Hörbahn der Katze, Pflügers Arch. ges. Physiol. 307, R135.
Plattig, K. H., Keidel, U. O., and David, E. (1967): Minderung der Ototoxizität des Kanamycins durch Pantothensäure, Deutsch. Med. Wschr. 92, 1391-1397.
Ranke, O. F. (1931): Die Gleichrichter-Resonanztheorie (Lehmann, München).
Ranke, O. F. (1950a): Theory of operation of the cochlea: a contribution to the hydrodynamics of the cochlea, J. Acoust. Soc. Amer. 22, 772-777.
Ranke, O. F. (1950b): Hydrodynamik der Schneckenflüssigkeit, Z. Biol. 103, 409-434.
Ranke, O. F. (1950c): Folgerungen aus der Theorie der Flüssigkeitsschwingungen in der Schnecke, Ber. ges. Physiol. 139, 183-184.
Ranke, O. F. (1952): Bedeutung der Adaptation für die Konstanz der Reizgestalt, Ber. ges. Physiol. 154, 279.
Ranke, O. F. (1953) in: Ranke, O. F., and Lullies, H., Gehör, Stimme, Sprache (Springer, Berlin).
Ranke, O. F., Keidel, W. D., and Weschke, H. G. (1952): Das Hören beim Verschluss des runden Fensters, Z. Laryngol. Rhinol. 31, 467-474.
Rasmussen, G. L. (1942): An efferent cochlear bundle, Anatomical Record 82, 441.
Ratliff, F., and Hartline, H. K. (1959): The responses of limulus optic nerve fibers to patterns of illumination on the receptor mosaic, J. Gen. Physiol. 42, 1241-1255.
Reichardt, W., and Mac Ginitie, G. (1962): Zur Theorie der lateralen Inhibition, Kybernetik 1, 155-165.
Siebert, W. M. (1962): Models for the dynamic behavior of the cochlear partition, M.I.T. Research Laboratory of Electronics, Quarterly Progress Report 64, 242-258.
Spoendlin, H. (1968): Ultrastructure and peripheral innervation pattern of the receptor in relation to the first coding of the acoustic message, in: Hearing Mechanisms in Vertebrates, A. V. S. de Reuck and J. Knight, Eds. (Churchill, London), pp. 89-119.
Spreng, M. (1967): Ueber die Messung der Frequenzgruppe und der Integrationszeit des menschlichen Gehörs durch vom Schall abhängige Hirnspannungen längs der Kopfhaut, Doctoral Dissertation, Technische Hochschule Stuttgart.
Stange, G., Kaufmann, F., Spreng, M., and Theissing, J. (1962): Influence of Kanamycin and Streptomycinsulfate to peripheral and cortical time course of adaptation to auditory stimuli in the cat, in: XXIInd International Congress of Physiological Sciences (Internat. Congr. Ser. 48, Excerpta Medica Foundation, Amsterdam), p. 1249.

Theissing, J. (1962): Wirkung des Kanamycins auf die Kurzzeitadaptation des Katzengehörs, Doctoral Dissertation, Universität Erlangen-Nürnberg.
Theissing, J., and Keidel, U. O. (1962): Zur Beeinflussung des Statoacusticus-Systems durch basische Antibiotica, Arzneimittelforschung *12*, 543-545.
Tonndorf, J., and Bergeyk, W. A. van (1958): Some principles of vestibular hydromechanics, Ann. Otol. Rhinol. Laryngol. *67*, 628-642.
Tonndorf, J., Khanna, S. M., and Fingerhood, B. J. (1966): The input impedance of the inner ear in cats, Ann. Otol. Rhinol. Laryngol. *75*, 752-763.
Ward, W. D. (1967): Adaptation and fatigue, in: Sensorineural Hearing Processes and Disorders, A. B. Graham, Ed. (Little, Brown & Co., Boston), pp. 113-121.
Ward, W. D. (1968): Susceptibility to auditory fatigue, in: Contributions to Sensory Physiology, W. D. Neff, Ed. (Academic Press, New York), Vol. 3, pp. 191-226.
Wien, M. (1905): Ein Bedenken gegen die Helmholtzsche Resonanztheorie des Hörens, Festschr. A. Wüllner (Verlag B. G. Teubner, Leipzig), pp. 28-35.
Worden, F. G., and Marsh, J. T. (1968): Frequency-following (microphonic-like) neural responses evoked by sound, Electroenceph. Clin. Neurophysiol. *25*, 42-52.
Zwislocki, J. (1948): Theorie der Schneckenmechanik, Acta Oto-Laryngol., Suppl. *72*.

DISCUSSION

Bosher: In our experiments measuring the endolymphatic potential by an approach through the stria vascularis, we have found a large negative voltage of the order of 40 mV which was associated with an increase in impedance of the electrode. I think, therefore, large negative potentials are very suspect because they may well be due to blocking of the tip of the electrode and consequent increase in its negative tip potential. We, ourselves, have never been able to find a zero potential at the tectorial membrane as Lawrence (1967) describes. We have investigated this by inserting electrodes at different angles. Either the electrode broke when it hit the bone of the inner sulcus and spiral limbus without showing any zero potential or we found a zero potential but then, as histological examination showed, only when we had pierced Reissner's membrane. So the electrodes must have been advanced through the tectorial membrane without finding a zero potential at that site.

Schwartzkopff: I have the same problem as Dr. Bosher. We never saw a jump in potential in our experiments with birds when we came through the tectorial membrane until we reached the level of the top of the hair cells. So we believe that the tectorial membrane is electronically transparent.

Keidel: I mentioned Dr. Lawrence's work because I think he obtained some results which are not in accord with the classical concepts of cochlear microphonics. I think we need much more detailed studies and we have to wait for further investigations by Dr. Lawrence to be sure whether there is a substantial influence of the neighbouring cells around the hair cell on the negative potential.

Schügerl: Do you think the sharpness of the envelope is the same at all places as Ranke's formula (1950) indicates or do you prefer Fletcher's curves (1951)

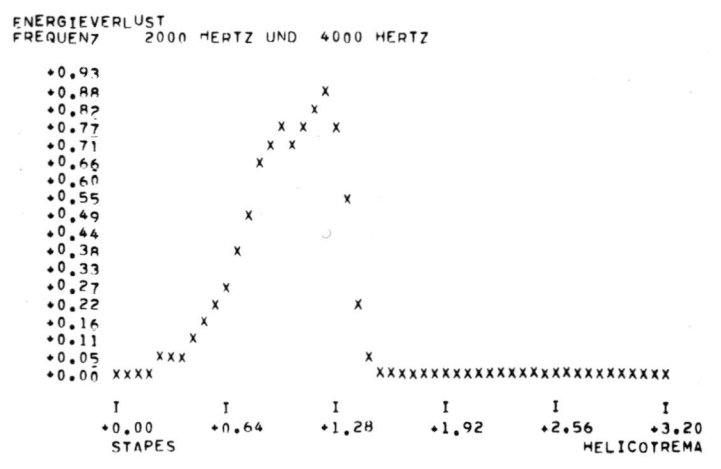

```
ENERGIEVERLUST
FREQUENZ    2000 HERTZ UND  4000 HERTZ
 +0.93
 +0.88                       X
 +0.82                       X
 +0.77                    X X   X
 +0.71                    X X
 +0.66                 X
 +0.60
 +0.55                             X
 +0.49              X
 +0.44
 +0.38              X
 +0.33
 +0.27           X
 +0.22           X           X
 +0.16          X
 +0.11         X
 +0.05       XXX
 +0.00  XXXX                 X     XXXXXXXXXXXXXXXXXXXXXXXXXX
        I       I       I       I       I       I
      +0.00   +0.64   +1.28   +1.92   +2.56   +3.20
      STAPES                                HELICOTREMA
```

Fig. 1 (Keidel). Energy loss along the cochlear partition for a combination of two sinusoids of 2 kHz and 4 kHz. Please compare with the envelope of displacement in Fig. 3b.

which show, according to von Békésy's (1960) measurements (for instance at 1600 Hz), a sharper envelope for higher frequencies?

Keidel: I agree that the envelopes of the higher frequencies are sharper, that the slopes are steeper, and I think Dr. Johnstone's results confirm this. But I would consider this as a secondary effect, it is not the main issue.

Schügerl: To explain the cochlear sharpening, it is not necessary to take all the temporal aspects into account. The resolution of a complex signal, composed of several harmonics, is already better if you take the root-mean-square of the excursion of the basilar membrane, calculated over the period of the signal, as a measure.

Keidel: Taking the root-mean-square as a measure instead of the envelope produces indeed a better resolution as is shown in Fig. 1. But I think there is also temporal information present and this information is probably used to improve the resolution beyond what the rms-measure would give.

REFERENCES

Békésy, G. von (1960): Experiments in Hearing (McGraw-Hill Company, New York).
Fletcher, H. (1951): On the dynamics of the cochlea, J. Acoust. Soc. Amer. *23*, 637-645.
Lawrence, M. (1967): Electric polarization of the tectorial membrane, Ann. Otol. Rhinol. Laryngol. *76*, 289-312.
Ranke, O. F. (1950): Hydrodynamik der Schneckenflüssigkeit, Z. Biol. *103*, 409-434.

MECHANICAL ASPECTS OF COCHLEAR FUNCTION

B. M. JOHNSTONE and K. TAYLOR

Department of Physiology
University of Western Australia
Nedlands, Australia

The role of the basilar membrane in frequency discrimination has been a matter of controversy for a century.

When von Békésy (1949) described the form of the travelling wave in the cochlea, doubts were expressed as to its significance for fine frequency analysis. The shape of the wave appeared too broad to account for the psychoacoustically known minimum observable frequency difference. To account for this, neural sharpening in the higher centres was suggested. This concept appeared to be confirmed in early neural recordings which showed a primary nerve tuning curve apparently shaped like the von Békésy travelling wave envelope and very sharp tuning curves from the cochlear nucleus and higher centres. Thus the coding in the cochlea appeared to be a simple following of a rather broad basilar membrane movement. This concept was overthrown when Kiang's (1965) more recent recordings from primary fibres showed very sharp tuning indeed. Tuning curves were nearly symmetrical and had slopes up to 100 dB/octave. As this was orders of magnitude sharper than the von Békésy wave, various types of peripheral neural tuning and lateral inhibition have been proposed. It is the purpose of this paper to suggest that in the light of new determinations of the basilar membrane motion (Johnstone and Boyle, 1967), neural interactions in the cochlea are minimal and the basilar membrane must be considered as the major frequency-selective element in the auditory system.

APPARATUS

Some explanation of our technique is necessary in order to appreciate the assumptions and possible artifacts in the results. The Mössbauer effect arises from recoil-free emission of gamma rays from suitable isotopes incorporated in solids and the subsequent resonance absorption of these, by nuclei similar to the daughter nuclei of the initial decays. The line widths for this process are very small so that small changes in energy (frequency) of the gamma rays can reduce the cross section for resonance absorption by measurable amounts. As the frequency shift (energy changes) increases, resonance absorption decreases. A Doppler shift due to a superimposed velocity (vibration) of the source, of a few mm per second, is a sufficient frequency change. If a small source, usually Co^{57} diffused into stainless steel foil, is vibrated and the gamma rays counted through a thin stainless steel absorber then the count rate will be a function of the motion of the source. Fig. 1 shows a calibration for this system in terms of gamma rays transmitted through the absorber, *versus* velocity amplitudes of the source for a sinusoidal vibration.

For simple amplitude measurements only two scalers are required. One to count the zero (no sound) rate, and one for counting with the sound on. The percentage count increase then gives the velocity directly. When phase information or waveform shapes are required, the counting system is more complicated and some sort of multichannel scaler or computer synchronized to the sound source must be used. We have devised a system enabling the counts within one input sound cycle to be stored into 64 channels. This system gives 1/2 μsec time resolution and enables phase determination of better than 6°. A typical direct output from this system is shown on Fig. 2. This curve must be transformed to recover the original waveform. Our most recent apparatus improves this resolution down to 1°.

As the Mössbauer technique is a velocity-sensitive system, small variations in animal position or low-frequency fluctuations are unimportant. Whilst this

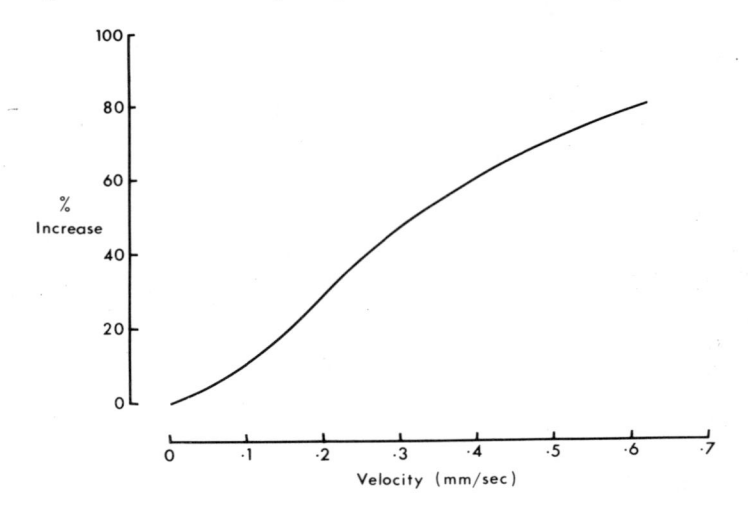

Fig. 1. Calibration of a zero isomer shift Co^{57} in stainless steel source and an enriched stainless steel absorber. The ordinate is percentage count increase above the zero velocity counts. The abscissa is the peak velocity for a sinusoidal motion.

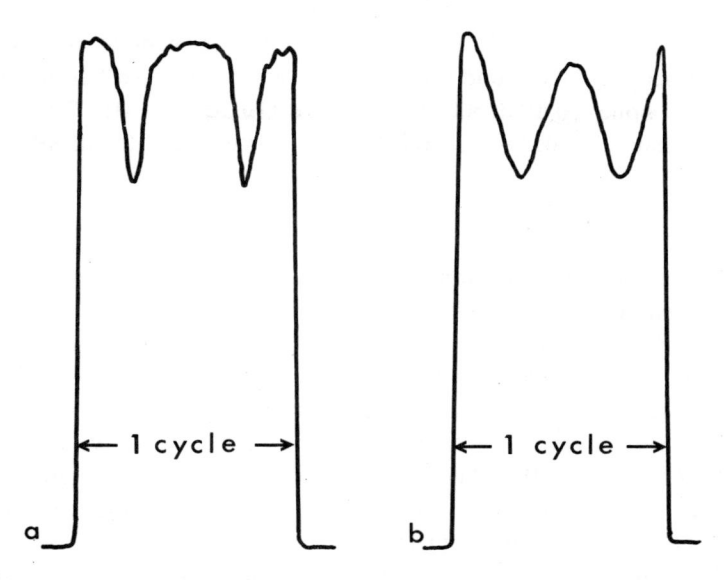

Fig. 2. The waveform output from the 64 channel counter. The abscissa is 1 cycle of the input sound, the ordinate is the counts per channel. Maximum counts corresponds to maximum velocity, minimum to zero velocity. When transformed, using the known filter (absorber) shape, the original velocity waveform is recovered. (a) shows a normal fairly high velocity waveform, (b) a moderate velocity with 2% 3rd harmonic distortion in the sound input. Note the lower maximum velocity of the centre peak, indicating a flattening of one peak of the sine wave (sound) input.

means we do not need a rigid mechanical set-up, it does mean that the results are sensitive to high-order harmonics. This is a disadvantage particularly at low frequencies, but this same problem can be turned to good use as it enables us to detect quite small amounts of 3rd harmonic distortion (Fig. 2).

TECHNIQUE

A small Mössbauer source of dimension about $40\mu \times 80\mu$ and $5\ \mu$ thick is placed on the guinea pig basilar membrane. The placing is difficult and the possibility of injury great. Injury is manifest as a flattening in the tuning curve and often as visible changes in the basilar membrane. Due to mechanical problems, we can measure only in the first turn, and then only over a restricted segment. At this place, the basilar membrane is about 160 μ wide, so the source covers about 1/4 to 1/2 of it. The 1st turn scala tympani is opened and drained, this is an unavoidable complication; however, on several occasions we have managed to replace the fluid and were unable to measure any significant difference whether the fluid was present or not.

As the absorber has a dynamic range of somewhat less than 20 dB the sound pressure level must be adjusted so that a count rate difference of between 10%

and 50% is recorded. That is, the measurements were made at approximately constant velocity and the sound pressure level varied from 70 dB at the resonant frequency to about 120 dB at 500 Hz and about 110 dB at 30 kHz. The measurements were than all scaled to 90 dB SPL; or in some cases, 100 dB SPL.

RESULTS

Bearing in mind the above limitations, we have been able to show a shift in the maximum point of stimulation from 19 kHz to 16.5 kHz for a shift in the source position from 1.5 to 2 mm from the stapes end. This result is in good agreement with calculations derived from von Békésy's data. The shape of the tuning curve for a single point is very different from von Békésy's (1949). Our best estimate to date gives a low-frequency slope of 12 dB/octave and a high-frequency slope of 95 dB/octave (Fig. 3). Naturally it is the high-frequency slope which is of interest and our measurement of it is subject to some uncertainty. The steepest slopes from 4 experiments were 100, 92, 88, 85 dB/octave, respectively. Our slowest from 4 similar experiments were 70, 65, 65, 62 dB/octave, respectively. The reasons for preferring the higher estimates are several: (1) they were obtained in later experiments and so are, hopefully, more reliable; (2) damage to the membrane always broadened the curve; (3) a slight drying of the membrane also broadened (reversibly) the curve; (4) the source occupies a finite length of basilar membrane, about equivalent to 250 Hz in frequency distribution. This latter will tend to spread out the slope by some amount, and so we feel justified in adding 5 dB to the mean high-frequency slope. All these factors lead to a best estimate of 95 dB/octave, although on at least one occasion, 100 dB/octave was measured.

The low-frequency slope of 12 dB/octave extends down to at least 3 kHz. Below this there is some evidence of a change to a flatter slope of about 6dB/octave. We have only one experiment in this area (the velocity is very small) and large SPL's must be used. Whether the change is real, and a function of frequency, or due to the high sound pressure, is under further study at the moment.

Although the general form of von Békésy's curves are similar to ours, his values are near 6 dB/octave for the low-frequency side and 20 dB/octave on the high side. We do not think this means there is a fundamental disagreement, after all von Békésy measured one end of the cochlea and we the other. Indeed a close inspection of von Békésy's curves for the guinea pig suggests a smooth graduation in tuning from the apex to the base (Fig. 5). This graduation is more striking when the high-frequency slopes only are considered. Such a progression in tuning has been noted by Kiang (1965), in the neural tuning curves of the cat.

It should be recognized that these figures all refer to a constant stapes amplitude. For a constant SPL at the ear drum, we may add an additional 9 to 12 dB/octave at the high frequencies to allow for the middle ear response.

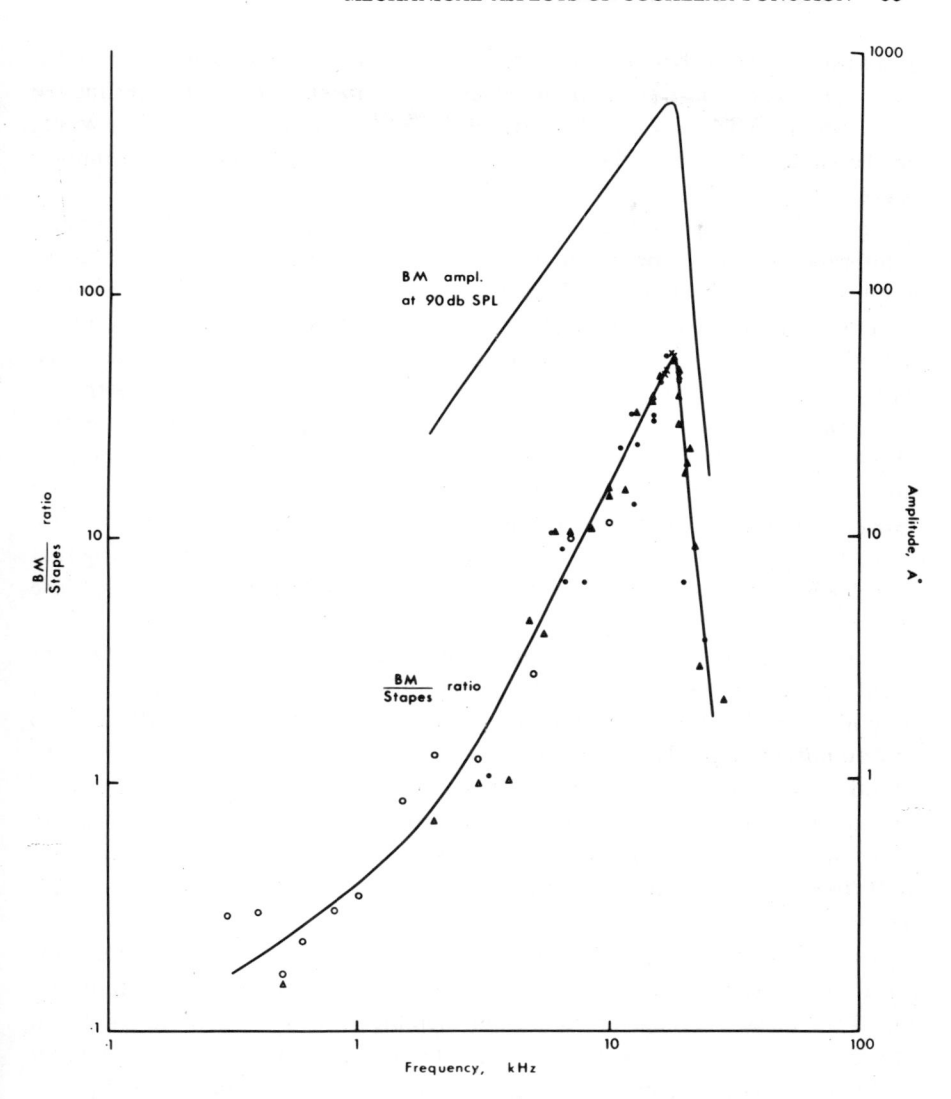

Fig. 3. Basilar membrane response in the guinea pig, of a point about 2 mm from the stapes end. There are 4 experiments. The change of slope at low frequencies must be regarded as tentative at the present time. The top curve is corrected for middle ear response and gives the basilar membrane amplitude for a 90 dB SPL at the ear drum. Later measurements gave more sensitive results, probably due to using better sound measuring apparatus (see text).

Hence we consider a slope of greater than 105 dB/octave to be possible in the mechanical response of the basilar membrane, referred to constant SPL at the ear drum (Fig. 3).

During the course of this work many checks of linearity were carried out.

The absorber has a dynamic range of rather less than 20 dB, so accurate checks were not possible; however, all movements were linear within our experimental error (about $\pm 10\%$ for counting and ± 1 dB for sound). The accuracy would not be sufficient to show the small nonlinearities of basilar membrane motion suggested by von Békésy (1960, p. 464).

Fig. 3 shows the result of some of our early experiments and the basilar membrane motion has been scaled to 90 dB SPL, at which level the peak has an amplitude of 650 Å. Later work using a more exact sound pressure measuring system gives an average figure of 550 Å for 100 dB SPL. This corresponds to 0.0055 Å at 0 dB SPL. It has been suggested that there are problems associated with such small movements (Naftalin, 1968). The main criticism has been that von Békésy carried out his work at very high sound pressure levels, in excess of 130 dB, and that an extrapolation of 6 or 7 orders of magnitude may be unjustified. Although scaled to 100 dB, in fact our peak measurements were carried out at 65 or 70 dB, and at these levels the gross linearity of the ear is not in question so such criticism does not apply to our results. It is interesting that von Békésy's figures for 0 dB work out at 0.01 Å (for 3 kHz), a value very close to ours.

Some experiments enabled a rough estimate of peak shift with SPL. On one occasion we measured 17 kHz (the peak) at 70 and 65 dB SPL. The relative amplitudes of basilar membrane vibration were close to 2:1 (≈ 6 dB) and so any amplitude-dependent frequency shift would seem to be small. Although we found no statistically significant shifts, on all occasions, the higher sound level measurements gave peaks at slightly higher frequencies. Since our error is about 10%, our findings do not disagree with von Békésy's who found a frequency shift (to a lower tone) of a little less than 10% for a 6 dB change in SPL (1960, p. 463).

We have recently expanded our apparatus to enable phase measurements (of about 6° accuracy) to be made. Fig. 2 shows the form of the output and Fig. 4 the phase response of the basilar membrane for a 19 kHz tuning point. This diagram must be taken as somewhat tentative at this stage, particularly at the high frequencies. However, two points are very clear: (1) the phase shift is a continuing function on the high-frequency side past the resonance, *i.e.* it does not asymptote to any final value, and this means Huxley's (1969) model cannot be applied to the basilar membrane; (2) the phase shift asymptotes to 90° lead at low frequencies. This latter is in contrast to von Békésy's result (1960, p. 462) of a zero phase asymptote. It is interesting to note that Flanagan (1962) disputed von Békésy's figure on theoretical grounds, and postulated that the phase must asymptote to 90° lead at low frequencies. Our findings certainly justify Flanagan's theories on this point.

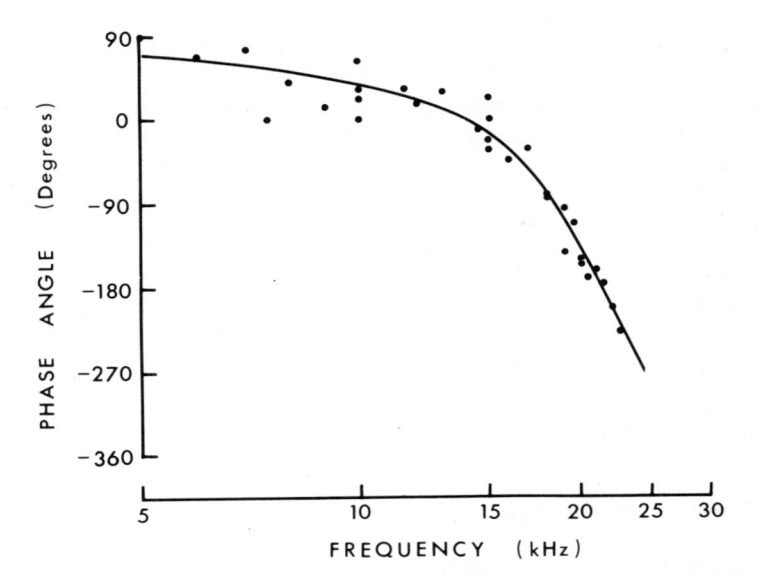

Fig. 4. Phase response of the basilar membrane for the same place as Fig. 3. There is a 90° ambiguity in our high-frequency measurements and so these results must be regarded as tentative. We are sure the low-frequency phase shift asymptotes to a 90° lead. Both the low-frequency asymptote of 90° and the high rate of phase change near resonance, are different to von Békésy's results (1960, p. 462).

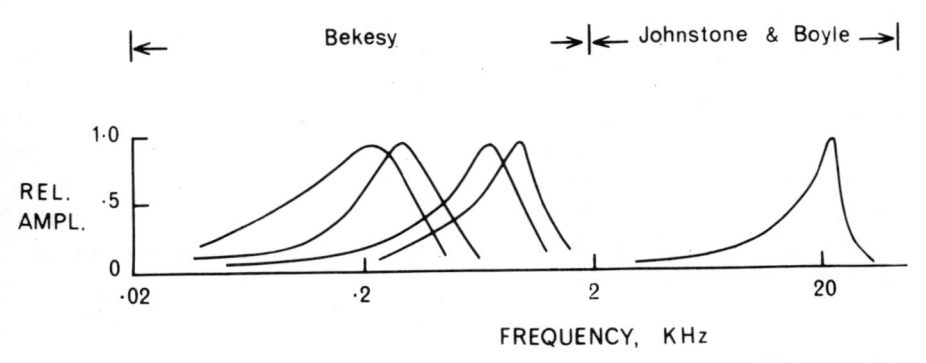

Fig. 5. Basilar membrane envelope of motion for various frequencies (guinea pig). The von Békésy curves are from reference (von Békésy, 1960, p. 504). The Johnstone and Boyle curve is redrawn from their reference (1967).

DETECTION OF THE TRAVELLING WAVE

Whilst the high-frequency slope of the travelling wave is similar to that of the neural curve, the low-frequency side presents a different picture. The neural curve is fairly symmetrical, whereas the mechanical motion is highly asymmetrical.

If nervous innervation is not confined to a very small point on the basilar membrane, but spread out over some finite length, say 150 μ (or 10 hair cells) then a simple slope detector is possible. The branching point of the nerve would act as a summing junction with the various lengths of dendrites supplying different hair cells, acting as delay lines. Such a system could have an optimum response at a particular amplitude and phase pattern over the hair cells, most likely where the phase change is large, *i.e.* just down from the peak amplitude on the high-frequency slope (Ratliff, 1969). The phase change on the low-frequency side is small and so detection efficiency will be poor. A combination of these effects yield a detector which will respond mainly to the fast (high-frequency) edge of the wave as it passes over the array, so a somewhat triangular response area results. This type of detector could sharpen the tuning curve; however, it should not be thought of as lateral inhibition in the normal sense, as there are no additional synapses.

DISTORTION IN THE COCHLEA

To date we have not been able to make satisfactory quantitative measurements of distorted motion, although preliminary experiments leading to a measure of 3rd harmonic distortion seem hopeful (Fig. 2).

One particular form of distortion, resulting in combination tones, has received much attention (Goldstein, 1967). The combination tone $2f_1 - f_2$ must be produced in the mechanical system as it can be cancelled by an external tone.

The amplitude of this tone is very frequency dependent; falling at 100 dB/octave of frequency separation. The right hand side of the tuning curve (Fig. 6) shows a rather steep longitudinal bending. It seems reasonable that distortion could most easily arise in this area. Intermodulation products would then be most easily produced where two such steep slopes from different frequencies overlap; such as shown by the dotted line in Fig. 6. This point of overlap will fall in amplitude at about 100 dB/octave frequency separation so it seems very likely that the combination tones are generated at this place.

A FINAL COMMENT

Returning to the difference between our curves and von Békésy's, several possible explanations come to mind:
(1) There is a real valid difference due to the different parts of the cochlea measured; this has already been discussed.
(2) Von Békésy used a higher sound pressure level of around 130 to 140 dB. We never used more than 120 dB and mostly 90 dB down to 70 dB. It is possible that the curve is flatter at very high sound pressure levels.
(3) An artifact in our measurements due to the finite length of the source. If the wavelength of the travelling wave becomes much shorter than our source

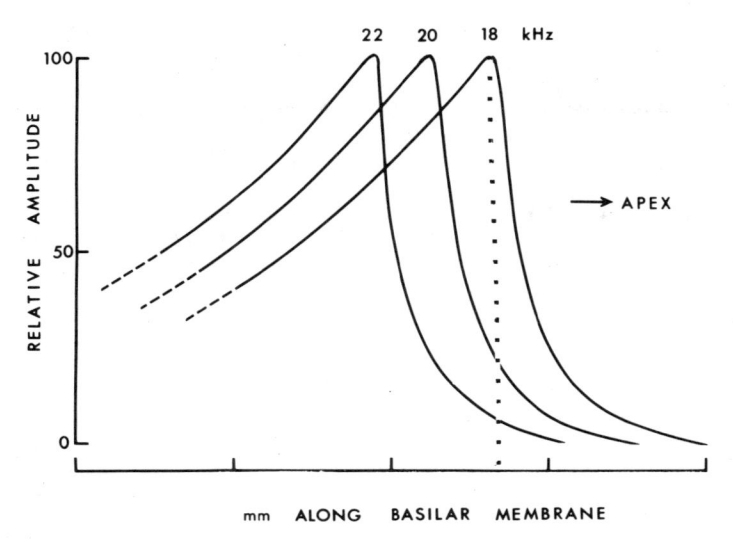

Fig. 6. Basilar membrane envelope for 3 frequencies near the stapes. The overlap of a low-frequency peak (18 kHz) with the high-slope side of a higher frequency (20 kHz) (dotted line) is a function of the position (separation) along the basilar membrane. The amplitude of the point of overlap on the high-frequency curve falls at about 100 dB/octave separation. The curves are diagrammatic and so no absolute scale of distance is presented.

then the source may bridge two wave crests and hence not move much at all, thus giving a faster rate of cut-off than the actual. This seems unlikely on at least 2 counts: (a) several sizes of sources were used, all gave similar tuning curves (but a very small source was not used and we are checking with such a source at the present); (b) during the measurement of basilar membrane phase, we recorded waveforms (Fig. 2) and at no stage did we see a frequency doubling or any untoward distortion, so it would appear no "rocking" motion of the source took place. Since a rocking motion would seem to be an essential prelude to the cresting effect, then it seems unlikely that the wavelength became as small as the source.

(4) An artifact in von Békésy's measurements. As far as we can find out, almost all von Békésy's measurements of the *shape* of the travelling wave were in fact done by observing Reissner's membrane, and a difference in travelling wave shape between this and the basilar membrane is quite possible.

(5) Most of von Békésy's work was on cadavers or partially dead animals. In our experience, death of the animal changes the shape of the tuning curve within 30 minutes (to a flatter shape). However, this point has not been investigated thoroughly as yet and also a dead cochlea is not the same thing as a preserved specimen.

It is obvious that much more will have to be done before these differences are finally resolved.

REFERENCES

Békésy, G. von (1949): The vibration of the cochlear partition in anatomical preparation and in models of the inner ear, J. Acoust. Soc. Amer. *21*, 233-245.
Békésy, G. von (1960): Experiments in Hearing (McGraw-Hill, New York).
Flanagan, J. L. (1962): Computational model for basilar membrane displacement, J. Acoust. Soc. Amer. *34*, 1370-1376.
Goldstein, J. L. (1967): Auditory nonlinearity, J. Acoust. Soc. Amer. *41*, 676-689.
Huxley, A. F. (1969): Is resonance possible in the cochlea after all?, Nature *221*, 935-940.
Johnstone, B. M., and Boyle, A. J. F. (1967): Basilar membrane vibration examined with the Mössbauer technique, Science *158*, 389-390.
Kiang, N. Y. -S. (1965): Discharge Patterns of Single Fibers in the Cat's Auditory Nerve (Research Monograph No. 35, M.I.T. Press, Cambridge, Mass.).
Naftalin, L. (1968): Acoustic transmission and transduction in the peripheral hearing apparatus, in: Progress in Biophysics and Molecular Biology, Butler and Noble Eds. (Pergamon Press, Oxford), pp. 1-27.
Ratliff, F., Knight, B. W., and Graham, N. (1969): On tuning and amplification by lateral inhibition, Proc. Nat. Acad. of Sci. U.S.A. *62*, 733-740.

DISCUSSION

Wilson: Dr. Johnstone makes the point that the high-frequency slope of the response area might reflect directly the mechanical excitation pattern on the basilar membrane. Although this idea is attractive it is not supported by two features of the data available. Firstly the mean value for auditory nerve fibres having characteristic frequencies (CF) above 2.5 kHz is about 200 dB/octave. This can be calculated from the data of Kiang (1965) and is also supported by our own data for cat. It is very significantly greater than the 70 to 105 dB/octave slope given by you for the mechanical motion of the basilar membrane of guinea pig. As it was thought possible that this discrepancy might be due to a species difference, E. F. Evans in our Department (unpublished data) measured response areas from the auditory nerve fibres of guinea pig (latency, spike waveform, and histology confirmed the location of the electrode). Some of these are reproduced in Fig. 1 together with Johnstone's mechanical results. It can be observed that the overall shapes are similar to, if not narrower than, those of cat. The mean value for the high slope is again in the region of 200 dB/octave with maximum slopes up to 500 dB/octave.

The second objection arises from the large variability of slopes obtained in the cochlea nerve data ranging from below 100 dB/octave to greater than 500 dB/octave, and observable in both cat and guinea pig. This appears to be true for all fibres with CF above about 2.5 kHz and there is no obvious general trend as a function of frequency within this range. The nature of this objection is most obvious when two fibres of the same CF in the same animal have different slopes. We have observed a few such examples. It should be pointed out that this variability greatly exceeds that which could be attributed to poor setting or reading of the oscillator even at the highest slopes.

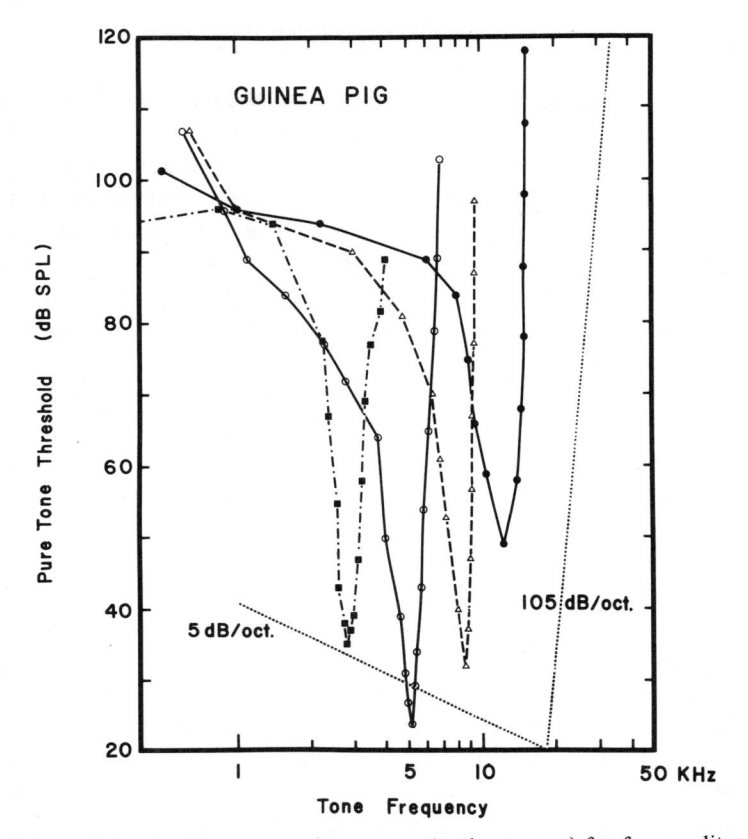

Fig. 1 (Wilson). Threshold response area curves (tuning curves) for four auditory nerve fibres of guinea pig compared with slopes of 5 and 105 dB/octave. Sound pressure was measured at the tympanic membrane.

It must be concluded then on present data that the high-frequency slope difference, as well as the more obvious low-frequency slope difference, requires some form of "sharpening" within the cochlea. How this is achieved remains to be demonstrated.

Johnstone: I am not sure that this sharpening cannot be explained by a simple detector that is based on the shift in phase along the steep slope. Indeed, there is a difference in the slopes I found and the slopes of the tuning curves measured in the single neuron, but I am not sure that it is so large that we should assume some rather complex lateral inhibition mechanism in the cochlea for the high frequencies involved.

Kohllöffel: I have my doubts about this speculation concerning a phase-sensitive detector as Dr. Johnstone proposed. We should not forget that, although the phase changes a lot along the slope, the amplitude drops drastically. I do not see how the nervous system will work on such a low amplitude. If lateral inhibition does occur, I would expect that it occurs where the amplitude is maximal.

Johnstone: There might be some semantic difficulty here. One should, perhaps, not think of lateral inhibition as distinct from a slope detector. These could be rather similar. Moreover, the amplitude along the slope is not so small.

Zwicker: May I try to relate Dr. Johnstone's findings to the results of psycho-acoustical measurements? We found that the slope of the excitation curve has, independently of frequency, a value of about 27 dB per critical bandwidth or Bark, which corresponds to about 20 dB/mm. In my opinion, expressing the slope in dB/octave is not so appropriate, because the frequency scale along the basilar membrane is more linear at low frequencies and more logarithmic at high frequencies. The slope we found corresponds at high frequencies to about 120 dB/octave, which is very similar to what you observed.

Another point I would like to mention is that there is at least a model for the detectibility of frequency changes since its value agrees with a change of 1 dB in amplitude. In this way, all frequency changes detectible by the human ear can be explained.

Johnstone: Yes, this amplitude change corresponds to a frequency change of a little less than 1% and I believe that the difference limen for frequency at 10 to 15 kHz is indeed of that order.

Rose: I would like to add a comment on the elegant measurements shown by Dr. Johnstone, particularly referring to the sharpening mechanism discussed before. There is a semantic difficulty about the term inhibition, so I will avoid this word. But, in fact, there is no evidence, whatsoever, for any neural interaction at the cochlear level that may account for this sharpening.

Johnstone: My only suggestion was that there may be some detector which takes into account the rapid change in phase along the slope. I would like to emphasize that I am not really sure myself about the possible role of this phase shift in sharpening the tuning curve. This shift would be enough to change the firing pattern if there is a phase-sensitive detector.

Whitfield: Ross (cited by Whitfield, 1968) did some work on the tuning curve of auditory nerve fibres within the cochlear nucleus and he came up with the finding that if the data are plotted on a scale representing the square root of frequency you get approximately the same shape for all characteristic frequencies. Did you find a similar result for the mechanical disturbances?

Johnstone: I am not sure about that but I cannot see why such a square-root relation should occur.

Terhardt: May I put again the question to which Dr. Zwicker referred: are the slopes of the vibration patterns different at different places or not?

Johnstone: I cannot answer that question because I could only measure at high frequencies. So, we have to compare our results with von Békésy's low-frequency measurements, and he found slopes much less steep than ours.

Plomp: We did some experiments on the aftermasking of a pure tone of 1000 Hz as a function of SPL. At low levels, the maximum in the masking pattern coincides with the 1000 Hz tone but at levels of about 100 dB a distinct shift of the whole pattern to higher frequencies was noticed. May I ask Dr. Johnstone whether he has found any similar shift of the vibration pattern as a function of SPL?

Johnstone: Our measuring technique makes it impossible to introduce SPL differences of more than about 10 dB. In these cases we did not observe any significant shift, that is to say, the shift, if present, is less than 10%. Only once we could measure over a range of 20 dB SPL and in that case there was some evidence for a slight shift.

REFERENCES

Kiang, N. Y.-S. (1965): Discharge Patterns of Single Fibers in the Cat's Auditory Nerve (Research Monograph No. 35, M.I.T. Press, Cambridge, Mass.), Ch. 7 (Figs. 7.3, 7.4, and 7.5).

Whitfield, I. C. (1968): The functional organization of the auditory pathways, J. Sound Vib. *8*, 108-117.

TEMPORAL AND SPATIAL DISTRIBUTION OF THE CM AND SP OF THE COCHLEA

VICENTE HONRUBIA
in collaboration with P. H. Ward

Department of Surgery/Head and Neck (Otolaryngology)
UCLA School of Medicine
Los Angeles, Calif., U.S.A.

Two generator potentials are produced in the organ of Corti during acoustic stimulation, the cochlear microphonics (CM) and the summating potential (SP). It is generally accepted that the CM reflect the mechanical events of the cochlear partition and are consequently linked to the stimulation process of the ear (Wever, 1966). In 1950 Davis *et al.* described a DC potential produced in the cochlea simultaneously with the appearance of the CM. This potential (the SP) was believed to reflect the summated stimulating activity of the CM. Limitations of our knowledge of the dependence of the SP on the parameters of the acoustic stimulus or its relationship to the CM make the SP the least understood of the cochlear potentials.

Recordings of the potentials from inside the cochlear duct are useful in determining the electroanatomy of the cochlea and produce little surgical damage to the cochlea (Tasaki, 1957). Because of this, we performed experiments with microelectrodes inserted inside the scala media to study the temporal and spatial distribution of the CM and SP in various animals.

METHOD

The experiments were conducted in 42 guinea pigs, 5 cats, 4 squirrel monkeys and 2 Macaca rhesus monkeys. In most of the guinea pigs, with the aid of an operating microscope and small sharp knives, the bony otic capsule was thinned and fenestrae 50-100 μ in width were made over the surface of the spiral ligament and the stria vascularis. Glass microelectrodes 1 to 3 μ diameter were introduced inside the scala media through the fenestrae. Potentials were recorded from the 4 turns of the cochlea in 16 guinea pigs. Based on the anatomical maps of von

Békésy (1960) and Fernandez (1952), the fenestrae were located in the first, second, third, and fourth turns at approximately 4.0, 9.5, 13.5, and 17.25 mm from the origin of the cochlea. In 5 guinea pigs and in all the cats and monkeys the recordings were made with microelectrodes inserted in the cochlear duct through the round window membrane.

Tone pips were used for stimulation. Calibrated sounds were delivered from an electrically shielded speaker 1 ft in diameter located 1 ft from the head of the animal in some of the experiments. In other experiments, as indicated in the text, a "closed" system designed after Kiang et al. (1965) was used for stimulation. Standard electrophysiological techniques were used for the synchronization of the stimulus, display, and recording systems (Honrubia and Ward, 1968, 1969a, 1969b).

The voltage of the CM indicates peak-to-peak values. The SP was measured on DC recordings and evaluated by measuring the voltage from the baseline of records to the center of the microphonics deflections.

RESULTS

Tuning curves of the CM along the cochlea were obtained in 8 guinea pigs. The animal's ear was stimulated with pure tones of the same frequency (250-6000 Hz) and sound pressure level (SPL) while recordings were made successively in each of the 4 turns of the cochlea. Input-output functions of the CM were obtained at each turn for the various frequencies. At low SPL the CM functions grew linearly with increased strength of stimulation with slopes usually slightly less than unity. The input-output functions start bending at lower SPL's for high frequencies than for lower frequencies.

The longitudinal distribution of the potentials was obtained from the plots of the input-output functions by joining with smooth curves the points that represent the voltage of the CM obtained at each turn with stimulation by the same SPL (Fig. 1). The accuracy of the interpolated values was substantiated by the fact that at intermediate points between the recording places the slope of the input-output functions of the CM was the same as in the recorded points. Nevertheless, envelopes for frequencies higher than 3000 Hz could not be drawn with confidence. The longitudinal attenuation of the voltage of the CM for the 300-Hz sound was 2 dB/mm, doubling in magnitude for a decade increment in the frequency. The slope of the longitudinal envelope on the proximal side remained the same for each frequency over the range of SPL's for which the input-output functions were linear. The points of both maximum CM voltage and maximum sensitivity were determined for every frequency. The point of maximum CM voltage was considered to be the location in the cochlea where the largest magnitude of the CM was measured for each SPL. For frequencies lower than 3000 Hz, the points of maximum CM voltage

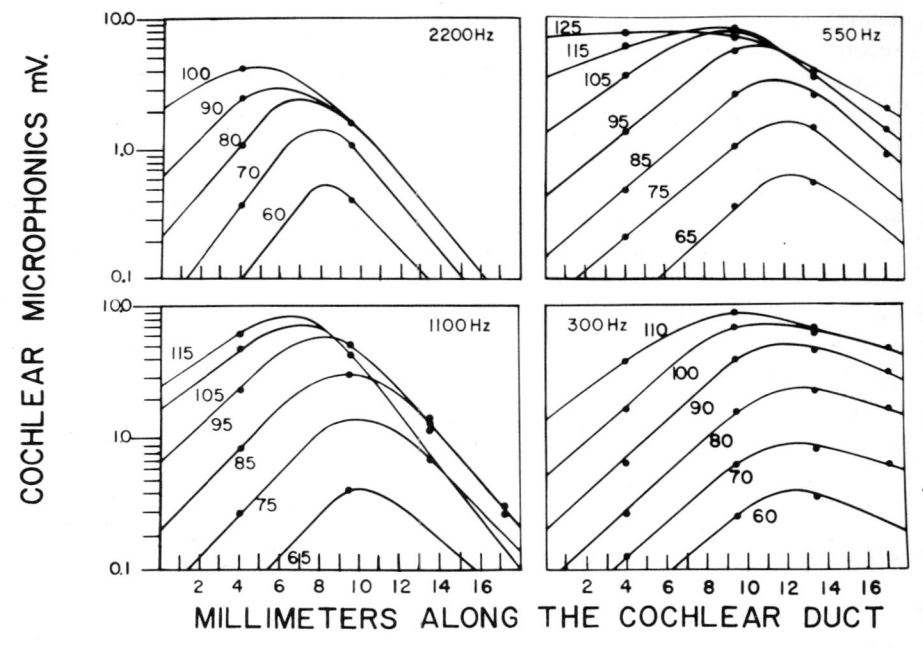

Fig. 1. Longitudinal distribution of the CM recorded inside the scala media. Responses obtained with 4 different frequencies at the indicated SPL's in dB.

moved toward the base of the cochlea as the SPL's were increased, and the displacement was greatest for the lower frequencies. This phenomenon was observed in all guinea pigs. The point of maximum sensitivity was determined as the place in the cochlear duct where the CM appeared with the lowest SPL of stimulation. The position along the cochlea of the maximum sensitivity points, as well as the points of maximum CM voltage, irrespective of the SPL, are shown by the dashed lines in Fig. 2. By changing the frequency of the sound

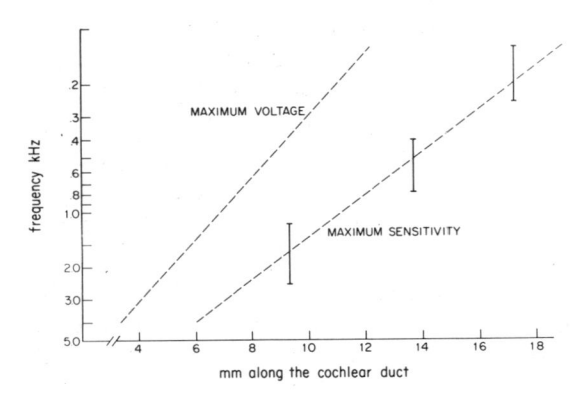

Fig. 2. The dashed lines, fit by the method of least squares, indicate the maximum voltage and maximum sensitivity of the CM along the cochlear duct of the guinea pig. The vertical bars indicate the range of frequencies at which the phase difference was 180° between the first and the second, third and fourth turns.

one octave, there is a shift of 2.6 mm in the location of the maximum sensitivity point and of 2 mm for the maximum amplitude.

The average slope of the linear segment of the CM input-output functions was 0.8 with a range of 0.75 to 0.9. The CM departed from linearity and reached saturation levels earlier at the frequency for which the specific turn showed the maximum sensitivity. The shift in the position of the maximum voltage points seems to be due to this phenomenon.

The SPL necessary for CM of equal magnitude at successive points along the cochlea was determined (Fig. 3). A consequence of the displacement of the maximum CM voltage toward the base of the cochlea is the creation of a steep cut-off response on the high-frequency side of the receptive fields. The hair cells located about 11 mm from the origin of the cochlear duct will be stimulated with frequencies successively lower than 1500 Hz by increasing the SPL. With higher frequencies (*i.e.*, >2500 Hz), this area will not be stimulated.

The phase difference between the CM recorded in the second, third and fourth turns and that of the CM in the first turn was measured in 9 guinea pigs. In 3 guinea pigs microelectrodes were introduced successively in the second, third and fourth turns with the microelectrode in the first turn remaining in place (Fig. 4). Recordings were made in groups of 2 guinea pigs from the first and second, first and third, and first and fourth turns. The combined results are shown in Fig. 2.

Tuning curves for the SP were constructed from the input-output functions from each turn in the same fashion as those of the CM (Fig. 5). In the first turn of the cochlea the SP were recorded — in reference to a ground electrode in the neck — with negative polarity for frequencies above 1500 Hz. For lower frequencies the SP was positive at low SPL and became negative at higher SPL's. It was observed that at each recording point in the cochlea there was a frequency

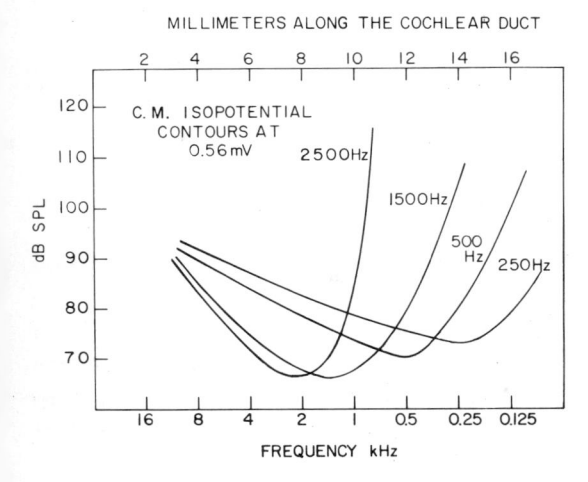

Fig. 3. SPL required to produce 0.56 mV of CM for the indicated frequencies at different positions along the cochlear duct (reproduced from Honrubia and Ward, 1968).

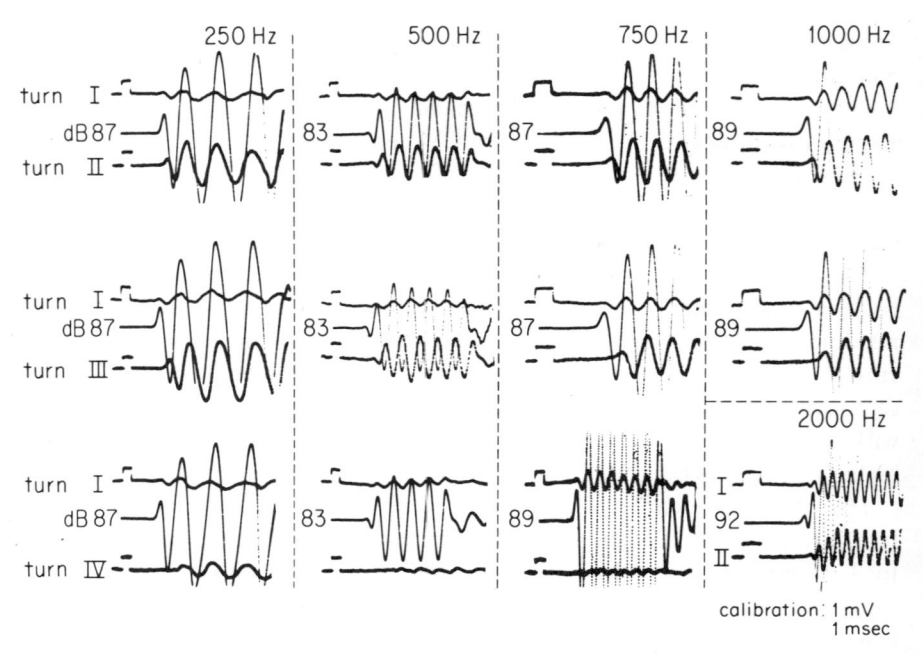

Fig. 4. The CM for various frequencies were recorded simultaneously in the first turn and the second, third and fourth turns of the cochlea of a guinea pig, successively. The recording at the top was made with the electrode in the first turn. The middle recording is the electrical output of the SPL-meter connected to a condenser microphone placed next to the animal's ear. The bottom record in each horizontal row was made from the electrode placed in the second turn (first row), third turn (second row) and fourth turn (third row).

for which the SP input-output functions showed a maximum steepness and for which the SP was recorded negative at lower SPL than anywhere else in the cochlea. For frequencies below that most sensitive one, the SP appeared positive at lower and negative at higher SPL. The lower the frequency, the higher the SPL necessary to produce negative SP. Comparison of the SP and CM tuning curves showed that the maxima of the negative SP occur on the distal envelope of the CM tuning curves. The arrows on Fig. 5 indicate the position of the maximum of the CM envelopes at 95 dB SPL for each of the frequencies in this guinea pig's cochlea. The magnitude of the SP reflects the change in the profile of the CM envelope.

The magnitude of the SP also depended on the stimulus duration. Recordings of CM and SP generated by stimulating the cochlea with 3400 Hz in the first turn are shown in Fig. 6. The calibrating signal seen before the recorded cochlear responses is always 1 mV and 10 msec. When the velocity of the sweep of the oscilloscope was decreased 2, 10 or 20 times, the calibrating signal had 1/2, 1/10 or 1/20 the length of that in the first row of pictures. Measurements of the SP at

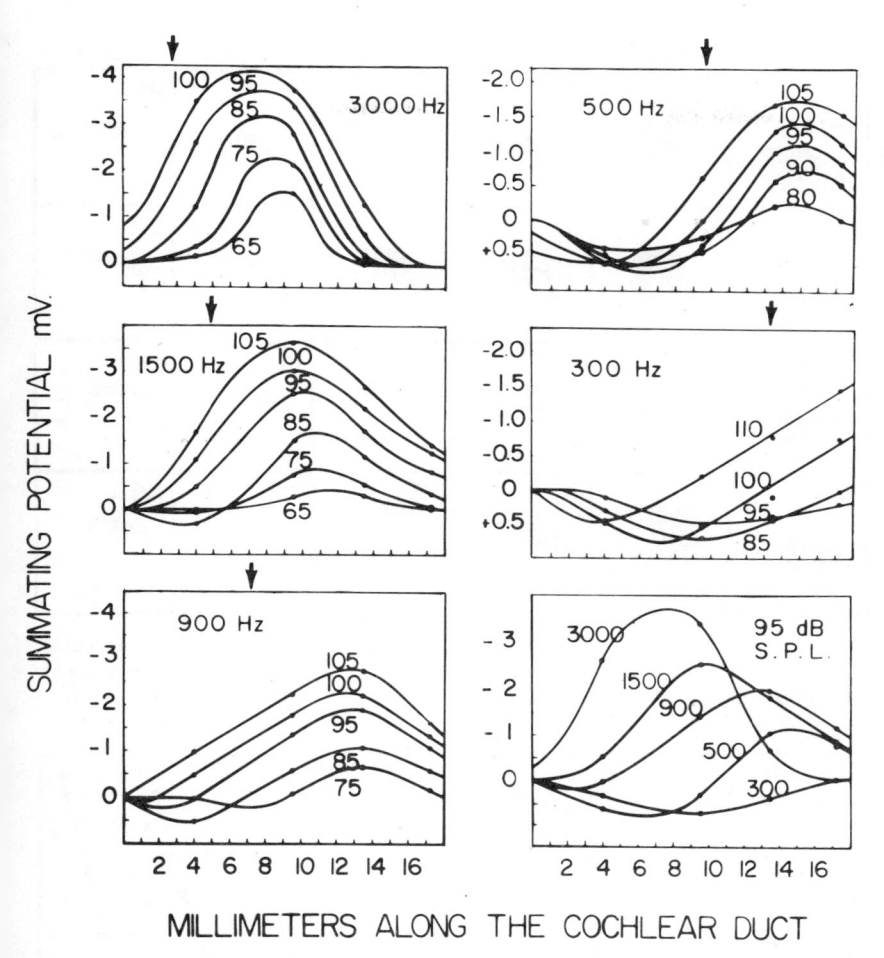

Fig. 5. Longitudinal distribution of the SP recorded inside the scala media. Responses obtained for 5 different frequencies at the indicated SPL's in dB. Right lower insert shows curves obtained at 95 dB SPL for the indicated frequencies (slightly modified from Honrubia and Ward, 1969c).

each SPL and duration are graphed in Fig. 7. The magnitude of the SP is proportional to the logarithm of the duration of stimulation. This relationship has been observed to be true in the first and second turns of the cochlea, for positive as well as for negative SP, for sounds of 40 msec to 2 sec duration, and for physiological SPL's.

Whenever the SP reversed the polarity with increase of the SPL of the stimulus, the transition of polarity occurred at the same SPL, regardless of the duration of the stimulus (Fig. 8). Therefore, although increments in SPL resulted in greater lengths of the organ of Corti covered by the SP envelope (Fig. 3), increments in duration produced only greater amplitudes of the SP

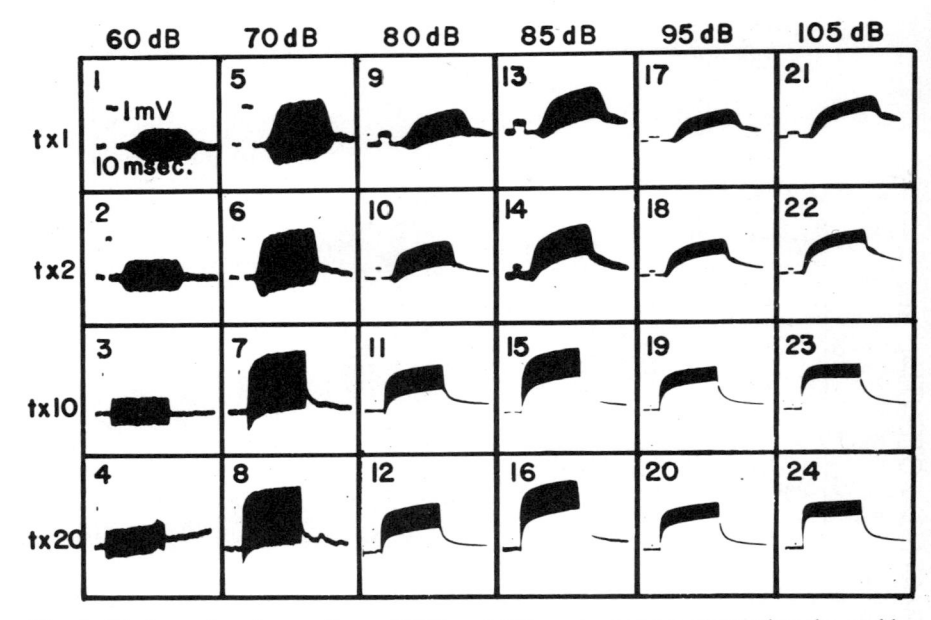

Fig. 6. Photographs of recordings of CM and SP's generated by stimulating the cochlea with 3400 Hz at the indicated SPL's and duration times. Further explanation is presented in the text. Calibrating signal 1 mV, 10 msec.

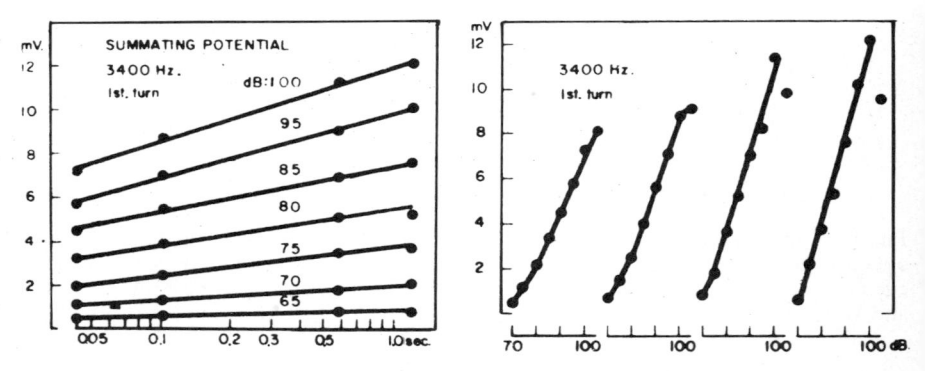

Fig. 7. Graphical representation of the SP recordings of Fig. 6. Left: relationship between the magnitude of the SP and the duration of the stimulus at the indicated SPL's. Right: input-output functions for the corresponding durations.

without recruitment of new basilar membrane segments to generate −SP.

A comparison of the cochlear potentials of the guinea pig, cat, squirrel monkey and Macaca rhesus monkey was made by recording the acoustically evoked potentials of the scala media at the level of the round window. The magnitude of the CM recorded at different frequencies at 95 dB SPL is shown in Fig. 9. For one squirrel monkey and one Macaca rhesus monkey, the SPL for which

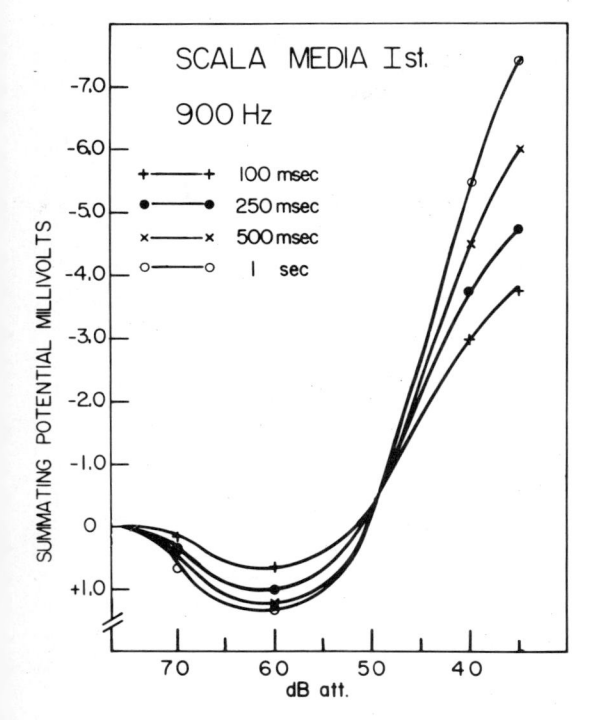

Fig. 8. Input-output function of the SP produced by a 900-Hz sound in the scala media of the first turn at the indicated durations of stimulation (modified from Honrubia and Ward, 1969c).

the CM are plotted was 105 dB.

The results of these measurements showed that with this SPL there was an optimum frequency at this level of the cochlea for which the output of the organ of Corti is maximum.

The cochlea showed the maximum sensitivity for 1640 Hz in the cat, 2500 Hz in the squirrel monkey, and 640 Hz in the guinea pig and rhesus monkey. In all animals it was possible to record measurable CM for the lowest frequency used (32 Hz). In the guinea pig the organ of Corti showed the greatest sensitivity for the lowest frequency (about 10 dB higher than that of the cat).

The magnitude of the CM is linearly related to the frequency of the stimulus. The cats showed the largest resolution power with a voltage change of about 8 dB/octave, followed by the squirrel monkey with 5 dB/octave and the rhesus monkey and guinea pig with approximately 4 dB/octave.

The behavior of the SP was similar in all 4 species. Tuning curves of both the CM and SP from one of the cats are shown in Fig. 10. The general properties of the SP at this level of the scala media were similar to those described for the first and second turns of the cochlea of the guinea pig. It was possible to measure the SP at lower SPL's for the higher than for the lower frequencies. At low frequencies the SP grew first positive and became smaller with increased

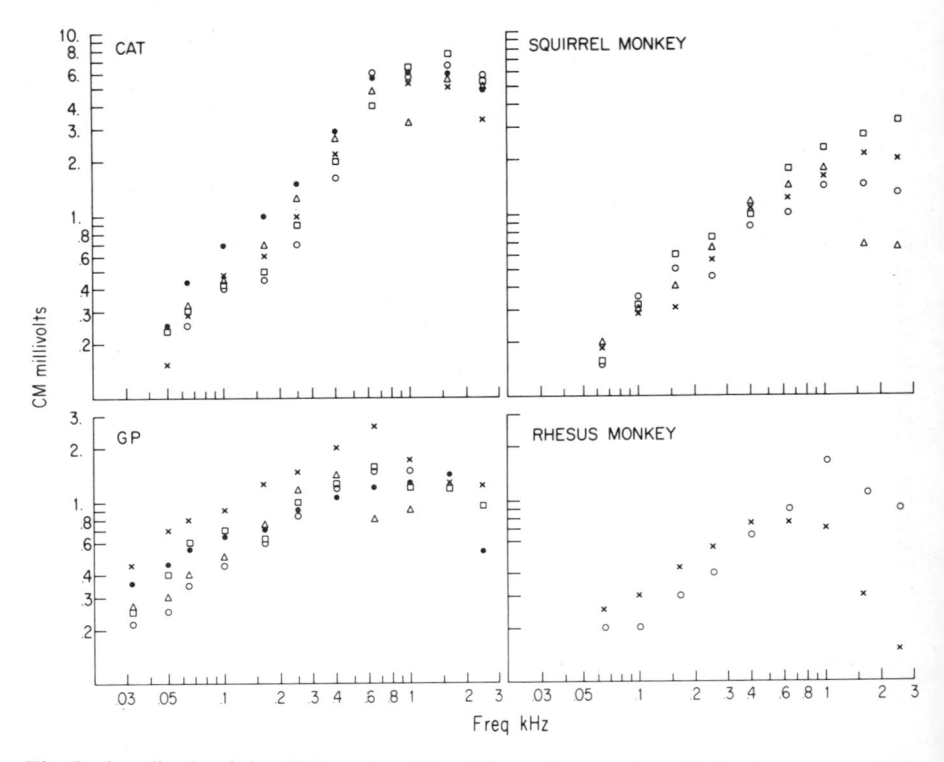

Fig. 9. Amplitude of the CM produced by different frequencies with 95 dB constant SPL at the ear drum. Recordings made inside the scala media at the level of the round window. Closed system stimulation. Each symbol represents a different animal (reproduced from Honrubia and Ward, 1969b).

SPL. At medium frequencies (400, 640, 1000 Hz) it was initially positive and then became negative. At the higher frequencies it was slightly positive or immediately became negative, increasing in magnitude with increased SPL of stimulation. The SPL at which the SP reversed polarity was lowest with the highest frequencies.

DISCUSSION

The cochlea represents the first station for the analysis of acoustic signals and also provides the nervous system with the mechanism for initiating nerve action potentials. Because of the traveling waves, different frequencies stimulate different segments of the basilar membrane. The effective stimulus is the tangential shear between the organ of Corti and the tectorial membrane. Two modes of shear, one radial and the other longitudinal, have been demonstrated in the guinea pig cochlea and in cochlear models and have been observed to be

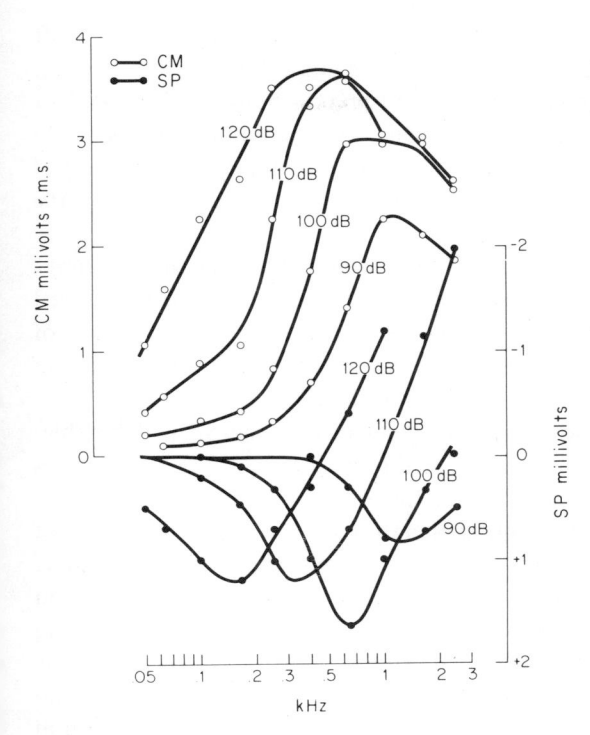

Fig. 10. Resonance curves for the CM and SP from scala media of cat A-24. Stimulation with closed system. SPL measured at the ear drum (reproduced from Honrubia and Ward, 1969b).

sharper than the traveling waves (von Békésy, 1960, pp. 497, 703; Tonndorf, 1960; Khanna et al., 1968).

By virtue of the existence of the potential gradient across the reticular lamina, the cochlea provides the energy sources for the production of the large electric current associated with the generation of CM. This electric current flows through the body of the hair cells and is believed to trigger the presynaptic events conducive to the stimulation of the eighth nerve fibers (Davis, 1968; Honrubia and Ward, 1969a).

A true representation of the temporal and spatial distribution of the CM inside of the cochlear duct affords an objective measurement of the mechanical events in the cochlea. Because of the electronic spread of currents along the core conductor, that is, along the cochlear partition, however, a discrepancy exists between the true CM and the potentials measured by electrodes placed inside the scala media. According to Misrahy et al. (1958), the current through the wall of the cochlear duct in the first turn becomes one-third of its magnitude every millimeter. For frequencies lower than 2000 Hz, the wave length is in excess of 10 mm, and the phase shift is less than 180° up to a point of maximum amplitude of the CM. Therefore, measurements for these frequencies on the proximal side of the maximum of the CM envelope should reflect the mechanical events in the cochlear partition with good approximation. The phase shift of the CM along the cochlea is proportional to the logarithm of the frequency, which indicates that the phase (wave) velocity is exponential and independent of the frequency of sound. This is in agreement with the findings of Zwislocki (1965), Tonndorf (1957) and Tasaki et al. (1952). Therefore, in the distal side of the CM envelope the measurements of the CM are affected greatly by the electronic spread of currents.

The CM by themselves cannot account for the production of tuning curves as sharp as those of the first-order neurons. The present experiments showed the existence of two phenomena that could collaborate in the production of sharp neural tuning curves. One is the displacement of the point of maximum stimulation (Fig. 2), the other is the presence of the SP.

The mechanism of production and the significance of the SP is debatable. The appearance of the SP is generally accepted to be the result of nonlinearities within the ear (Whitfield and Ross, 1965; Engebretson and Eldredge, 1968). Their two models provide a reasonable explanation for the production of SP of one polarity. In the cochlea, however, the SP may be either positive or negative.

Direct observations in the human cochlea and in cochlear models (von Békésy, 1960; Tonndorf, 1960) showed that in the distal region of the traveling wave the longitudinal bending of the cochlear partition was accompanied by rectification of the vibratory movement. Proximal to the point of maximum mechanical stimulation, the displacement is larger toward the scala vestibuli and toward the scala tympani in the distal part of the cochlea. According to these observations, the polarity of the SP should be positive on the distal and negative on the proximal side of the traveling wave, in accordance with the findings of Tasaki et al. (1954), Teas et al. (1962), Butler and Honrubia (1963), and Kiang et al. (1965). Although the mechanism of production is debatable, the functional implications of the SP follow from their polarity. The bias created in the scala media by the positive SP on the proximal side of the CM envelope, in all likelihood tends to inhibit the excitatory effect resulting from the outward flow of current in the hair cells during the negative half-wave of the CM. This results in a sharpening of the receptive fields of the cochlea. The negative SP has long been claimed to act as a stimulus for the eighth nerve, especially for high frequencies for which the CM are inefficient as a stimulator for the nerve fiber (Davis et al., 1958).

The appearance of the round-window tuning curves points to the existence of pronounced differences among the different groups of animals. At present it is difficult to evaluate their significance. The only comparison between the cochlear potentials that is valid is that of the slope of their input-output function, which is the same in all species measured (Wever, 1966). The absolute magnitude of the CM, as well as their longitudinal distribution, is affected by electrical characteristics of the cochlear partition, of which little is known.

REFERENCES

Békésy, G. von (1960): Experiments in Hearing (McGraw-Hill, New York).
Butler, R. A., and Honrubia, V. (1963): Responses of cochlear potentials to changes in hydrostatic pressure, J. Acoust. Soc. Amer. *35*, 1188-1192.
Davis, H., Fernandez, C., and McAuliffe, D. R. (1950): The excitatory process in the cochlea, Proc. Nat. Acad. Sci. *36*, 580-587.
Davis, H. (1968): Mechanism of the inner ear, Ann. Otol. Rhinol. Laryngol. *77*, 644-655.
Engebretson, A. M., and Eldredge, D. H. (1968): Model for the nonlinear characteristics of cochlear potentials, J. Acoust. Soc. Amer. *44*, 548-554.
Fernandez, C. (1952): Dimensions of the cochlea (guinea pig), J. Acoust. Soc. Amer. *24*, 519-523.
Honrubia, V., and Ward, P. H. (1968): Longitudinal distribution of the cochlear microphonics inside the cochlear duct (guinea pig), J. Acoust. Soc. Amer. *44*, 951-958.
Honrubia, V., and Ward, P. H. (1969a): The dependence of the cochlear microphonics and the summating potential on the endocochlear potential, J. Acoust. Soc. Amer. *46*, 388-392.
Honrubia, V., and Ward, P. H. (1969b): Cochlear potentials inside the cochlear duct at the level of the round window, Ann. Otol. Rhinol. Laryngol., *78*, 1154.
Honrubia, V., and Ward, P. H. (1969c): Properties of the summating potential of the guinea pig's cochlea, J. Acoust. Soc. Amer. *45*, 1443-1449.
Khanna, S. M., Sears, R. E., and Tonndorf, J. (1968): Some properties of longitudinal shear waves: a study by computer simulation, J. Acoust. Soc. Amer. *43*, 1077-1084.
Kiang, N. Y., Watanabe, T., Thomas, E. C., and Clark, L. F. (1965): Discharge Patterns of Single Fibers in the Cat's Auditory Nerve (Research Monograph No. 35, M.I.T. Press, Cambridge, Mass.).
Misrahy, G. A., Hildreth, K. M., Shinabarger, E. W., Clark, L. C., and Rise, E. A. (1958): Electrical properties of wall of endolymphatic space of the cochlea (guinea pig), Amer. J. Physiol. *194*, 396-402.
Tasaki, I., Davis, H., and Legouix, J. P. (1952): The space-time pattern of the cochlear microphonics (guinea pig) as recorded by differential electrodes, J. Acoust. Soc. Amer. *24*, 502-519.
Tasaki, I., Davis, H., and Eldredge, D. H. (1954): Exploration of cochlear potentials in guinea pig with a microelectrode, J. Acoust. Soc. Amer. *26*, 765-773.
Tasaki, I. (1957): Hearing, Ann. Rev. of Physiol. *19*, 417-438.
Teas, D. C., Eldredge, D. H., and Davis, H. (1962): Cochlear responses to acoustic transients: an interpretation of whole nerve action potentials, J. Acoust. Soc. Amer. *34*, 1438-1459.
Tonndorf, J. (1957): Fluid motion in cochlear models, J. Acoust. Soc. Amer. *29*, 558-568.
Tonndorf, J. (1960): Shearing motion in scala media of cochlear models, J. Acoust. Soc. Amer. *32*, 238-244.
Wever, E. G. (1966): Electrical potentials of the cochlea, Physiol. Rev. *46*, 102-127.
Whitfield, I. C., and Ross, H. F. (1965): Cochlear microphonic and summating potentials and the outputs of individual hair cell generators, J. Acoust. Soc. Amer. *38*, 126-131.
Zwislocki, J. J. (1965): Analysis of some auditory characteristics, in: Handbook of Mathematical Psychology, R. D. Luce, R. R. Bush, and E. Galanter, Eds. (John Wiley & Sons, New York), Vol. III, pp. 1-97.

DISCUSSION

Schwartzkopff: Is it possible that the change in polarity of the SP (see Fig. 8 of the paper) is caused by other current sources coming into effect? We did many experiments with birds and found a change of polarity only in 2 cases out of 50. I realize, of course, that the bird has a more simple cochlea than mammals.

Honrubia: It is possible that the generators for the $+SP$ and the $-SP$ are different, but I don't know why the change of polarity is not seen in the cochlea of birds.

Whitfield: In recording from the two ends of the cochlea, we have found that when the activity is at the basal end there is a small positive SP at the apical end and *vice versa*, whereas there is a negative electrode potential near the activity. Now, I don't know how this might be related, but I remember Davis' original observations (1958) showing a positive SP for small amplitudes.

Honrubia: If you refer to the polarity of the SP at the point of maximum sensitivity, I saw this phenomenon only once in some 75 experiments with different frequencies of stimulation.

Whitfield: Another point concerns the difference with time. With a stimulus lasting more than a few tens of msec, there is a positive after-effect at the end of the stimulus when the polarity during the stimulus is negative. Its size depends upon the duration of the stimulus and it behaves just like a neuronal afterpotential. Now it seems possible, if there is such an after-effect, that you may get the corresponding effect of a persistent depolarization during the stimulus so that, indeed, a part of the negative SP may be a synaptic depolarization added to the other effect.

Honrubia: Within the range of stimuli I have used I never saw this afterdepolarization inside the scala media. We are planning to study the SP of deinnervated cochleas to see whether there are differences with respect to the SP of normal cochleas.

Whitfield: I don't know whether that will help you.

Honrubia: The second point is that, for the interpretation of the potential, one has to take into account that the generation is in the reticular lamina. Because of both capacitance and resistance, the potential is different whether it is recorded inside the scala media or outside, in the scala tympani or scala vestibuli. Not only is there an attenuation of the potential, but also the time constant of the membranes is an important factor.

Whitfield: When you move the electrode from the scala media into the scala vestibuli, there is no polarity reversal of that part of the potential produced at the synaptic ends of the hair cells. This will complicate the interpretation.

REFERENCES

Davis, H., Deatherage, B. H., Eldredge, D. H., and Smith, C. A. (1958): Summating potentials of the cochlea, Amer. J. Physiol. *195*, 251-261.

COCHLEAR MICROPHONICS DISTRIBUTION AND SPATIAL FILTERING

L. U. E. KOHLLÖFFEL

Neurocommunications Research Unit
University of Birmingham
Birmingham, England

INTRODUCTION

The basilar membrane/hair cell transducer mechanism is still the subject of much speculation, not only as regards the intimate mechanism of the hair cell action but more broadly with respect to the mechanoelectric behaviour of the cochlear partition itself. Although there is some experimental evidence establishing a qualitative correspondence between a mechanical travelling wave on the basilar membrane and the responses of nerve fibres arising at points along it, the transfer functions cannot at present be studied quantitatively because the behaviour of the basilar membrane is much less well measured than that of the nerve fibres.

Owing to the very small displacements involved, it is exceedingly difficult to study directly the motion of the membrane except at rather high stimulus intensities, where the system is probably highly nonlinear (Whitfield and Ross, 1965). Such studies were of course attempted in the classical experiments of von Békésy (1947) and more recently by Johnstone and Boyle (1967). However, displacement of the cochlear partition leads to a change in the potential across it, and the distribution of these changes, which manifest themselves as the cochlear microphonic (CM) and summating potentials (SP) have been widely used in attempts to infer the corresponding distribution of mechanical movement along the membrane. The experiments of von Békésy (1953) using vibrating needles applied to the partition, and those of Tasaki *et al.* (1954) in which microelectrodes penetrated the membrane, suggest that the source of the CM is located at the hair-bearing ends of individual hair cells. The potential generated at any given point might be expected, therefore, to be an index of the relative displacement of that particular point. If the whole system were placed in a medium of longitudinally low conductance, then an electrode placed close to the partition at any point along it should record an attenuated voltage which would be a measure of the amplitude of vibration at that point, and the phase angle would be a measure of the phase delay at the point. Much of the work which has been done on CM measurements has been interpreted on the basis that this assumption is approximately true. However, what evidence there is suggests that the assumption is very far from true, and that the attenuation factor along the cochlea is of the order of only 6 or 7 dB/mm (von Békésy, 1951; Tasaki et al., 1952). As pointed out by Whitfield and Ross (1965) this means that the signals arising from individual hair cell operators spread out and interfere with each other to a large extent in the surrounding inner ear fluids and the activity recorded by an individual electrode is a weighted average of the whole hair cell array, rather than the activity of a small number of hair cells in the immediate vicinity of the electrode. Mathematically this hypothesis takes the form of a convolution integral:

$$CM(x_o) = \int H(x) \cdot W(x_o - x) \mathrm{d}x = H(x) * W(x) \tag{1}$$

where $H(x)$ denotes hair cell output along x and $W(x_o - x)$ is a weighting function accounting for the attenuation of a signal between its point of generation x and the recording site x_o. The CM distribution appears thus as the spatially filtered image of $H(x)$. It is evident that because of the shorter wavelength of the travelling wave, the interactions will be much more noticeable for high frequency stimuli than for low ones, and outputs will show the greatest departure from the 'local electrode' assumption. As a result of the cancellations from out-of-phase generators we might predict qualitatively that there will be amplitude minima in the microphonic at one or more points along the cochlear scalae and that these may be accompanied by abrupt changes of phase between neighbouring points.

MEASUREMENT OF CM AND DATA ANALYSIS

In order to study the problem, it is clearly necessary to record simultaneously from a number of closely spaced points along the hair cell array. In the experiments to be described an array of twelve closely spaced electrodes was inserted into the wall of the basal turn of scala tympani in the guinea pig. Two slightly different techniques were used. In one, twelve separate holes were drilled in the bony wall, into each of which a 50 μ insulated stainless steel wire was sealed. In the other technique, an array of twelve electrodes was constructed by embedding similar insulated steel wires in polystyrene. The inter-electrode distance in the ribbon-like array so formed was 150 μ. This electrode assembly was inserted into the cochlea through a slit 2 mm long and 0.5 mm wide drilled parallel to the basilar membrane in the wall of the scala tympani. The position of the most basal electrode of the array was approximately 3 to 3.5 mm from the basal end of the basilar membrane. The presence of the slit does not seem to interfere with the proper functioning of the inner ear; both the 'slit' and the 'closed hole' techniques gave substantially the same results (Kohllöffel, 1970). The multi-electrode array, however, gives more detail over a shorter length of membrane than do the more widely spaced electrodes of the 'closed hole' technique.

The basic pattern of the results obtained is illustrated by the example of Fig. 1. This figure shows a typical set of amplitude and phase curves for stimulus frequencies between 4 and 16 kHz and a SPL in the range of 70 to 90 dB. The systematic variation in the general amplitude patterns confirms that with increasing frequency the active region of hair cells moves towards the round window. This is indicated by the maintained drop in CM amplitude at electrodes most distant from the round window as 14 kHz is exceeded. Note, however, the existence of minima in the responses (well shown in the 6.5 kHz record), an occurrence predicted by the spatial filter hypothesis. Even more startling are the phase patterns. With increasing frequency the phase changes very suddenly from lag to lead; this takes place at about 13 kHz in the example shown. This, too, is at complete variance with the hypothesis that a given electrode records activity over only a small number of hair cells. If it were so then one would expect the CM distribution to exhibit a phase pattern similar to that which characterizes the travelling wave on the basilar membrane, *i.e.*, a progressive continuous phase lag from the round window towards the apex. Clearly this is not the case. On the other hand, a simple interpretation of the phase shift phenomenon can be given, if the 'local electrode' assumption is dropped. In Fig. 2 the data of Fig. 1 for 11.5 kHz and 15 kHz have been redrawn in vector form. Each vector represents the amplitude and phase of a particular electrode output, and their endpoints are joined in each case through a trajectory. With increasing frequency the trajectory gradually approaches and crosses the origin of the vector plane. After it has crossed the origin (which takes place between 12.8 kHz and 13 kHz), we observe that vectors distant from the round window lead, in phase, those closer to the round window. If we were to add to all electrode outputs in a case like Fig. 2a a small vector of large and constant phase lag, but with an amplitude which declines progressively at each electrode as the round window end of the array is approached, the effect on the trajectory would be as shown in Fig. 2c. In other words, the trajectory is constrained to cross the origin, and to show a phase lead comparable to the pattern found in the 15 kHz case (Fig. 2b). In the real situation something quite analogous to this may be expected to happen. With rising frequency, the wavelength of the travelling wave shortens and the hair cells more distant from the round window operate with greater phase lag than before

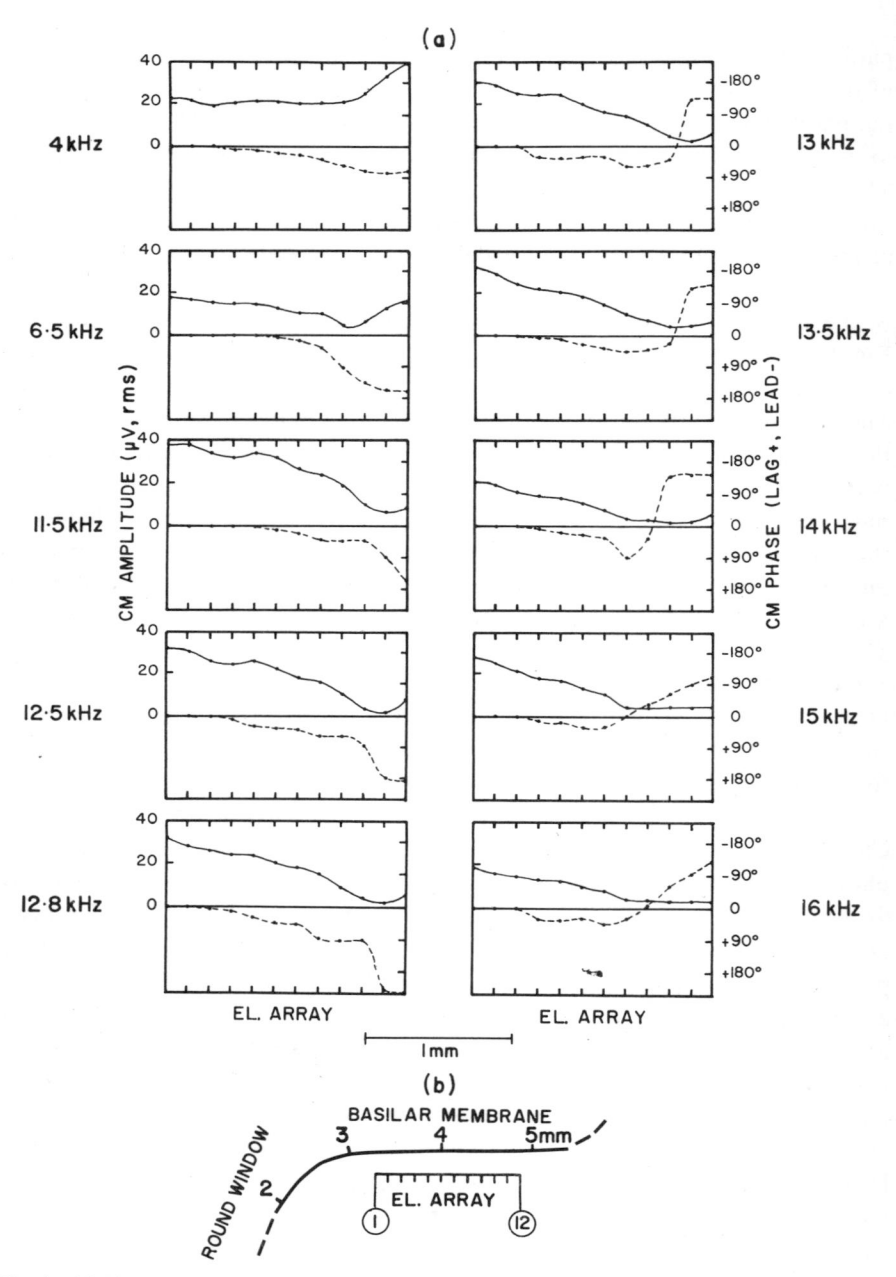

Fig. 1. (a) Typical example of a series of CM amplitude and phase curves obtained from one guinea pig. Note that large phase changes occur in the region of minimum amplitude. At frequencies beyond 12.8 kHz outputs distant from the round window lead in phase more proximal ones. Amplitude —, phase - - -
(b) Electrode array position in scala tympani shown in relation to longitudinal dimensions of basilar membrane (from Kohllöffel, 1970).

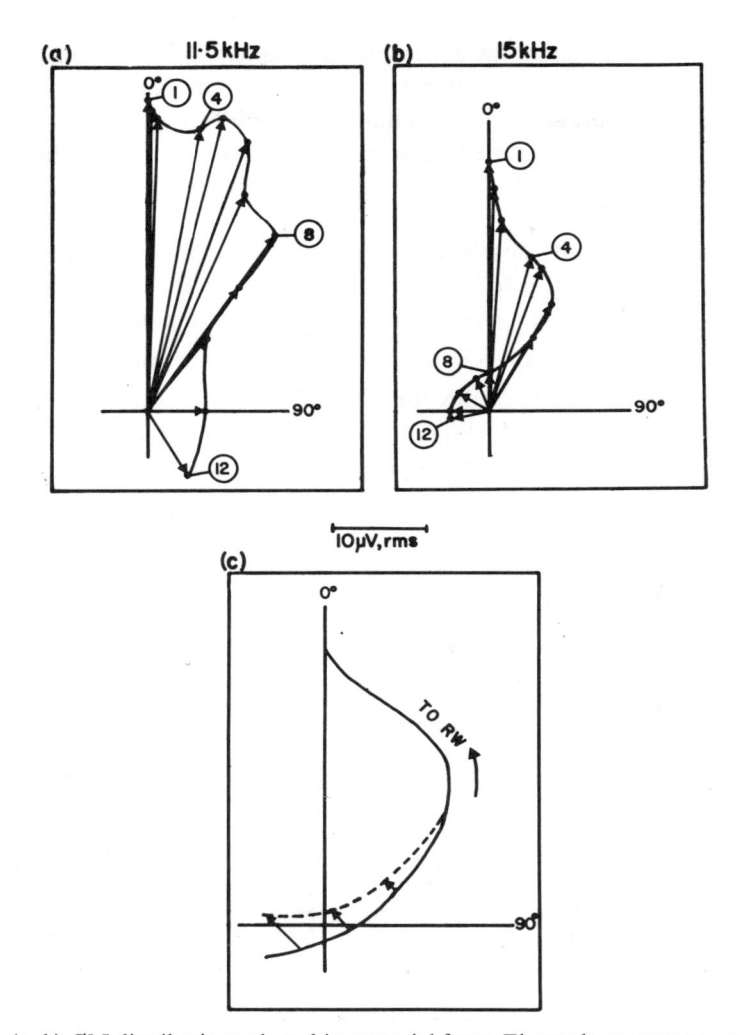

Fig. 2. (a, b) CM distributions plotted in vectorial form. Electrode outputs are numbered from 1 to 12 beginning with the most basal electrode. Phase reversal within the vector bundle occurs after the trajectory has crossed the origin of the vector plane.
(c) Phase reversal artificially induced by addition of a small vector whose amplitude declines towards the round window (from Kohllöffel, 1970).

their influence being of course felt progressively less at electrodes towards the round window end of the array.

The above treatment provides a qualitative explanation of the observed phase reversal. Quantitative treatment of the data leads to essentially the same conclusion, but can be extended to throw light on some possible limits for the configuration of the mechanical disturbance of the membrane itself. In the first instance, let us take the amplitude and phase curves as measured by von

Békésy (1947) and adapt them for higher frequencies by translating them towards the round window and compressing their longitudinal distribution. If we assume as a working hypothesis that the electrical output would be directly proportional to the membrane displacement, we then have a hypothetical hair cell output distribution $H(x)$ on which we can execute the spatial filtering operation.

Two examples of this distribution are shown in Fig. 3(II); the right-hand example (b) is compressed to half the length of membrane compared with (a), and therefore represents a higher frequency. The weighting function is given as:

$$W(x) = \exp(-\lambda |x|) \qquad \lambda = 0.69;\ x \text{ in mm}. \tag{2}$$

The corresponding graphs of this function $W(x_0-x)$ are sketched in Fig. 3(I) for an arbitrarily chosen x_0. Equation (2) takes account of the exponential attenuation of a signal along the cochlear canal. Attenuation in the direction

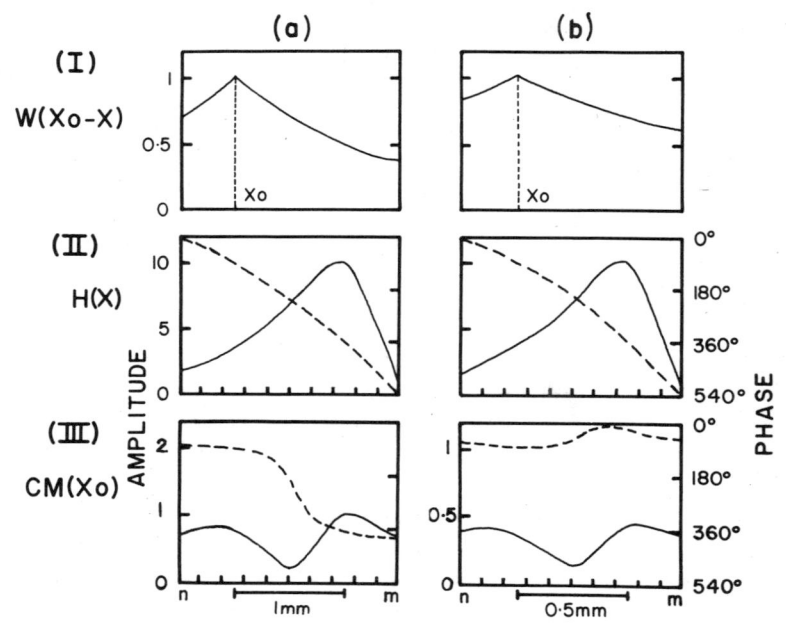

Fig. 3. Spatial filtering of 'von Békésy pattern' $H(x)$.
(I) Weighting function $W(x_0-x) = \exp(-0.69 |x_0-x|)$.
(II) Hypothetical hair cell output distribution $H(x)$.
(III) Theoretical microphonic distribution resulting from graphical integration of $CM(x_0)$
$= \int_n^m H(x) \cdot W(x_0-x)\,dx$. The interval (n, m) is in case (a) 2 mm in the 'higher frequency' example (b) only 1 mm long.
(a, III) Note the large variation in phase in the region of the amplitude minimum. This is consistent with features of real distributions shown in Fig. 1a.
(b, III) In the 'higher frequency' case the more distant portion of $CM(x_0)$ leads the more proximal part in phase.
Amplitude ———, phase – – – (from Kohllöffel, 1970).

perpendicular to the basilar membrane is neglected as a first approximation. The numerical value for $\lambda = 0.69$ corresponds to a potential drop of 6 dB/mm, a figure based on von Békésy's data (1951). Assuming a real λ means that the capacitive properties of cochlear tissues and fluids are neglected. We may be justified in so doing since von Békésy (1951) and Tasaki et al. (1952) found that frequency had only a minor influence on λ.

With the functions $H(x)$ and $W(x)$ thus given $CM(x_o)$ was calculated by graphically integrating the equation

$$CM(x_o) = \int_n^m H(x) \cdot W(x_o - x)\, dx.$$

The resultant values of the cochlear microphonic phase and amplitude distributions $CM(x_o)$ are shown in Fig. 3(III). It is to be noted again that whereas in (a) the distribution exhibits a progressive phase lag as we move away from the basal end of the integration interval, in the 'higher frequency' example of (b) the more apical part of $CM(x_o)$ leads the more basal one.

We may now try to find the shape of $H(x)$ which corresponds to a real CM distribution. In Fig. 4d the CM patterns from three different guinea pigs are shown for stimulus frequencies of 13.25, 13.5, 14 kHz, respectively. From Greenwood's data (1961) relating frequency to distance along the membrane, one would infer that the envelope of the travelling wave has a maximum at these frequencies somewhere in the region of 3 to 5 mm from the stapes end of the basilar membrane, i.e., in the region where the electrode array is located. We may therefore assume for these frequencies a hair cell output distribution like the one in Fig. 3a(II); the interval (n, m) is then taken as extending from 3 to 5 mm along the membrane. Fig. 3a(II) along with Fig. 3a(III) has been redrawn in Fig. 4a.

Comparison of the $CM(x_o)$ curve with the real distributions shows at once that the quantitative fit is rather poor and suggests that the 'von Békésy shape' of curve is not the correct one for high frequencies. There are two distributions which we can modify in an attempt to obtain a better fit, those of phase and amplitude. We have tried modifying only the amplitude distribution and have left the phase distribution unchanged. As can be seen from Fig. 4, making the basal limb of the amplitude curve relatively flatter and steepening the apical limb, produces theoretical CM phase and amplitude curves which approximate the observed data much more closely than those obtained from the classical displacement curve. The suggestion of a more steeply falling amplitude curve on the 'low frequency' side conforms with the conclusions of Johnstone and Boyle (1967) obtained by direct measurement using the Mössbauer effect.

INVERSION OF THE SPATIAL FILTER OPERATION

The mathematical procedure demonstrated in Fig. 3 and Fig. 4 turned out to be rather cumbersome and ill suited to deal with the great mass of data collected with the electrode array. Assuming a likely pattern of $H(x)$ and systematically modifying its shape until agreement between the resulting $CM(x_o)$ and an experimental CM pattern is achieved is not an elegant way of data analysis. Theoretically there also remains the possibility that functions $H(x)$ other than those in Fig. 4 exist whose spatially filtered image shows sufficient similarity with real distributions. It is therefore highly desirable to express $H(x)$ as a function of $CM(x)$ explicitly; such a function will allow us to deduce directly from any given $CM(x)$ the corresponding $H(x)$.

The derivation of the expression for $H(x)$ from Eqs. (1) and (2) is outlined below.

Convolution integral: $\quad CM(x) = H(x) * W(x).$ \hfill (1)

Weighting function: $\quad W(x) = \exp(-\lambda |x|).$ \hfill (2)

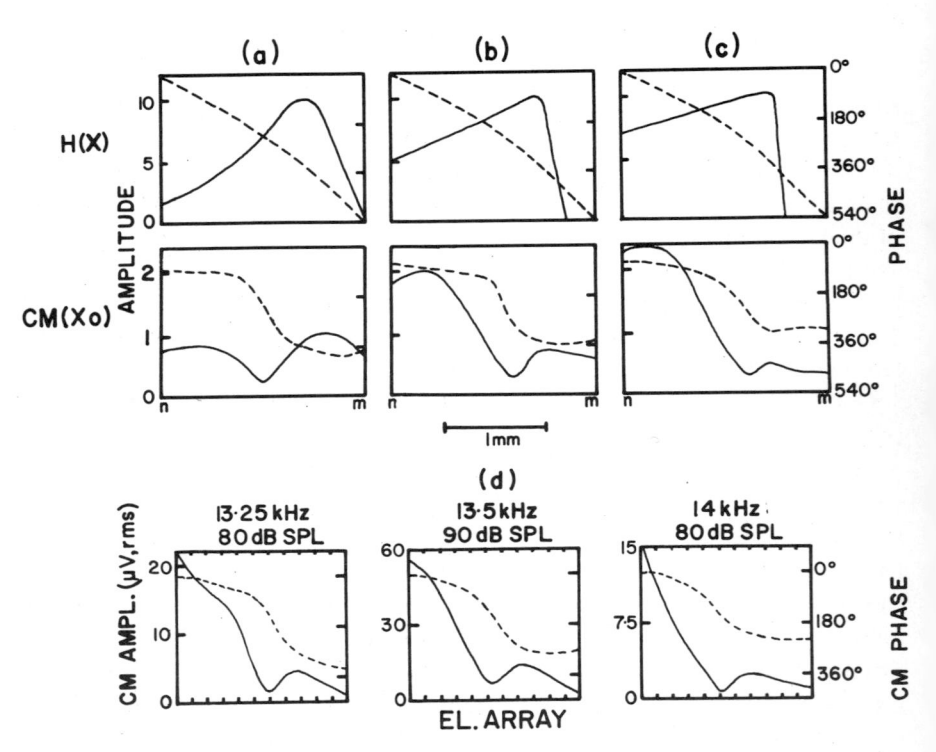

Fig. 4. (a, b, c) Modifying the amplitude evelope of $H(x)$ leads to a better fit between $CM(x_o)$ and experimental CM curves. The limits of the interval (n, m) correspond to distances 3 and 5 mm respectively, that is, counting from the basal end of the membrane. This locates the interval in a region corresponding to the position of the electrode array. (d) CM curves from three different guinea pigs are given for comparison. Amplitude ———, phase – – – (from Kohllöffel, 1970).

By taking the Fourier transform of (1) we get:

$$F_{CM} = F_H \cdot F_W; \quad F_H = \frac{F_{CM}}{F_W}. \tag{3}$$

The Fourier transform of (2) has the form:

$$F_W = \frac{2\lambda}{\lambda^2 + v^2} \tag{4}$$

where v is the independent variable in the Fourier domain. Inserting (4) into (3) yields:

$$F_H = \frac{\lambda}{2} F_{CM} + \frac{v^2}{2\lambda} F_{CM}. \tag{5}$$

The inverse Fourier transform of (5) will be:

$$H(x) = \frac{\lambda}{2} CM(x) - \frac{1}{2\lambda} \frac{d^2 CM(x)}{dx^2}. \tag{6}$$

In Eq. (6) $H(x)$ appears as the linear combination of $CM(x)$ and its second spatial derivative. This has important implications as regards data collection. It is not enough to get a fair idea of the longitudinal distribution of CM but this distribution must be measured with sufficient accuracy to allow the calculation of its second spatial derivative. Indeed, this demands a great deal from CM measurements. For consider the CM distribution of a high-frequency stimulus. The very short wavelengths of the excitation pattern on the membrane will cause slight ripples in this CM distribution. They will be strongly enhanced by virtue of the second derivative and the correct $H(x)$ may be restored from CM data. However, the second derivative will also magnify any irregularity in the CM pattern. Hence it is clear that the distinction must be made between a 'meaningful ripple' and a 'meaningless irregularity'. Systematic examination of many experimental CM patterns with Eq. (6) may reveal criteria and guidelines which in turn could help to make such a distinction. Only further research will provide the answer.

CONCLUSION

The properties of the measured CM distributions, namely the occurrence of an amplitude minimum and the peculiar phase variation along the array, were found to be explicable on the basis of the spatial filter hypothesis. It was shown that spatial filtering executed on the 'von Békésy pattern' of disturbance produced essentially the same effects in the resulting theoretical CM distributions; however, their quantitative agreement with experimental CM curves turned out to be rather poor. Flattening the high-frequency branch and steepening the low-frequency branch of the classical displacement pattern markedly improved this. The amplitude envelopes of the modified functions $H(x)$ of Fig. 4 show great similarity to the envelope patterns one might expect from frequency response curves of the basal basilar membrane measured by Johnstone and Boyle (1967).

The spatial filter operation can be inverted and it is shown that $H(x)$ depends on the CM distribution and its second spatial derivative. The high sensitivity of the second derivative to irregularities in the CM pattern poses a great problem. It is hoped that further investigation may lead to their elimination.

ACKNOWLEDGEMENT

The author wishes to thank Dr. I. C. Whitfield for guidance in this project and for help in the preparation of this manuscript. The work is supported by the Science Research Council.

REFERENCES

Békésy, G. von (1947): The variation of phase along the basilar membrane with sinusoidal vibrations, J. Acoust. Soc. Amer. *19*, 452-460.
Békésy, G. von (1951): The coarse pattern of the electrical resistance in the cochlea of the guinea pig (electro-anatomy of the cochlea), J. Acoust. Soc. Amer. *23*, 18-28.
Békésy, G. von (1953): Shearing microphonics produced by vibrations near the inner and outer hair cells, J. Acoust. Soc. Amer. *25*, 786-790.
Greenwood, D. D. (1961): Critical bandwidth and the frequency coordinates of the basilar membrane, J. Acoust. Soc. Amer. *33*, 1344-1356.
Johnstone, B. M., and Boyle, A. J. F. (1967): Basilar membrane vibration examined with the Mössbauer technique, Science *158*, 389-390.
Kohllöffel, L. U. E. (1970): Longitudinal amplitude and phase distribution of the cochlear microphonic (guinea pig) and spatial filtering, J. Sound Vib. *11*, 325-334.
Tasaki, I., Davis, H., and Legouix, J. P. (1952): The space-time pattern of the cochlear microphonics (guinea pig), as recorded by differential electrodes, J. Acoust. Soc. Amer. *24*, 502-519.
Tasaki, I., Davis, H., and Eldredge, D. H. (1954): Exploration of cochlear potentials in guinea pig with a microelectrode, J. Acoust. Soc. Amer. *26*, 765-773.
Whitfield, I. C., and Ross, H. F. (1965): Cochlear-microphonic and summating potentials and the outputs of individual hair-cell generators, J. Acoust. Soc. Amer. *38*, 126-131.

DISCUSSION

Anderson: Have you tried to apply a matrix formulation to the problem? If the vector of microphonic potentials is expressed as the product of a symmetric matrix describing the assumed weighting function and a vector of the potentials $H(x)$, the amplitude and phase distributions of $H(x)$ can then be obtained by the premultiplication of both sides of the equation by the inverse of the weighting matrix. Thus, the use of the second derivative is avoided.

Kohllöffel: We considered several analyzing techniques; matrix inversion is, in fact, the method we thought of first. Expressing $H(x)$ explicitly as a function of $CM(x)$ has the advantage that it points out how accurate CM measurements have to be.

Dallos: Your technique is beautiful, but in comparing it to more conventional recording methods we must keep in mind that it makes quite a difference in spatial resolution whether the single or the differential electrode technique is used. For example, when recording from the third turn, the single electrode will show a cut-off rate of, perhaps, 30 dB per decade on the high-frequency side; a well-balanced differential pair will give a cut-off rate in excess of 100 dB per decade.

Kohllöffel: The fact is that potentials generated within the organ of Corti do spread in the inner ear canals. Even differential electrodes are not able to suppress potentials from adjacent points. Whichever electrode technique you apply, the spatial spread has to be taken into account.

Honrubia: I agree with your interpretation of the differential electrode details. There is a claim in literature that only a differential recording technique is good. But the idea that a differential pair will look to only 2 mm and one electrode will look to the whole cochlea just does not make any sense. They look to the same thing, and you just made the right interpretation of the differing results. Differential electrodes cancel out only cross talk of CM between turns but not the potentials generated along the cochlear partition and arriving in opposite phase through scala vestibuli and scala tympani. We try to record with a single electrode from the scale media because we think the spread of potentials will be different there. Though your technique may help a lot, I think we should primarily use an appropriate choice of stimulus parameters (*i.e.*: low frequencies) to minimize the handicap of the conductivity of the cochlear fluids.

Kohllöffel: I agree with you that this technique does not solve all problems. But I would like to emphasize that, although it is difficult, the technique is worth-while.

EXPERIMENTS ON COCHLEAR ANALYSIS FOR TRANSIENTS IN THE GUINEA PIG

J. P. LEGOUIX

Laboratoire de Neurophysiologie Générale
Collège de France
Paris, France

INTRODUCTION

Short transients provoke movements of the basilar membrane which take the appearance of travelling waves. Such waves have been described by von Békésy (1933) and studied by several authors on dimensional models (Peterson and Bogert, 1950; Tonndorf, 1960; Flanagan, 1962). In the present paper, some observations are reported on the cochlear microphonic (CM) response of guinea pig to several types of transients. This study raises many problems concerning the mechanical processes involved in the cochlear analysis of transients. It considers also the important question of the relationships of the auditory nerve action potential with the motion of the basilar membrane, and more generally, the question of the neural coding of the physical characteristics of the stimulus.

PROCEDURE

Differential electrodes of the type described by Tasaki *et al.* (1952) were used to record the cochlear microphonic. This method consists of inserting two fine metal wire electrodes on opposite sides of the cochlear partition, one in scala vestibuli, the other in scala tympani. Several of such pairs of electrodes were placed in the four turns of the cochlea and connected to separate differential amplifiers. These electrodes record the cochlear potentials from the narrow region of the organ of Corti which lies between them (about 1 mm), without interference from potentials of remote sources. Hence, they provide information about the local movements at the site of the electrodes.

RESPONSES TO STEP FUNCTION SIGNALS

Step signals are the simplest form of transient which might be applied to the ear. From a practical point of view, they appeared the easiest to make. Since it is not possible to transmit such stimuli by air conduction, they were directly applied to the stapes by a mechanical device made of a metal rod attached to the membrane of a magnetic earphone. The piston-like motion of the rode provoked a forced movement of the cochlear partition of short duration followed by free oscillations. The source impedance of the mechanical transducer was higher than that of the ossicles. This difference did not appear to modify the cochlear properties. Two components might be described in the microphonic responses to step signals (Fig. 1):
(1) An early deflexion which is the effect of direct driving of the stapes and of the corresponding hydraulic pressure on the basilar membrane.
(2) After-oscillations, which occur when the stapes is at rest.

The early deflexion

The shape of the early deflexion, as Tonndorf (1960) showed on models, represents the first derivative of the input signal. We observed that its aspect varies according to the place where it appears:
(1) Its amplitude is larger in the basal part of the cochlea and smaller toward the apex.
(2) Its rise time increases progressively from the base to the apex.
(3) It appears with a delay which increases progressively toward the apex; 1 msec at the 3rd turn, and 3 msec at the 4th turn.
(4) The shape of this deflexion is altered when the slope of the signal is modified. With slower slope, the size and the form of the deflexion become similar in the various parts of the cochlea.

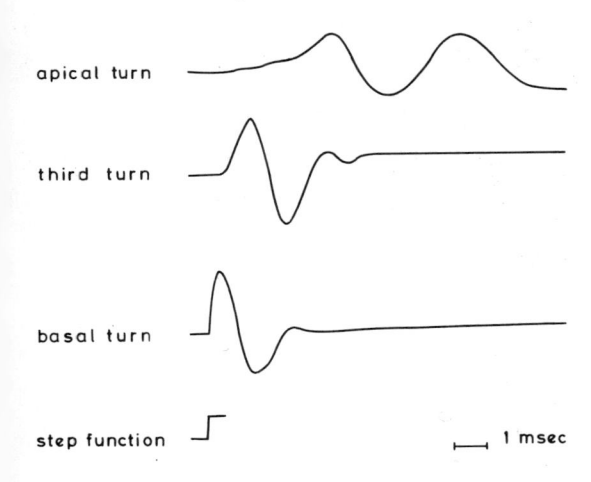

Fig. 1. Cochlear microphonics in response to a step function signal.

As a whole, these characteristics appear to reflect the progressive increase of elasticity and of inertia from the base to the apex.

The after-oscillations

The cochlear structures which have stored potential energy in the early deflexion release this energy in the form of free oscillations. It is noted that, with this form of stimulus, the direct driving of the cochlear partition is very short and the main part of the response is due to a free oscillation. This oscillation shows different characteristics according to the location of the electrode. In the basal turn, the period is about 2 msec and the damping is very high; the logarithmic decrement is 1.7. At the fourth turn, the period is much longer, around 10 msec, but the logarithmic decrement is much less; about 1.

However, when rise time of the step signal increases, the late oscillations observed in the fourth turn show a greater amplitude. This fact seems to indicate that, in the distal regions, the movement is not a simple damped oscillation, but a true travelling wave, some energy being transmitted along the basilar membrane from more proximate points to these regions (Wever *et al.*, 1954).

RESPONSES TO SHORT AIR-CONDUCTED TONE BURSTS

Air-conducted tone bursts, if they are very short, act as impulse stimuli and exert a very brief forced motion which initiates relatively large free oscillations in the various parts of the cochlea, particularly in the apical regions, where the damping is lower. When the duration of the tone burst is increased, the forced motion begins to predominate and, since the spectrum is narrowing, the responses spread in more restricted regions (Fig. 2).

In some cases, at the beginning of a pure tone, or of a tone burst, an onset transient, generated in the cochlea, appears in the microphonic response. It

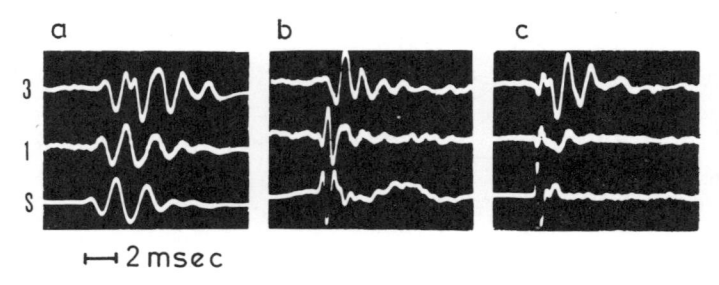

Fig. 2. Cochlear microphonic responses to short tone bursts.
S: signal recorded by a condenser microphone.
1: microphonic potential from the basal turn.
3: microphonic potential from the third turn.
In (a) the stimulus duration is long enough to give a forced movement, except in the 3rd turn where the response is less damped. In (b) and (c) the signal is shorter and the free oscillations are dominant.

seems to be an after-oscillation of the type just mentioned. Such transients are more obvious in the regions that are not tuned to the stimulus frequency, and when the SPL is high. With tone bursts, it is possible, when the stimulus parameters are appropriate, to observe the transformation of the onset transient into a slow deflexion which may be identified as the classical summating potential (SP). The latter shows greater amplitude when the stimulus contains a frequency higher than the characteristic frequency at the electrode location.

ACTION POTENTIALS PROVOKED BY TRANSIENTS

The differential electrodes which were used pick up a mixture of cochlear microphonics (CM) and action potentials (AP) from the auditory nerve. Thanks to a differential network (Tasaki *et al.*, 1952) it is possible to record in the same time, on separate amplifiers, the CM and the AP.

When the stimulus is a short transient, rather well synchronized volleys of unit impulses are elicited in the nerve trunk. These impulses, since they are coming from different fibres, have not all the same latency and are slightly scattered in the time. However, they group in a complex manner which provides, with all types of transients, a rather stereotyped form of response which presents three negative peaks, usually named N_1, N_2, N_3. When slow transients were mechanically applied to the stapes, the motion transmitted to the ossicles could be relatively simple, and this made it easier to relate the form of the neural response to the displacement. It was possible to verify that the AP is provoked by the displacement of the cochlear partition from the scala tympani toward the scala vestibuli. The AP is initiated when the CM gets to a critical level, but its amplitude is growing when the slope of microphonics is increasing (Fig. 3). This fact can be explained by a smaller dispersion in time of the individual impulses when the slope is steeper; then more fibres are stimulated at one instant.

When the transient presents several periods, separate sets of N_1, N_2, N_3 deflexions are elicited each time the cochlear partition moves toward the scala vestibuli. They show definite relations to the phase of the microphonic response and form corresponding patterns. In particular, when the phases of the stimulus are not exactly symmetrical, the pattern of the AP is modified when the polarity is reversed, because the stimulating deflexions are thereby changed.

Many complex stimuli show waveforms which are similar to the above described transients. This is the case for plosive consonants and vowels which, to some extent, might be compared to repeated tone pulses. The CM provoked by these speech sounds present important low-frequency oscillations in the apical part of the cochlea, similar to those observed with impulse stimuli. These waves show typical features, for instance, they are more prominent with "p" than with "t". With "k", they appear more complicated, and often look as if two successive transients were superimposed.

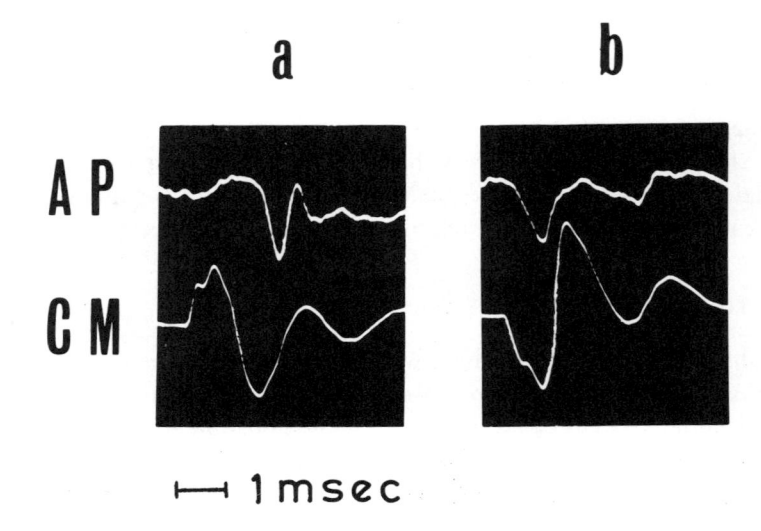

Fig. 3. Lower sweep: microphonic potentials from the basal turn, provoked by a slow transient applied mechanically to the stapes. The negativity in scala vestibuli is recorded downward on the graph.
Upper sweep: whole nerve action potential of the auditory nerve.
In (a) the polarity of the signal is opposite to that in (b). The time pattern of the volleys of impulses is modified when the polarity of the signal is reversed.

The action potentials in response to these sounds were studied in the auditory nerve as well as in higher centres. They present temporal patterns which are closely related to the waveform of the microphonic response and, for this reason, might be supposed to carry particular clues about the waveform of the stimulus.

CONCLUSIONS

The movements of the basilar membrane produced by acoustic transients and also by short mechanical pulses are similar to the responses of a set of low-pass filters with distributed frequency characteristics, damping and bandwidth. Several studies of electrical analogues with corresponding parameters have been made (Zwislocki, 1948; Perterson and Bogert, 1950). The validity of such models is supported by the fact that they show impulse responses which are similar to the physiological responses described in this paper. Morever, some particular cochlear characteristics may also be reproduced with such models, as onset transients. Responses, analogous to SP, can be observed, according to the recent study of Engebretson and Eldredge (1968), in a low-pass filter combined with a nonlinear network. More simply, SP-like responses may be obtained when an electric signal, corresponding to a tone burst, is applied to a low-pass filter, if the intensity is high enough to drive the inductive components of the filter in a nonlinear part of their characteristics (Legouix, 1968). Because in such filters the time constant determines the form of the transient response, it may be concluded that the waveform of the cochlear response depends mainly upon the properties of the receptor itself. Then, apparently, the frequency analysis for these short signals is not likely made in the time domain but, more probably, it depends upon the Fourier analysis which depends upon the localization process.

However, several of the data reported in this paper suggest that the response waveform in the cochlea, that follows the Fourier analysis, is encoded in the temporal feature of the few volleys of impulses which are elicited by short transients, and that it has some interesting significance for the recognition of the signal. For instance the volleys of impulses in the auditory nerve follow closely the microphonic oscillations and may give rise to a pitch sensation derived from their periodicity. Moreover, within a single volley of nerve impulses, there exists also a possibility of temporal coding at least for slow transients. As it appears, with slow mechanical stimuli, the phase and the rise time of the signal is encoded in the time pattern of impulses. More generally, the propagation of travelling waves from the base to the apex requires some time, so that the successive points of the organ of Corti are stimulated with an increased delay. For this reason, as it was pointed out by Teas et al. (1962), a slow transient will excite nerve impulses in different fibres, continuously over a period of 1 msec or more. Then it might be supposed that the variable delays of the impulses may also carry information about the time characteristics of the signals (Davis, 1957). How these neural messages are used by the brain centres, is a matter of investigation.

REFERENCES

Békésy, G. von (1933): Ueber den Knall und die Theorie des Hörens, Phys. Z. *34*, 577-582.
Békésy, G. von (1942): Ueber die Schwingungen der Schneckentrennwand beim Präparat und Ohrenmodell, Akust. Z. *7*, 173-186.
Davis, H. (1957): Biophysics and physiology of the inner ear, Physiol. Rev. *37*, 1-49.
Engebretson, A. M., and Eldredge, D. H. (1968): Model for the nonlinear characteristics of cochlear potentials, J. Acoust. Soc. Amer. *44*, 548-555.
Flanagan, J. (1962): Computational model of basilar membrane displacement, J. Acoust. Soc. Amer. *34*, 1370-1377.
Legouix, J. P. (1968): Réponses microphoniques cochléaires provoquées par différents types de transitoires, C.R. Soc. Biol. *162*, 45-49.
Peterson, L., and Bogert, B. (1950): A dynamical theory of the cochlea, J. Acoust. Soc. Amer. *22*, 369-381.
Tasaki, I., Davis, H., and Legouix, J. P. (1952): The space-time pattern of the cochlear microphonics as recorded by differential electrodes, J. Acoust. Soc. Amer. *24*, 502-519.
Teas, D. C., Eldredge, D., and Davis, H. (1962): Cochlear responses to acoustic transients: an interpretation of whole-nerve action potentials, J. Acoust. Soc. Amer. *34*, 1438-1460.
Tonndorf, J. (1960): Response of cochlear models to aperiodic signals and to random noises, J. Acoust. Soc. Amer. *32*, 1344-1355.

Wever, E. G., Lawrence, M., and Békésy, G. von (1954): A note on recent developments in auditory theory, Proc. Nat. Acad. Sci. *40*, 508-512.
Zwislocki, J. J. (1948): Theorie der Schneckenmechanik, Doctoral Dissertation, Eidg. Tech. Hochschule, Zürich.

DISCUSSION

Zwislocki: Dr. Legouix, your slides show that CM responses to transients are highly damped. This does not agree with the sharp tuning curves Dr. Johnstone found (p. 85). Are you able to find an explanation for this discrepancy?

Legouix: We have worked with differential electrodes and so there may be an effect of spatial integration as Mr. Kohllöffel showed (p. 107). Apart from this, our technique cannot decide whether we obtain an exact value of the damping of the cochlear partition. We are dealing with a travelling wave which will bring some additional energy so that the observed oscillation might not be a simple damped oscillation. On the other hand, it is not likely that the mechanical device increases the damping inside the cochlea, and I would expect that any defect in the technique would make the damping appear too low rather than too high. These considerations give me some confidence in the results.

Johnstone: It is obvious that my measurements of the mechanical frequency selectivity of the basilar membrane, together with the damping found by Dr. Legouix, do not agree with the properties of a simple LC-circuit. The damping factor is a missing element in every model including Ranke's (1950). Our own preliminary and somewhat imperfect results are in line with the classical idea that there is substantial damping. I believe the measurements of Dr. Legouix are very accurate.

Zwislocki: The width of resonance and the damping are related in a linear system, no matter what sort of model is used. So, it seems to me that the question of the selectivity of the cochlea is a real and important problem. As far as the discrepancies are concerned, we have seen that the methods of recording CM can be improved a lot. On the other hand, I would like to ask Dr. Johnstone if, perhaps, the sharpness of his curves could be partly due to relative movements of the probe. Do you have any evidence that the probe does not move relative to the scala media?

Johnstone: There is no direct evidence, but I believe the probe did not move.

Honrubia: I think there are two problems in recording CM. One is the interference of the fast-changing phase of potentials along the basilar membrane. The other problem is the occurrence of cross talk between turns. The technique of differential electrodes suppresses the potentials due to cross talk because they come in phase to points that are relatively close to each other. Do you think, Dr. Legouix, that the appearance of humps before the main deflection or discontinuities as shown in Fig. 2 of the paper could be due to insufficient suppression of the cross talk?

Legouix: The differential recording is a phase-sensitive technique. It greatly improves the space discrimination as it cancels out microphonic responses from remote sources. With respect to the humps you were mentioning, I must say that the technique of applying a mechanical impulse to the stapes is rather difficult. Many alterations of the curve are thought to be due to the mechanical device itself, which always has some shortcomings of damping and frequency response.

REFERENCES

Ranke, O. F. (1950): Theory of operation of the cochlea, J. Acoust. Soc. Amer. *22*, 772-777.

THE SP IN CONNECTION WITH THE
MOVEMENTS OF THE BASILAR MEMBRANE

R. KUPPERMAN

Laboratory of Labyrinthology, ENT-Department
University Hospital
Utrecht, The Netherlands

INTRODUCTION

Direct observation of the movements of the basilar membrane by von Békésy (1952, 1953) gave new impetus to the study of the membrane mechanics. He observed a place of maximal movement corresponding to a specific stimulus frequency, the tonotopic projection for a high frequency was in the basal part, the low frequencies were situated in the apical part of the cochlea. To observe the movements, a rather high sound pressure level (SPL) of 120 dB was used in the frequency range of 300-1500 Hz and the displacement of the basilar membrane was of such an order that distortion phenomena easily could occur. Recently Johnstone and Boyle (1967; pp. 81-93 of this volume) measured the movements of the basilar membrane with the Mössbauer technique. They could use only the basal turn of the cochlea. The SPL they used was 90 dB, the membrane displacement at 18 kHz was 900 Å. The methods used before have a certain disadvantage. First of all the cochlear distortion is already present for a SPL exceeding 60 dB. Further on the measurements of movements are in von Békésy's work limited to a frequency range of 300-1000 Hz, in the work of Johnstone and Boyle they are limited to the first turn of the guinea pig's cochlea having a frequency range from 12 kHz to 20 kHz. Another method to get information about the tonotopic relation of the movements of the basilar membrane is to use the place of generation of the different cochlear potentials.

Tasaki *et al.* (1952) measured the cochlear microphonics (CM) in different turns of the cochlea with a differential electrode technique. They found some envelope in the amplitudes of the CM using a constant frequency. Teas *et al.* (1962) used acoustic clicks of a well defined form to measure space-time patterns in the cochlea; thus, in combination with the CM data, they were able to postulate some suggestions about the exact form of the envelope of the movements of the membrane.

Because an alternating potential spreads widely in the perilymphatic space, Honrubia and Ward (1968) measured the CM in the scala media where the electrical spread of the CM is small, so that the CM is better related to the place of its origin. In 1966 we found (Kupperman, 1966) that the positive and the negative summating potential (+SP and −SP) measured in the scala tympani provide us with a better way for measuring the form of the envelope during stimulation with tone bursts of different frequencies. It was found that the +SP was strictly related to the place of the cochlear partition that moved during stimulation; the maximal +SP value corresponding to the maximal mechanical movement. The present study was designed to collect more information about both the +SP and −SP related to the movements of the cochlear partition. The different places of origin of both potentials in the organ of Corti were studied under masking conditions.

METHOD

Potentials were recorded from the scala tympani of the cochlea of healthy medium weight guinea pigs. Silver/silverchloride electrodes with a diameter of 50 μ were used. These electrodes were connected to a specially designed low-noise amplifier (Philbrick) with an amplification factor of 1000. The mass electrode, made of the same material, made contact with the neck muscles. The amplified bio-electric response was monitored on a four-beam oscilloscope together with the electric and acoustic tone burst. The responses were summed with a CAT 1000 computer. Because the tone bursts were not synchronized in their beginning the CM was cancelled out and only the DC component in the total electric response of the guinea pig was recorded. The signal stored in the memory of the computer was plotted on an X-Y writer.

The tone bursts could be varied in duration, amplitude and tone frequency. With sequential coupled keys, the total tone burst could be built up from different frequencies and noise sources. The acoustic signal was delivered by a condenser telephone with a flat frequency response characteristic between 0.1 and 30 kHz. The SPL in the outer ear canal was measured during the experiments by a calibrated subminiature condenser microphone.

RESULTS

The space pattern of the +SP

The +SP was measured in the four turns of the cochlea. The SPL of the 25 msec tone pulse was 64 dB. The maximal value of the +SP occurred in the part of the cochlea that corresponded to von Békésy's (1952) tonotopical localization of frequency. As the tone pulse frequency was raised, moving the travelling wave envelope closer to the basal end of the cochlea, the SP suddenly changed from positive to negative. By measuring in two turns of the cochlea simultaneously, the change in polarity of the SP could be nicely demonstrated as the frequency of the tone stimulus was varied from 500 Hz to 10 kHz.

The frequency at which the polarity in the third turn and basal turn changed from + to − was in the third turn about 2000 Hz and in the basal turn about 8000 Hz (Fig. 1). The polarity change of the SP is strictly related to the electrode location; a different location in a turn will introduce a difference in turnover frequency. In Fig. 2 the change is demonstrated in the third turn in another animal, the electrode location was some millimeters more in the direction of the oval window than the electrode location in Fig. 1.

Remote masking of the +SP and −SP

To change the neural conditions of the cochlea just before producing a SP a 40 msec burst of white noise was presented at 70 dB, 3 msec before the tonal stimulus of 25 msec. The 3 msec interval is long enough to damp vibrations of the cochlear partition entirely. We found no influence on the +SP, but the −SP was partially suppressed. If a pure tone of the same duration and SPL is sub-

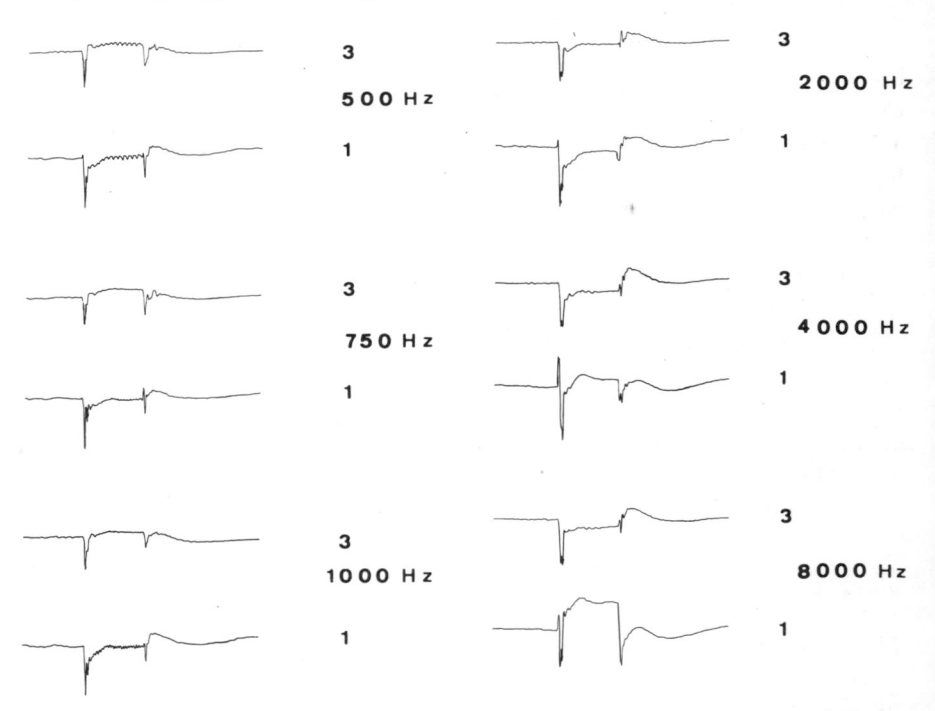

Fig. 1. The +SP and −SP measured in the third and basal turns of the cochlea for different stimulus frequencies.

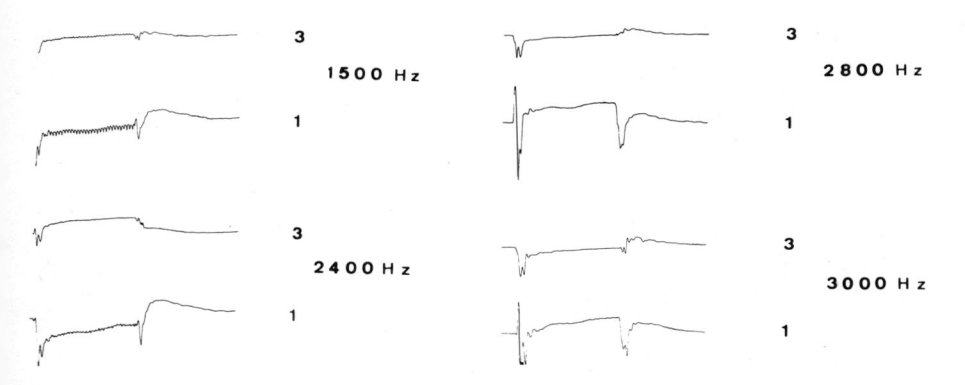

Fig. 2. Change in polarity from −SP to +SP in the third turn of the cochlea. The upper track is the third turn, the lower track is the basal turn, where the +SP is already present.

stituted for the noise, little or no influence is seen on the +SP. The −SP can be affected when the frequency of the masking sound preceding the tone stimulus is over 2000 Hz. The masking effect on the −SP never reaches the same value as when a white noise masking sound is used.

DISCUSSION

These experiments show that there is a specific tonotopic relation between the maximal movements of the basilar membrane for a given frequency and the maximal value of the +SP. The −SP has no relation with any exact point of vibration of the cochlear partition. The first point to prove is that the behaviour

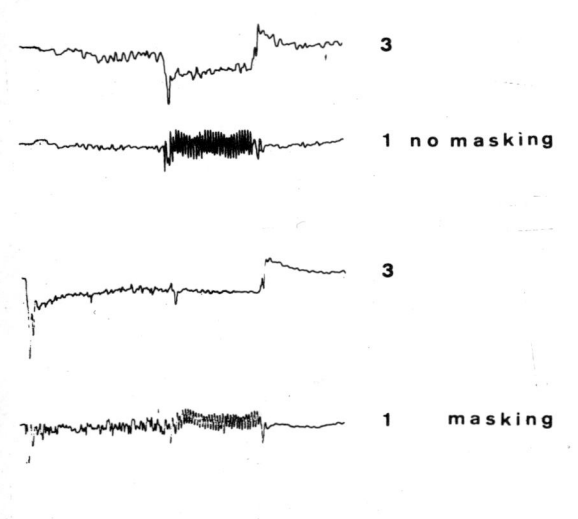

Fig. 3. Remote masking with a white noise of 40 msec. The tone pulse has a frequency of 6600 Hz. The lowest track indicates time.

of the +SP as related to the frequency of the tone stimulus is at least that of the CM; and the second point is that there is a different place of origin of the −SP and +SP. Only then the function of the +SP can be made clear.

To examine the relation between the +SP and CM envelopes the Q-factors (resonance frequency divided by the width of the half-power band) of the electrically measured vibration curves were compared. The Q-factor for +SP was calculated from the curves shown in Fig. 4. In turn 1 and 4 it is of the order of magnitude of 3.0; in turn 2 and 3 it amounts to 7 ± 1. The Q-factor computed from von Békésy's (1952) direct observation of the vibrations of the cochlear partition is in the order of 1.6; Johnstone and Boyle (1967) found a Q-factor of 2.5 with the aid of the Mössbauer effect. From the tuning curve of a single-neuron preparation a Q-factor can be found in the order of 10-12. We see that the +SP result lies between; probably the +SP represents an intermediary function during the stimulation of the hair cells after a possible mechanical shearing due to the movements between hair cell and the overlying tectorial membrane, as Davis and Eldredge (1959) suggested. The final sharpening of the frequency resolution could have a neural origin (lateral inhibition, according to von Békésy). The CM represents, in this hypothesis, the electrical side effect of the gross movement of the cochlear partition. This behaviour of the +SP mentioned before could be found only in the scala tympani, probably as the result of an electric leakage from the highly positive cochlear duct through the organ of Corti. For that matter, the +SP can be seen as a side effect proportional to the real stimulation of the hair cell, that is the bending of the hairs of the hair cell (Davis, 1958).

Still the function and place of origin of the −SP is not quite clear. Measurements by the present author (1966) of −SP during anoxia and of its latency gave

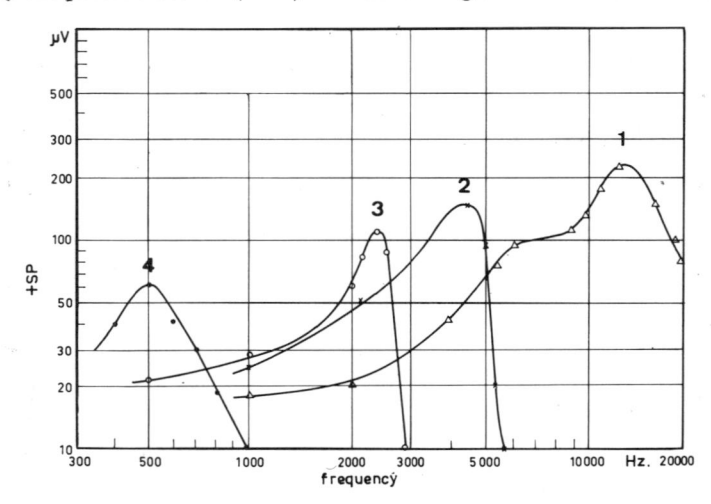

Fig. 4. Values of the +SP measured in four turns of the cochlea.

the impression that the −SP must have a neural nature. Continuation of these experiments with masking noise showed that −SP can be masked but +SP cannot. This points to the neural nature of −SP and the different origin of +SP. As −SP and the action potential showed the same behaviour in being masked, the name "summating potential" as derived from summation of action potentials is still valid for the −SP.

REFERENCES

Békésy, G. von (1952): Direct observations of the vibration of the cochlear partition under a microscope, Acta Oto-Laryngol. *42*, 197-201.
Békésy, G. von (1953): Shearing microphonics produced by vibrations near the inner and outer hair cells, J. Acoust. Soc. Amer. *25*, 786-790.
Davis, H. (1958): A mechano-electrical theory of cochlear action, Ann. Otol. Rhin. Laryngol. *67*, 789-801.
Davis, H., and Eldredge, D. H. (1959): An interpretation of the mechanical detector action of the cochlea, Trans. Amer. Otol. Soc. *47*, 28-37.
Honrubia, V., and Ward, P. H. (1968): Longitudinal distributions of the cochlear microphonics inside the cochlear duct (guinea pig), J. Acoust. Soc. Amer. *44*, 951-958.
Johnstone, J. R., and Johnstone, B. M. (1966): Origin of summating potential, J. Acoust. Soc. Amer. *40*, 1405-1413.
Johnstone, B. M., and Boyle, A. J. P. (1967): Basilar membrane vibration examined with the Mössbauer technique, Science *138*, 389-390.
Kupperman, R. (1966): The dynamic DC potential in the cochlea of the guinea pig, Acta Oto-Laryngol. *62*, 465-480.
Takasi, I., Davis, H., and Legouix, J. P. (1952): Space-time pattern of the cochlear microphonic as recorded by differential electrodes, J. Acoust. Soc. Amer. *24*, 502-519.
Teas, D. C., Eldredge, D. H., and Davis, H. (1962): Cochlear responses to acoustic transients. An interpretation of whole-nerve action potentials, J. Acoust. Soc. Amer. *34*, 1438-1459.

DISCUSSION

Whitfield: What you call the positive SP does not seem to be affected in any way by the neural response and yet certainly the after-potential of opposite polarity does seem to be affected. I think that your polarities are reversed with respect to the original descriptions of Davis *et al.* (1958).

Kupperman: Yes, I believe so. I discussed it with him. I like to call an SP negative if the deflection is in the same direction as the action potentials. This negative SP, which can be masked, behaves like a summing up of action potentials.

Johnstone: I am very pleased to see the steep slopes in your Fig. 4. Moreover, there is a nice increase in slope up to nearly 80 dB/octave at 5 kHz, which is very interesting. I agree very much with you that the negative SP (your positive SP in scala tympani) is definitely a hair cell potential, produced by a circulating current from the hair cells. I also support your idea that the other SP is not being generated in the same way, though I am not prepared to say whether it comes from the neuron or from another source in the hair cell.

Schwartzkopff: I wonder whether you, or anyone else, tried to differentiate the components of the SP by oxygen deprivation. We found a rather poor influence of nitrogen in the bird. We concluded that the SP we recorded in birds is produced essentially on the contact zone between the tectorial membrane and the hairs. We have not yet seen much of the neural component in birds. Perhaps you are able to find both aspects.

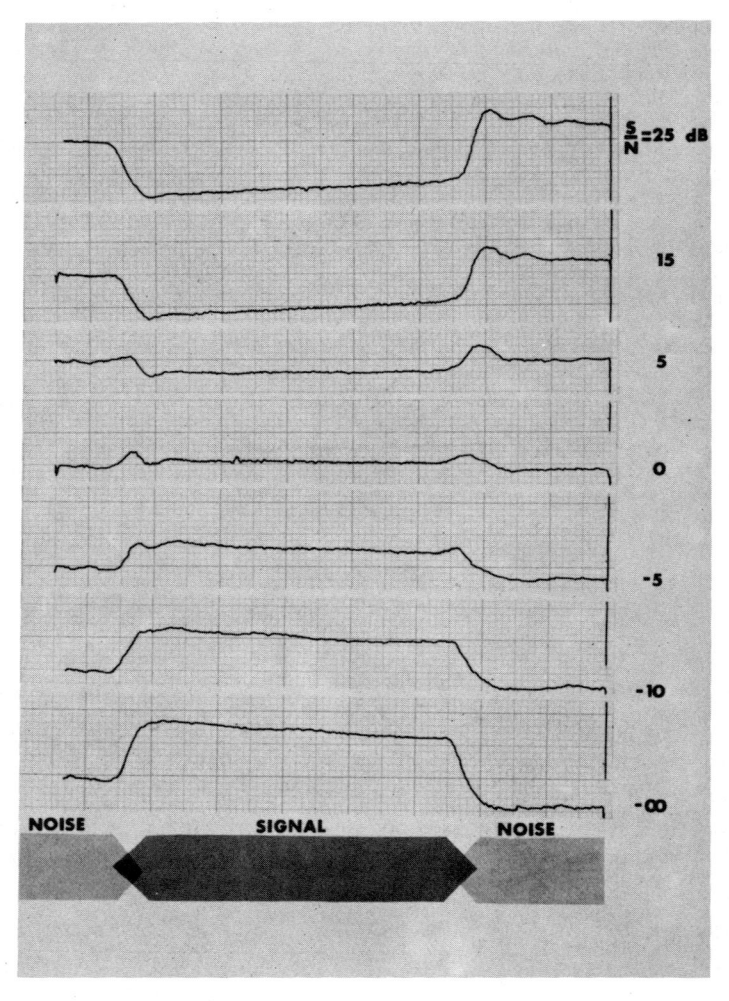

Fig. 1 (Dallos). Chinchilla, basal differential recording. Signal: 4000 Hz tone, 37 msec duration. Noise: narrow-band noise with $f_o = 4000$ Hz, 100 msec duration. Signal-to-noise ratio in decibels is the indicated parameter.

Kupperman: I tried this kind of experiment but up to now with little success. The procedure is very tricky; it is very hard to realize a stable state of metabolism in the animal. I have some results with a rough experiment in which the blood supply was clamped. The negative SP (negative in scala tympani) disappears at the same high speed as the action potentials. The positive SP (positive in scala tympani) stays much longer and then diminishes up to the rate of decrease of the CM. Afterward, a residual CM remains and a negative potential occurs.

Dallos: It appears to me that one of your strong arguments for a neural origin of the negative SP (negative in scala tympani) is that it can be masked. I think that what you call masking is not really that. As it is clearly demonstrated in the series of traces shown (Fig. 1, above), the so-called masking stimulus itself generates an SP, raising the level of the baseline to the level of the SP that is generated by the stimulus. Thus what happens is that the baseline is raised to the level of the stimulus-generated SP, and not the SP diminished to the level of the baseline. As the picture shows, this process can be carried to its extreme; with very unfavourable signal-to-noise ratios the SP generated by the noise can become dominant.

Kupperman: Yes, of course the noise produces a positive SP, but this SP disappears immediately after switching off the noise. I never measured after-effects as long as 10 or 20 msec and you are able to see this in Fig. 3 of my paper. The after-effects depend, however, on SPL. No troubles will occur at levels of 40 to 50 dB. High levels will lead to the problem you are talking about.

De Boer: If the +SP and CM are both potentials from hair cells, why has one a steeper frequency response curve than the other?

Kupperman: This is simply due to the different nature of CM and SP, CM being an alternating current and SP being a direct current. There is, for instance, good reason to expect a cross leakage of the alternating CM currents between parts of the cochlea: this smoothes the frequency response curve as compared with the SP curves which have less electric leakage.

REFERENCES

Davis, H., Deatherage, B. H., Eldredge, D. H., and Smith, C. A. (1958): Summating potentials of the cochlea, Amer. J. Physiol. *195*, 251-261.

Section 3

THE AUDITORY PATHWAY

It seems to be beyond dispute that frequency coding is preserved in the auditory pathway from the cochlea up to the cortex. Also, for low stimulus frequencies, there is general agreement on the preservation of temporal information in the timing of nerve discharges up to certain levels of the auditory pathway. But the relative extent to which frequency coding and pulse interval coding contribute to pitch perception is still a matter of controversy as this section shows.

CENTRAL NERVOUS PROCESSING IN RELATION TO SPATIO-TEMPORAL DISCRIMINATION OF AUDITORY PATTERNS

I. C. WHITFIELD

Neurocommunications Research Unit
University of Birmingham
Birmingham, England

INTRODUCTION

In considering the neurophysiological status of place and periodicity theories of pitch perception, the most satisfactory point at which to start is that of the central process of the auditory nerve. Sufficient data are now available to give a reasonably complete picture of the kind of behaviour we may expect in response to a given sound stimulus, especially in the case of single tonal stimuli. The experiments of Kiang (1965) provide extensive data about the frequency response areas of single fibres in the cat over a major portion of its frequency range, and the transformation suggested by Ross (Whitfield, 1968a) enables us to convert these data into a picture of the spatial distribution of activity over the entire fibre array for any given frequency and intensity of stimulus. The results both of this calculation and of the behavioural studies following lesions carried out by Schuknecht (1960) demonstrate that a considerable proportion of the total array is activated by a single tone. Although the fibres activated are different for each frequency, nevertheless, there is a considerable overlap in the active arrays related to two adjacent discriminable tones.

Study of auditory nerve fibres also leaves no doubt that for low-frequency stimuli there is a correlation between the period of the stimulus and the intervals between successive nerve impulses in an activated fibre. This phenomenon, postulated in the 19th century (Rutherford, 1886), was actually first observed as long ago as the 1930's (*e.g.*, Derbyshire and Davis, 1935), and formed the basis of Wever's (1949) volley theory. With the advent of computers to process the data it has received ample detailed confirmation (see Rose *et al.*, 1968 for references).

If we now look at the transducer mechanism giving rise to the neural pattern, we can readily find the appropriate mechanisms for both these phenomena. Different stimulus frequencies give rise to disturbances in different parts of the basilar membrane, and these vibration envelopes are fairly broad (von Békésy, 1949; Schuknecht, 1960). The mechanical behaviour of the transducer structure, therefore, correlates at least qualitatively with the corresponding spatial pattern of activity in the auditory nerve. It has been shown too, that when the processes of a vestibular hair cell are deflected the tendency for an impulse to be discharged in its associated nerve fibre is high when the hairs are deflected towards the kinocilium and low when they are deflected away from it (Lowenstein et al., 1964). Bearing in mind current ideas on shearing forces in the cochlea (von Békésy, 1953), it is to be expected therefore that discharge of nerve impulses would be correlated with the movement of the membrane in one direction and hence with a particular phase of the stimulus. The time course of the action potential, and of its generation at the hair cell junction, might lead one to expect that this relationship would become progressively weaker for stimuli with periods much shorter than the duration of the nerve impulse, and this proves to be the case. There is, not surprisingly, no abrupt upper limit, but an approximate figure might be 4000-5000 Hz (Rose et al., 1968).

We have present in the auditory nerve, therefore, potential information about stimulus frequency (at least over a limited range) both in the form of a contrast between those fibres which are active and those which are inactive, and in the form of the intervals between successive impulses in a group of such fibres. We cannot decide whether either (or both) of these possibilities are made use of by looking at the transducer; neither can we decide by studying the auditory nerve patterns themselves, even with any of the sophisticated instruments now available to us. What we must seek to determine is the extent to which the central nervous system can utilize the data; if it cannot, anything we can extract with an oscilloscope or a computer is quite irrelevant.

I have dealt extensively elsewhere (Whitfield, 1967) with possible mechanisms of processing place data, and shall confine myself here, since this is the subject of our conference, primarily to a consideration of temporal processing. In the first place it must, I think, be obvious that at some point within the nervous system information carried in terms of pulse periodicity would have to be converted to probability of response of a neurone or neurones. This necessity arises, in the limit, from the observation that the frequency of a sound stimulus does not appear in an associated motor response. It is instructive, therefore, to investigate how far the periodicity seen in the auditory nerve can be traced within the central nervous system.

The earliest data are those provided by Kemp et al., (1937) who found synchronous responses up to 2500 Hz at the trapezoid body, and up to 1000 or 1500 Hz at the inferior colliculus. More recent studies of single units in the superior olive (Moushegian et al., 1964) indicate at that level a reasonable degree of synchrony for frequencies up to 1500 Hz and a negligible amount at 2500 Hz.

FEATURES OF A PULSE INTERVAL SYSTEM

It must be evident that if the system is to take account of pulse interval in the way required, it is necessary that there be within the nervous system some standard timing device or 'clock'. What evidence is there for the existence of such a device? The most ubiquitous and fundamental timing device we know of involves the time constants of cell membranes. These time constants lead, of course, to the phenomena of spatial and temporal summation. With suitable disposition of excitatory and inhibitory endings, the relative time of arrival of impulses along two input channels can strongly influence the probability of firing of the target neurone.

A now classic example of a neural timing device is seen in the accessory olive. Cells in this nucleus can be influenced by stimuli applied to either ear, and the relative timing of binaural clicks is a crucial factor in the firing of the cell. A change of ± 250 μsec on each side of simultaneity can change the probability of firing of a typical neurone from 40% at the one limit to 70% at the other, according to Hall's (1965) data. Again, Rose et al. (1966) have observed that binaural time disparities of as little as 10 or 20 μsec can affect the firing of neurones at the inferior colliculus. There is thus no question that the nervous system can in certain circumstances "identify" time intervals and convert them to probability of response. However, in both the cases we have just considered the system has been detecting a disparity in the timing of two impulses in two different fibres. The essence of the pitch problem is to determine the interval between two successive pulses in the *same* fibre (Fig. 1). Thus to detect the difference between 500 Hz and 505 Hz it is necessary not to detect a difference of 20 μsec but to detect the difference between an interval of 2000 μsec and one of 1980 μsec—a very different proposition. Most people can give a reasonably accurate guess at the interval between two clicks a second apart, but few could reliably distinguish an interval 60 seconds long from one 59 seconds long.

Eccles showed many years ago (1953) that temporal summation at a single synaptic ending is comparatively unimportant. For a train of impulses arriving along a single fibre, if the first two fail to elicit an output then succeeding members of the train are unlikely to do so. This fact arises from the temporal characteristics of the post-synaptic membrane threshold. Not only, therefore, has the ability of a neurone to detect differences in the intervals between pulses which arrive at it along a single fibre not been demonstrated, it seems inherently unlikely that it is physiologically possible.

There are two ways in which the one-channel problem could be solved in

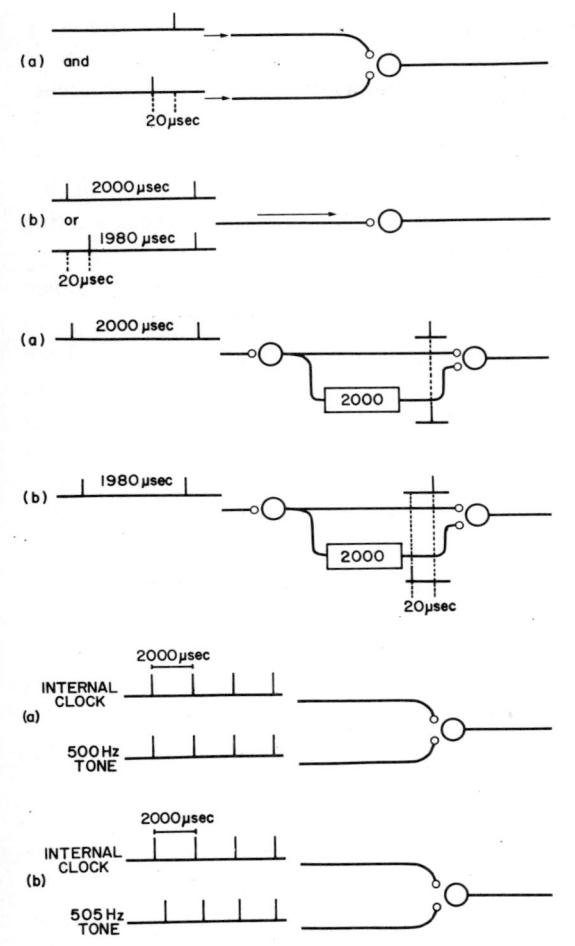

Fig. 1. (a) The binaural or two-fibre situation; (b) the periodicity or one-fibre situation. Although in (a) the cell can give an output which will distinguish 20 μsec from 0 μsec or 40 μsec the percentage accuracy required is not high. In (b) the much longer pulse intervals would need an accuracy of 1% in the detector to distinguish them.

Fig. 2. The delay line solution. A 2000 μsec delay line reduces the problem of Fig. 1b to that of Fig. 1a for this particular frequency discrimination. A large number of such delay lines would be required.

Fig. 3. The clock pulse solution. If the cell is a coincidence counter, it will give an output in (a) but not in (b). Either many clocks would be required, or a single fast-running clock with scaling circuits. In either event a high degree of short term stability would be needed.

terms of the two-channel problem, and both involve an extra timing device. Clearly, a parallel delay line which would deliver a replica of the original pattern but delayed by, in our example, 2000 μsec would enable the 500 Hz/505 Hz discrimination to be made in terms of a known property of the nervous system — the ability to discriminate a 20 μsec *versus* 0 μsec disparity in two fibres (Fig. 2). This could be carried out in principle simply by having nerve fibres of different lengths, but such a system would need a considerable number of accurately adjusted lines, and although such models had been postulated many years ago, no actual examples have been demonstrated in the nervous system.

The possibility of such a delay line system being the mechanism underlying the directional responses of auditory cortical units was examined by Whitfield (1969). Since the responses observed were dependent on direction of frequency change, but hardly at all on the rate of change, it was concluded that an inhibitory rather than a delay-line mechanism was the more economical hypothesis.

An alternative way of transforming the single channel output would be to provide a second channel containing a regular clock pulse (Fig. 3). Coincidence or lack of coincidence of this train with the input pattern could then be made to determine the output of the target neurone. Again, the existence of fairly regularly firing neurones has been reported in the central nervous system, but nothing on the scale which would be required to operate a discriminating mechanism of the range found in the auditory system.

If we accept that we have still to look for and identify some such mechanism before we can regard pulse periodicity as having physiological validity as a pitch discrimination mechanism, where ought we to look? Since, as has already been stated above, the maximum 'following' rate declines progressively as we ascend the system, presumably the earlier we look the better. The earliest point in the system at which a relationship between binaural pulse position and probability of firing has been demonstrated is at the medial olivary nucleus (Hall, 1965), but this may to some extent reflect the fact that this is the lowest level at which reasonably direct connections from the two sides interact on the same neurone. (The influences which have been shown to be exerted by the contralateral ear on the cochlear nucleus (Pfalz, 1962) and on the cochlea itself (Fex, 1962) are probably too indirect and diffuse to function in the way required.) In the periodicity case binaural action is not required, and the cochlear nucleus is the earliest point at which we might search. Just what we should look for is difficult to say. The time delays which would be required are, as we have seen, of the order of 1 to 10 msec. Such delays would need to be obtained with 1 or 2% accuracy, so that multiple synapses are unlikely to be a good answer. The obvious alternative is nerve conduction time, and in order to achieve the necessary delay by this method the path would probably have to travel outside the nucleus. Such loops do exist, and the path from the ventral cochlear nucleus to the lateral olivary nucleus and back to the cochlear nucleus is an example. Furthermore, the centrifugal section of this pathway has a latency of the order of 30 msec or more (Whitfield and Comis, 1968). However, this latency has been found to be highly dependent on the strength of stimulation of the olivary nucleus (Comis and Whitfield, 1968), an observation which renders the pathway unlikely as a candidate for a delay line.

The alternative 'clock pulse' hypothesis would require that some central clock would feed regular pulses into the cochlear nucleus, and one or other of the centrifugal pathways which terminate there would presumably have to be the route by which they enter. Not all of these pathways have been examined, but those which have (Comis and Whitfield, 1968; Whitfield, 1968b) seemed to be concerned with exerting a tonic influence on neuronal excitability and show no evidence, at least in the anaesthetized animal, of any regular clock-like activity.

ELECTRICAL EXCITATION OF THE AUDITORY SYSTEM

One of the difficulties in deciding between position and periodicity as the vehicle for information carriage in the auditory nerve is that both occur together in the normal system. It would be instructive if we could somehow separate the two sources of information and study the subjective result.

Two groups of workers have carried out experiments involving stimulation with electrodes directly on the 8th nerve in human subjects. Simmons (1966) introduced a multiple electrode array into the modiolus of a deaf patient. The effects both of applying electrical stimulation to different electrodes, and of varying the stimulus parameters (including the frequency of stimulation) were investigated. Although this is probably the most extensive study which has been done, unfortunately the subject appears to have been a very poor 'witness' when it came to reporting the sensations produced, and it is almost impossible to decide whether there were true changes of pitch, or whether these were not rather complex changes in the character of the sound. Even if the subject's reports of 'highness' or 'lowness' be taken at their face value, there was nevertheless no systematic relationship between stimulus frequency and pitch. Furthermore, the presence of the cochleae, even though true hearing was apparently absent on the side under test, is a complicating factor in view of the known possibilities of electrophonic hearing (Stevens, 1937; Jones *et al.*, 1940). This latter effect seems to be a response of the intact cochlea to electrical stimulation. It covers nearly the whole frequency range of normal hearing, and the resultant tones will interact with frequencies introduced into the cochlea by the normal acoustic route. An earlier investigation by Simmons (1964) on a different patient who had an intact cochlea suffers from the same ambiguity. The range and discrimination of pitches found in this case accord more with electrophonic hearing than with the results of other nerve stimulation experiments. Square pulses were used as the stimuli and it is notable that the subject reported that the 'tones' had the quality of square-wave acoustic stimuli rather than sinusoidal ones. This suggests that cochlear, rather than auditory nerve, stimulation was the basis of the effects.

A more satisfactory experimental situation from the point of view of the investigation is that reported by Djourno *et al.* (1957). These authors had a completely deaf patient with bilateral cochlear destruction, and they implanted electrodes on the remaining central stump of the 8th nerve on one side. The electrodes could be remotely activated via a coil implanted subcutaneously in the temporalis muscle. They carried out tests using impulsive stimuli at 100 pps and alternating currents of a few cycles per second up to 'musical frequencies'. With bursts of 100 pps the sensation was reported as like the chirp of a cricket at low stimulus intensities, and like blasts on a 'pea whistle' at high intensity. With alternating currents the 'pea whistle' effect was obtained with a 'roughness' which diminished as the stimulus frequency rose. The qualitative change, which was very limited, did not seem at all to be in terms of pitch. Vowels spoken through a microphone were perceived only as different loudness of the same note. In a second paper (Djourno and Kayser, 1958) the results on two patients are compared. In the one patient stimulation always produced sensations of a 'sharp metallic quality' while in the other one they had a deep, low-pitched timbre, like the 'soughing' of the wind. The authors concluded that it was the relative placement of the electrodes in relation to the nerve trunk which was responsible for the differences. This in turn suggests that it is the particular pattern of fibres activated and the relative strength of their activation which is all important, and that the actual periodicity of the impulses does not affect the sensation.

The direct evidence then suggests that in man periodicity of impulses in the auditory nerve is not a dominant factor in the perception of the signal. There remains however the possibility that it could be of importance in circumstances where place information is absent or ambiguous.

PERIODICITY PITCH IN FISH

The possibility of periodicity being used as the means of pitch perception has been urged in the case of fish, since these do not have the elaborate cochlear analytic mechanism found in mammals. It appears, nevertheless, that differential frequency response between fibres does exist and therefore there must presumably be some differential receptor sensitivity. Enger (1963) analysed the activity of single auditory neurones in the sculpin (Cottus scorpius) and found four types of fibre classified with respect to their response to sound; (1) units with a regular discharge which did not respond to sound; (2) units responding to low frequencies only (below 200 Hz); (3) units which responded to all frequencies which the animal presumably can hear; (4) units which responded to frequencies up to about 300 Hz with a discharge synchronized to the stimulating sound.

In an example quoted by Enger (1968) a single auditory fibre of the sculpin needed sound intensities which varied progressively from 10 dB at 100 Hz to 30 dB at 250 Hz to produce the same discharge rate in the fibre. It is evident, therefore, that the possibility exists even in fishes for frequency to be coded in terms of the differential responses of different fibres, and it is not open to us to hypothesize that their discrimination is based upon periodicity simply because we have no other explanation. Here, too, positive evidence is required, and is so far lacking.

NERVE IMPULSE PERIODICITY AND PITCH

If we actually record the nerve impulses in for example the cochlear nucleus of the cat and examine their periodicity, we find a number of discrepancies between these observations and the human subjective pitch sensation. If we employ a complex tone consisting say, of components of 1.2 and 1.4 kHz, we are as likely to get a periodicity corresponding to one of the components as to 200 Hz (Fig. 4a). However, as the frequencies of the components are increased, the locking to 200 Hz tends to become better and this effect extends beyond the existence region (Fig. 4b). Indeed excellent 200 Hz locking can be obtained with components of 10 kHz and 9.8 kHz. A perhaps more cogent point is that the

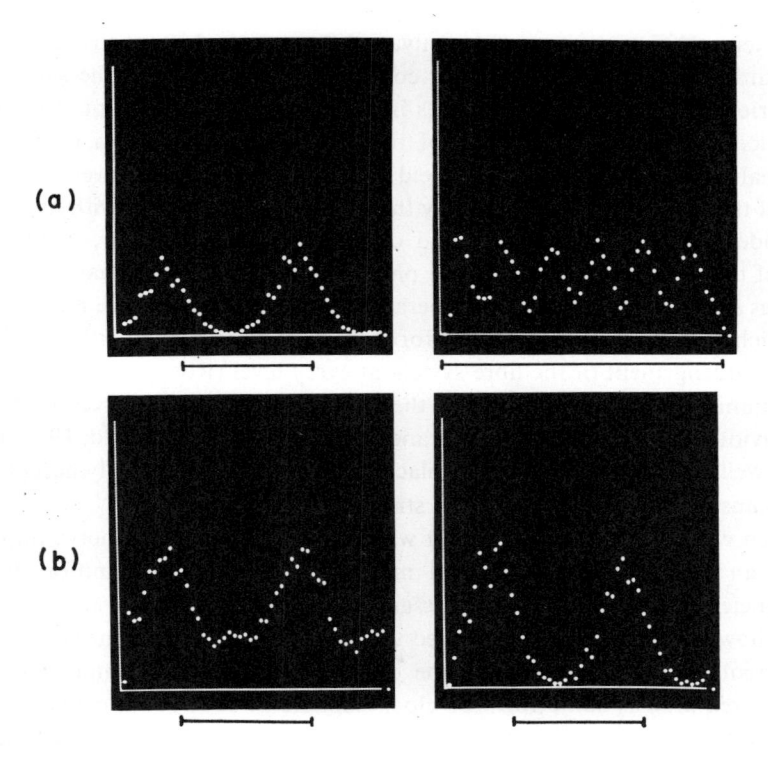

Fig. 4a. Responses of two different units in the cat cochlear nucleus with characteristic frequencies around 1.4 kHz to a complex tone with equal components at 1.2 and 1.4 kHz. Left: histogram shows periodicity corresponding to 200 Hz. Right: histogram shows periodicity corresponding to individual peaks of the waveform (1200 Hz).
Fig. 4b. Left: response of a CN unit to a complex tone with equal components at 3.6 and 3.8 kHz. Right: response of a CN unit to a tone with equal components at 1100 and 1300 Hz.

The two components of the tone were phase-locked in each case. The points on the histograms represent the total number of spikes occurring at a given time after a marker pulse locked to an arbitrary point on the stimulus waveform. This pulse recurred every 10 msec (200 μsec per point) in all the records except the top right hand one where the analysis rate was doubled (100 μsec per point) for greater clarity. Only spike periodicities integrally related to the trigger repetition rate give rise to peaks in the histogram. Time bars = 5 msec.

nerve fibre periodicity remains locked to 200 even when two components differing by 200 Hz but not harmonics of that frequency (for example 1100 and 1300 Hz) are used. The apparent pitch of this complex does not, of course, correspond to 200 Hz (see papers by Ritsma and Smoorenburg, pp. 250 and 267).

DISCUSSION

Many experiments have demonstrated physiologically that information about stimulus frequency is potentially coded in two different ways for low frequencies. Different frequencies activate different members of the total array of auditory nerve fibres, albeit with considerable 'overlap'. At the same time the periodicity of the nerve impulses in the nerve bears a relationship to the periodicity of the stimulus. Both of these statements appear to be true for mammals, birds (Stopp and Whitfield, 1961) and fishes. However, although each of these codes can be utilized by the experimental observer with his microelectrodes, oscilloscopes, clocks and computers, it is necessary, if either or both of them are to form part of the physiological mechanism, that the central nervous system should be able to operate on the information. The mechanisms by which the place coding of the information is preserved and enhanced — the orderly arrangement of the fibre system at every level (Rose et al., 1959, 1963; Tsuchitani and Boudreau, 1966) and the mechanisms which preserve the identity of individual fibre groups (Allanson and Whitfield, 1955; Whitfield, 1956, 1965) — are well known, and indeed the place mechanism is very widely accepted as the means by which high frequency stimuli are processed.

When we come to the question of whether the periodicity of nerve impulses forms an alternative or additional mechanism of pitch recognition at low frequencies, we find that the necessary physiological mechanisms have not been shown to exist. It is emphasized that the problem is a different one from that involved in measuring the time difference of dichotic stimuli, and that what is required is some form of clock capable of measuring comparatively long intervals of several milliseconds to a 1% accuracy. No such clock has so far been demonstrated, and it will be necessary to locate such a clock and its mechanism in order to validate the theory.

The most cogent evidence against the proposal that pitch is coded in terms of pulse periodicity is provided by experiments in which periodic pulses were artificially excited in auditory nerve fibres without regard to their position in the total array. In so far as the resultant sensation had any pitch, this seemed to be entirely unrelated to the pulse repetition rate of the stimulus. On the other hand, there was some suggestion that the position of the stimulating electrodes, and hence the particular fibres excited, could affect the pitch. What is clearly required now is a subjective experiment of the Woolsey and Walzl (1942) type in which apical cochlear fibres are differentially stimulated, while simultaneous behavioural tests are carried out.

If the case for nerve impulse rate as a means of transmitting the pitch of single low-frequency tones must be rated as "not proven" from a physiological point of view, what of Periodicity Pitch (the 'missing fundamental' and related phenomena)? Here the same objections and lack of mechanisms apply. Furthermore, it may turn out to be quite possible to embrace these phenomena in a classical place theory. According to this theory a complex tone gives rise to a series of active groups of nerve fibres in the central auditory pathway, separated by groups of inactive fibres (Whitfield, 1967) (Fig. 5). Although each of these groups corresponds to one of the partials of the complex, we do not normally hear these as such, unless we train ourselves to do so. What we do hear is a *unitary* experience of a complex sound to which we assign a pitch. The total pattern in the auditory pathway array corresponds to this experience. As with many other sensory experiences it is eminently possible that the presence or absence of a small part of this pattern, or its distortion, will not entirely destroy the normal sensation; the interpretation put on the available data is that which is 'most likely'. Of course, there is at present no direct physiological evidence for this aspect of the theory, just as there is none for the pitch carrying role of pulse interval in the auditory nerve. The point to be made is that we cannot adopt either theory simply on the basis of there being no other. A great deal more psychophysiological experiment is needed.

I should like to acknowledge the assistance of Dr. S. D. Comis and Mr. H. F. Ross and the financial support of the Science Research Council in some of the experiments described herein.

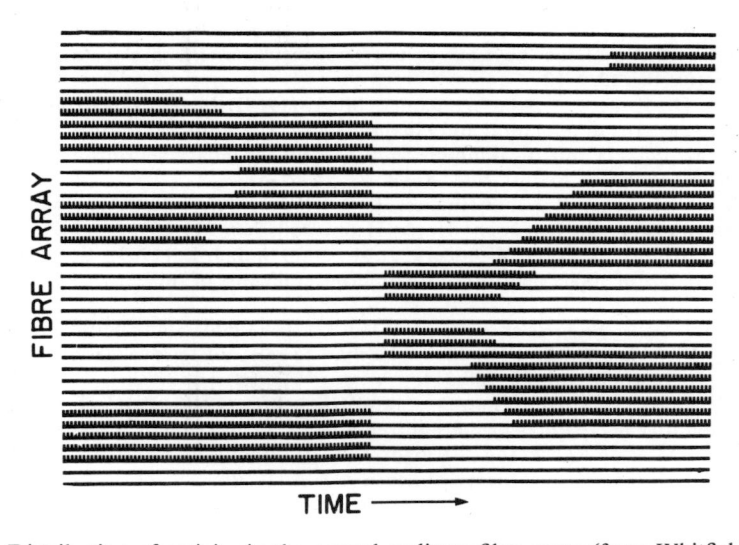

Fig. 5. Distribution of activity in the central auditory fibre array (from Whitfield, 1967).

REFERENCES

Allanson, J. T., and Whitfield, I. C. (1955): The cochlear nucleus and its relation to theories of hearing, Third London Symposium on Information Theory, C. Cherry, Ed. (Butterworths, London), pp. 269-286.
Békésy, G. von (1949): On the resonance curve and the decay period at various points on the cochlear partition, J. Acoust. Soc. Amer. 21, 245-254.
Békésy, G. von (1953): Description of some mechanical properties of the organ of Corti, J. Acoust. Soc. Amer. 25, 770-785.
Comis, S. D., and Whitfield, I. C. (1968): Influence of centrifugal pathways on unit activity of the cochlear nucleus, J. Neurophysiol. 31, 62-68.
Derbyshire, A. J., and Davis, H. (1935): The action potentials of the auditory nerve, Amer. J. Physiol. 113, 476-504.
Djourno, A., Eyriès, C., and Vallancien, B. (1957): De l'excitation électrique du nerf cochléaire chez l'homme, par induction à distance, à l'aide d'un micro-bobinage inclus à demeure, C.r. Soc. Biol. 151, 423-425.
Djourno, A., and Kayser, D. (1958): Perspectives nouvelles en matière de prothèse sensorielle par action directe sur les fibres ou les centres nerveux, C.r. Soc. Biol. 152, 1433-1434.
Eccles, J. C. (1953): The Neurophysiological Basis of Mind (Clarendon Press, Oxford).
Enger, P. S. (1963): Single unit activity in the peripheral auditory system of the teleost fish, Acta Physiol. Scand. 55, Suppl. 210.
Enger, P. S. (1968): Contribution to discussion in: Hearing Mechanisms in Vertebrates, A. V. S. de Reuck and J. Knight, Eds. (Churchill, London) p. 15.
Fex, J. (1962): Auditory activity in centrifugal and centripetal cochlear fibres in cat, Acta Physiol. Scand. 55, Suppl. 189.
Hall, J. L. (1965): Binaural interaction in the accessory superior olivary nucleus of the cat, J. Acoust. Soc. Amer. 37, 814-823.
Jones, R. C., Stevens, S. S., and Lurie, M. H. (1940): Three mechanisms of hearing by electrical stimulation, J. Acoust. Soc. Amer. 12, 281-290.
Kemp, E. H., Coppée, G. E., and Robinson, E. H. (1937): Electric responses of the brain stem to unilateral auditory stimulation, Amer. J. Physiol. 120, 304-315.
Kiang, N. Y.-S. (1965): Discharge Patterns of Single Fibers in the Cat's Auditory Nerve (Research Monograph No. 35, M.I.T. Press, Cambridge, Mass.).
Lowenstein, O., Osborne, M. P., and Wersäll, J. (1964): Structure and innervation of the sensory epithelia of the labyrinth in the Thornback ray (Raja clavata), Proc. Roy. Soc., Series B, 160, 1-12.
Moushegian, G., Rupert, A., and Whitcomb, M. A. (1964): Medial superior-olivary-unit response patterns to monaural and binaural clicks, J. Acoust. Soc. Amer. 36, 196-202.
Pfalz, R. K. J. (1962): Centrifugal inhibition of afferent secondary neurons in the cochlear nucleus by sound, J. Acoust. Soc. Amer. 34, 1472-1477.
Rose, J. E., Galambos, R., and Hughes, J. R. (1959): Microelectrode studies of the cochlear nuclei of the cat, Johns Hopk. Hosp. Bull. 104, 211-251.
Rose, J. E., Greenwood, D. D., Goldberg, J. M., and Hind, J. E. (1963): Some discharge characteristics of single neurons in the inferior colliculus of the cat, J. Neurophysiol. 26, 294-320.
Rose, J. E., Gross, N. B., Geisler, C. D., and Hind, J. E. (1966): Some neural mechanisms in the inferior colliculus of the cat which may be relevant to localization of a sound source, J. Neurophysiol. 29, 288-314.
Rose, J. E., Brugge, J. F., Anderson, D. J., and Hind, J. E. (1968): Patterns of activity in single auditory nerve fibres of the squirrel monkey, in: Hearing Mechanisms in Vertebrates, A. V. S. de Reuck and J. Knight, Eds. (Churchill, London), pp. 144-157.
Rutherford, W. (1886): A new theory of hearing, J. Anat. Physiol. 21, 166-168.
Schuknecht, H. F. (1960): Neuroanatomical correlates of auditory sensitivity and pitch discrimination in the cat, in: Neural Mechanisms of the Auditory and Vestibular Systems, G. L. Rasmussen and W. F. Windle, Eds. (Charles C. Thomas, Springfield), pp. 76-90.

Simmons, F. B. (1964): Electrical stimulation of acoustical nerve and inferior colliculus, Arch. Otolaryngol. *79*, 559-568.
Simmons, F. B. (1966): Electrical stimulation of the auditory nerve in man, Arch. Otolaryngol. *84*, 2-54.
Stevens, S. S. (1937): On hearing by electrical stimulation, J. Acoust. Soc. Amer. *8*, 191-195.
Stopp, P. E., and Whitfield, I. C. (1961): Unit responses from brain stem nuclei in the pigeon, J. Physiol. *158*, 165-177.
Tsuchitani, C., and Boudreau, J. C. (1966): Single unit analysis of cat superior olive S-segment with tonal stimuli, J. Neurophysiol. *29*, 684-697.
Wever, E. G. (1949): Theory of Hearing (Wiley, New York).
Whitfield, I. C. (1956): Electrophysiology of the central auditory pathway, Brit. Med. Bull. *12*, 105-109.
Whitfield, I. C. (1965): 'Edges' in auditory information processing, XXIII Internat. Physiol. Cong. (Tokyo), pp. 245-247.
Whitfield, I. C. (1967): The Auditory Pathway (Monographs of the Physiological Society, No. 17; Arnold, London).
Whitfield, I. C. (1968a): The functional organization of the auditory pathways, J. Sound Vib. *8*, 108-117.
Whitfield, I. C. (1968b): Centrifugal control mechanisms of the auditory pathway, in: Hearing Mechanisms in Vertebrates, A. V. S. de Reuck and J. Knight, Eds. (Churchill, London), pp. 246-254.
Whitfield, I. C. (1969): The response of the auditory nervous system to simple time-dependent acoustic stimuli, Ann. New York Acad. Sci. *156*, 671-677.
Whitfield, I. C., and Comis, S. D. (1968): A reciprocal gating mechanism in the auditory pathway, in: Cybernetic Problems in Bionics, Oestreicher and Moore, Eds. (Gordon and Breach, New York), pp. 301-312.
Woolsey, C. N., and Walzl, E. M. (1942): Topical projection of nerve fibers from local regions of the cochlea to the cerebral cortex of the cat, Johns Hopk. Hosp. Bull. *71*, 315-344.

DISCUSSION

Keidel: How would you combine the demultiplication at higher levels of the auditory pathway with the delay line models you discussed? Do you accept that these mechanisms are able to discriminate between at least 1000 μsec and 1020 μsec?

Whitfield: I'm not putting these mechanisms forward as models but as things which would be needed. As you may have gathered, I don't think that this is the way it works. As you go up to higher levels in the nervous system, you get less relationship to the periodicity of the signal until at the cortex, which is a good example of reduction in the amount of activity, there is no relationship at all. I agree that a mathematical-like count-down will work but that is not what happens.

Keidel: Are you certain about that?

Whitfield: I am not able to find a relationship between the periodicity of the stimulus and the response of the cortex.

Keidel: There is some indication in the geniculate body.

Boerger: I think I can give some experimental evidence that the place principle holds also for residue pitch. In cats we looked for fibres with a characteristic frequency (CF) near 500 Hz. Measurements from one fibre are given in Fig. 1, which shows the firing rate as a function of the SPL for four different stimuli. In one case the stimulus was a complex tone, consisting of the 3rd to 6th harmonics of the fibre's CF. Even at 10 dB SPL, the response to this complex (circles) was greater than to a pure tone at the CF (crosses). However, bear in mind that the partials are added in phase producing a sharp time course. This means that the peak level is about 15 dB above the individual levels. I think that these results support the place theory. This does not mean that the time structure of this signal has no significance. We are of the opinion that these time structures are converted by the basilar membrane into activity at a frequency dependent place, perhaps by a motion which has not yet been investigated, in connection with the different directional sensitivity of the hair cells.

On the other hand, we have measured response curves by means of such a high-frequency complex (Fig. 2). The frequency of the absent fundamental was at the range of the CF.

Whitfield: In your last figure, what did you record for a pure tone?

Boerger: We did not measure it exactly but the curves are qualitatively the same. This response curve is obtained from a signal which has no component in the CF region. The lowest component was about 1500 Hz.

Whitfield: These components were all outside the tuning curve for the single frequency components?

Boerger: Yes. At the eardrum the SPL at CF was at least 40 dB below the

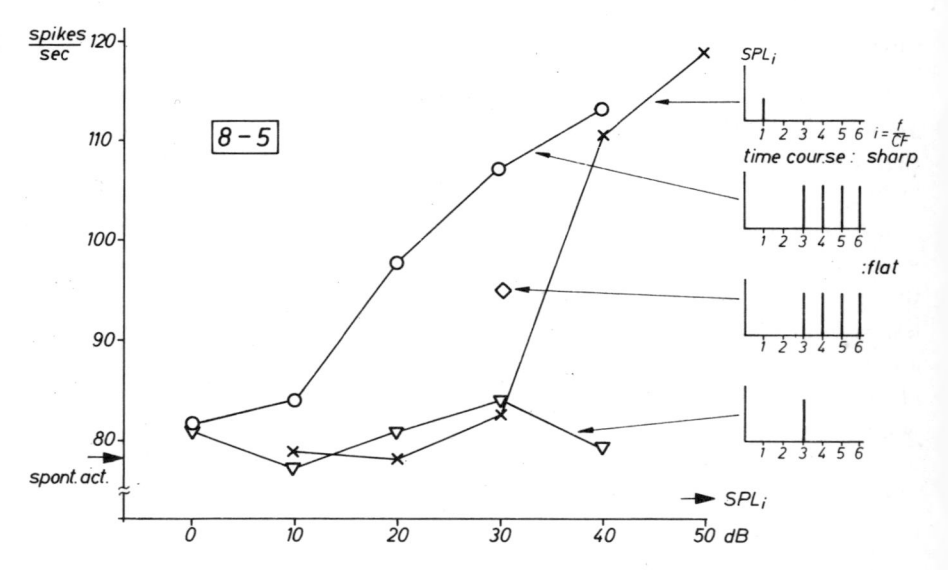

Fig. 1 (Boerger). Auditory nerve single fibre discharge rate at different stimulus configurations (see inserts). The neuron with CF=420 Hz responds to a complex containing no energy within its response area more sensitive than to CF−stimulation.

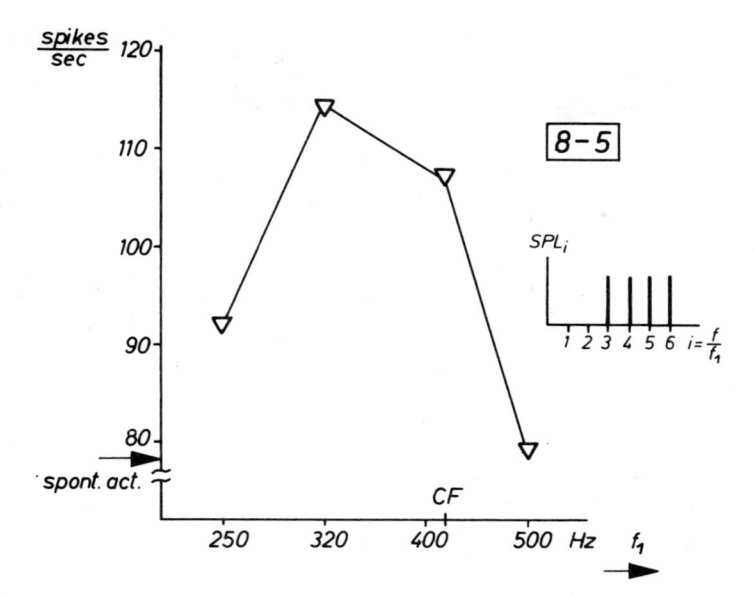

Fig. 2 (Boerger). Response curve of the same fibre produced by variation of the fundamental in the vicinity of the CF. Stimulus: see insert; time course: sharp; SPL of each component = 25 dB.

individual components. As to be seen from Fig. 1 the lowest component alone produces no response. Those who are familiar with the combination tones would object that the $2f_1-f_2$ combination tone, which may originate from the 3rd and the 5th components, is involved. By removing the 3rd component a possible cubic distortion tone could no longer coincide with the fundamental. In this case the response of the fibre remained qualitatively the same.

Smoorenburg: Are you certain of the location of the electrodes?

Boerger: Yes, I think we know very well we had the primary fibres.

Klinke: Dr. Zwislocki, can you imagine a sort of mechanical event which is able to explain Dr. Boerger's finding that a fibre can be activated by a harmonic complex without the first and second partial and with a periodicity corresponding to the CF? May the place on the basilar membrane corresponding to the CF be activated by a stretch movement of the membrane introduced by the envelope of such a residue?

Zwislocki: I do not have a ready answer; I think it was said that some kind of distortion will probably answer the question.

Goldstein: It is a very intriguing possibility that a periodic stimulus with no energy at the fundamental stimulates a fibre with a CF near the missing fundamental. This result apparently contradicts the psychophysical experiment in which one can not mask the residue with energy at the fundamental frequency.

De Boer: I think it is not so very important to look at time detecting mechanisms for pulses that are carried along *one* fibre, because we know that whatever signal you present, the individual nerve fibres will always tend to skip one or more cycles. So it is probably a kind of volleying effect which should be taken into account.

The basic problem which Dr. Whitfield pointed to is the absence of any delay mechanism in the early stages of the auditory system. When you look at the latency of the responses of the primary auditory nerve fibres (Kiang, 1965, p. 26) and you take a number of units with the same CF you will find that the latency times vary. Do you think that this phenomenon may be due to a delay line mechanism which could be a serious candidate?

Whitfield: It is possible of course. The difficulty, I think, is that it would have to occur in the high-frequency fibre region. As you can vary the pitch of the residue while keeping the harmonics in the same frequency area you need all the corresponding delay lines in that area. So you need a lot of delays in the cochlea that have no apparent natural purpose other than just being able to produce the residue. This seems unlikely, but it is a possibility.

Schwartzkopff: Why is it unlikely?

Whitfield: I think one has to demonstrate it.

Goldstein: Concerning the question of Dr. de Boer, I think that the actual latency times involved range between 1.5 msec for $f_{CF} > 3$ kHz up to about 4 msec for f_{CF} about 300 Hz (Kiang, 1965). It just seems that these are not the right numbers in order to measure pitches corresponding to 100 Hz to 1000 Hz. You don't have 10 msec latency times in the region of 3 kHz!

De Boer: In the cat.

Goldstein: Yes, in the cat. But there are some indications that the auditory nerve responses in the cat should be similar to those found in other mammals. This is of course an extrapolation.

With respect to the cochlear nucleus responses you showed, Dr. Whitfield, how can you assume that the auditory system is looking where you look? Could it not, in effect, combine information from different units?

Whitfield: Oh yes, you mean the case where you can hear the residue and where we found no periodicity. This does not prove anything because the mechanism can be somewhere else.

Rose: Dr. Whitfield, you mentioned the experiments of Simmons and of Djourno *et al.* If a cochlea is destroyed the fibres will degenerate. So what has been stimulated by introducing electric signals?

Whitfield: He claimed that the nerve existed. I do not know how far down.

Rose: The spiral ganglion cells are destroyed.

Whitfield: In spite of the damage there is no doubt that sensations of sound were obtained.

Rose: Yes, but I do not think this can be an argument one way or another. The important thing is the total pattern and when you destroy some part, the pattern is destroyed.

Zwislocki: From the point of view of periodicity pitch, will it make any difference to the neural system at what level the stimulation occurs?

Rose: I do not expect a bonafide sensation from electric stimulation.

Whitfield: I agree that this experiment has its shortcomings, but as far as I know a pitch related to the repetition frequency of an electric stimulus has not been demonstrated in anybody up to now.

Plomp: If pitch is related to place you would expect that a shift in the stimulation pattern with SPL introduces a corresponding pitch shift. I think that we have three levels at which such a shift in the stimulation pattern has been observed: Dr. Honrubia demonstrated it in the CM, Dr. Rose showed it for the single nerve responses, and I noticed it in the aftermasking pattern of a simple tone. Psychophysical measurements show, however, some pitch shift but the shift is too small to be related to the shift found in the different types of experiments I just mentioned. Those shifts in stimulation pattern amount to half an octave or even more.

Whitfield: I do not think the CM can give you an estimate of the pitch shift. The CM are too complicated.

Cardozo: I would like to ask Dr. Whitfield if he would go along with my trying to formulate his position by saying that, although he has heard the residue, he has measured the necessary information even in the cat with an adequate stimulus, he has *still* some difficulty in finding or formulating a physiological mechanism at this moment which could account for the observed phenomena of the periodicity pitch.

Whitfield: I would go along with the second part that I can not formulate a physiological mechanism. But I do not think I can go along with the first part because the electrophysiological results are in contradiction with the psychophysical ones. Especially in case of the set of high harmonics because it seems that they lock even better than lower ones. If the impulses are randomly ordered, instead of regularly, one might get some differences in quality. But this will be independent of the particular periodicity; there should be no differences in pitch in that case.

Schouten: If you are able to hear a pitch of the residue, the information should be in the auditory system whether or not you find it. On the other hand, it can be there without being measured by the auditory system. Your example of a harmonic stimulus with a repetition frequency of 200 Hz consisting of a group of partials around 3000 Hz is situated at the border of the existence region of the tonal residue. So the periodicity detector needed for this pitch may be lacking, but this does not mean that corresponding periodicities should be lacking.

Whitfield: Your assumption is: you must have it. I say I do not see the necessity. I have an alternative hypothesis. From the day you are born you learn about the relationships between harmonics of sounds. If you get a fragment of this pattern you tend to consider this as the whole.

Schouten: You still need a measuring apparatus.

REFERENCES

Kiang, N. Y.-S., *et al.* (1965): Discharge Patterns of Single Fibers in the Cat's Auditory Nerve (Research Monograph No. 35, M.I.T. Press, Cambridge, Mass.), Ch. 4.

TWO DIFFERENT NEURONAL DISCHARGE PERIODICITIES IN THE ACOUSTICAL CHANNEL

S. KALLERT
in collaboration with E. David, P. Finkenzeller and W. D. Keidel

*I. Physiologisches Institut der Universität Erlangen-Nürnberg
Erlangen, Germany*

The spike discharges of single units from the cat's inferior colliculus and medial geniculate body were recorded by means of glass microelectrode techniques. A LINC-computer was used to find out the temporal spike discharge pattern of both spontaneously active and stimulus-activated units and to describe it in terms of PST histograms, interval histograms and joint-interval histograms (Rodieck et al., 1962). The cats were anaesthetized with pentobarbital sodium. The anaesthetic level was as light as possible but deep enough to prevent motion of the preparation. The stimuli were tone bursts and noise bursts of 200 msec in duration and 70 dB SPL. All bursts had 6 msec rise and fall times. The stimulating and recording techniques used in this study are described by David et al. (1967).

The majority of unit response features was identical with the findings already described in the literature (Galambos and Davis, 1943; Tasaki, 1954; Galambos et al., 1959; Rupert et al., 1963; Rose et al., 1966, 1967; Moushegian et al., 1967; Hind et al., 1967; Kiang et al., 1965; Pfeiffer, 1966; Aitkin et al., 1966; David et al., 1968, 1969). But a few units showed properties which we think are interesting enough to be reported here.

(1) There is spontaneous activity showing a permanent discharge periodicity.
   Fig. 1 shows the interval and joint-interval histogram of such a spontaneously active unit in the medial geniculate body. The time of analysis is here 100 msec and one can see that the period is about 20 msec. Fig. 2 shows the histogram of another spontaneously active unit, the basic period is again about 20 msec but here are also periods of twice the duration of the basic period. We found spontaneously active units with discharge periods in the range of about 8 to 25 msec.
(2) There are stimulus-driven periodic spike discharges especially in the

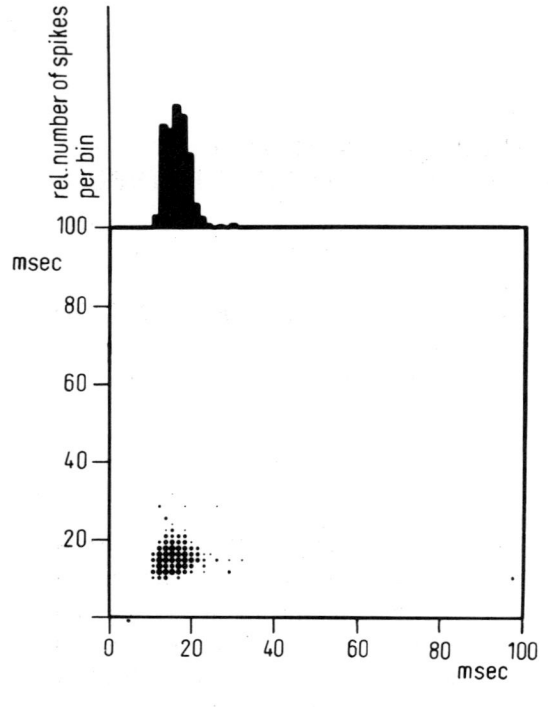

Fig. 1. Interval and joint-interval histograms of spontaneous discharge of a unit in the medial geniculate body showing a spike-discharge period in the order of 20 msec. Each time bin = 1.6 msec.

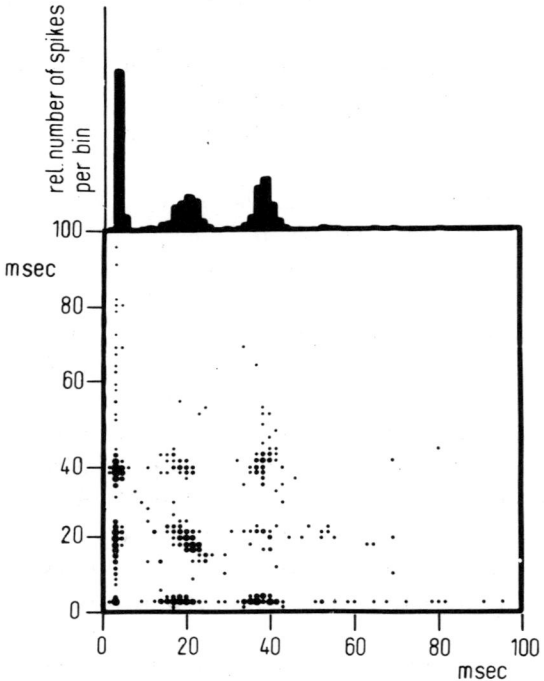

Fig. 2. Interval and joint-interval histograms of another spontaneously active unit of the medial geniculate body. The basic period is nearly the same as in Fig. 1, but here also periods of twice the duration of the basic period are existent. The period cannot be influenced by any tone stimulus (B-unit). Each time bin = 1.6 msec.

colliculus inferior with periods equal to the tone period or to integral multiples of the tone period (A-units). But it is to emphasize that for high stimulus-tone frequencies even the basic discharge period is a multiple of the tone period.

Fig. 3 shows the interval and joint-interval histogram of a stimulus-driven unit. The time of analysis here is 200 msec. The stimuli were tone bursts of 200 msec in duration and the tone frequency was 3.5 kHz which means a tone period of about 0.285 msec. One can see, however, that the basic spike discharge period, *i.e.* the shortest discharge period, is in the order of 20 msec. This can be explained by demultiplication mechanisms as described by Keidel (1956).

Fig. 4 shows the PST histogram of another unit; at the top driven by noise bursts, at the bottom driven by tone bursts. Below each histogram the stimulus duration is indicated. It is the well-known picture consisting of onset discharge, silent period and sustained discharge during the remaining stimulus time. One can see that the second part of the discharge represents the stimulus quality, if the histograms for noise and tone stimulus are compared. And while the tone period is in the order of 0.14 msec, corresponding to 7.0 kHz, the discharge period is again in the order of 20 msec.

(3) There are units in the geniculate body showing frequency response curves

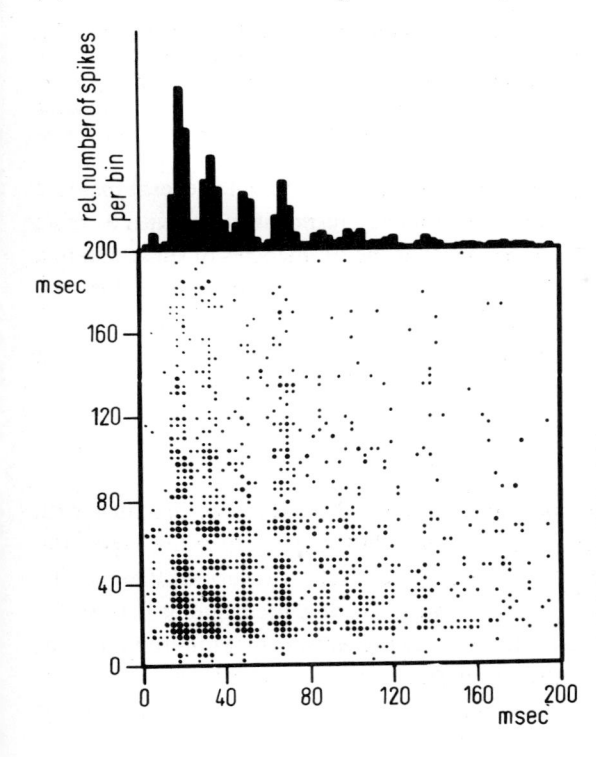

Fig. 3. Interval and joint-interval histograms of a stimulus-driven unit of the colliculus inferior. This unit did not show any spontaneous activity. Stimulus tone bursts of 200 msec. Repetition rate: 1 per 2 sec; tone frequency: 3.5 kHz; each time bin = 3.2 msec. The histogram is based on discharges which occurred during 20 presentations of the stimulus.

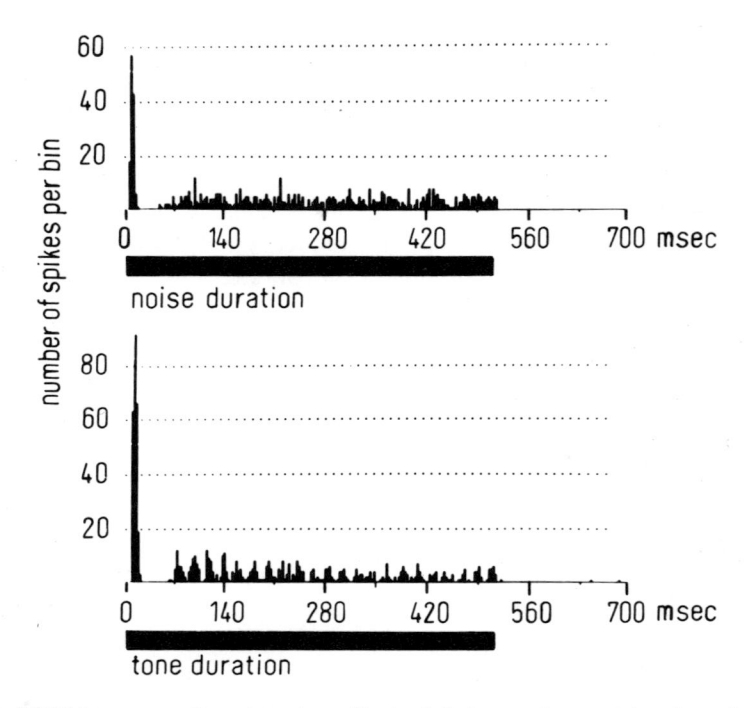

Fig. 4. PST-histograms of a unit in the colliculus inferior; at the top driven by noise bursts, at the bottom driven by tone bursts. Below each histogram the stimulus duration is indicated. This unit did not show any spontaneous activity. Stimulus tone bursts of 500 msec; repetition rate: 1 per 3 sec; tone frequency: 7.0 kHz; each time bin = 2.7 msec. Each histogram is based on discharges which occurred during 20 presentations of the stimulus.

(number of spikes as a function of the tone frequency) with several maxima in opposition to the usual form of a frequency response curve with a single maximum. It is particularly interesting that the frequencies of these maxima are in integral numerical ratios to each other, as it is known for the sensation of consonance in man.

Fig. 5 shows such a frequency response curve. The clear maxima are at about 1.65, 3.3, and 5.0 kHz.

CONCLUSION

It is shown that there are spontaneous spike-discharge periodicities with periods in the range of 8-25 msec in the geniculate body. There are especially in the colliculus inferior stimulus-driven spike-discharge periodicities of the same order even when the tone period is much shorter, for instance 0.14 msec. And there are frequency response curves showing several maxima for different frequencies which are in an integral numerical ratio to each other.

These findings suggest that the spontaneously active units serve as a sort of

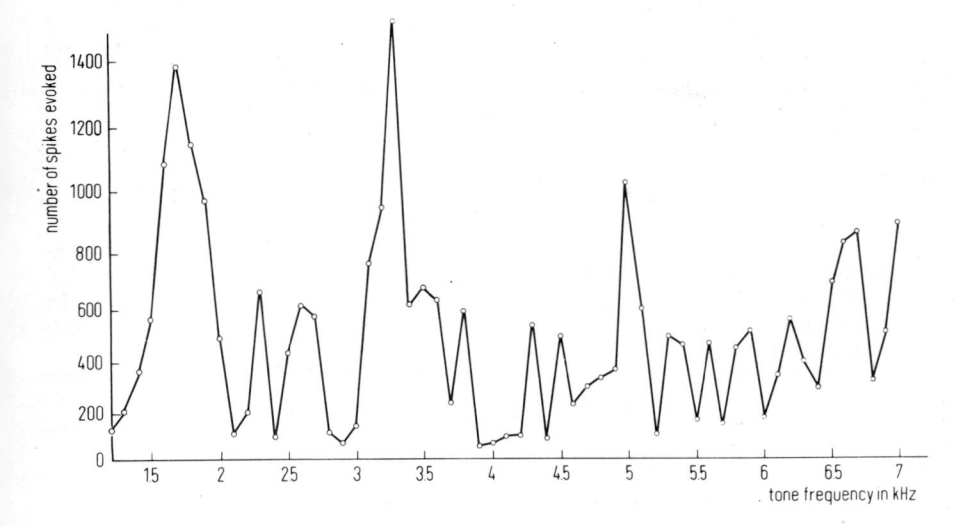

Fig. 5. Frequency response curve of a unit in the medial geniculate body. Abscissa: tone frequency; ordinate: number of spikes evoked by 50 tone bursts of 200 msec respectively.

clock. The output of these clocks and the output of the stimulus-driven units may both terminate on a unit which is able to measure coincidence between these two signals, and in this way the multi-maxima curve for frequency response could result. Fig. 6 shows a scheme of the suggested mechanism: the spontaneously active unit serving as a clock, the stimulus-driven unit and the coincidence unit.

It is conceivable that such frequency response curves with several maxima

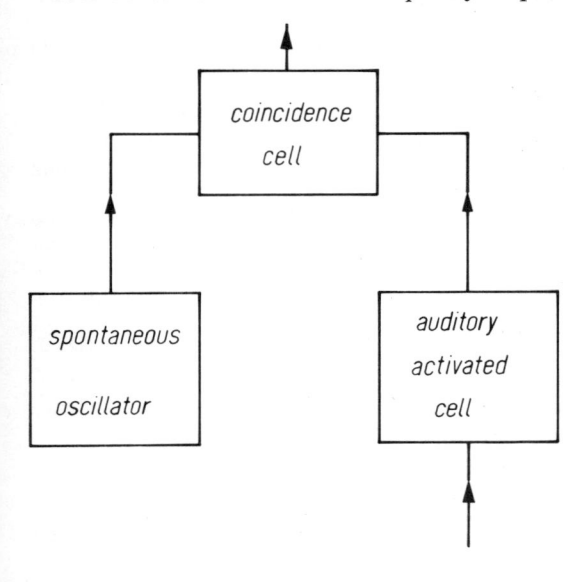

Fig. 6. Schema for the measurement of coincidence, described in the text.

are the electrophysiological correlate to the sensation of consonance, if we are allowed to relate measurements on the cat to sensation known by man. Of course, it is also imaginable that multi-peak frequency response curves result from convergence of units with single-peak response curves.

REFERENCES

Aitkin, L. M., Dunlop, C. W., and Webster, W. R. (1966): Click-evoked response patterns of single units in the medial geniculate body of the cat, J. Neurophysiol. 29, 109-123.

David, E., Finkenzeller, P., Kallert, S., and Keidel, W. D. (1967): Die Bedeutung der temporalen Hemmung im Bereich der akustischen Informationsverarbeitung, Pflügers Arch. ges. Physiol. 298, 322-335.

David, E., Finkenzeller, P., Kallert, S., and Keidel, W. D. (1969): Reizfrequenzkorrelierte ableitbare Reaktion einzelner Elemente des colliculus inferior und des corpus geniculatum mediale auf akustische Reize verschiedener Form und verschiedener Intensität, Pflügers Arch. ges. Physiol. 299, 83-93.

David, E., Finkenzeller, P., Kallert, S., and Keidel, W. D. (1969): Reizfrequenzkorrelierte "untersetzte" neuronale Entladungsperiodizität im Colliculus inferior und im Corpus geniculatum mediale, Pflügers Arch. ges. Physiol. 309, 11-20.

Galambos, R., and Davis, H. (1943): The response of single nerve fibers to acoustic stimulation, J. Neurophysiol. 6, 39-58.

Galambos, R., Schwartzkopff, J., and Rupert, A. L. (1959): Microelectrode study of superior olivary nuclei, Amer. J. Physiol. 197, 527-536.

Hind, J. E., Anderson, D. J., Brugge, J. F., and Rose, J. E. (1967): Coding of information pertaining to paired low-frequency tones in single auditory nerve fibers of the squirrel monkey, J. Neurophysiol. 30, 794-816.

Keidel, W. D. (1956): Vibrationsrezeption. Der Erschütterungssinn des Menschen. Erlanger Forschungen, Reihe B, Naturwissenschaften, Bd. 2 (Universitätsbund, Erlangen).

Kiang, N. Y.-S., Pfeiffer, R. R., Warr, W. B., and Backus, A. S. (1965): Stimulus coding in the cochlear nucleus, Ann. Otol. Rhinol. Laryngol. 74, 463-484.

Moushegian, G., Rupert, A. L., and Langford, T. L. (1967): Stimulus coding by medial superior olivary neurons, J. Neurophysiol. 30, 1239-1261.

Pfeiffer, R. R. (1966): Classification of response patterns of spike discharges for the units in the cochlear nucleus: toneburst stimulation, Exp. Brain Res. 1, 220-235.

Rodieck, R. W., Kiang, N. Y.-S., and Gerstein, G. L. (1962): Some quantitative methods for the study of spontaneous activity of single neurons, Biophys. J. 2, 351-368.

Rose, J. E., Gross, N. B., Geisler, C. D., and Hind, J. E. (1966): Some neural mechanisms in the inferior colliculus of the cat which may be relevant to localization of sound source, J. Neurophysiol. 29, 288-314.

Rose, J. E., Brugge, J. F., Anderson, D. J., and Hind, J. E. (1967): Phase-locked response to low-frequency tones in single auditory nerve fibers of the squirrel monkey, J. Neurophysiol. 30, 769-793.

Rupert, A., Moushegian, G., and Galambos, R. (1963): Unit responses to sound from auditory nerve of the cat, J. Neurophysiol. 26, 449-465.

Tasaki, I. (1954): Nerve impulses in individual auditory nerve fibers of guinea pig, J. Neurophysiol. 17, 97-122.

DISCUSSION

Schwartzkopff: I once found that stimulating the cat with a click produced some kind of periodicity in the spike activity at the level of the superior olive (Schwartzkopff, 1958; Galambos et al., 1959). We had some ideas on the origin of this periodicity, but these were turned down by Kiang's finding (1965) that, for a click, the basilar membrane produces a damped vibration which can account for the periodic discharges. As we had no computer available, we were not able to test the spontaneous activity of these units for the presence of periodicities. In the light of Mr. Kallert's results, I would like to pick up again the idea that some kind of oscillators which need first to be "pushed" in order to oscillate according to their characteristic frequency are present at different places in the auditory system.

Keidel: Regarding Dr. Schwartzkopff's point, we should have in mind that there are normal units, called A-units in Mr. Kallert's paper, which follow the periodicity of the stimulating tone. On the other hand, there are B-units whose periodicity is not influenced by the stimulus. So, we can not escape the conclusion that we have in the medial geniculate body at least some neurophysiological evidence for clock-type units, as discussed by Dr. Whitfield. Whether they are related to the sensation in man is quite a different problem. However, one may speculate that those neurons play a part in our differentiation between consonance and dissonance in music.

Ward: It sounds to me that you are trying to explain something that we don't really believe in any more. The experiments of Plomp and Levelt (1965) and others have shown pretty well that interval ratios are not important for the consonance between two pure tones. The question is simply whether they are within a critical band of each other. These findings of Mr. Kallert are certainly pertinent to periodicity pitch perception but I don't think that they are pertinent to consonance.

Keidel: I was not speaking about the consonance of two tones sounding simultaneously, but of successive tones. If you listen to a sinusoidal tone moving from low to high frequencies, you have the sensation of some sort of similarity when you pass the octave.

Whitfield: Can you tell us how stable these oscillators are and under what conditions the recordings were made?

Keidel: It is important to have a very light anaesthesia. If you have a medium or deep anaesthesia, no responses can be obtained.

Kallert: The stability of the periods was so good that we could not see any change in the time histograms at the beginning and at the end of an experiment, after about 30 minutes.

Keidel: Allow me to add that the accuracy is in the order of 1 to 2%.

Zwicker: If I look at the histogram (see Fig. 2 of the paper), I see a rather wide interval distribution.

Keidel: I was referring to the stability of the tops in the interval distribution; the accuracy of their octave relation is in the order of 1 to 2%.

Zwicker: This is not the point I have in mind. I am asking: what is the accuracy with which, after a particular action potential, the firing time of the next one can be predicted? On the basis of Fig. 2, I conclude that it is much worse than 1 to 2%.

I have a second remark on the same figure. It shows that spontaneous activity, stimulation by a sinusoidal tone of 10,000 Hz and by another tone of 1800 Hz result all in about the same histograms. To me, it looks like these histograms can be used only to make a decision what type of neuron you have picked on your electrode.

Keidel: I would only like to say that this type of neuron, about 3% of the total number of units, has a bandwidth of 20 to 70 Hz, so, more than one octave, and this is all you need for this coincidence.

Møller: What happens when such a unit with a spontaneous activity of 20 per second is stimulated with a sound of about the same periodicity? Do you get time-locked responses or do you get beats?

Kallert: We did not investigate this point.

Møller: I also have a comment relevant to Dr. Zwicker's question. It may be difficult to determine the stability from the interval histograms. There is obviously a serial correlation between pulses here. The autocorrelation function or expectation density would give you a direct measure of the stability.

REFERENCES

Galambos, R., Schwartzkopff, J., and Rupert, A. (1959): Microelectrode study of superior olivary nuclei, Amer. J. Physiol. *197*, 527-536.
Kiang, N. Y.-S. (1965): Stimulus coding in the auditory nerve and cochlear nucleus, Acta Oto-Laryngol. *59*, 186-200.
Plomp, R., and Levelt, W. J. M. (1965): Tonal consonance and critical bandwidth, J. Acoust. Soc. Amer. *38*, 548-560.
Schwartzkopff, J. (1958): Ueber nervenphysiologische Resonanz im Acusticus-System des Wellensittichs (Nelopsittacus undulatus Shaw), Z. Naturforsch. *13b*, 205-208.

THE INFLUENCE OF THE FREQUENCY RELATION IN DICHOTIC STIMULATION UPON THE COCHLEAR NUCLEUS ACTIVITY

R. KLINKE

Physiologisches Institut der Freien Universität Berlin
Berlin, Germany

G. BOERGER and J. GRUBER

Heinrich-Hertz-Institut
Berlin-Charlottenburg, Germany

The receptor cells of the organ of Corti, as well as the secondary neurons of the cochlear nucleus show efferent innervation (Rasmussen, 1946, 1953, 1960; Rossi and Cortesina, 1962; and electronmicroscopic findings of many investigators, see Iurato, 1967). The efferent fibres running to one cochlea can be activated by sound stimulation of the contralateral ear (Fex, 1962, 1965). On the other hand, electrical stimulation of efferent fibres of the crossed olivo-cochlear bundle leads to augmentation of cochlear microphonics (Fex, 1959; Desmedt, 1962; Sohmer, 1965), to diminution of the compound action potential of the auditory nerve (Galambos, 1956; Fex, 1962; Wiederhold and Peake, 1965, 1966; Sohmer, 1965, 1966) and to inhibition of the activity of primary auditory fibres (Chamberlain *et al.*, 1968). Electrical stimulation of the efferent fibres of the uncrossed olivo-cochlear bundle does not affect the cochlear microphonics (CM) but only the compound action potential of the cochlear nerve (Desmedt and La Grutta, 1963; Sohmer, 1966; Fex, 1967, 1968).

The spontaneous activity of secondary neurons of the de-afferented cochlear nucleus can be inhibited by acoustic stimulation of the intact contralateral ear (Pfalz, 1962; Dunker *et al.*, 1964; Dunker and Grubel, 1965; Dunker and Wachsmuth, 1965). This inhibition is due to efferent innervation of secondary neurons. It depends on the frequency applied to the contralateral ear.

The interaural acoustic attenuation in cats amounts to at least 40 dB (Boerger *et al.*, 1968a, 1968b) if the animal is stimulated by means of specially adapted earphones. Thus if the SPL does not exceed 40 dB above threshold the stimulation of one ear remains sub-threshold for the other. It is thus possible to activate efferent fibres by monaural sound stimulation, as Fex (1962, 1965) did, and to check the influence of the activated efferent fibres on sound-evoked neuronal activity of the other side without fear of direct acoustic influence by interaural cross talk.

METHOD

The experiments were performed on slightly anaesthetized cats (Sodium-pentobarbital) in a sound-proof chamber. The cochlear nucleus of one side was stereotaxically located with tungsten micro-electrodes (this side will be henceforth called the ipsilateral side).

As soon as an active unit was found, its characteristic frequency (CF) was determined. The ipsilateral ear was first stimulated with tone bursts of the CF and, in general, of 400 msec duration. The SPL did not exceed 40 dB above threshold, thus supra-threshold interaural cross talk could be excluded. The action potentials of single neurons were amplified and stored on magnetic tape. In every case the neuronal responses to at least 100 stimuli were recorded. Following this initial procedure the contralateral ear was simultaneously stimulated with tone bursts of 400 msec duration and a SPL maximally 40 dB above the auditory threshold. The frequency of the contralaterally applied sound was varied. The SPL's on both ears were held constant throughout the experiment. Also, in the case of stimulating both ears (dichotic stimulation), the responses to 100 individual stimuli were recorded. The neuronal activity was evaluated off-line with a LINC-8-computer.

RESULTS

Forty-one units could be recorded for a sufficient length of time to permit a comparison of their activity in response to both monaural ipsilateral and dichotic acoustic stimulation. Of these, 27 showed diminished activity with dichotic stimulation in comparison to ipsilateral monaural stimulation alone.

In 17 of these neurons the recordings could not be maintained long enough to permit more than only a few frequencies to be employed on the contralateral ear. So a complete frequency response curve for inhibition could not be determined. In the remaining 10 neurons the dependency of the average discharge rate per stimulus on the frequency of the contralaterally applied sound could be demonstrated. The inhibition was weak when the contralateral ear as well as the ipsilateral ear were stimulated with the CF of the neuron in question. The inhibition was stronger when higher or lower frequencies, adjacent to the CF, were applied to the contralateral ear. With widely displaced frequencies, the inhibition was once again weak. Thus, W-shaped curves resulted when the neuronal activity was plotted as a function of the frequency applied to the contralateral ear. Fig. 1 is an example of such an evaluation. A maximum for inhibition was found when the frequency of the contralaterally applied sound was shifted away from the CF of the neuron in question from $\pm 10\%$ to $\pm 30\%$. There the inhibition observed could reduce the activity of the neuron under study by more than 50%.

The latency of the observed inhibition was usually 30-50 msec. The inhibition lasted 50 to 100 msec beyond the cessation of the stimulus to the contralateral ear and was followed by a post-inhibitory activation.

Of the remaining neurons 12 were not influenced by contralaterally applied sound and 2 showed a slight increase in activity. Many of the neurons which were not influenced by contralateral stimulation under our experimental conditions were inhibited, however, when the SPL on the contralateral side was raised. These neurons were not further evaluated, because supra-threshold cross talk could not be excluded. Nevertheless, this finding suggests that efferent inhibition exists for a much greater proportion of neurons of the cochlear nucleus than were experimentally observed.

DISCUSSION

We used the stereotaxic method of recording from the cochlear nucleus to ensure that no efferent pathway was destroyed. As the recordings were made

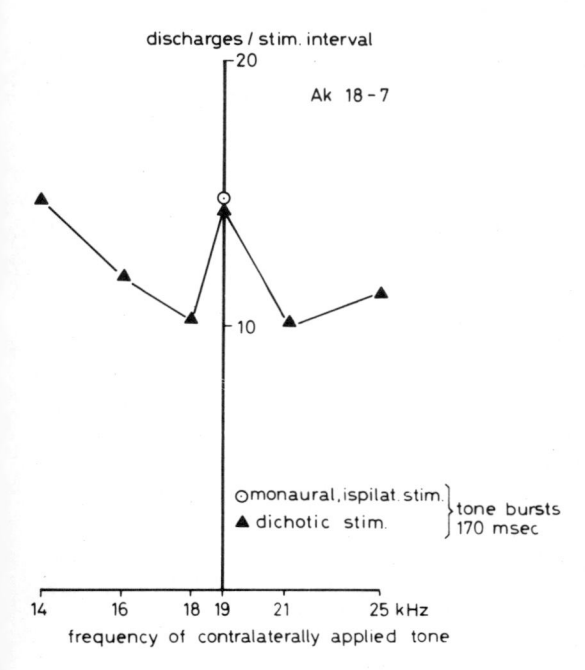

Fig. 1. Number of discharges per stimulus interval as a function of the frequency of the contralaterally applied tone. Each point represents an average of 100 responses. The ipsilateral ear was stimulated with 19 kHz, the characteristic frequency of the neuron and, with approximately 40 dB SPL, constant throughout the experimental procedure. When present, the SPL of the contralateral tone was also 40 dB.

from secondary neurons, the observed effect can be due to efferent influence upon the receptor cells, or to efferent influence upon the secondary neurons, or to a synergistic influence upon both stations. This question cannot be answered at present. We are inclined to think that the efferent innervation of the cochlear nucleus is at least important for the effect described here, because Dunker et al. (1964) described frequency dependent inhibition of spontaneous activity of the de-afferented cochlear nucleus, when the intact contralateral ear was stimulated by sound.

Experiments similar to those described above are in progress, in which recordings are being made from single fibres of the auditory nerve. Preliminary results show that, in primary neurons, a similarly strong inhibition does not exist, but further investigations are necessary.

The W-shaped curve of frequency dependent inhibition suggests that the afferent mechanism may sharpen frequency discrimination, because the curve is similar to those found with the visual system in which lateral inhibition is known to sharpen visual acuity. Thus frequency discrimination might not only be enhanced by monaural mechanisms such as "two-tone inhibition" as described by Sachs and Kiang (1968) and others, but also by a binaural mechanism as described here. This seems to be reasonable from the functional point of view because in the natural environment both ears receive practically the same acoustic signal with only slight differences in intensity and phase.

The functional significance of the efferent innervation for frequency discrimination was elucidated by Capps and Ades (1968) in squirrel monkeys. They demonstrated a significant decrease of frequency discrimination, following sectioning of the crossed olivo-cochlear bundle. In contrast to our findings and those of the authors just mentioned, psychophysical experiments do not yield an indication for an improvement of frequency discrimination in binaural listening compared with monaural application of the stimulus.

This work was supported by grants of the Deutsche Forschungsgemeinschaft.

REFERENCES

Boerger, G., Gruber, J., and Klinke, R. (1968a): Interaurales Übersprechen bei der Katze, Naturwissenschaften 55, 234.
Boerger, G., Gruber, J., and Klinke, R. (1968b): Zulässiger Schalldruck bei binauraler Reizung ohne überschwelliges interaurales Übersprechen, Experientia 24, 1223-1224.
Capps, M. J., and Ades, H. W. (1968): Auditory frequency discrimination after transsection of the olivo-cochlear bundle in squirrel monkeys, Exp. Neurol. 21, 147-158.
Chamberlain, S. C., Moxon, E. C., and Wiederhold, M. L. (1968): Efferent inhibition of electrically stimulated response in cat auditory-nerve fibers, M.I.T. Research Laboratory of Electronics, Quart. Progr. Rep. 90, 266-270.
Desmedt, J. E. (1962): Auditory-evoked potentials from cochlea to cortex as influenced by activation of the efferent olivo-cochlear bundle, J. Acoust. Soc. Amer. 34, 1478-1496.
Desmedt, J. E., and La Grutta, V. (1963): Function of the uncrossed efferent olivo-cochlear fibres in the cat, Nature 200, 472-474.

Dunker, E., and Grubel, G. (1965): Zur Funktionsweise des efferenten auditorischen Systems. II. Änderung von Variabilität der Spike-Intervalle und Hemmungs-Empfindlichkeit deafferentierter Cochlearis-Neurone nach Durchschneidung ventral im Brückenhirn kreuzender Bahnen (Hund), Pflügers Arch. ges. Physiol. *283*, 270-284.

Dunker, E., and Wachsmuth, D. (1965): Zur Funktionsweise des efferenten auditorischen Systems. III. Analyse der Fluktuation von Spike-Intervallen deafferentierter Einzelneurone des Nucleus cochlearis dorsalis bei hemmend und fördernd beeinflußter Aktivität, Pflügers Arch. ges. Physiol. *284*, 347-359.

Dunker, E., Grubel, G., and Pfalz, R. (1964): Beeinflussung von spontanaktiven, deafferentierten Einzelneuronen des Nucleus cochlearis der Katze durch Tonreizung der Gegenseite, Pflügers Arch. ges. Physiol. *278*, 610-623.

Fex, J. (1959): Augmentation of cochlear microphonics by stimulation of efferent fibres to the cochlea, Acta Oto-Laryngol. *50*, 540-541.

Fex, J. (1962): Auditory activity in centrifugal and centripetal cochlear fibres in cat, Acta Physiol. Scand. *55*, Suppl. 189, 5-68.

Fex, J. (1965): Auditory activity in uncrossed centrifugal cochlear fibres in cat. A study of a feedback system, II, Acta Physiol. Scand. *64*, 43-57.

Fex, J. (1967): Calcium action at an inhibitory synapse, Nature *213*, 1233-1234.

Fex, J. (1968): Efferent inhibition in the cochlea by the olivo-cochlear bundle, in: Hearing Mechanisms in Vertebrates, A. V. S. de Reuck and J. Knight, Eds. (Churchill, London), pp. 169-186.

Galambos, R. (1956): Suppression of auditory nerve activity by stimulation of efferent fibers to cochlea, J. Neurophysiol. *19*, 424-437.

Iurato, S. (1967): Submicroscopic Structure of the Inner Ear (Pergamon Press, Oxford).

Pfalz, R. (1962): Einfluß schallgereizter efferenter Hörbahnteile auf den de-afferentierten Nucleus cochlearis (Meerschweinchen), Pflügers Arch. ges. Physiol. *274*, 533-552.

Rasmussen, G. L. (1946): The olivary peduncle and other fiber projections of the superior olivary complex, J. Comp. Neurol. *84*, 141-219.

Rasmussen, G. L. (1953): Further observations of the efferent cochlear bundle, J. Comp. Neurol. *99*, 61-74.

Rasmussen, G. L. (1960): Efferent fibers of the cochlear nerve and cochlear nucleus, in: Neural Mechanisms of the Auditory and Vestibular Systems, G. L. Rasmussen and W. F. Windle, Eds. (Charles C. Thomas, Springfield, Ill.), pp. 105-115.

Rossi, G., and Cortesina, G. (1962): The efferent innervation of the inner ear, Panminerva Med. *4*, 478-500.

Sachs, M. B., and Kiang, N. Y.-S. (1968): Two-tone inhibition in auditory-nerve fibers, J. Acoust. Soc. Amer. *43*, 1120-1128.

Sohmer, H. (1965): The effect of contralateral olivo-cochlear bundle stimulation on the cochlear potentials evoked by acoustic stimuli of various frequencies and intensities, Acta Oto-Laryngol. *60*, 59-70.

Sohmer, H. (1966): A comparison of the efferent effects of the homolateral and contralateral olivo-cochlear bundles, Acta Oto-Laryngol. *62*, 74-87.

Wiederhold, M. L., and Peake, W. T. (1965): Dependence of efferent inhibition of auditory nerve responses on intensity of acoustic stimuli. M.I.T. Res. Laboratory of Electronics, Quart. Progr. Rep. *77*, 347-353.

Wiederhold, M. L., and Peake, W. T. (1966): Efferent inhibition of auditory-nerve responses: dependence on acoustic-stimulus parameters, J. Acoust. Soc. Amer. *40*, 1427-1430.

DISCUSSION

Keidel: Do you think, with respect to latency times of the order of more than 60 msec as found by Pfalz (1962a, 1962b, 1962c) in our laboratory, that the inhibitory activity comes from a lower level of the auditory pathway?

Klinke: I do not know. Actually, the latency ranges between 30 and 50 msec and the inhibition lasts 50 to 100 msec beyond the cessation of the contralateral stimulus.

Keidel: Do you think that latency times of the order of 30 msec may have to do with inhibition effects from low levels?

Klinke: The latency of 30 to 50 msec corresponds to findings of Fex (1962). Because of the long latency higher levels might be involved.

Rose: Where were these neurons located in the cochlear nucleus?

Klinke: We have done some histological control, according to this we had neurons from both the ventral and the dorsal cochlear nucleus.

Rose: What were the relative SPL's at both ears? Was the inhibition effect visible for contralateral SPL's from threshold on as you went up in level?

Klinke: The stimuli were both about 35 to 40 dB above threshold. The inhibition starts at contralateral levels of about 20 to 30 dB above threshold. So only 10 to 20 dB are left to determine the intensity function, which is not enough.

Scharf: Chocholle and Saulnier (1962, 1964) measured frequency discrimination in one ear in the presence of a sound in the other ear. Discrimination was poorer when the contralateral sound was a pure tone of the same frequency as in the measured ear or was a white noise; discrimination was unaffected by a contralateral tone very different in frequency.

Johnstone: Dr. Klinke, you think that this contralateral inhibition may be a sharpening mechanism for pitch perception. Before an additional filter can sharpen a signal, it obviously has to be as sharp if not sharper than what is coming in. From the rough sketch of the notch of your W-shaped curve, it seems to me that this is not the case. Can you give us the sharpness of the tuning curve without inhibition to be able to compare this with the frequency selectivity of the inhibition.

Klinke: We did not measure the complete tuning curve without inhibition. Our primary interest was directed to the contralateral influence.

Zwislocki: I do not think that Dr. Johnstone's remark holds generally. A broad inhibiting activity may still suppress part of a localized excitation producing an even sharper localization. This occurs in vision.

REFERENCES

Chocholle, R., and Saulnier, C. (1962): Le seuil différentiel de fréquence: l'effet de la présence simultanée d'un son ou d'un bruit blanc sur l'oreille opposée, Journal de Physiologie, 54, 315-316.

Chocholle, R., and Saulnier, C. (1964): Les seuils différentiels de fréquence en présence d'un son sur l'oreille contralatérale, Acustica 14, 35-44.

Fex, J. (1962): Auditory activity in centrifugal and centripetal cochlear fibres in cat, Acta Physiol. Scand. 55, Suppl. 189, 5-68.

Pfalz, R. (1962a): Einfluss schallgereizter efferenter Hörbahnteile auf den de-afferentierten

Nucleus cochlearis (Meerschweinchen), Pflügers Arch. ges. Physiol. *274*, 533-552.
Pfalz, R. (1962b): Erregung efferenter Einzelneurone innerhalb der Cochlea durch kontralaterale Beschallung mit Klickfolgen (Meerschweinchen), Z. Biol. *113*, 215.
Pfalz, R. (1962c): Centrifugal inhibition of afferent secondary neurons in the cochlear nucleus by sound, J. Acoust. Soc. Amer. *34*, 1472-1477.

TWO DIFFERENT TYPES OF FREQUENCY SELECTIVE NEURONS IN THE COCHLEAR NUCLEUS OF THE RAT

AAGE R. MØLLER

Department of Physiology
Karolinska Institutet
Stockholm, Sweden

Recordings from single units in the cochlear nucleus of the rat reveal that some neurons differ radically in their response pattern from others. Most units will respond to continuous stimulation by tones or noise with a sustained train of nervous discharges. The response of these units has been described earlier (Møller, 1969a) in detailed plots of their response areas as a function of frequency and SPL (tuning curves). Thus we have a concept of the spectral selectivity of these units. A few units will, however, respond to tone bursts with only a single discharge evoked immediately after the onset of the tone. Recording from these "transient" units further reveals that they respond to repetitive click sounds of low repetition rate with a train of discharges. One discharge is evoked by each click. However, these units will suddenly cease firing when the click repetition rate is increased above a certain value. When clicks of high repetition rate are presented in bursts, each burst evokes only a single immediate discharge. The difference in response pattern of these transient units and ordinary units is illustrated in Fig. 1 which shows a dot display of the discharges evoked by bursts of clicks with different repetition rates. While the response of the ordinary unit fuses above a certain click rate, the response of the transient unit shows perfect synchrony with respect to the individual clicks up to the critical rate where the unit fails to respond. This specific click rate varies among the transient units (200 to 800 pps). Fig. 2 shows in a more quantitative way the difference between these types of units in response to click sounds. The graphs show spike counts as a function of repetition rate of clicks presented in 50 msec bursts. The upper two graphs are from a transient unit and the lower graph from an ordinary unit. The dashed line in Fig. 2a gives a one-to-one relationship between stimulus and response. Similar results were obtained when the sounds

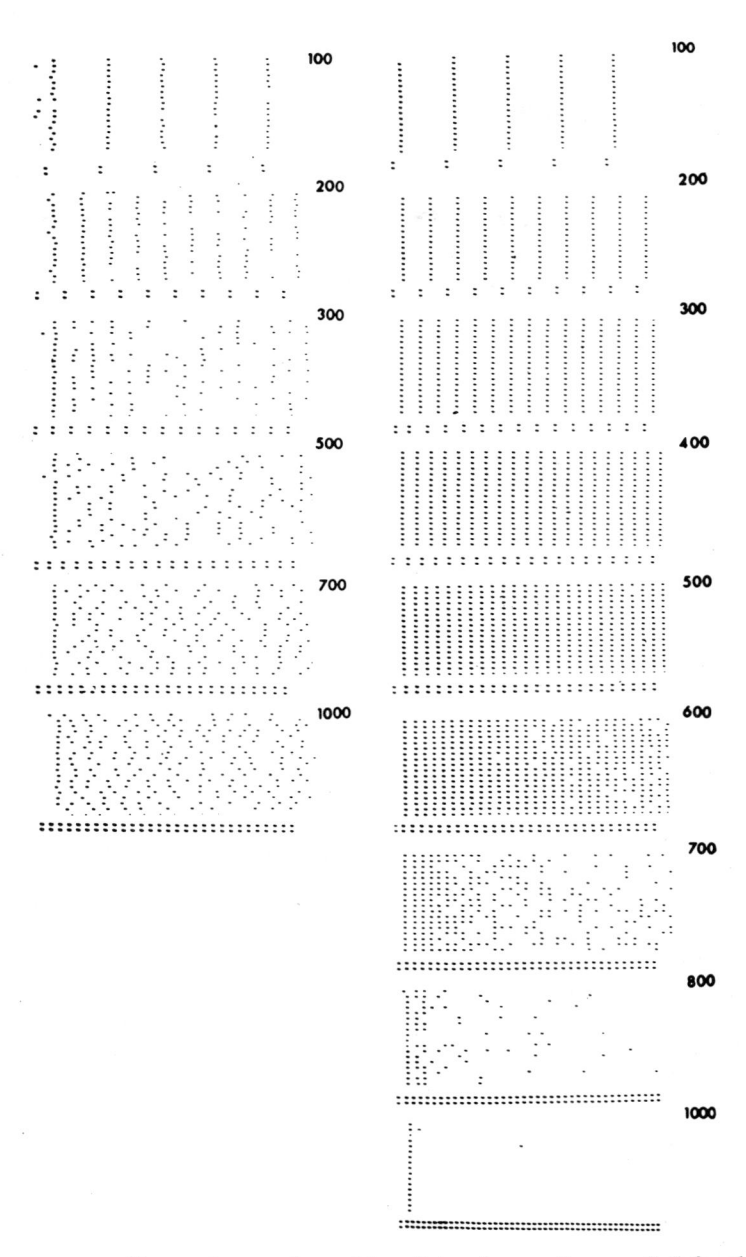

Fig. 1. Responses to 50 msec bursts of repetitive clicks of an ordinary unit (left column) and a transient unit (right column). Each dot shows the occurrence of a nervous discharge. The repetition rate in clicks per sec is indicated by the inserted numbers and the individual clicks are shown by the double dots below each recording sequence. Click duration was 30 μsec.

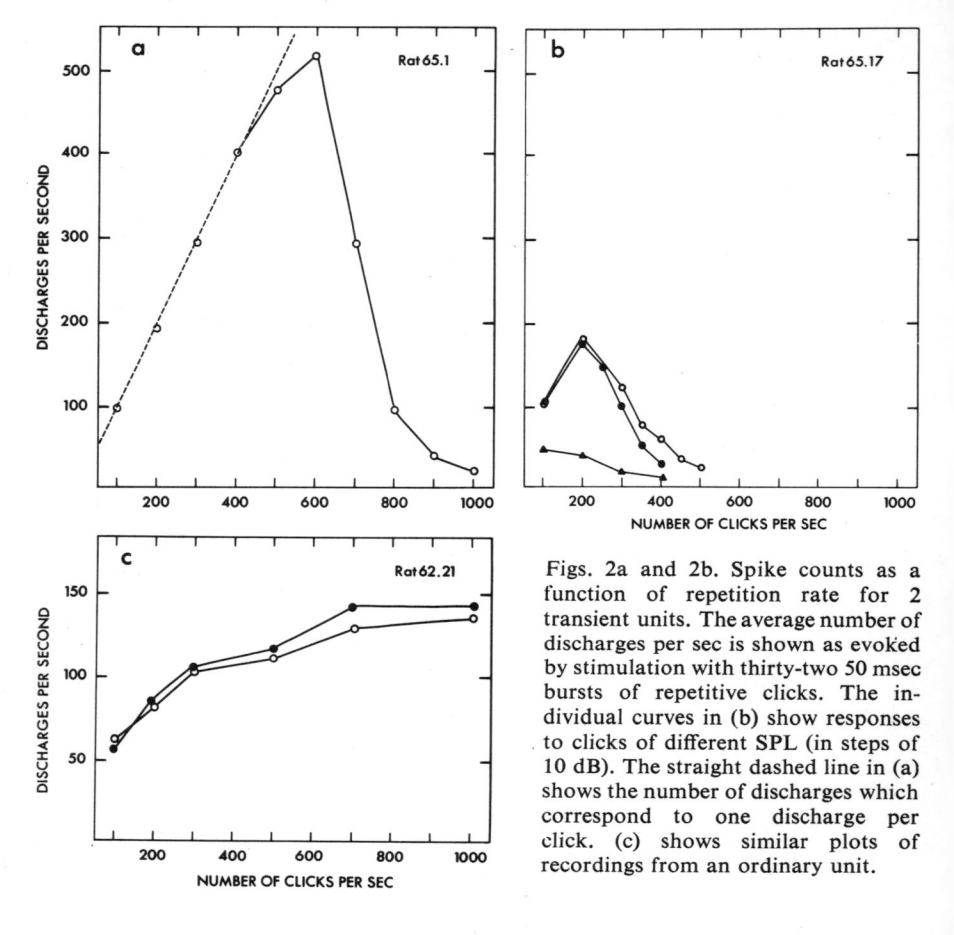

Figs. 2a and 2b. Spike counts as a function of repetition rate for 2 transient units. The average number of discharges per sec is shown as evoked by stimulation with thirty-two 50 msec bursts of repetitive clicks. The individual curves in (b) show responses to clicks of different SPL (in steps of 10 dB). The straight dashed line in (a) shows the number of discharges which correspond to one discharge per click. (c) shows similar plots of recordings from an ordinary unit.

were presented as a continuous train of clicks instead of presenting the clicks in bursts (Møller, 1969b).

Further study of these transient units revealed that they were not selective to repetition rate *per se* but that they had a specificity to the duration of silence between two sounds. A sound which followed another sound could evoke a second response only if there had been a sufficiently long period of silence between the two sounds. This is illustrated in Fig. 3a which shows the response by a transient unit to short tone bursts presented repetitively in the same way as the click sounds. Different durations of sound were used. It is seen that the repetition rate at which the unit's firing rate went down depends on the duration of the bursts. The data of Fig. 3a are replotted in Fig. 3b to show the number of discharges evoked by each sound burst as a function of the silent interval between the sounds. The response is seen to be independent of the duration of the single bursts and the probability of firing then depends only on the duration

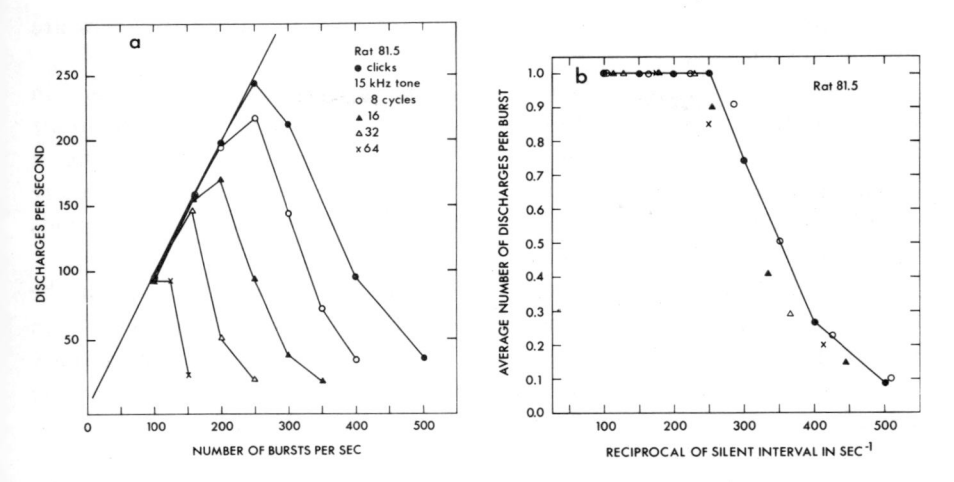

Fig. 3. (a) Responses to tone bursts (15 kHz) of various durations compared with those of clicks. The sounds were presented in the same way as in Fig. 2 and average discharge frequency is shown as a function of repetition rate.
(b) The same data as in (a) but the average number of discharges per sound is shown as a function of the reciprocal of the length of the silent interval between the individual sounds. The same symbols as in (a) are used. The data for click stimulation are connected by straight lines.

of silence between the sounds. Thus it is shown that a minimum silent interval is required, being about 4 msec in the particular unit in Fig. 3. In addition to this feature these units showed a broad spectral sensitivity. Compared to tuning curves of ordinary units, the response of transient units to pure tones indicates that the response area is much broader than that of the normal tuning curve, see Fig. 4, and the threshold is rather high. The range of best frequency (7-15 kHz) is far above the unit's critical repetition rate for clicks. Similar "tuning curves" were obtained by using band-pass filtered clicks in which the threshold was designated as a function of the centre frequency of the band-pass filter.

The responses of transient units to clicks can be completely inhibited by noise and by pure tones in certain frequency ranges. More specifically, pure

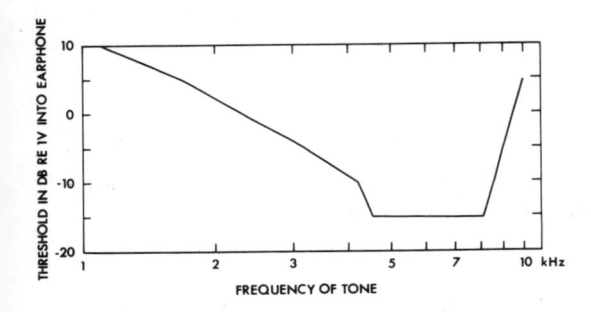

Fig. 4. Response area of a transient unit to stimulation with tone bursts as a function of tone frequency.

tones with a frequency immediately below the best excitatory spectrum are usually most effective.

Low-frequency sinusoids (below 1000 Hz) could not elicit any response in these units nor could a response be elicited by tones which were amplitude modulated with a low-frequency sine wave. In connection with these findings it should be noted that the maximal firing rate of some of these units was more than 700 discharges per sec while units which responded to sustained sounds never discharged more than 450 per sec.

These transient units are probably similar to or the same as those found in the cat cochlear nucleus and described by Kiang (1965) and Pfeiffer (1966). These authors named the neurons "on units". As they did not systematically study the response, they did not observe the specificity of discharge to the repetition rate of transient sounds (or rather to the interval between individual sounds).

Observations on the response pattern of rat transient units are valuable since they may elucidate the mechanisms underlying our ability to discriminate the pitch of certain sounds. Numerous psychoacoustic investigations indicate that pitch discrimination of some certain sounds cannot be explained by spectral analysis. Thus, for instance, removal of the fundamental frequency of a train of repetitive clicks does not change the judged pitch of the sound (Schouten, 1960; see also p. 41 of this volume). Furthermore, recent psychoacoustic investigations have shown that the pitch of periodic clicks of a certain duration is, under some circumstances, related to the duration of silence between these pulses rather than the periodicity of the pulses (McClellan and Small, 1965; Rosenberg, 1965; Zwislocki, 1967). Although the duration of silence may have some weak spectral correlates, other factors make it difficult to explain the phenomenon on the basis of a spectral analysis similar to that performed by the basilar membrane.

Of additional interest to our findings is that this pitch phenomenon is present only at low repetition rates. In this regard, the rat transient units display a response pattern similar to the psychoacoustic results. The repetition rate selectivity of rat transient units is not a result of spectral analysis but is due to a time-pattern analysis. This statement is supported by the fact that reversing the polarity of every second click in a train consisting of clicks of the same polarity did not change the response pattern. The precise timing of the transient units (Fig. 1) makes it further plausible that these units play an active role in the perception of the periodicity of sounds.

The work reported in this paper was supported by grants from the Swedish Medical Research Council (Grant B68-14X-90-03), Therese and Johan Anderssons Minne and from Magnus Bergvalls Stiftelse.

REFERENCES

Kiang, N. Y.-S. (1965): Stimulus coding in the auditory nerve and cochlear nucleus, Acta Oto-Laryngol. *59*, 186-200.
McClellan, M. E., and Small, A. M. (1965): Time-separation pitch associated with correlated noise bursts, J. Acoust. Soc. Amer. *38*, 142-143.
Møller, A. R. (1969a): Unit responses in the cochlear nucleus of the rat to pure tones, Acta Physiol. Scand. *75*, 530-541.
Møller, A. R. (1969b): Unit responses in the rat cochlear nucleus to repetitive, transient sounds, Acta Physiol. Scand. *75*, 542-551.
Pfeiffer, R. R. (1966): Classification of response patterns of spike discharges for units in the cochlear nucleus: tone-burst stimulation, Exp. Brain Res. *1*, 220-235.
Rosenberg, A. E. (1965): Effect of masking on the pitch of periodic pulses, J. Acoust. Soc. Amer. *38*, 747-758.
Schouten, J. F. (1960): Five Articles on the Perception of Sound (1938-1940), Instituut voor Perceptie Onderzoek, Eindhoven.
Zwislocki, J. J. (1967): Audition, Ann. Rev. Psychol. *18*, 407-436.

DISCUSSION

Goldstein: When you were looking in the cochlear nucleus, what proportion of the units that you picked up had the transient characteristic?

Møller: That is difficult to say. I made no statistics about the percentage of these units, but it was not difficult to find them. I first found them in the dorsal cochlear nucleus and, looking more carefully for them, I found these units also in the ventral cochlear nucleus. Of course, I could determine the percentage from my note books but I don't think this would be very relevant because I am afraid that a particular microelectrode is rather selective with regard to a certain type of neuron. All that I can say is that of some 300 units I had about 60 of the transient type.

Whitfield: We have seen something similar to this with frequency modulation of a pure tone. Changing the frequency in one direction resulted in a response and it did not matter how fast the frequency varied. On the other hand, a sinusoidal continuous frequency modulation did not elicit responses for modulation frequencies above about 20 Hz. As you said, one gets an initial response and then nothing. This result is not due to the faster changing frequency but to the repetition itself. We have postulated that the mechanism is an inhibitory one and that one has, as it were, to wait for the decay of inhibition.

I was wondering if you have anything comparable in your situation. You used clicks which presumably stimulate all fibres successively down the basilar membrane. Have you also tried using shaped tone pulses without abrupt onset and decay?

Møller: I have used band pass filtered clicks that look like damped oscillations, similar to what I think you have in mind. At low repetition rates the units will respond to this type of sound but they fail at high rates. The critical rate depends on the bandwidth of the filter, and thus, probably, on the steepness of the onset. If I diminish the bandwidth, then this critical rate for which the responses fail goes down.

I have also used other types of signals, for instance, tones of which the frequency was swept rapidly in a triangular pattern. The transient neurons responded very nicely if the frequency variation was above a certain rate. Obviously, the excitation of the nervous system has to be fast enough to give a response.

Wilson: You point out that these transient units are extremely precise in their temporal characteristics. Have you any quantitative measure of this? I noticed that they occupy only 2 bins in your post stimulus time histogram.

Møller: That's right. A post stimulus time histogram of the responses of these units to click sounds shows that the time jitter of the discharges may be less than 200 μsec. I have not made much of an effort to look for this precision because the method I used is not well suited for doing that. One should take certain things into consideration, for instance, the way in which the nerve discharges are analyzed. In my case, electric pulses are generated when the amplitude of the nerve spikes exceeded a certain value. These pulses are then displayed in the dot displays or they are fed into a histogram analyzer. The influence of any variation of the base line is reduced by high pass filtering of the spikes with a cut-off frequency of about 1000 Hz. However, one can not avoid a certain amount of jitter due to small changes in amplitude. One must be very careful with things like that, therefore I have not given any figure of the jitter.

Scharf: Do you have any notion about how silent the intervals between the sounds have to be to elicit the response?

Møller: No, it would be very interesting to know. I have to try that.

Section 4

PERIPHERAL NERVE-FIBRE DISCHARGES

Studies of the single-fibre discharges in the peripheral auditory system have contributed much to a better understanding of the first stages of signal processing. It has been shown that the single fibres carry information both about the frequency and the time structure of the stimulus. Frequency coding is revealed in the tuning curves and response areas of single fibres; temporal coding is observed in the synchrony between neural discharges and the time structure of the stimulus. This latter feature receives special attention in this section.

DISCHARGES OF SINGLE FIBERS IN THE MAMMALIAN AUDITORY NERVE

J. E. ROSE

Laboratory of Neurophysiology
University of Wisconsin
Madison, Wisconsin, U.S.A.

I shall consider here the available evidence that bears on the question how the discharges in the auditory nerve fibers are aroused and timed. I shall start with the findings pertaining to the response areas.

RESPONSE AREAS OF THE AUDITORY FIBERS

All workers are agreed that the frequency response of an auditory neuron is, within limits, a function of stimulus strength and, hence, the frequency-intensity domain that activates the neuron is traditionally called its *response area*. The original observations to that effect were made by Galambos and Davis (1943) on the neurons of the cochlear complex, but similar findings have been obtained for other synaptic regions of the auditory system and for the auditory nerve fibers as well. Some generalizations seem in order. There is always a sound pressure level at which only a very narrow band of frequencies activates the neuron; the center point of such a band is called its *best*, or its *characteristic*, frequency. While the best frequency is constant for a given neuron, apparently any frequency in the audible spectrum can be the best frequency, with the probable exception of those that are quite close to the limits of the spectrum.

As the sound pressure level of a tonal stimulus is raised, the effective frequency band broadens (Fig. 1). For fairly intense tones the effective band may constitute a substantial segment of the audible spectrum. The widening of the effective band under these conditions is asymmetrical in respect to the best frequency if the latter is higher than about 2 kHz. There is always only a limited spread towards higher frequencies but the spread towards lower frequencies may be great. Neurons differ greatly both in the threshold for the best frequency and in the width of the frequency band which may activate them at comparable sound pressure levels.

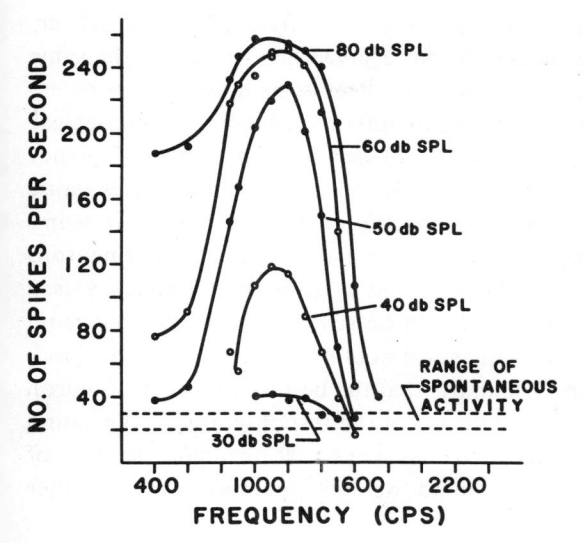

Fig. 1. Response area for unit 65-48-1 showing the number of spikes produced at different sound pressure levels. Dashed lines indicate the range of spontaneous activity. Here, and in all other figures, the frequency is given in cycles/sec (cps). Interspike interval distributions for this fiber for stimuli at 80 dB SPL are shown in Fig. 2 (from Rose et al., 1967).

The functional significance of the response areas has been the subject o many conjectures. Katsuki et al. (1958) believed that the threshold tuning curves become (at least up to some levels) progressively sharper as one examines the successive regions in the auditory synaptic chain. However, further work (Kiang et al., 1965, Chapter 7; Rose et al., 1967) does not, I believe, sustain this view. There is no good evidence that the response areas of the auditory nerve fibers differ in any obvious way from those in the central stations. Any sharpening of the response area would thus have to take place in the cochlea. Despite diligent search, there is, I think, no evidence for, and considerable evidence against, the notion that lateral inhibition in the cochlea plays a role in the determination of the response areas such as are obtained in an anesthetized animal. For the present, the inference is that the response area of an auditory nerve fiber essentially results from excitation caused by mechanical motions in the cochlea. I may add that, when two tones interact, a drastic reduction in spike count which often occurs (as compared with that produced by one of the component tones acting alone) is also almost certainly due to mechanical events rather than, as commonly believed, to a genuine neural inhibition (Katsuki et al., 1959; Nomoto et al., 1964; Kiang et al., 1965, Chapter 10; Sachs and Kiang, 1968).

It has long been assumed, though rather rarely explicitly stated, that the best frequency of a fiber reflects and, in a sense, is given by its connections with only a limited segment of the cochlear partition. An indirect support for such a notion has been offered by Kiang et al. (1965, Chapters 4 and 5). They have shown that responses of a fiber to clicks tend to group around several modal values if the best frequency of the fiber is low. The latent period to the first modal peak is, within limits, longer for lower best frequencies. Moreover, separations between the modal peaks are, for a given neuron, about the same and equal to the period of the best frequency. Clearly, such results can be expected if the individual fibers are connected to different segments of the cochlear partition, and if they respond to the vibrations of such segments with a frequency that is maximally effective at this segment. Since different frequencies cause maximal elevations at different points along the cochlear partition, it seems obvious that the best frequency is always determinable, that different neurons have different best frequencies and, finally, that a given best frequency is a constant attribute of a given fiber. It should be stressed, however, that from such an understanding of the best frequency it does not necessarily follow (and, in fact, it is not always true experimentally) that the best frequency of a neuron always remains the most effective frequency in the response area if the sound pressure level of the stimulus is increased. Assuming that our understanding of the best frequency is correct, how should we interpret responses to other frequencies when the stimulus intensity is raised?

TIME STRUCTURE OF LOW-FREQUENCY RESPONSES

The answer to this question is unequivocal when low-frequency tones are employed (Rose et al., 1967, 1968). Typical results are shown in Fig. 2. The discharges are always locked to the cycles and occur at intervals which group around the integral multiples of the period of the stimulating frequency. For the time structure of the discharges it matters not at all what the best frequency of the neuron may be; the sound pressure level of the stimulus is also irrelevant as long as the stimulus is above threshold. For a given neuron, therefore, it is the sound pressure level which determines what frequency band will be effective; in this sense, and only in this sense, one can speak of the frequency specificity of a given fiber. It is the position of a given frequency within the response area which determines the discharge rate at a given sound pressure level; and it is solely the cadence of the stimulating cycles which governs the time structure of the responses.

The observation that in response to a low frequency, the discharges may be phase-locked, has been made in the past by many workers (Galambos and Davis, 1943; Tasaki, 1954; Katsuki et al., 1962; Rupert et al., 1963; Nomoto et al., 1964; Kiang et al., 1965, Chapter 6). It is only recently, however, that this

Fig. 2. Neuron 65-48-1. Periodic distributions of interspike intervals when a pure tone of variable frequency activated the fiber. Stimulus frequency is indicated in each graph. Intensity of all stimuli: 80 dB SPL. Tone duration: 1 sec. Responses to 10 stimuli constitute the sample upon which each histogram is based. Abscissa: time in msec; each bin = 100 μ sec. Dots below abscissa indicate integral values of the period of each frequency employed. N = number of interspike intervals in the sample. N is given as two numbers. The first indicates the number of plotted intervals; the second is the number of intervals whose values exceeded the maximum value of the abscissa (from Rose et al., 1967).

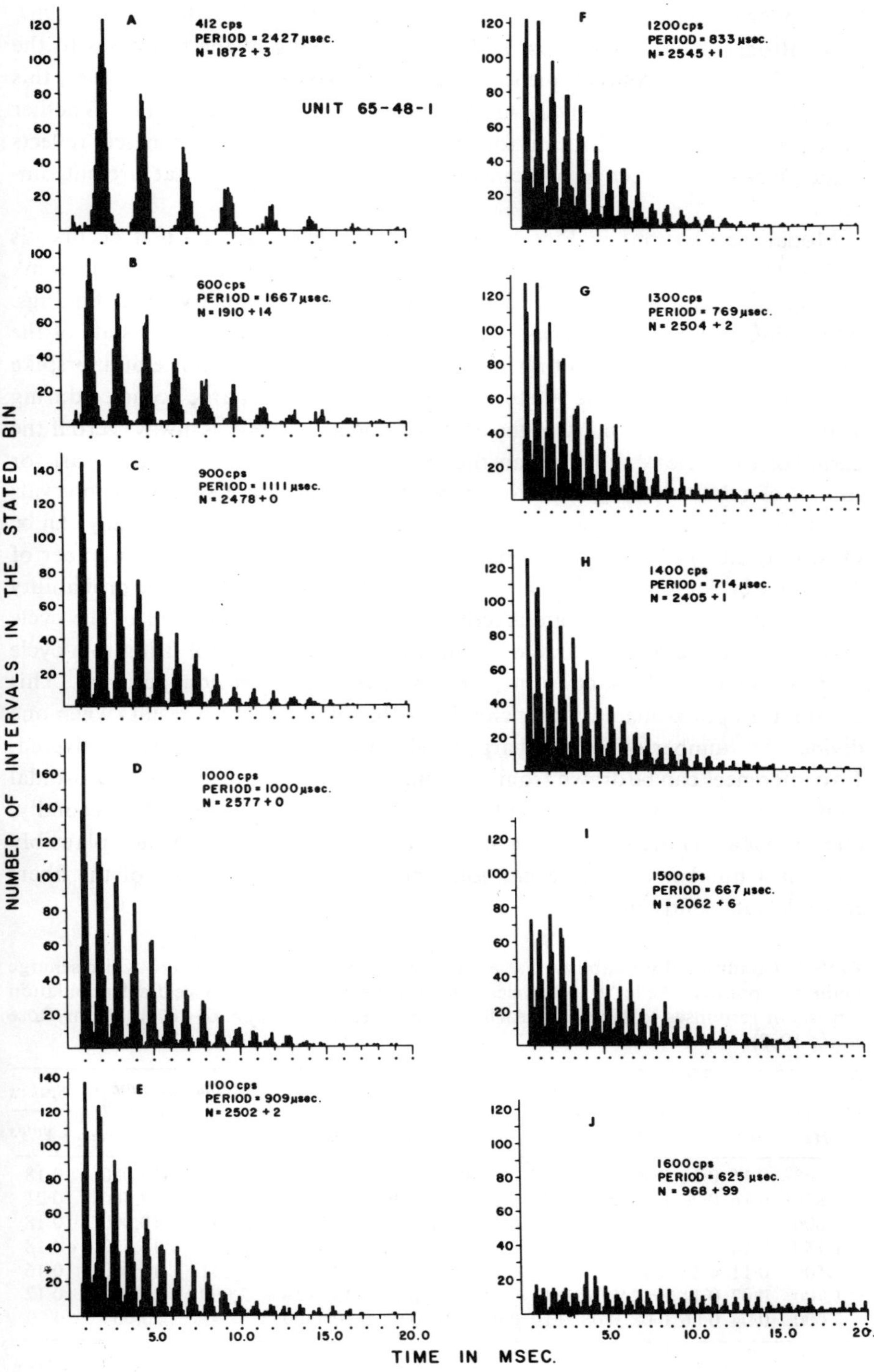
A
412 cps
PERIOD = 2427 μsec.
N = 1872 + 3
UNIT 65-48-1
B
600 cps
PERIOD = 1667 μsec.
N = 1910 + 14
C
900 cps
PERIOD = 1111 μsec.
N = 2478 + 0
D
1000 cps
PERIOD = 1000 μsec.
N = 2577 + 0
E
1100 cps
PERIOD = 909 μsec.
N = 2502 + 2
F
1200 cps
PERIOD = 833 μsec.
N = 2545 + 1
G
1300 cps
PERIOD = 769 μsec.
N = 2504 + 2
H
1400 cps
PERIOD = 714 μsec.
N = 2405 + 1
I
1500 cps
PERIOD = 667 μsec.
N = 2062 + 6
J
1600 cps
PERIOD = 625 μsec.
N = 968 + 99
NUMBER OF INTERVALS IN THE STATED BIN
TIME IN MSEC.
5.0
10.0
15.0
20.0

problem has been examined in some detail (Hind et al., 1967; Rose et al., 1967, 1968; Brugge et al., 1969; Rose et al., 1969). Locking of the responses to the cycles of the stimulating frequency occurs up to about 5000 Hz. However, this relation becomes progressively more blurred above some 2500 Hz. Whether the observed desynchronization for higher frequencies is real or merely reflects inadequacies in correct measurements of spike occurrence is, at present, uncertain.

Modal distributions in Fig. 2 indicate that, while a spike – if it occurs – is locked to the cycle, not every cycle is an effective stimulus. One could think that the refractoriness of the fiber is the primary reason for such findings. Interestingly, this is not the case (Rose et al., 1967). The refractory state of the fiber obviously can be a limiting factor and prevent the occurrence of interspike intervals shorter than some 500 to 700 μsec. Thus, if a discharge occurred during a given cycle, the probability of the next cycle being effective is always zero if the period of the cycle is shorter than the refractory period. However, at least for stimuli of relatively long duration, most of the spikes are spaced at intervals which are very much longer than the refractory period. In fact, it usually can be shown (Table I) that the conditional probability of firing after any number of ineffective cycles is nearly the same, and only slightly higher than, the probability of firing immediately after an effective cycle (provided the period of the cycle exceeds the value of the refractory period). The probability of firing to a cycle is, thus, very nearly a constant for any given stimulus configuration. This constant is approximated rather closely by the quotient which results when one divides the number of the discharged spikes by the number of cycles delivered. Two inferences can be drawn from such findings. First, one can treat a sinusoidal stimulus as consisting effectively of as many stimuli as there are cycles; secondly, one can view the occurrence of a spike as a probabilistic event which plausibly arises if a number of local excitations (perhaps in the terminals of the fiber) coincide sufficiently in time.

Table I. Conditional probability that a unit which discharged at time zero will discharge again in response to the 1st to 10th cycles following the time-zero discharge. Each population consists of responses to 10 stimuli each lasting 1 second (neuron 65-107-1) (from Rose et al., 1968).

Frequency Hz	Cycle										Number of spikes discharged	Number of cycles delivered	Spikes/Cycles
	1st	2nd	3rd	4th	5th	6th	7th	8th	9th	10th			
408	0·16	0·16	0·19	0·20	0·17	0·19	0·18	0·18	0·25	0·25	746	4,080	0·18
850	0·16	0·20	0·22	0·23	0·21	0·22	0·23	0·30	0·22	0·26	1,798	8,500	0·21
1,000	0·16	0·16	0·18	0·20	0·20	0·19	0·18	0·24	0·18	0·18	1,832	10,000	0·18
1,150	0·14	0·16	0·18	0·18	0·21	0·20	0·20	0·21	0·21	0·19	2,077	11,500	0·18
1,500	0·11	0·15	0·15	0·17	0·16	0·17	0·17	0·19	0·19	0·15	2,351	15,000	0·16
1,700	0·07	0·11	0·12	0·14	0·13	0·14	0·13	0·12	0·13	0·16	2,099	17,000	0·12
2,000	0·04	0·06	0·08	0·08	0·10	0·09	0·10	0·09	0·10	0·09	1,791	20,000	0·09

Despite some reports to the contrary (Rupert et al., 1963), it seems right to believe that responses to low-frequency stimuli are always locked to the cycles of the stimulating tone. What, then, is the mechanism which causes phase-locking and what functional significance should be attached to it?

The assumption that auditory nerve fibers are excited only by unidirectional movements of the cochlear partition would, in a simple way, account for phase-locking because only half of the periodic waveform could then be an effective stimulus. Actually, such an assumption has been made in the past for other reasons (Davis et al., 1950). The significance of the time structure of the discharges is a matter for conjecture. It is clear from Fig. 2 that the time structure can be viewed as a code which could be used by an observer to determine the frequency of the stimulating tone. The question is, is this code actually utilized by the central nervous system? I shall assume that it is, and shall consider some experimental data which seem consistent with that idea.

THE STIMULATING WAVEFORM

Let us first specify what such a time code implies. If the spikes are discharged according to successive unidirectional elevations of the cochlear partition, and if the low-frequency information is extracted from spacings between the spikes, then spacings between the unidirectional elevations of the stimulating waveform must be critical for the frequency information which is being delivered to the central nervous system.

In the simplest case of a single pure tone all the peaks of the waveform are equidistant and are of equal amplitude. Hence, all interspike intervals can be expected to group around the integral multiples of the stimulus period and the frequency information transmitted would pertain to the frequency of the stimulating tone.

When two pure tones, not harmonically related, are sounded together and no constant phase relation between them exists, then the time structure of the responses has been shown to be in one of three modes (Hind et al., 1967). The responses are locked to the cycles of: (1) the first primary, or (2) the second primary, or (3) both primaries. Which mode prevails is a function of the respective strength of the primary tones. Figs. 3A and 3B show interspike interval distributions generated by discharges of a fiber to two non-harmonically related tones. For Fig. 3A the relation between the sound pressure levels was such that the responses were locked essentially only to the cycles of the lower primary, even though the higher primary tone was highly effective when presented alone (Fig. 3C). Thus, the higher primary can be said to have been masked.

With an attenuation of the lower primary by 10 dB, a quite irregular distribution results (Fig. 3B). A closer inspection reveals that this pattern appears

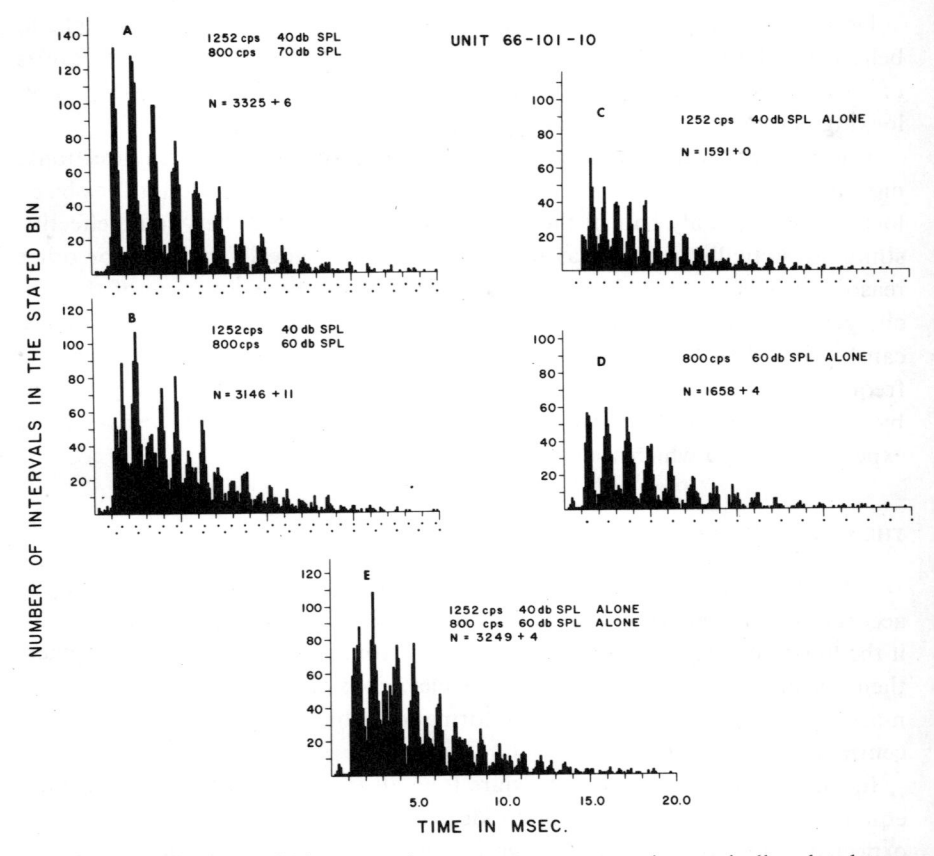

Fig. 3. Interspike interval histograms in response to two non-harmonically related tones presented together (A and B) and separately (C and D). Plot E is a bin-by-bin sum of histograms C and D. All stimuli were 8 sec in duration. N in A and B = number of interspike intervals in two presentations; N in C and D = number of interspike intervals during one presentation. Each bin = 100 μsec (slightly modified from Hind et al., 1967).

because each component frequency causes locking of responses to its cycles. That this is so is shown rather convincingly by the striking similarity of the distributions in Fig. 3B to those in Fig. 3E. However, Fig. 3E is merely a graphic combination of the histograms for each component alone (Figs. 3C and 3D), added bin by bin.

If, thus, interspike intervals are significant for low-frequency information, then in a response to a non-harmonically related two-tone combination, a single fiber will generate frequency information pertaining to one or the other primary as a function of the respective strength of the stimulus. However, if the waveforms of both primaries interact in the relevant sector of the cochlear partition, the frequency information generated will be irregular and not easily defined, since it presumably is the function of momentary interactions of the two primary waveforms.

An entirely different situation prevails when the stimulus consists of two harmonically related tones which are firmly locked to each other. A stable, complex periodic waveform is generated which can be systematically affected by an amplitude or phase shift of the component frequencies. It is useful to consider the data from such experiments (Brugge et al., 1969; Rose et al., 1969) from two points of view. First, one can relate the discharges to the period of the complex sound by constructing histograms modulo this period (period histograms), and, thus, gain some insight as to the way the discharges are produced. Secondly, one can study the interspike interval distributions to evaluate the information that is actually delivered to the central nervous system.

Fig. 4 shows a set of period histograms for a fiber stimulated by two tones at 50 and 60 dB SPL, respectively. These tones were locked in a ratio of 1:2, and the phase of the higher primary was shifted by the indicated amount in respect to the stimulus configuration for Fig. 4C. The phase shifts had only a small, if any, effect on the discharge rate but caused systematically significant changes in the timing of spikes. When one synthesizes a waveform that presumably arises on the cochlear partition by simply adding the component sinusoids, choosing an arbitrary amplitude ratio and an arbitrary phase relation until a reasonable fit is obtained for Fig. 4C, then — assuming that the spikes are actually discharged according to the unilateral elevations of the stimulating waveform — all other histograms should be fitted by waveforms which are obtained if in the equation for the waveform in Fig. 4C one merely shifts the phase angle of the higher primary by exactly the same amounts as was done during the experiment. All the curves in Figs. 4D to 4L were obtained in this way. The results, while by no means exact, are, I think, rather striking. They suggest that the period histogram for an auditory fiber can be viewed, to a first approximation, as reflecting a rectified profile of the stimulating waveform.

Since similar results have been shown to hold true for various frequency and amplitude ratios (Brugge et al., 1969), the evidence seems strong that: (1) the auditory nerve fibers are in fact excited by deflections in only one direction of the cochlear partition; (2) discharges occur at times which correspond to the unidirectional elevations of the stimulating waveform; (3) the stimulating waveform which arises in the relevant segment of the cochlear partition can be reasonably approximated — for stimulus intensities up to at least 100 dB SPL — by a curve resulting from simple addition of the component sinusoids. It may be added that recent experiments of Suga, Arthur and Pfeiffer (personal communication) are in harmony with these statements. Moreover, it could be expected that discharges should follow any waveform regardless of the way such a waveform is generated. Recent findings of Goblick and Pfeiffer (1969) suggest that this holds true in their study of the responses to clicks.

184 J. E. ROSE

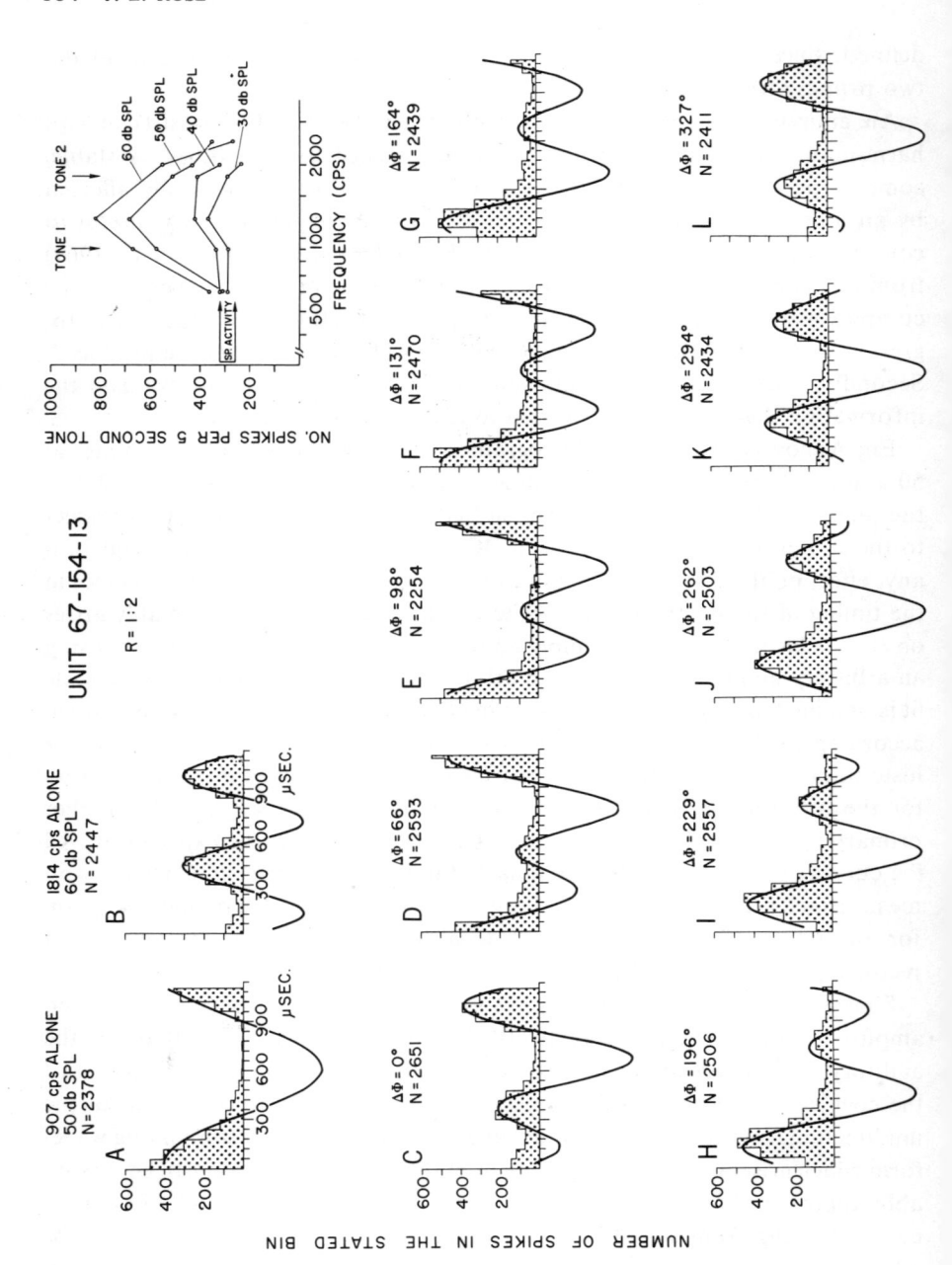

Interspike intervals which must arise when a complex periodic sound is presented are of special interest. Of course, if the amplitude of one or the other primary dominates the waveform, the interspike intervals will group around the integral multiples of the period of the dominating primary. If, on the other hand, the two waveforms interact strongly interspike intervals must occur which do not correspond to the frequency of either primary tone. Fig. 5 illustrates that, for a frequency ratio of 2:3, these expectations are fulfilled (Rose et al., 1969). For the stimulus configuration in Fig. 5A the interspike intervals group around the integral multiples of the period of the lower primary; they group around the integral multiples of the higher primary in Fig. 5F. For intermediate stimulus configurations certain intervals occur much more frequently than others. Thus, in Figs. 5B to 5D the numerically dominant intervals (mode 3) group around the integral multiples of the period of the complex sound. The frequency that corresponds to these intervals must, of course, always be that of the fundamental; in this case (frequency ratio of 2:3), it is also that of a difference tone. In addition to these intervals, some other modular intervals (modes 1 and 2), which do not correspond to the periods of either primary, are present in Figs. 5D and 5E. It will be noted that the values of these intervals are critically dependent on the phase and amplitude relation between the stimuli. By contrast, the interspike intervals that correspond to the integral multiples of the period of the fundamental frequency must arise (though with different probability) for any phase and amplitude relation of the primary tones.

It has been suggested (Rose et al., 1969) that the interspike intervals that appear when complex periodic sounds are presented may be instrumental in

Fig. 4. Variations in discharge patterns as a function of the phase relation between component tones. Tones locked in a ratio of 1:2. Frequencies used: 907 Hz at 50 dB SPL and 1814 Hz at 60 dB SPL. Each period histogram is based on discharges which occurred during two 10 sec presentations of the stimulus. Abscissa: time in μsec; each bin = 60 μsec. Period of the complex sound = 1102 μsec. A and B: period histograms for each primary tone presented alone. C to L: period histograms in response to a complex periodic sound for various phase relations between the component tones. $\Delta\Phi$ indicates a phase shift in degrees of the higher-frequency tone in relation to the configuration shown in C. Response area for the neuron for four sound pressure levels is shown in upper right corner. Vertical arrows (tone 1 and tone 2) indicate, respectively, the position of the 907 and 1814 Hz tone within the response area of the fiber. Horizontal arrows bracket the range of spontaneous activity. Two sine waves with frequency ratio of 1:2 were summed to produce the curves fitted to period histograms C to L. The amplitude and phase relation between the component sinusoids have been adjusted until the fit shown in C resulted. Curves in D to L have the same parameters as in C but the phase angle of the higher frequency has been shifted by the amount equal to the shift introduced during the experiment. For clarity, the abscissas and ordinates for the fitted curves have been omitted. The center line of the fitted curves always coincides with the abscissa of the period histograms. Period histograms A and B have been fitted by a single sine function. Note that in B the abscissa is the period of the lower frequency and, thus, the spikes are shown distributed over two cycles of the higher frequency (from Brugge et al., 1969).

186 J. E. ROSE

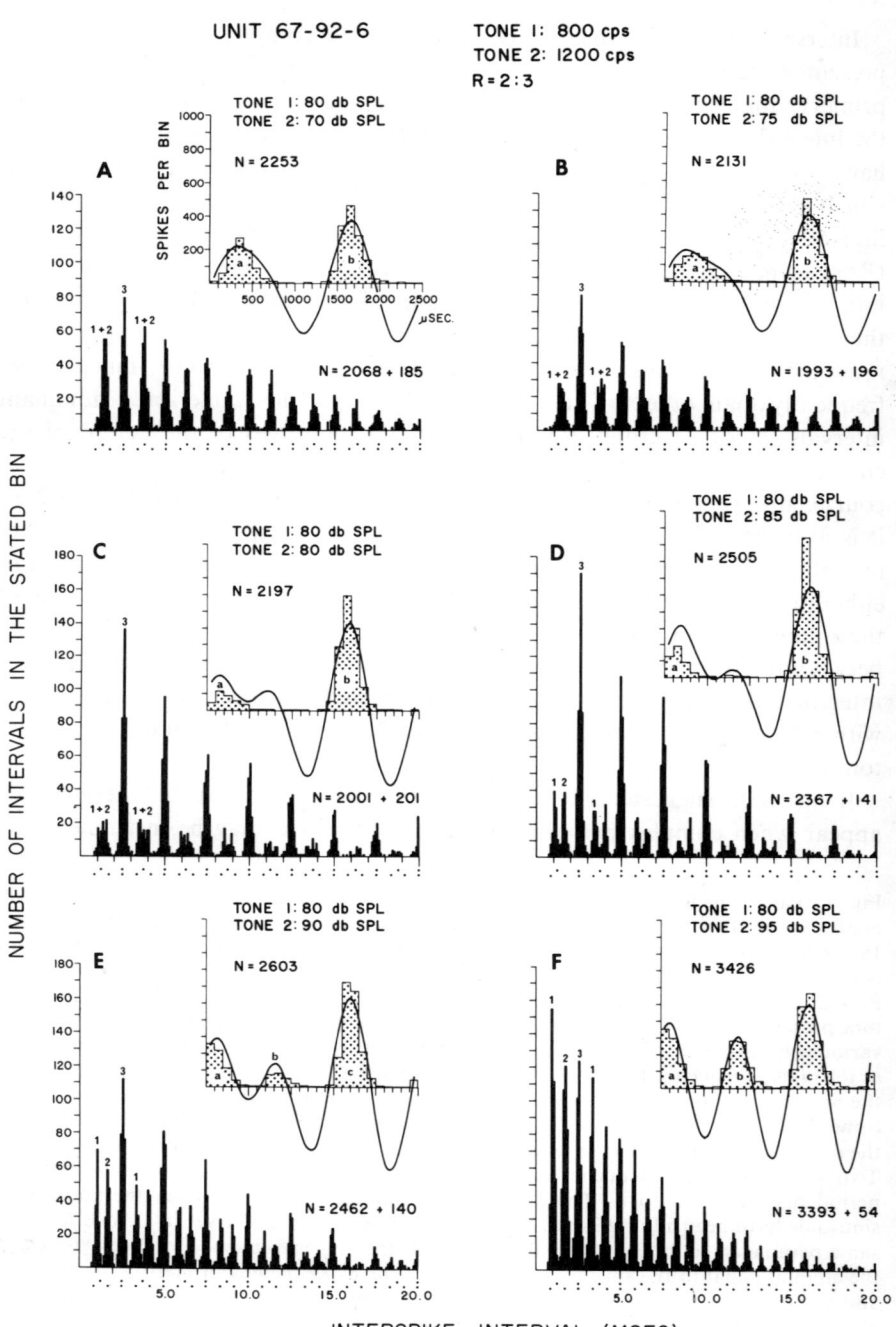

generating information about, at least, some combination tones and particularly about the fundamental frequency of the complex sound. Whether such a suggestion has merit remains to be established.

This study was supported by National Institutes of Health Program Project Grant NB-06225.

REFERENCES

Brugge, J. F., Anderson, D. J., Hind, J. E., and Rose, J. E. (1969): Time structure of discharges in single auditory nerve fibers of the squirrel monkey in response to complex periodic sounds, J. Neurophysiol. 32, 386-401.
Davis, H., Fernández, C., and McAuliffe, D. R. (1950): The excitatory process in the cochlea, Proc. Natl. Acad. Sci. 30, 580-587.
Galambos, R., and Davis, H. (1943): The response of single auditory-nerve fibers to acoustic stimulation, J. Neurophysiol. 6, 39-57.
Goblick, T. J., Jr., and Pfeiffer, R. R. (1969): Time-domain measurements of cochlear alinearities using combination click stimuli, J. Acoust. Soc. Amer. 46, 924-938.
Hind, J. E., Anderson, D. J., Brugge, J. F., and Rose, J. E. (1967): Coding of information pertaining to paired low-frequency tones in single auditory nerve fibers of the squirrel monkey, J. Neurophysiol. 30, 794-816.
Katsuki, Y., Sumi, T., Uchiyama, H., and Watanabe, T. (1958): Electric responses of auditory neurons in cat to sound stimulation, J. Neurophysiol. 21, 569-588.
Katsuki, Y., Watanabe, T., and Suga, N. (1959): Interaction of auditory neurons in response to two sound stimuli in cat, J. Neurophysiol. 22, 603-623.
Katsuki, Y., Suga, N., and Kanno, Y. (1962): Neural mechanism of the peripheral and central auditory system in monkeys, J. Acoust. Soc. Amer. 34, 1396-1410.
Kiang, N. Y.-S., Watanabe, T., Thomas, E. C., and Clark, L. F. (1965): Discharge Patterns of Single Fibers in the Cat's Auditory Nerve (Research Monograph No. 35, M.I.T. Press, Cambridge, Mass.).
Nomoto, M., Suga, N., and Katsuki, Y. (1964): Discharge pattern and inhibition of primary auditory nerve fibers in the monkey, J. Neurophysiol. 27, 768-787.
Rose, J. E., Brugge, J. F., Anderson, D. J., and Hind, J. E. (1967): Phase-locked response to low-frequency tones in single auditory nerve fibers of the squirrel moneky, J. Neurophysiol. 30, 769-793.
Rose, J. E., Brugge, J. F., Anderson, D. J., and Hind, J. E. (1968): Patterns of activity in single auditory nerve fibres of the squirrel monkey, in: Hearing Mechanisms in Vertebrates, A. V. S. de Reuck and J. Knight, Eds. (Churchill, London), pp. 144-157.
Rose, J. E., Brugge, J. F., Anderson, D. J., and Hind, J. E. (1969): Some possible neural correlates of combination tones, J. Neurophysiol. 32, 402-423.

Fig. 5. Relation between interspike intervals and distribution of spikes in the period histograms. A tone of 800 Hz was locked to a tone of 1200 Hz in a ratio of 2:3. A to F: interspike interval and period histograms obtained when the lower frequency was at 80 dB SPL while the strength of the higher frequency varied as indicated. Each sample is based on responses to two 10 sec presentations. Interspike interval histograms: abscissa: time in msec, each bin = 100 μsec. Ordinate: number of interspike intervals in each bin. Period histograms: abscissa: time in μsec; each bin = 100 μsec; period of the complex sound = 2500 μsec. Ordinate: number of spikes in each bin. Each period histogram is fitted by a curve which is the sum of two sine functions. Legends as in Figs. 2 and 4 (from Rose et al., 1969).

Rupert, A., Moushegian, G., and Galambos, R. (1963): Unit responses to sound from auditory nerve of the cat, J. Neurophysiol. 26, 449-465.
Sachs, M. B., and Kiang, N. Y.-S. (1968): Two-tone inhibition in auditory-nerve fibers, J. Acoust. Soc. Amer. 43, 1120-1128.
Tasaki, I. (1954): Nerve impulses in individual auditory nerve fibers of guinea pig, J. Neurophysiol. 17, 97-122.

DISCUSSION

Smoorenburg: I am afraid I do not understand your definition of conditional probability in Table I of the paper. Do you mean by "the probability that a unit which discharged at time zero will discharge again in response to the nth cycle" that the unit may or may not fire at the 1st to $(n-1)$th cycles? The data represent the serial correlation coefficients if the unit may fire in the mean time. In case of no discharge in response to the 1st to $(n-1)$th cycles we are dealing with the probability that an interval n will occur and we expect the sum of the probabilities to be equal to 1. In fact, we expect the data to correspond with Fig. 2 of the paper. Evidently, you are not talking about the latter.

Rose: It is neither your first nor your second supposition. The probabilities shown in Table I are conditional probabilities of firing during the 1st, the 2nd, the 3rd, etc., stimulus cycle which follows the original discharge at time zero. The condition is that the fibre did not yet discharge again until the time interval under consideration. The sum of the conditional probabilities so calculated will always exceed 1. A concrete example will perhaps be useful. Consider populations for 3 intervals. Let us assume (which is not true for the actual data) that all 3 are of the same size. The unconditional probability of firing will be then nearly 0.33 for each interval and the sum of probabilities will be 1. The conditional probability, as defined, will be still 0.33 for the 1st interval since there cannot be any discharges that occurred earlier than the 1st interval. The conditional probability will be, however, 0.5 for the 2nd interval because exactly half of all discharges which did not occur during the 1st interval will occur during the 2nd. The conditional probability for the 3rd interval will be 1 since all discharges which did not occur during the 1st and 2nd intervals will now occur. In any finite sample the conditional probability (under condition stated) must be 1 for the last interval. Whether this has any meaning depends, of course, on the size of the sample for that interval. The significant point made by the figures in Table I is that the probability of firing of a fibre is about the same regardless whether the fibre rested for 1 to 2 or as long as 10 msec.

Smoorenburg: I wonder whether you tried more advanced fitting procedures. The idea of unilateral deflections is, I think, a bit simplified. This approach cannot account for spontaneous activity in the fibre and for activity at downward deflections. Have you tried to incorporate noise?

Rose: I believe that one can, and probably should, introduce a DC shift in fitting the waveforms for tones at high SPL's. We did not work with noise.

Ward: I am glad to see that we finally have quite a bit of the data that we have been hoping for as evidence to support the importance of phase relations among the components of a complex wave. I just have one question. What interpretation do you have for the fact that Fig. 4L of your paper, which represents the phase relation between the two tones of 907 Hz and 1814 Hz where you get the maximum double humping, looks, at least to me, completely indistinguishable from B which is the 1814 Hz tone alone. These are responses of only one unit, but do you think that this implies that for this particular phase relation the second harmonic will be more easily heard than, for instance, in the situations F and G where you get very little of a second hump?

Rose: There are two aspects as I see it. The first concerns the way in which the unit responds. The fibre responds to the unilateral deflection of the cochlear partition and this holds whether or not one thinks that the timing has anything to do with what we hear. The second question is, of course, whether timing is indeed significant. Even if we assume that it is, I am not certain that Fig. 4L allows us to conclude that the time interval corresponds to the second harmonic. The period histograms are plotted on a rather small scale, so the agreement is a little deceptive. If the distance between the peaks in Fig. 4L is not identical to the one in B, the frequency to which this distance corresponds could be far from being a second harmonic.

Whitfield: What you just said in favour of timing is not in conflict with the ideas on what the nerve is forced to do by the nature of the receptor mechanism. The findings for two-tone stimuli are in agreement with the way we would expect this mechanism to respond. So, we are still left with the same question whether timing is relevant in hearing or not. This is a matter of interpretation, of how we look at the data, not of the data as such. We have still the problem that we can produce time periodicity in the nerve responses in situations where we don't hear a corresponding tone.

Rose: I fully agree that our findings may reflect properties of the transducer. Obviously we do not have proof that the time structure of the period histograms in Figs. 4 and 5 has any significance for hearing. I am only saying that it is worthwhile to explore the relation of the period histograms to the waveform of the stimulus because this *might* be significant.

Whitfield: It *could* be, yes.

Schügerl: Dr. Rose, do you think that there is any relation between your results for two-tone stimuli in Fig. 4 and the changes in timbre, which can be heard during a slowly beating period of a mistuned octave?

Rose: Well, I think, it is a little dangerous to attach too much importance to this. I am rather surprised that there is such an agreement between the electrophysiological data and perception. After all, what we hear is not likely to be always a simple function of what the cochlea does. It is difficult to judge whether or not timbre differences can be explained on the basis of the differences in period histograms. I would not stress any simple interpretation because what one hears must be a function of activity of all fibres and this is very difficult to predict from the responses of a single fibre.

Plomp: I agree with you that we have to be careful in comparing psychophysical results with data on single nerve responses. However, as far as I can see, the beats we get for mistuned consonances—two sinusoidal tones with a frequency ratio slightly different from a simple frequency ratio—cannot be explained by nonlinearities in the mechanics of the cochlea, whereas their origin is clear if we accept that the hearing organ is sensitive to the differences of the time structure of the nerve impulses you have shown. A second interesting psychophysical result is the sweep tone faintly audible, for instance, for a complex stimulus consisting of two simple tones with frequencies of 200 and 599 or 601 Hz (Plomp, 1967). The pitch of this inharmonic tone varies in a sawtooth-like manner over a range corresponding to about 600-740 Hz, with the slow phase going up for 200+599 Hz and going down for 200+601 Hz. In my opinion, this sweep tone supports the hypothesis that the hearing organ is able to use time intervals between nerve impulses for pitch perception.

Zwislocki: I would like to put forward two points but, first, may I congratulate Dr. Rose for the beautiful results he has shown. My first point is that, as it seems to me, these results present the necessary but not sufficient conditions for pitch perception on the basis of time periodicities. In this connection, I would like to call your attention to the fact that, in order to preserve time information for lateralization, synchronized single-unit responses would be necessary even if the information were not utilized for pitch perception. In humans, phase sensitivity ends around 1300 Hz, whereas the peripheral neural time locking goes as high as 5000 Hz as you showed us. I think, this example shows that the phase locking is not always utilized, provided that the animal ear may be compared with the human ear in this respect.

My second point, which I think was completely overlooked because it is not the main point of your paper, concerns the curves of Fig. 1. These curves are rather flat in frequency distribution. They give the illusion to be steep because the firing rate is plotted on a linear scale but if a logarithmic scale had been used, the curves would have been very flat. They are much flatter than, for instance, Kiang's response areas where the SPL required to get a certain increase of firing rate is plotted as a function of frequency.

Rose: Well, Kiang's curves are threshold tuning curves. We get the same steep slopes for tuning curves. However, the response becomes much flatter for higher SPL's.

Goldstein: I have a question and a comment. My question is related to the interesting results presented in Fig. 4. The graphs show very nicely that the neural response follows the phase of the stimulus. I wonder, however, what happens when the SPL of one of the two tones is changed. Does the amplitude of this tone in the complex waveform change in the same proportion in the new best fit to the data?

Furthermore, I would like to make a comment on the relationship between psychophysics and these physiological results. I agree with the former speakers who made similar comments on the difficulties in relating psychophysics to physiology. There are many situations in psychophysics where you do get phase sensitivity and one wonders whether all these situations are associated with the single fibre responses; conversely, there are situations where you get phase sensitivity in these responses while it is known that you don't get it in the psychophysical situation. Let me give an example. In listening to complexes consisting of two tones that are close in frequency but also close to a harmonic ratio, for instance, 15:16, one does not hear a phase effect at all; for some ratios, e.g. 4:5, a phase effect is heard only at high SPL's. Plomp (1967) published some psychophysical data on this. Yet it is known from the physiology that one will get a phase sensitivity in the neural response. This is a problem, and I think the only hope for arriving at a solution is by comparing the influence of specific parameters in one modality, physiology, with the influence of specific parameters in the other modality, psychophysics. If they match, then you could guess better than otherwise.

Rose: If SPL of one tone is changed, the amplitude of this tone must be changed also in order to obtain the best fitting for the complex waveform. However, the amount of change is often not at all in the same proportion as the SPL change and occasionally, at least, the phase angle must be changed as well.

I am not a psychoacoustician and I do not know what is the prevalent opinion as to the capacity of the ear to detect changes in phase of one of the components of a complex sound. However, I think, that our data suggest that such a phase sensitivity should be quite restricted. While it is true that a single fibre is very sensitive to a phase shift of a component frequency the total effect in an ensemble of fibres can be expected to be small or altogether negligible.

Schouten: In your Fig. 2, it strikes me that, independent of frequency, the decay times of all interspike interval distributions have about the same value of, roughly, 8 to 12 msec. Is that observation trivial or do we get some additional general information from it about the behaviour of the neural units?

Rose: I think that your statement is very important. This decay time is indeed rather constant for different stimulus frequencies. It depends, however, on the SPL of the stimulus.

De Boer: In one of his latest reports, Kiang (1968) maintained that the most sensitive point of the nerve fibre is always quite close to the threshold curve which means that, when he has a good preparation, he only finds very sensitive neurons. Do you have the same experience?

Rose: What we find is a distribution of sensitivity over a rather large range in SPL.

Honrubia: My observations on the tuning curves of the cochlear microphonics (CM) could be of significance in reference to the shape of your curves in Fig. 1. Because of the shift of the maximum of the CM toward the high frequencies as the SPL is increased, the CM isopotential curves have steeper high-frequency slopes than CM tuning curves themselves (see p. 97 of this volume, Fig. 3). The effect is similar to the one obtained if your curves of Fig. 1 would be plotted with the rate of spikes as the parameter instead of SPL's. That is why there is no contradiction between the shape of your curves and the threshold tuning curves.

Rose: I also do not think that there is a contradiction.

REFERENCES

Kiang, N. Y.-S. (1968): A survey of recent developments in the study of auditory physiology, Ann. Otol. Rhinol. Laryngol. *77*, 656-675.
Plomp, R. (1967): Beats of mistuned consonances, J. Acoust. Soc. Amer. *42*, 462-474.

TWO-TONE MASKING EFFECTS IN SQUIRREL MONKEY AUDITORY NERVE FIBERS

J. E. HIND
in collaboration with J. E. Rose, J. F. Brugge and D. J. Anderson

Laboratory of Neurophysiology
University of Wisconsin
Madison, Wisconsin, U.S.A.

INTRODUCTION

A number of microelectrode studies have shown that the discharge rate of primary auditory nerve fibers in response to a pure tone may be diminished when a second tone of suitable frequency and intensity is sounded simultaneously. This effect has been observed in the monkey by Nomoto, Suga and Katsuki (1964), in the bat by Frishkopf (1964), and in the cat by Kiang *et al.* (1965, Chapter 9), by Sachs and Kiang (1968), and by Sachs (1969). Driven by the harsh demands of the "place" theory of pitch perception and encouraged by the demonstration of lateral inhibition in the retina, auditory neurophysiologists have generally viewed the reduction in discharge rate as an inhibitory phenomenon and little attention has been directed toward the fine time structure of the trains of spikes. In contrast, recent work in our laboratory indicates that two-tone interaction in squirrel monkey auditory nerve fibers, at least for the lower frequencies, is intimately related to the close correspondence we have observed between the effective stimulating waveform and the temporal distribution of discharges.

OBSERVATIONS WITH NON-HARMONICALLY RELATED TONES

Our initial two-tone studies employed stimuli composed of two low frequencies which were not harmonically related. The resulting waveform thus changed continuously with the shifting phase relation between the component sinusoids. For each paired-tone stimulus the temporal patterns of discharge were analyzed with a LINC computer. Periodicities in the spike trains were revealed in the interspike interval histogram which displays the distribution of time intervals between successive discharges. Phase-locking of the discharges to the individual cycles of the two component frequencies was studied by means of the period histogram which depicts the distribution of spikes throughout the cycle of a periodic stimulus. Examination of these histograms as the sound pressure levels (SPL's) of the components were varied revealed that the discharges may be phase-locked to either one tone or the other or to both tones simultaneously (Hind et al., 1967). Which of these modes prevails is determined by the SPL's of the two tones and their positions with respect to the "response area" of the fiber in the frequency-intensity domain. When phase-locking occurs to only one tone of a pair, each of which is effective when acting alone, the fine time structure of the response may be indistinguishable from that which occurs when that tone is presented alone. Moreover, when one component of a pair thus dominates the phase-locked response it usually follows that the discharge rate tends to approach the value produced by that component acting alone, even though this rate may be substantially less than that evoked by the other tone when presented alone. Under this circumstance the dominant tone can with good justification be considered to have masked the other component. Thus the masking tone dominates the fiber not by causing a saturation rate of discharge but by preventing the masked tone from producing its characteristic phase-locked response.

USE OF COMPLEX PERIODIC STIMULI

In order to pursue the relation between waveform and response pattern, experiments were performed with stimuli composed of two tones whose frequencies were locked precisely in the ratio of small integers. Since the waveform of such stimuli is periodic, the period histogram can be arranged to display the distribution of discharges throughout the period of the complex wave. On the basis of a substantial number of observations in which systematic variations were made in the SPL's and phase relations of the component tones, the following conclusions were drawn (Rose, p. 183 of this volume; Brugge et al., 1969): (1) auditory nerve fibers are excited by deflections of the cochlear partition in only one direction; (2) the discharges occur at times which correspond to the unilateral elevations of the stimulating waveform; (3) the effective stimulating waveform can be approximated by addition of the component sinusoids, although the required amplitude and phase relations usually cannot be taken directly from the actual stimulus parameters.

DISTRIBUTION OF TWO-TONE INTERACTION THROUGHOUT THE RESPONSE AREA

In order to relate our findings to the concept of two-tone "inhibition" which has dominated earlier studies we have examined our material with an emphasis on the reduction in discharge rate which can be observed when a second tone of appropriate frequency and SPL is added to a tone at or near the best frequency of the fiber. Since our experiments were directed toward study of the relation between waveform and the temporal distribution of discharges, our usual procedure for each fiber involved detailed observation of a limited number of frequency ratios. Although the consistency of our findings indicates that the correlation between waveform and discharge pattern holds for all frequencies and SPL's capable of affecting a fiber, we have no single case for which a complete survey was made of the effective frequency range.

Our most complete coverage for one fiber was obtained for unit 67-135-7, the response area of which is shown in Fig. 1. The curves illustrate the number of discharges obtained in response to a pure tone stimulus of 10 sec duration at each of the indicated frequencies and SPL's. For tones at a level of 40 dB only frequencies between approximately 1 and 2 kHz were effective. When the SPL was raised to 100 dB the low-frequency limit dropped sharply to the vicinity of 100 Hz while the upper limiting frequency increased only slightly to about 2.5 kHz. This asymmetrical broadening of the effective frequency band with increasing SPL is exhibited by a majority of the fibers we have studied and is,

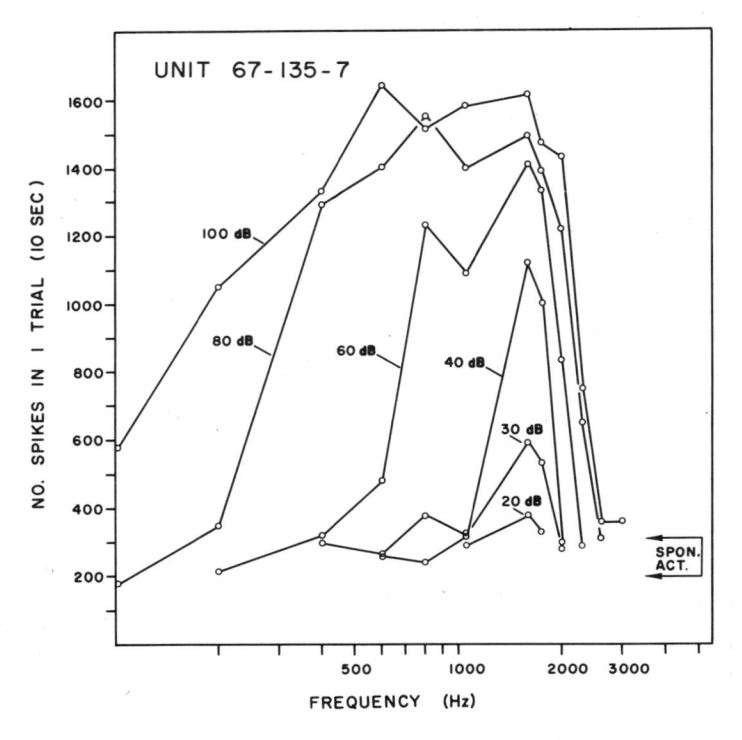

Fig. 1. Response area for unit 67-135-7 to a single pure tone.

we believe, a fundamental factor in determining the response to paired tonal stimuli.

Fig. 2 illustrates period histograms obtained for unit 67-135-7 when an invariant tone (IT) at a frequency close to the best frequency and a fixed SPL of 60 dB was paired successively with variant tones (VT) at four sub-harmonically related lower frequencies at the indicated SPL's (columns A, C, E, and G). The variant frequencies were locked precisely in the ratios of 1:7, 1:5, 1:3 and 1:2 relative to the invariant frequency, corresponding respectively to frequencies of 198, 277, 463, and 676 Hz. As shown at the top of the figure, the invariant frequency also varied slightly in the four frequency ratios but the amount of variation was negligible in relation to the response area of the fiber. The results are also shown when each variant tone was presented alone (columns B, D, F, and H) and when the invariant tone was similarly presented alone (A_5). The length of the abscissa of each histogram is equal to the period of the variant tone and is thus also equal to the period of the complex wave formed by addition of the two sinusoids. The horizontal time scale is maintained constant at 100 μsec per bin; each horizontal scale division represents 500 μsec. Thus as the frequency of the variant tone increases, the length of the histograms shortens proportionally. The vertical scale is also the same throughout, each scale division indicating a value of 80 spikes per bin. All histograms are based on the

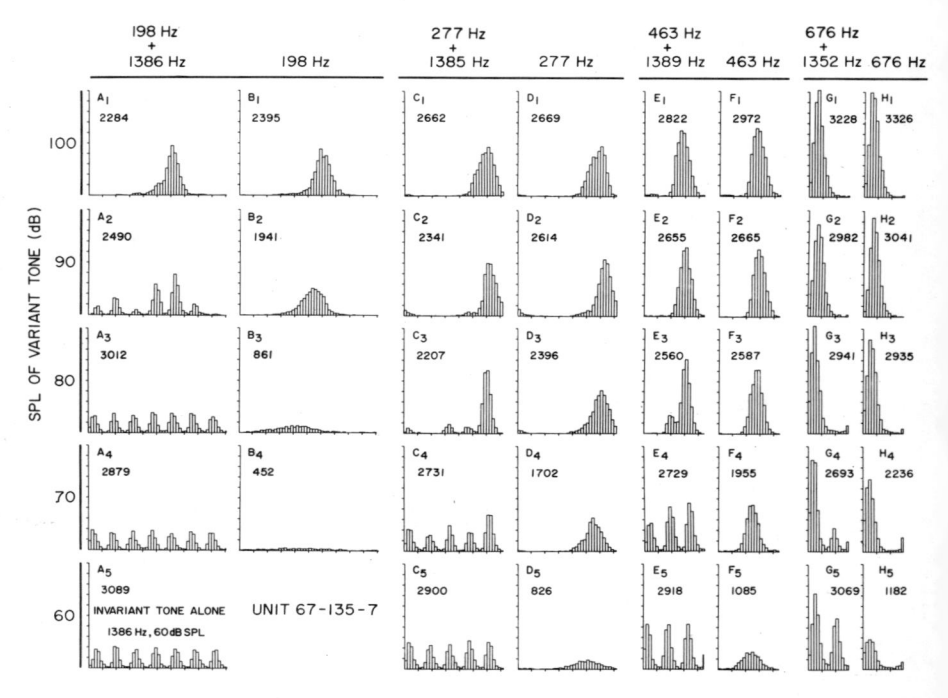

Fig. 2. Period histograms for unit 67-135-7.

responses to two stimulus presentations, each 10 sec in duration. The number on each histogram indicates the total number of discharges in the sample.

When a variant tone of 198 Hz at 70 dB SPL was combined with the IT, the temporal distribution of discharges (A_4) was indistinguishable from that obtained when the IT was presented alone (A_5). Furthermore, while the number of discharges to the paired tones was somewhat less than to the IT acting alone, this amount of difference, about 7%, can reasonably be attributed to sampling variability. Even when the 198 Hz component was raised to 80 dB the pattern and rate (A_3) did not differ significantly from those associated with the IT. However, when the VT was raised to 90 dB, a marked change occurred: the profile of the period histogram then reflected both of the component frequencies and the number of discharges dropped significantly (A_2). A further 10 dB increase in the VT resulted in a further decline in discharge rate and virtual domination of the period histogram by the invariant frequency (A_1); domination by the VT was not absolute, for the distribution was slightly more peaked than that obtained for the VT alone (B_1), suggesting that the IT still exerted a detectable influence.

A similar progression of events was observed for a VT of 277 Hz (C_1-C_5). In this case the variant component achieved complete domination of the period histogram at the 100 dB level (C_1), both in terms of profile and in number of discharges which differed by only 7 from the value obtained with the VT acting alone (D_1). Comparable transitions in dominance from the IT to the VT stimulus were observed for the variant frequencies of 463 Hz (E_1-E_5) and 676 Hz (G_1-G_5). As the variant frequency came closer to the best frequency of the fiber, the VT became more effective and its influence was noticeable even at the lowest level shown, 60 dB (G_5). The greater effectiveness of the 676 Hz VT was also evident at the 100 dB level where the number of discharges slightly exceeded that of the IT acting alone (G_1).

Thus for all variant frequencies a similar progression of events is observed. For the lowest SPL's of the VT the IT dominates the response and determines the discharge rate. As the level of the VT is raised, a relatively restricted range of SPL's is reached throughout which the profile of the period histogram reveals modulation of one sinusoidal waveform by the other; in this range the discharge rate is always less than for the IT acting alone and may even be somewhat less than for the VT acting alone (C_2 and C_3). The sound pressure level of the range throughout which such waveform interaction is observed depends upon the location of the variant frequency with respect to the response area of the fiber. Upon further increase in VT level the variant tone comes to dominate the period histogram and the discharge rate approaches a value similar to that observed when the VT is presented alone, this value again being determined by the location of the variant frequency with respect to the response area. For variant frequencies far removed from the best frequency the number of spikes discharged at the highest SPL's may be significantly less than the number elicited by the weaker invariant tone acting alone, while for frequencies close to the best frequency, discharge rate may equal or exceed that of the IT.

The variations in discharge rate for this unit are in good accord with the results of our previous studies using both harmonically and non-harmonically related tones. The number of spikes produced in response to a two-tone stimulus is generally not substantially larger than that in response to the more effective tone alone and usually not smaller than to the less effective tone.

RELATION OF OUR RESULTS TO THE CONCEPT OF AN INHIBITORY AREA

In Fig. 3 the spike counts from Fig. 2 have been normalized and arranged on a plot with a logarithmic frequency scale. The number of discharges for each paired stimulus is expressed as a decimal fraction of the number of spikes (3089) evoked by the invariant tone acting alone. Additional data for frequency ratios of 1:6 and 1:4 have also been added. The location of the invariant tone is indicated by a cross. In accord with the definition of inhibitory area used by Sachs and Kiang (1968), circles have been drawn around values which are 20% or more below the rate evoked by the IT acting alone. (The region of statistically significant depression of rate for this unit is actually larger than delimited by the 20% criterion.)

The region of 20% or greater depression in response is quite similar to the low-frequency inhibitory area of Sachs and Kiang (1968) which adjoins and partly overlaps the low-frequency border of the response area. A comparison is most directly drawn by reference to Fig. 6 of their paper which illustrates the low- and high-frequency inhibitory areas for unit 365-18 which had a characteristic (best) frequency of 1130 Hz and an upper limiting frequency for the low-frequency inhibitory area of slightly less than 400 Hz. Our unit 67-135-7 had a

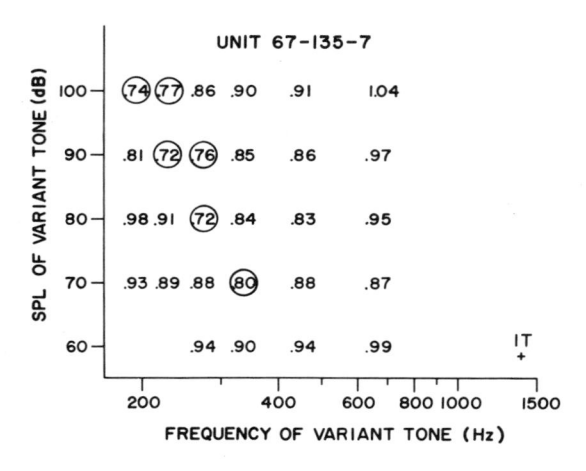

Fig. 3. Normalized discharge rates for unit 67-135-7.

best frequency which was higher by several hundred Hz; the highest frequency at which we observed a 20% reduction in rate was 338 Hz but the next higher tested frequency, 463 Hz, still produced a highly significant reduction of 17% at 80 dB SPL. Considering the difference in species and the limitations imposed by our crude sampling of the frequency spectrum, the agreement is impressive. Further, although our data points are relatively sparse, the extent of our region of 20% reduction is generally consistent with the shape of the Sachs and Kiang low-frequency area. It is unfortunate that we do not have data for unit 67-135-7 in the region corresponding to the high-frequency inhibitory area of Sachs and Kiang. However, in all of our experimental material the same relationships hold between waveform and discharge rate, regardless of whether the frequencies are above or below the best frequency. We thus are led to infer that the boundaries of the high-frequency region of reduced discharge are similarly determined by modulation effects and through domination by variant tones which rapidly decrease in effectiveness as their frequency is raised above the best frequency.

DISCUSSION

Of the various properties we have observed in the response of auditory nerve fibers to paired tonal stimuli we consider that the one most likely to prove relevant to psychoacoustic masking is the "capture effect" by which one component comes to dominate the excitatory process as its sound pressure level is raised relative to a second. We are less impressed with the reduction in discharge rate which often accompanies transitions in dominance and are reluctant to interpret this reduction as a mechanism for achieving significant sharpening of the tuning curve of an individual fiber. We believe that our findings argue strongly that most of the reductions in discharge rate merely reflect the dependence of the excitatory process upon the effective stimulating waveform. While true neural inhibitory processes may operate in addition to the waveform mechanism, no convincing experimental evidence is presently available to support this idea. Thus the properties of the so-called inhibitory areas seem to follow from considerations of waveform and the nature of the relationship between discharge rate, frequency, and SPL for a given fiber as expressed in its response area.

While all of our observations on the correlation between waveform and response pattern were made with harmonically related tones, we are convinced that the significance of our findings is not limited to such stimuli. In all of our experiments in which systematic phase shifts between the two components were introduced, the period histograms were compatible with the corresponding changes in waveform, and the discharge rate rarely varied more than 10% throughout all phase relations. Thus our earlier results with non-harmonically related paired tones may be interpreted by the plausible assumption that each fiber responds to the instantaneous stimulating waveform even though this waveform may be changing in time.

We think it prudent to leave open the important question of whether the correlation between waveform and response pattern holds for the higher frequencies. We do know that a low-frequency tone can modulate the discharge pattern of a fiber responding to a high frequency at which our techniques are unable to demonstrate phase-locking. Moreover, as emphasized by Sachs and Kiang (1968), the rate-reduction features of what these authors refer to as two-tone inhibition are remarkably constant throughout the entire frequency spectrum; such evidence tempts one to speculate that a common mechanism may be responsible.

This investigation was aided by grants NB 06225 and FR 00249 from the National Institutes of Health.

REFERENCES

Brugge, J. F., Anderson, D. J., Hind, J. E., and Rose, J. E. (1969): Time structure of discharges in single auditory nerve fibers of the squirrel monkey in response to complex periodic sounds, J. Neurophysiol. *32*, 386-401.

Frishkopf, L. S. (1964): Excitation and inhibition of primary auditory neurons in the little brown bat, J. Acoust. Soc. Amer. *36*, 1016(A).

Hind, J. E., Anderson, D. J., Brugge, J. F., and Rose, J. E. (1967): Coding of information pertaining to paired low-frequency tones in single auditory nerve fibers of the squirrel monkey, J. Neurophysiol. *30*, 794-816.

Kiang, N. Y.-S., Watanabe, T., Thomas, E. C., and Clark, L. F. (1965): Discharge Patterns of Single Fibers in the Cat's Auditory Nerve (Research Monograph No. 35, M.I.T. Press, Cambridge, Mass.).

Nomoto, M., Suga, N., and Katsuki, Y. (1964): Discharge pattern and inhibition of primary auditory nerve fibers in the monkey, J. Neurophysiol. *27*, 768-787.

Sachs, M. B. (1969): Stimulus-response relation for auditory-noise fibers: two-tone stimuli, J. Acoust. Soc. Amer. *45*, 1025-1036.

Sachs, M. B., and Kiang, N. Y.-S. (1968): Two-tone inhibition in auditory-nerve fibers, J. Acoust. Soc. Amer. *43*, 1120-1128.

DISCUSSION

De Boer: In my opinion, it is an illusion to think of two-tone inhibition as having anything to do with the stimulating waveform. The main reason is that two-tone inhibition is also found in units with a high characteristic frequency in which one cannot observe a relation with the stimulating waveform.

Hind: I agree that we surely cannot prove a relation between waveform and two-tone suppression of discharge at high frequencies. But we are unwilling to accept the conclusion that this proves that two-tone suppression has nothing to do with the stimulating waveform. The correlations at low frequencies are very good and the nature of the so-called inhibitory area seems remarkably similar throughout the frequency spectrum; there is no obvious discontinuity from low to high frequencies. Moreover, if one is willing to include the possibility of a DC component in the concept of effective stimulating waveform, the relationship between waveform and two-tone suppression might be postulated to extend to the highest frequencies, including those cases where the variant tone does not itself excite the fibre. Although not discussed by Dr. Rose in this symposium, we have found that some of our two-tone period histograms can be fitted satisfactorily only by assuming the existence of a DC component. Such a DC component might correspond to the asymmetry in vibration of the cochlear partition which has been observed in physical models of the cochlea by Tonndorf (1958) and which has been suggested as an explanation for the summating potential by Whitfield and Ross (1965).

Smoorenburg: I wonder whether you examined a situation that is not so appropriate to bear out the general idea. In this particular situation, the firing rate of the invariant tone alone is almost the maximum firing rate possible. This means that an increase of amplitude will result in an increment of firing rate smaller than the decrement of firing rate with a similar decrease of amplitude. This may explain that the introduction of the variant tone, which introduces a modulation, gives all together a lower firing rate. The inhibitory areas of Sachs and Kiang (1968), however, are also found in conditions where the invariant tone alone excites the fibre at half the maximum firing rate.

Hind: I agree with your interpretation of the case illustrated in this paper. The invariant tone (1386 Hz at 60 dB) produced 3089 spikes in two 10-sec trials which, from Fig. 1, can be seen to be close to the saturation rate; modulation with other tones is more likely to chop away from the effective excitation rather than to increase it significantly. However, in general when one component of a pair dominates the phase-locked response, the discharge rate tends to approach the value produced by that component acting alone, a value which depends upon the frequency and SPL coordinates of that component with respect to the response area and which therefore may be either substantially more or less than the rate produced by the other component.

Goldstein: In your paper you describe the inhibition phenomenon as the result of a modulation mechanism. Would you care to elaborate on that?

Hind: Not really. We are presently trying to fit the period histograms for two-tone stimulus configurations and, as Dr. Rose indicated in his remarks, it is not a simple matter to derive amplitude factors. I think we see some hope but we do not have anything which we can share with people.

Zwicker: I like to point out something a little bit dangerous. In Fig. 1 of your paper we see very clearly a large nonlinearity. On the other hand you are thinking in linear terms with respect to the modulation and the amplitude and phase relations as discussed by Dr. Rose too. Looking at these things together, I think there is some kind of contradiction.

Hind: There is no doubt that nonlinear mechanisms are interposed between the stimulus waveform and the discharge. In fact, our present job is trying to elucidate the nature of these nonlinearities.

Zwicker: Can you go a little bit further into what you would mean by inhibition?

Hind: I prefer not to use the term in this situation. It has been used by previous workers who tended to view two-tone suppression in terms of neural inhibition. Our position is that true neural inhibition does not need to be involved at the level of the first auditory fibre to explain our results. True neural inhibition may, in fact, occur but we simply do not have evidence that it occurs at this stage. There is no doubt that true neural inhibition exists at higher levels of the auditory pathway as, for example, in the cochlear nucleus.

Whitfield: I very much like to support your idea that this effect should not be called inhibition. I think you found an elegant relationship which is very different from what one finds in, for example, the cochlear nucleus where you can inhibit a response with a tone that is maybe 30 dB down on the stimulating tone. You do not see this type of thing, I gather.

Zwislocki: I would like to keep open the case of inhibition; I mean real inhibition that includes spontaneous activity. And I would like to do that by stating the difference between the experiments in vision and in hearing. In vision, especially in the eye of the limulus, it is possible to stimulate a receptor unit without stimulating any other receptor units. This offers the opportunity to observe the inhibitory effect of one unit on the other units. In the ear, this is not possible because, whatever stimulus you introduce in the cochlea, it spreads over many units so that they cannot be isolated. Thus, it is possible that a stimulated unit inhibits other units that are also excited. There is an interaction between the two effects which, so far, have not been separated. I do not think that we can decide whether there is inhibition or not.

Whitfield: May I just come back to that. I'm not disputing that there may be inhibition. I am saying that we should not call this effect inhibition until we can demonstrate that there is something corresponding to what, in the central nervous system, we mean by inhibition. This has not been demonstrated. So let us not colour it by calling it inhibition.

De Boer: I would like to prefer the term two-tone depression.

Goldstein: As far as I know, Kiang has always used the term inhibition of auditory nerve responses to describe specific data rather than to propose underlying mechanisms (*e.g.* Sachs and Kiang, 1968). In my own view every research worker has to discipline himself to neutralize word-labels given to phenomena by other workers and instead strive to understand the phenomena. Therefore, it is not clear to me that one should necessarily strive for the most neutral label rather than use a borrowed term in the effort to suggest analogy or even mechanism.

REFERENCES

Sachs, M. B., and Kiang, N. Y.-S. (1968): Two-tone inhibition in auditory-nerve fibers, J. Acoust. Soc. Amer. *43*, 1120-1128.

Tonndorf, J. (1958): Harmonic distortion in cochlear models, J. Acoust. Soc. Amer. *30*, 929-937.

Whitfield, I. C., and Ross, H. F. (1965): Cochlear-microphonic and summating potentials and the outputs of individual hair-cell generators, J. Acoust. Soc. Amer. *38*, 126-131.

SYNCHRONY BETWEEN ACOUSTIC STIMULI AND NERVE-FIBRE DISCHARGES

E. DE BOER

Physical Laboratory, ENT.-Department
Wilhelmina Hospital, University of Amsterdam
Amsterdam, The Netherlands

WEISS' MODEL OF COCHLEAR ACTION

Two phenomena have provided clues for the simplest model of the initiation of nerve impulses in the cochlea. The first is that nerve fibres respond differentially to various frequencies and the second is that the firing instants are partially synchronous to the acoustic stimulus. In setting up a model one assumes that there is a quantity that goes up and down just as the stimulus waveform does, and that the nerve fibre is most likely to fire when this quantity is maximal. In view of the frequency selectivity that neural responses exhibit one may come to regard this excitatory quantity as a *filtered version of the acoustic stimulus*. For purposes of easy analysis it is most appropriate to regard the filter involved as a *linear* one. By linear we mean the property that, if the input signal $x(t)$ is the sum of two parts, $x_1(t)$ and $x_2(t)$, and the responses to each of the parts alone are $y_1(t)$ and $y_2(t)$, then the response $y(t)$ to $x(t)=x_1(t)+x_2(t)$ is the sum of $y_1(t)$ and $y_2(t)$: $y(t)=y_1(t)+y_2(t)$.

Fig. 1 shows the model proposed by Weiss (1964, 1966) built up in three steps. In Fig. 1A we see the stimulus $x(t)$ transformed into a signal $y(t)$ by the linear circuit C. A nerve spike is produced whenever $y(t)$ passes a pre-set threshold. The signal $x(t)$ is acoustic in nature but the transformed signal $y(t)$ (although originally meant to represent something like the deflection of the basilar membrane) may be anything: perhaps a complex shearing force on a hair cell or the amount of a chemical substance emitted. The main points are that $y(t)$ is assumed to be a linear transform, and that the circuit C represents all of the frequency selectivity that the neural responses show.

Weiss immediately realized several shortcomings of his model and these led him to include additional features: Fig. 1B shows the addition of an auxiliary

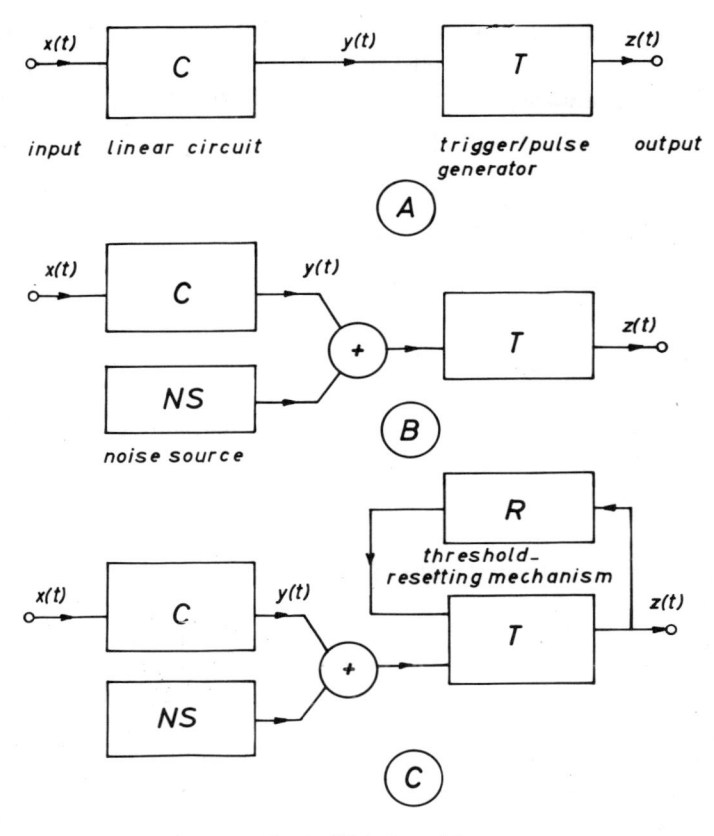

Fig. 1. Weiss' model.

noise source and Fig. 1C introduces a threshold-resetting mechanism. The noise source causes irregular spontaneous discharges, such as have been observed experimentally, and it causes the firings to be irregular in time even when a perfectly regular stimulus, like a sinusoid, is presented. The threshold-resetting mechanism simulates refractoriness of the fibre: after a fibre has fired it cannot immediately respond again.

It is the purpose of this paper to discuss several aspects of this functional model of cochlear action. The model still shows several shortcomings; the question is: can it be improved materially by a proper choice of parameters and, what is more important, does it indeed represent cochlear action in a functional manner? One problem in Weiss' original formulation was that by letting the linear filter stand for cochlear hydrodynamics one cannot explain experimentally obtained tuning curves for sinusoidal stimuli. If all frequency selectivity is considered as concentrated in the linear circuit C, the frequency response of C should be shaped like the experimental tuning curve. Cochlear hydrodynamics just provides too little frequency selectivity for this purpose.

But this point would seem just a matter of choosing proper filter parameters; it has nothing to do with the model itself. Another point at which the model falls short of explaining experimental evidence is the observed irregularity of spike trains for very intensive impulsive stimuli. It may be necessary to include a source of time jitter as involved in the timing of spikes (de Boer, 1969a) but a physiological correlate cannot be ascertained at present. Other improvements can be suggested but these all concern minor items.

What is most needed is a re-evaluation of the model itself, or, an independent check of its validity. To this end we will explore the phenomena of synchrony between stimulus and neural response somewhat more deeply, both from the experimental side and in view of modeling. In doing so we should discuss physiological and model experiments for several types of stimuli: tones, double tones, harmonic and inharmonic tone complexes, periodic series of clicks and white noise. Because of space limitations we can only touch upon a few — seemingly little related — aspects.

SYNCHRONY BETWEEN STIMULUS AND RESPONSE FOR NOISE STIMULI

Let us start with noise as the stimulus. Fig. 2 shows what can be expected in Weiss' model: Fig. 2A shows the input signal $x(t)$, Fig. 2B the transformed version $y(t)$: in this case band-limited noise. It is easy to state that the firing instants will tend to synchronize with the maximal excursions of $y(t)$ but this does not help us very much. The signal $y(t)$ is a hypothetical one and, as yet, completely inaccessible to the experimenter. Still there is a way out of this dilemma. The waveshapes of $x(t)$, situated around the instants that the nerve fibre fires, must have one hidden property in common, namely they should all,

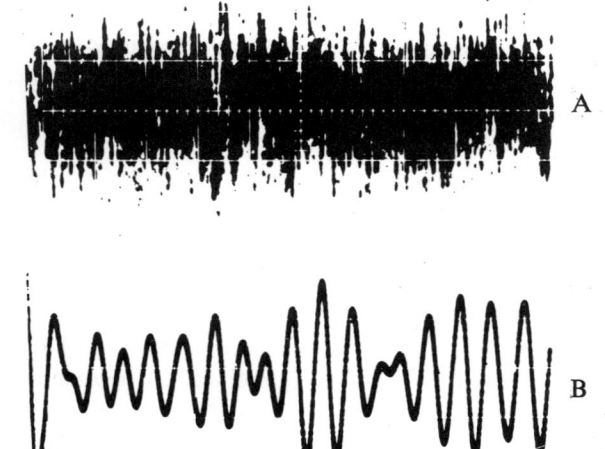

Fig. 2. White noise $x(t)$ and filtered noise $y(t)$.

via the transformation, lead to a y-signal with large excursions. Inversely, when in a physiological experiment the noise fragments that immediately precede the firing instants of a nerve fibre are recorded, it should be possible to bring this hidden property to light. This principle has been worked out in the theory of triggered correlation. The mathematics can be found in a paper by de Boer and Kuyper (1968). A heuristic introduction, designed toward the non-mathematically oriented reader, has been given in two papers (de Boer, 1968, 1969b). It appears that from a close study of the above-mentioned noise fragments quantitative conclusions can be drawn regarding the process involved in cochlear frequency resolution. Thus the parameters of the model can, at least in principle, be inferred from a physiological experiment.

The procedure is as follows. A recording is made of the action potentials in one primary auditory fibre. The method is similar to that reported by Kiang et al. (1965). The ear of the animal (cat) is stimulated with white noise. All waveshape fragments of the stimulus that immediately precede nerve spikes are isolated and added. Usually a period of 15 msec terminating at the firing instant was chosen. The addition can be done with an average response computer (de Boer, 1967; de Boer and Jongkees, 1968). If Weiss' model holds, the theory states that several thousands of noise fragments added together yield a waveform that very closely resembles the impulse response of the linear system C. Thus it is, at least in principle, possible to measure parameters of the model from a physiological experiment with white noise as stimulus. The computations to be done involve only input and output signals of the cochlea: the noise stimulus and the nerve spikes.

The resultant correlogram manifests itself as a damped sinusoid, the average frequency being equal to the characteristic frequency of the neuron under study. But the damping is much less than one would expect on the basis of cochlear hydrodynamics. Hence, according to these results, the linear system C in a model that portrays these results must be a resonant circuit with relatively little damping. By Fourier transformation the frequency response of the model filter C can be obtained. A few neurons could be kept long enough in contact with the recording electrode so that the correlogram as well as the tuning curve could be determined. The effective frequency response of the circuit C was found to agree reasonably well in shape with the measured pure-tone tuning curve of the same neuron (de Boer, 1969a, 1969b). Both differ greatly from the response that the basilar membrane exhibits on hydrodynamical grounds (Zwislocki, 1953). The effective frequency response, derived from the correlogram, as well as the tuning curve of a neuron show the largest steepness at frequencies above the characteristic frequency. The former has a slope of 80 dB per octave in the 1 kHz region, the latter of 100 dB per octave (cochlear movement patterns would account for only 20 dB per octave). It is to be noted that Johnstone and Boyle (1967) measured a steepness of 75 dB per octave of the microscopical movement pattern, although in a different animal (guinea pig) and in a different frequency region (18 kHz).

These data lead to the conclusion that Weiss' model appears to explain physiological findings quite well provided its linear circuit is assumed to have greater frequency selectivity than has hitherto been supposed. The model then serves to explain and to tie together experimental results obtained with diverse types of stimuli, notably sinusoids and white noise. Even the general shape of Post Stimulus Time (PST) histograms measured from click stimuli can be readily explained since the impulse response of the linear filter shows just the required number of oscillations (*cf.* de Boer, 1969a). But because of the large influence of refractoriness here it is difficult to study this point quantitatively.

One other point that appears to support the validity of the modified model is the following. It was observed that the correlogram derived from experiments with white-noise stimuli approximately retains its shape when the SPL of the stimulus is varied over 20-40 dB. This implies that the frequency response of the filter in the model remains approximately the same under intensity variations (as indeed it should). In physiological terms this means that a nerve fibre keeps sending the same type of information about its stimulating signal to the brain under different intensities of stimulation. The relative weights with which various frequency regions contribute to neural firings remain the same no matter whether the SPL of the stimulus is high or low. This has been termed the "principle of specific coding" (de Boer, 1969b) for complex stimuli.

From the foregoing discussion it can be concluded that the process of frequency resolution in the cochlea can rather well be described as a linear process. Experimental data with tones, noises and clicks as stimuli can be tied together under this notion. Although several discrepancies exist, it seems safe to state that nonlinear processes (as undoubtedly are active in the cochlea) do not materially contribute to the frequency resolution that is ultimately attained by the cochlea.

SYNCHRONY BETWEEN STIMULUS AND RESPONSE FOR DOUBLE TONES

We next take up a completely different matter: the response of nerve fibres to non-random complex stimuli. In this case the theory of triggered correlation cannot be applied to the model without modification (see the original papers for the conditions under which the theoretical results hold). We thus may expect the model to have completely different and unexpected properties, and, as we shall see, there is in this case at least one explanation for the appearance of higher-order combination tones.

Let us limit ourselves to double tones as stimuli. In a recent paper Brugge *et al.* (1969) show how the time distribution of nerve firings can, at least in principle, be related to a waveform consisting of two sinusoids with the same frequencies as the tones presented. The probability of firing goes up and down as this hypothetical waveform. But it is shown that it does not necessarily do so in a proportional manner, nor is this hypothetical waveform a linear transform of the stimulus. This waveform thus cannot be identical to the signal $y(t)$ in the model of Fig. 1; it is necessary to include nonlinearities in the model. In first approximation these nonlinearities can be regarded as *static* ones, so that all frequency selectivity that responses of nerve fibres exhibit remains represented by the original linear circuit in the model. A closer study of these nonlinearities is needed to find out whether this approximation is warranted or not.

Hind et al. (1967) reported that under special circumstances the firing instants can be partially synchronous to either of the two stimulating sinusoids. This we can explain in very simple terms. Let us assume that two tones of frequencies f_1 and f_2 are presented. Let us furthermore assume that *these frequencies are harmonically related*, i.e. that both are integral multiples of a fundamental frequency f_0. Later on this restriction will be removed. As before, we suppose that firing is controlled by a waveform (yt) that is a linear transform of the stimulus (any interposed static nonlinearity is immaterial in this connection). See the upper half of Fig. 3 for this waveform. Most of the firings will occur at the instants that $y(t)$ is maximal. At these instants each input component assumes a specific phase (*e.g.* its phase may be near 0°) and this situation occurs again and again, each time $y(t)$ is maximal. Hence it is easy to see that the response is phase-locked to either of the two constituent tones. Up till here the frequencies were assumed to be harmonically related. The results reported by Hind et al. (1967) pertain as well to the case in which signals were not harmonic. In order to extend our reasoning we next assume that our two tones, while still harmonically related, both acquire the same phase shift of φ_0 radians. The resultant waveform $y(t)$ will still be periodic, the envelope has remained the same but the fine structure (the nearly sinusoidal oscillations between upper and lower envelope) shows a phase shift φ_0 (*cf.* de Boer, 1961). The preferred moments of firing will be shifted over a period Δt that corresponds to the phase φ_0. This time shift will be $\Delta t = \varphi_0/2\pi f_m$ in which f_m is the average of f_1 and f_2. If we now consider one firing instant it seems as if the two original tones acquired phase shifts of $\varphi_1 = \varphi_0 \cdot f_1/f_m$ and $\varphi_2 = \varphi_0 \cdot f_2/f_m$, respectively.

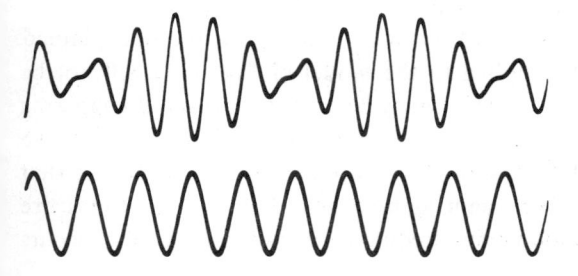

Fig. 3. Upper part: harmonic two-tone signal, shown at the stage of the effective excitation signal $y(t)$. Lower part: combination-tone signal, used as a time reference.

Since we can assume f_1 and f_2 to be close together, φ_1 and φ_2 will be approximately equal to φ_o, the actual phase shift that each of the two components acquired. Hence the synchrony that existed previously has not disappeared. The phase of each primary tone that corresponds to a firing instant will appear as only slightly shifted (the ratios f_1/f_m and f_2/f_m are not exactly unity) but this will be hardly noticeable. This holds for any value of φ_o.

This procedure can be extrapolated by letting φ_o increase linearly with time at a rate of $\Delta f = d\varphi_o/dt$. The effect is that the original frequencies f_1 and f_2 become $f_1 + \Delta f$ and $f_2 + \Delta f$ so that they are no longer harmonic. Since φ_o now varies from 0 to 2π radians, the preferred phases of the two tones corresponding to the firing instants will appear as slightly smeared out. When f_1 and f_2 are close together (as they should be, lest the filtering process interferes) this effect will be small. We may conclude that, in the general case, firing instants should be partially synchronous to each of the constituent stimulus tones, and this provides a simple explanation of Hind's results. Note that the main point of the discussion is that only a small portion, namely the part with maximum excursions, of $y(t)$ is considered to be contributing to nerve firings. If the threshold of firing were assumed to vary randomly over the full range of y-values no synchrony would be expected at all. In the second place the two primary frequencies must be close together. But this condition is automatically met when one studies a situation where each of the two tones contributes to neural excitation.

COMBINATION TONES

An important study about nerve-fibre discharges to two-tone stimuli has been presented by Goldstein and Kiang (1968). They presented two tones to the animal's ear and studied a very specific temporal relation in the discharges of a nerve fibre. To this end they generated electronically a distortion product with frequency $2f_1 - f_2$, in the sequel to be called "combination tone". This signal was *not* contained in the stimulus. It was used solely as a time reference for the investigation of synchrony. In many cases Goldstein and Kiang found a distinct synchrony between the nerve discharges and this auxiliary tone. Some of the properties that audible combination tones in psychophysical experiments are known to exhibit were parallelled by physiological findings.

We will now show that such a synchrony can, at least in principle, be explained by a similar type of argument, applied to the model of Fig. 1. It will not be necessary to invoke any additional nonlinearity beyond the threshold-triggering mechanism involved in the generation of the nerve spike. We begin with two tones with frequencies f_1 and f_2, leading to a stimulating waveform like that shown in Fig. 3, upper part. We again assume first that the two tones are harmonically related. The fundamental frequency is called f_o. Firing occurs when $y(t)$ passes a pre-set value; this threshold may be set so high that only the part of $y(t)$ in which the excursions are maximal, contributes to the firing. We have already established that the firing instants are partially synchronous with either of the constituent tones. Similarly, the discharges will be partially synchronous with any signal that has a frequency close to f_1 and f_2 and that is an integral multiple of the fundamental frequency f_o. The combination tone with frequency $f_c = 2f_1 - f_2$ belongs to this class: when f_1 and f_2 both are integral multiples of f_o, so is f_c, when f_1 and f_2 are close together, f_c is close to f_1 and f_2 as well. Hence in the harmonic case the firing instants will correspond to a specific phase of the combination tone $2f_1 - f_2$.

We now shift the phase of both primary tones by φ_o. As stated earlier, the firing instants will shift over a time $\Delta t = \varphi_o/2\pi f_m$ in which f_m is the mean of f_1 and f_2. The electronically generated combination tone with frequency $2f_1 - f_2$ also acquires the phase shift φ_o. It will appear to be shifted over a time $\Delta t_c = \varphi_o/2\pi f_c$. When f_1 and f_2 are close together, the firing instants still occur at approximately the same phase of the combination tone. This holds for any value of φ_o. If we now let φ_o increase linearly with time, both primary tones get a frequency shift of $\Delta f = d\varphi_o/dt$, and they are no longer harmonic. Still the firing instants will be concentrated in a particular phase region of the combination tone. The synchrony will be slightly less than in the pure harmonic case but this small difference probably goes unnoticed. It is to be noted that this reasoning is based entirely on the model of Fig. 1 with as its main parts a linear filter (C) and a triggering mechanism (T) with a fixed threshold. The argument shows that it is not strictly necessary to modify the model in order to account for experiments in which nerve impulses are found to be partially synchronous to a combination tone even if the latter is not contained in the stimulus. For a general discussion of this topic see de Boer *et al.* (1969). It remains to be seen how much of the experimental data can be explained in this way; *other* experimental facts may point toward basic additional nonlinear phenomena in the cochlea, of a type that involves frequency selectivity of its own.

GENERAL CONCLUSIONS

From the above exposition it follows that the simplest type of model for the cochlea is capable of explaining many observations. It explains the synchrony between a tone and the resulting nerve spikes, it explains the micro-synchrony between noise stimuli and nerve spikes (but only when the model is equipped with a fairly sharp filter), it explains fairly well the shape of the tuning curve for pure tones and the PST histogram for click stimuli. The same simple model also explains qualitatively why one may observe synchrony with specific combination tones, not present in the stimulus. These findings seem to justify the model as the simplest picture of cochlear action. Modifications will certainly be needed to account for all experimental evidence. But it may be safe to assume for the moment that the effect of the modifications will be limited. Then the model — in the form equipped with a fairly sharp filter — may serve for a while to link together experimental evidence and to guide our thinking about the coding of auditory information that is carried out in the cochlea.

REFERENCES

Boer, E. de (1961): A note on phase distortion and hearing, Acustica *11*, 182-184.
Boer, E. de (1967): Correlation studies applied to the frequency resolution of the cochlea, J. Audit. Research *7*, 209-217.
Boer, E. de (1968): Reverse correlation. I. A heuristic introduction to the technique of triggered correlation, with applications to the analysis of compound systems, Proc. Kon. Akad. Wetensch. (Amsterdam), Series C, *71*, 472-486.
Boer, E. de (1969a): Encoding of frequency information in the discharge patterns of auditory nerve fibres, Internat. Audiol. *8*, 547-556.
Boer, E. de (1969b): Reverse correlation. II. Initiation of nerve impulses in the inner ear, Proc. Kon. Akad. Wetensch. (Amsterdam), Series C, *72*, 129-151.
Boer, E. de, and Jongkees, L. B. W. (1968): On cochlear sharpening and cross-correlation methods, Acta Oto-Laryngol. *65*, 97-104.
Boer, E. de, and Kuyper, P. (1968): Triggered correlation, IEEE Trans. Biomed. Engin. BME-*15*, 169-179.
Boer, E. de, Kuyper, P., and Smoorenburg, G. F. (1969): Proposed explanation of synchrony of auditory-nerve impulses to combination tones, J. Acoust. Soc. Amer. *46*, 1579-1581.
Brugge, J. F., Anderson, D. J., Hind, J. E., and Rose, J. E. (1969): Time structure of discharges in single auditory nerve fibers of the squirrel monkey in response to complex periodic sounds, J. Neurophysiol. *32*, 386-401.
Goldstein, J. L., and Kiang, N. Y.-S. (1968): Neural correlates of the aural combination tone $2f_1-f_2$, Proc. IEEE *56*, 981-992.
Hind, J. E., Anderson, D. J., Brugge, J. F., and Rose, J. E. (1967): Coding of information pertaining to paired low-frequency tones in single auditory nerve fibers of the squirrel monkey, J. Neurophysiol. *30*, 794-816.
Johnstone, B. M., and Boyle, A. J. F. (1967): Basilar membrane vibration examined with the Mössbauer technique, Science *158*, 389-390.
Kiang, N. Y.-S., Watanabe, T., Thomas, E. C., and Clark, L. F. (1965): Discharge Patterns of Single Fibers in the Cat's Auditory Nerve (Research Monograph No. 35, M.I.T. Press, Cambridge, Mass.).
Weiss, Th. F. (1964): A model for firing patterns of auditory nerve fibers, M.I.T., Research Laboratory of Electronics, Cambridge, Mass., Techn. Report No. 418.
Weiss, Th. F. (1966): A model of the peripheral auditory system, Kybernetik *3*, 153-175.
Zwislocki, J. J. (1953): Review of recent mathematical theories of cochlear dynamics, J. Acoust. Soc. Amer. *25*, 743-751.

DISCUSSION

Zwislocki: Dr. Rose's paper (p. 177, Fig. 1) showed tuning curves, if I may call them so, that increased in width with intensity. Does this have any meaning in terms of your model?

De Boer: The shape of the frequency response curve of the equivalent filter is asymmetrical, it spreads more to the lower frequencies than to the higher ones. As I showed in my paper, this shape is about the same as that of the tuning curve. So, also from my results on reverse correlation, I can deduct that, with increasing intensity of a pure tone, a neuron should get progressively more sensitive to lower frequencies. But this holds only for pure tones, not for noise.

Bosher: Dr. H. Davis used to tell his students that the nervous system works on the all-or-none principle, though it was designed to hide this as far as possible. I think much the same can be said here as the linear model can only be made to function satisfactorily by the addition of nonlinearities. A nonlinearity which interests me is one which Dr. de Boer (1969) has himself stressed, namely the discrepancy between the increase in average firing rate compared to the increase in stimulus intensity. This, he has said, suggests that the threshold of the triggering circuit must depend in some way on stimulus intensity. Lynn (1969) and Sayers at Imperial College in London have tried to compute Kiang's PST histograms from the available data on the hydrodynamics of the cochlea and have failed to do so. They found they could only succeed if they inserted just such a device: *i.e.* by making the threshold level of the triggering circuit dependent on the magnitude of the initiating stimulus. I feel this fits in very nicely with Dr. de Boer's idea but it does emphasize the need for modification of the initially linear system.

De Boer: I would like to state explicitly—this is better for the discussion—that by nonlinearities you can mean nearly everything. What I imply here is that for frequency selectivity there are no frequency selective nonlinearities involved, whereas the dependence of the setting of the threshold on the intensity is of course a nonlinear affair. This threshold level is set to some averaged value of intensity with a time constant of, let us say, over 20 msec. I am talking here, however, about something which occurs within one period of the characteristic frequency. This is quite a different type of nonlinearity.

Bosher: Your actual analysis is between the noise input and the spike output and between these two occur the nonlinearities. One can say this part is linear and the next is nonlinear but surely this is only hypothetical.

De Boer: No, the point is if the nonlinear mechanism in the model were frequency dependent, then the method would show up the linear part of it. That means, then, it should show up von Békésy's wave pattern which it does not.

Zwicker: Most of what I would like to say is already said and I agree with what Dr. de Boer has put forward. So I may restrict myself to two comments. First, on the basis of psychoacoustical experiments we tried to build a model (Zwicker, 1962; Terhardt, 1966) and came out with something almost the same as you have shown. Second, this means that we have some kind of frequency selectivity built in the ear and some nonlinearity, but this nonlinearity operates *after* the filtering process. Then, there comes the triggering of the nerve impulses and this is a very strong nonlinearity. This conclusion from psychoacoustical measurements agrees very well with your model.

Møller: I think I got you right when you said that the frequency response curve, which was based on your recovered impulse responses, is actually what you recover. Then, your results would mean that the bandwidth is independent of intensity for noise stimuli.

De Boer: That's what we found in varying the sound pressure level over about 40 dB. I had one occasion in which I could change the SPL over 60 dB and in that case the shape of the frequency response curve was changed a little.

Møller: This constancy is just what a linear filter would do. So, your remark that this finding is not in agreement with the tuning curve, or isorate curve, or whatever we call it, is not really true. If we make a frequency response curve of a linear filter, then we get a wider curve at higher levels, but of course the bandwidth remains the same, so it is just a matter of saying.

De Boer: No, I may have expressed myself in the wrong way. I meant to say that in terms of tuning curves different neurons will respond when you increase intensity. But this happens only when you present *one* sinusoid. If it is a part of a complex sound, this is no longer true because the neuron always picks out those components near its characteristic frequency in a relative proportion which is independent of intensity.

Møller: This is exactly what a linear filter would do. You can also say that the range of a linear filter is larger at higher intensities, but it is never said that way.

De Boer: Yes, indeed. But we have to keep in mind that we assume that the threshold level of the triggering mechanism is shifted for higher intensities. Otherwise, this mechanism would be overloaded.

Møller: I agree. My next point is the following. You said that your measuring technique is insensitive to nonlinearities and this holds indeed for nonlinearities of the non-memory type. But what happens when the nonlinearity is of the memory type, let me say some sort of a low-pass filter or a process with some time constant? Then it does not hold any longer that the crosscorrelation between input and output is a good representation of the system impulse response.

De Boer: Well, in fact it does for a system where you have coincidence. If you have a multitude of pulses fed into a coincidence network which selects only certain of those impulses to be carried along, and that is what Ranke (1955) supposes to be the case in the cochlea, you can prove that the method of reverse correlation will show up the linear part that precedes the generation of the pulses. So, if the sharpening in the cochlea would be attained by such a nonlinear system that does have memory, the measuring technique would show up that linear part.

Møller: No, I am sorry, I mean a memory *after* the filtering process. If you have, for instance, a very long time constant compared to the signal, then you would not get any correlation at all. The system impulse response that you recover from your technique would not be the true response in that case.

De Boer: That is true. But then there would not be any correlation for sinusoids either.

Møller: So my question is: what is the frequency limit of your method? To what frequency with regard to characteristic frequency of the unit can you go?

De Boer: It is the same limit as for the preservation of periodicity of pure tones, 4 to 5 kHz. It is correct to say that the method has limits in the case you stated but we have not yet explored those limits systematically. We are going to do so.

Anderson: In deriving results of your correlation technique you have assumed a threshold model. Dr. Geisler (1968) of our laboratory has performed simulation experiments on such models and has obtained responses for low-frequency sinusoidal stimuli with somewhat disappointing results particularly when a wide intensity range was used. Have you made an attempt to analyze your correlation technique for models other than those of the threshold type?

De Boer: I wrote in my paper that, actually, threshold crossings rather than peaks may trigger the nerve impulses. If we then have uncorrelated noise added to the signal, the probability of firing is some function of that signal in the neighbourhood of the threshold crossing and you get the same model as I discussed. So, I don't think there is a real problem.

Anderson: The same noise source will not work, however, at all intensities.

De Boer: That is right.

Johnstone: I would like to return to Dr. Møller's remark. From a physiologist's point of view these models consisting of integrating time constants, linear filters, nonlinearities, etc. look rather unreal. What we actually have in the ear is a mechanical filter followed by an electric transducer followed by a chemical transmitter, which itself is solely one way, but certainly followed by an integrating receptor. So in principle we have a rectifier followed by an integrator which would smear out your correlogram.

De Boer: In effect this is represented by the jitter source in the model. As long as the jitter does not smear out the signal over a good part of one period of the characteristic frequency, it is all right, and the method works.

Whitfield: Is there any relation between your findings concerning the effect of the average level and the experiments of Dewson (1968), who found that discrimination between speech sounds by monkeys in the presence of noise was significantly worse after section of the olivo-cochlear bundle? In the absence of noise there was no such loss of discrimination post-operatively.

De Boer: I cannot say anything about that.

REFERENCES

Boer, E. de (1969): Reverse correlation. II. Initiation of nerve impulses in the inner ear, Proc. Kon. Akad. Wetensch., Series C, 72, 129-151.

Dewson, J. H. (1968): Efferent olivo-cochlear bundle: some relationships to stimulus discrimination in noise, J. Neurophysiol. 31, 122-130.

Geisler, C. D. (1968): A model of the peripheral auditory system responding to low-frequency tones, Biophysical J. 8, 1-15.

Lynn, P. A. (1969): Processing of signals in peripheral auditory system in relation to aural perception, Doctoral Dissertation, University of London.

Lynn, P. A., and Sayers, B. McA. (1970): Cochlear innervation, signal processing, and their relation to auditory time-intensity effects, J. Acoust. Soc. Amer. 47, 525-533.

Ranke, O. F. (1955): Die Fortentwicklung der Hörtheorie und ihre klinische Bedeutung, Arch. Ohr usw. Heilk. u. Z. Hals usw. Heilk. 167, 1-15.

Terhardt, E. (1966): Beitrag zur automatischen Erkennung gesprochener Ziffern, Kybernetik 3, 136-143.

Zwicker, E. (1962): Ueber ein einfaches Funktionsschema des Gehörs, Acustica 12, 22-28.

Section 5

ORIGIN OF COMBINATION TONES

In recent years combination tones have obtained renewed attention because of their enigmatic character, some being audible even at low stimulus levels. This section touches upon the psychoacoustically revealed properties of combination tones and reports the search for physiological correlates in cochlear microphonics and single-fibre responses.

COMBINATION TONES IN COCHLEAR MICROPHONIC POTENTIALS

PETER DALLOS

Auditory Research Laboratory
Northwestern University
Evanston, Illinois, U.S.A.

INTRODUCTION

This paper is addressed to two seemingly unrelated problems: the origin and mechanism of intermodulation distortion (combination tone) generation in the cochlea, and the role of cochlear microphonic (CM) potentials in hearing. It is our contention that the solution of the former problem, along with the consideration of some psychoacoustical observations on combination tone percepti on, provides important clues to the understanding of the significance of microphonic potentials. We hope to show that the behavior of microphonic distortion potentials (MDC) and the properties of audible combination tones are radically different, the implication being that there might not be a direct causal relationship between auditory perception and microphonic potentials.

Some manifestations of distortion by the ear were noted as early as the beginning of the 18th century, but even today the mechanism of distortion generation is subject to considerable disagreement. Studies of MDC's proved that at least up to extremely high SPL's the distortion primarily originates in the cochlea and not in the middle ear (Wever and Lawrence, 1954). While the cochlea is clearly implicated as the major source of distortion, there are two apparently conflicting views concerning the mechanism of generation of classic distortion components (harmonics and intermodulation components). Some maintain that the site of nonlinearity is in the final, mechano-electrical transduction process (Wever and Lawrence, 1954), whereas others contend that the principal nonlinearities are to be sought in the mechanical vibration of the cochlear partition or in the hydrodynamic properties of the cochlear fluids (Tonndorf, 1958). Some of our results (Dallos and Sweetman, 1969; Sweetman and Dallos, 1969) as well as those of de Boer and Six (1960) showed promise that the two contradictory views could be reconciled, in that, depending on the SPL of the primary stimulus, either the mechano-electrical or the mechano-hydraulic mechanism of distortion generation can predominate. In the following we intend to demonstrate that: (1) At moderate SPL's the MDC's are not distributed according to the place principle. (2) At moderate SPL's the MDC's originate in the mechano-electric conversion process. (3) These moderate-level MDC's do not appear to have an audible correlate.

METHOD

Young guinea pigs anesthetized with urethane were used as experimental animals. All recording of CM potentials was done with intracochlear differential electrodes (from scalae vestibuli and tympani), two sets of electrodes in two different turns of the cochlea at one time. In some of the experiments, stimulating electrodes were also used to pass a direct current across the cochlear partition. Sound stimuli, delivered to the bony meatus in a closed system, were monitored at the eardrum with a calibrated probe-tube microphone. The stimuli used to elicit MDC's consisted of two pure tones (primaries). Single sinusoidal stimuli were also used when comparisons were desired between MDC's and pure tones. During cancellation experiments, the canceling sinusoidal stimulus was delivered with bone conduction. Details of our experimental procedures and performance criteria are described in Sweetman and Dallos (1969); Dallos (1969a); and Dallos et al. (1969a).

RESULTS

Spatial distribution of MDC's

By delivering a sinusoidal stimulus at the frequency of and phase-locked with the MDC we attempted to cancel the CM distortion potential throughout the cochlea by adjusting the SPL and phase of the pure tone (Sweetman and Dallos, 1969). Fig. 1 summarizes the results. In panel (A) the CM magnitude of a quadratic intermodulation distortion component (simple difference tone: f_2-f_1) is shown from turns I and III as the function of the phase of a pure tone having identical frequency with the difference tone, and whose amplitude was so adjusted that optimal cancellation of the MDC occurred in turn I. Note that while the MDC can be eliminated in turn I, the canceling stimulus has little effect on the MDC in turn III. In panel (B) the converse is shown: cancellation of the MDC in turn III does not remove it from turn I. In panel (C) a control experiment is shown where an air-conducted sinusoid is canceled in both turns simultaneously by a properly adjusted bone-conducted sinusoid of the same frequency. These experiments (and others on different orders of intermodulation distortion components) indicate that the distribution patterns of MDC's and

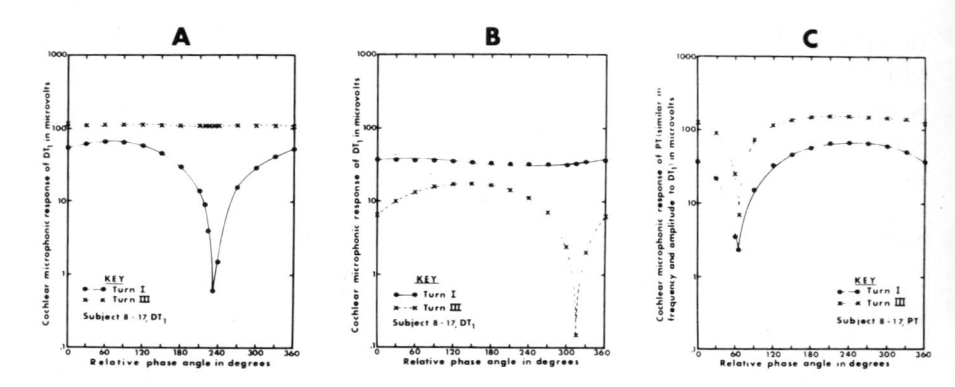

Fig. 1. Cancellation of CM as a function of the phase of the bone-conducted sinusoid. (A): cancellation of f_2-f_1 is optimized in turn I. (B): cancellation of f_2-f_1 is optimized in turn III. (C): cancellation of CM elicited by a pure tone having the same frequency as f_2-f_1. The primary frequencies and SPL's are: $f_1=6000$ Hz, 90 dB; $f_2=7050$ Hz, 90 dB; pure tone: 1050 Hz, 67 dB in the presence of $f_2=7050$ Hz, 90 dB (from Sweetman and Dallos, 1969).

CM potentials in response to sinusoidal stimuli of the same frequency are quite different, because it is impossible to create destructive interference between them over an extended length of the cochlea.

By measuring the relative magnitudes of distortion components and of sinusoids one generally finds that at low and moderate levels the MDC shows a spatial distribution pattern similar to that of the higher frequency eliciting primary, while at very high levels, it partially mimics the distribution of the CM in response to a sinusoid of the same frequency as the MDC (Worthington, 1969). Thus MDC's at moderate levels are prominent in the same vicinity where their higher frequency primary generates maximal disturbance.

Origin of MDC's

After ascertaining this distribution pattern we investigated whether the low-level distortion originates in mechano-hydraulic or mechano-electric processes. We used DC polarization of the cochlear partition (Tasaki and Fernandez, 1952; Legouix and Chocholle, 1957) to investigate this problem (Dallos et al., 1969b).

Our experiment was guided by the following hypothesis. If the relative change in a distortion component as the result of DC polarization is the same as for the CM responses to the primaries and a pure tone that simulates the distortion component in question, then it is reasonable to assume that the distortion is mechanical or hydraulic in origin. This is so, for mechano-hydraulically generated distortion components would be transduced into microphonic potentials by the hair cells via the same mechanism as would be sinusoids. Thus, there is no reason to assume that a change in the biasing of the mechano-electric transducer would differentially affect the transduction of mechanically generated nonlinear components and pure tones. On the other hand, if the distortion originates in the mechano-electric transducer then any change in the transducer's operating point brought about by polarization should differentially affect pure tones and nonlinear components.

In Fig. 2 two pairs of curves are seen for the CM response of one of the primaries and for the MDC (difference tone: f_2-f_1), at moderate and at high SPL's. The relative change in microphonic component is plotted as the function of the magnitude and polarity of the DC bias. Note that the functions for the primary at both SPL's and also for the difference tone at high levels are like those reported in the literature (Tasaki and Fernandez, 1952). That is, we see a roughly proportional increase in microphonic with current strength when the scala vestibuli is positive, and similar proportional decrease when the scala vestibuli is negative. In contrast is the curve for the MDC at the moderate sound level. The behavior is exactly reversed: positive polarity creates a decrease whereas negative polarity creates an increase in this nonlinear component.

These results indicate that cochlear distortion, as manifested by combination frequencies in CM, is at least a two-stage process with signal level playing the critical role. At low levels the distortion is primarily a hair cell phenomenon, and it is most prominent at the location of the maximal primary microphonic potential. At high levels a mechanical type of nonlinearity predominates, and it is manifested by components which are probably distributed within the cochlea by traveling waves. The genesis of these ideas was illustrated in this

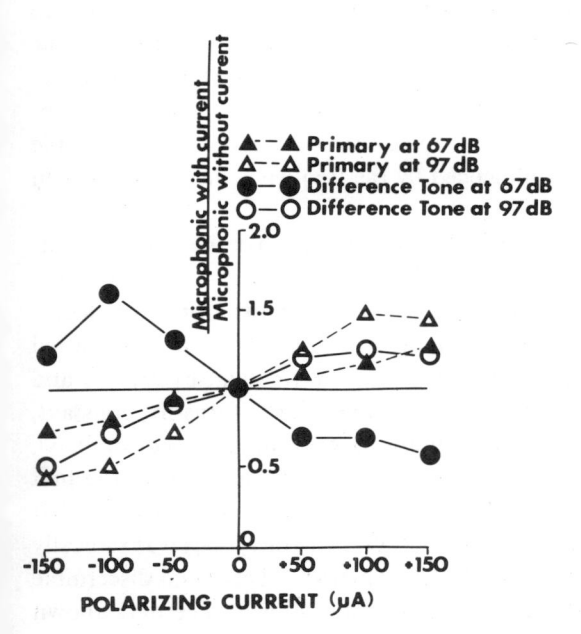

Fig. 2. Ratio of MDC (f_2-f_1) magnitude with and without polarization. Measurements are made at various current levels (abscissa), and at two SPL's of the primaries (67 and 97 dB). The two primaries (4000 and 4500 Hz) are presented at equal SPL's (from Dallos et al., 1969b).

paper by showing relationships that apply to the quadratic distortion component, f_2-f_1 (difference tone). Other experiments indicate that generalization of these ideas to other intermodulation distortion components in CM (summation tone and various orders of combination tone) is quite legitimate.

Comparison of f_2-f_1 and $2f_1-f_2$

In this section the behavior of the quadratic (f_2-f_1) and a particular cubic ($2f_1-f_2$) MDC is compared, and also some comparisons are made between these microphonic distortion potentials and the subjective experience (as reported in the literature) that corresponds to similar distortion components (f_2-f_1 and $2f_1-f_2$). The motivation of the comparisons is that recent psychophysical experiments by Zwicker (1955), Plomp (1965), and Goldstein (1967, also pp. 230-247 of this volume) have clearly shown that the aural combination tone $2f_1-f_2$ behaves in a significantly different manner than do other types of combination tones. Goldstein's (1967) findings are clearly contrary to the classic view of distortion generation in the auditory system which originated with von Helmholtz, and according to which the peripheral auditory system has a saturating type of nonlinearity.

Goldstein's experiments conclusively showed that the combination tone $2f_1-f_2$ is the most prominent subjective distortion component, furthermore the relative loudness of this component was found to be independent of the SPL of the stimulus, and it could be perceived at very low sensation levels. All three of these observations are contrary to the polynomial distortion model and they compelled Goldstein to postulate that the cochlea possesses an essential nonlinearity. An additional significant observation indicated that the magnitude of the $2f_1-f_2$ component is extremely dependent upon the frequency separation of the two primaries, in fact, it decreased at a rate of approximately 100 dB per octave with f_2/f_1. In contrast, the loudness of a simple difference tone f_2-f_1 was barely influenced by the frequency disparity between the primaries.

We wish to give some data now which demonstrate that the $2f_1-f_2$ component in CM behaves just like any other MDC and that it conforms to the polynomial nonlinearity model. In Fig. 3 data are shown for experiments that were performed by fixing f_1 and changing f_2 so that the ratio f_2/f_1 varied between 1.05 and 1.8. Microphonic potentials at the frequencies $2f_1-f_2$ and f_2-f_1 were recorded from both the first and third cochlear turns for constant, and equal, SPL's of the primaries. Presentations were made at constant 60, 80, and 100 dB SPL. In the left panel the potentials measured from the first turn are shown, while the third turn data are given in the right panel. The left panel reveals that neither distortion component ($2f_1-f_2$ or f_2-f_1) was radically influenced by the frequency separation of the primaries. There is no discernible difference between the behavior of the two types of MDC's. The data shown

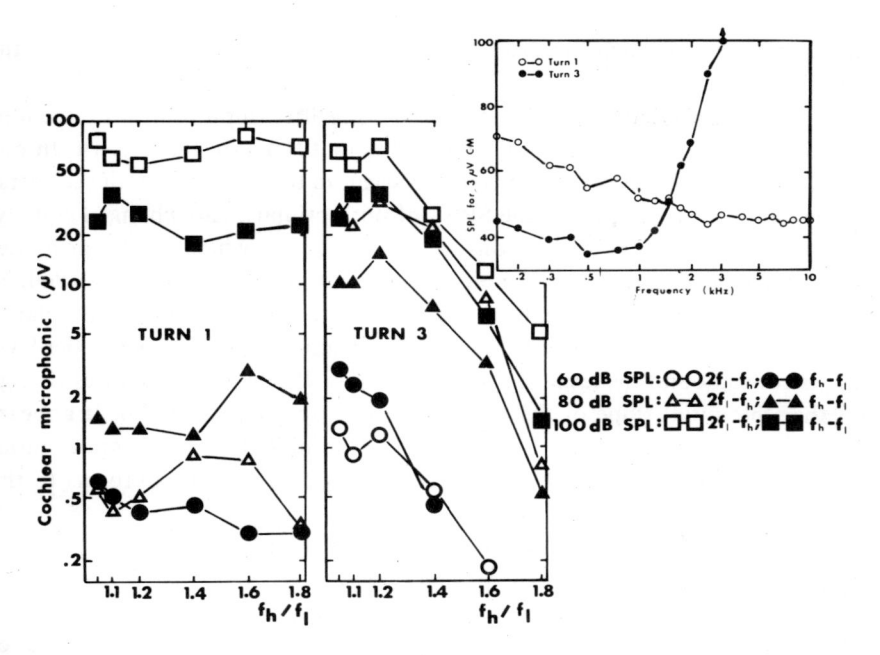

Fig. 3. Magnitude of MDC's f_2-f_1 (closed symbols) and $2f_1-f_2$ (open symbols) at 60, 80, and 100 dB SPL of the primaries as a function of the frequency separation of the latter (f_2/f_1). Recordings are from cochlear turns I and III. The insert shows the 3 μV isopotential curves for this animal for both recording sites. The lower primary frequency $f_2=1000$ Hz (from Dallos, 1969b).

in the right panel show that both types of MDC's, at all three SPL's, appear to be strongly influenced by the primary frequency ratio. The difference between the flat curves from the basal turn and the rather precipitous drop in microphonic from the third turn can be explained by noting the relative sensitivity of the two electrode locations to the primary frequencies. To facilitate the comparison, the insert shows 3 μV isopotential curves for both electrode pairs for this animal. From these curves one can conclude that the apparent diminution of the distortion components with increasing f_2/f_1 is simply due to a decrease in the magnitude of excitation by the higher frequency primary.

The dependence on SPL of the magnitude of the combination tone $2f_1-f_2$ was approximated by Goldstein (1967) by the function $L=a^2b/(a+b)^2$, where a and b are the amplitudes of f_1 and f_2, respectively. This description indicates that if both a and b are changed equally then the relative distortion stays constant. In contrast, according to a polynomial nonlinearity the magnitude of the distortion component $2f_1-f_2$ should increase linearly if the intensity of f_1 is held constant while f_2 is changed; furthermore, the increase should be proportional to the square of the magnitude of f_1 if this primary is variable while the other is held constant. Finally, if both primaries are increased equally and simultaneously then the distortion product should rise as the cube of the intensity.

In Fig. 4 representative CM magnitude *vs.* SPL (input-output) functions are plotted for the MDC: $2f_1-f_2$. Three different functions are shown. In one the SPL of f_1 is held constant while f_2 is varied, in the second case the converse is true, while in the third case the SPL's of both primaries are changed equally. The initial slope of these functions is determined by which of the primaries is the variable. Clearly defined unity slopes were seen in the cases examined by us when the SPL of f_1 was held constant and f_2 was the variable. Similarly, the initial slope of the $2f_1-f_2$ microphonic function is double when the level of f_2 is constant and the level of f_1 is the variable. When both primaries change together the initial slope of the CM function is approximately 3.0. It appears valid to conclude that the slope of the CM combination tone input-output functions, including the function for the $2f_1-f_2$ component, conforms to the polynomial nonlinearity model.

DISCUSSION

It is concluded that cochlear distortion is a two-stage process, mechano-electrical at low levels and mechano-hydraulic at high levels. In addition, all MDC's behave alike, irrespective of their order. In other words, combination tones of the type $2f_1-f_2$ as recorded from cochlear microphonic potentials

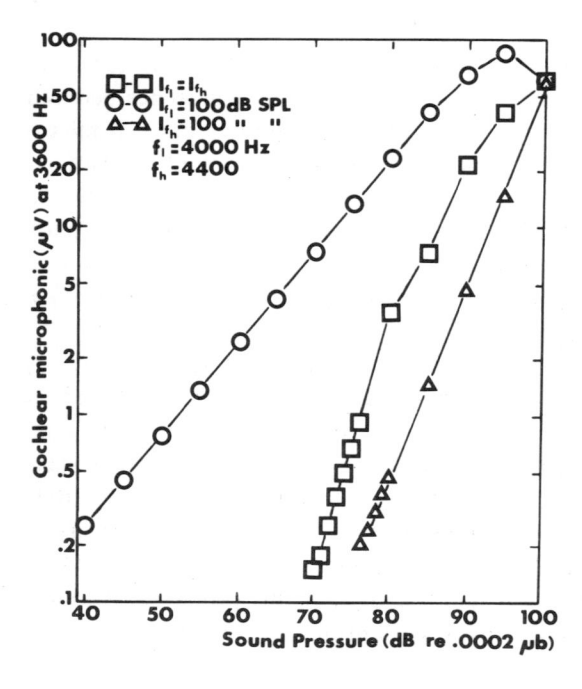

Fig. 4. CM input-output functions for the component $2f_1-f_2$ ($f_1=4$ kHz, $f_2 = 4.4$ kHz), with the SPL's of both primaries the same, or if either is held at a constant 100 dB (from Dallos, 1969b).

do not in any respect behave contrary to the expectations based on a simple polynomial nonlinearity model. This finding is highly significant when one considers that Goldstein (1967) and Goldstein and Kiang (1968) came to diametrically opposed conclusions concerning the behavior of this combination tone, based on psychoacoustical experiments with human observers and on physiological studies of single unit responses in the cat's auditory nerve. Quite clearly the behavioral and neural data on the one hand and the microphonic data on the other do not seem to correlate, thus an explanation of the discrepancies is in order.

One of Goldstein's most significant observations was that the relative $2f_1 - f_2$ distortion was independent of the stimulus level, and that this distortion was observable at very low sensation levels of the primaries, around 20 dB. The magnitude of the distortion increased at the same rate as that of the primaries, staying only 15-20 dB below the latter. The simple difference tone, $f_2 - f_1$, does not become audible until the primaries reach the approximate sensation level of 50 dB (Plomp, 1965; Goldstein, 1967), $f_2 + f_1$ is not audible at all. In contrast with these psychoacoustical observations, our results indicate that the microphonic component that corresponds to $2f_1 - f_2$ does not appear in significant proportions at low SPL's. When both primaries are the same SPL then this combination tone component does not reach the 0.1 μV level below 50 dB SPL. In contrast, the quadratic MDC's, $f_2 - f_1$ and $f_2 + f_1$, are usually detectable at lower levels, often as low as 30 dB SPL.

It appears to us that audible, psychoacoustically measurable, nonlinear components are most likely to arise from two, probably unrelated, sources. The first of these is tied to the high-level second stage of peripheral distortion, while the second process produces the $(n+1)f_1 - nf_2$ components which were studied by Zwicker (1955) and Goldstein (1967). It is to be emphasized again that the low-level distortion, so apparent in CM, which is well describable by a classic polynomial nonlinearity, evidently does not have an audible correlate. Conversely, the peculiar distortion components of order $(n+1)f_1 - nf_2$, and particularly the $2f_1 - f_2$ cubic product, do not seem to have a microphonic correlate which would match their properties. These observations lead us to suggest that, first, the hearing of a particular pitch might not necessarily imply that a microphonic potential of appropriate frequency is present in the cochlea, and second, the presence of a particular cochlear microphonic potential might not inevitably be accompanied by hearing sensation.

These ideas are very much in line with some of Plomp's suggestions (1965); they indicate that CM potentials might not necessarily form an essential link in the peripheral transduction chain.

ACKNOWLEDGMENTS

Many of these results were obtained in cooperation with M. A. Cheatham, Z. G. Schoeny, R. H. Sweetman, and D. W. Worthington. This research was supported by grants from the National Institutes of Health, USPHS.

REFERENCES

Boer, E. de, and Six, P. D. (1960): The cochlear difference tone, Acta Oto-Laryngol. *51*, 84-88.
Dallos, P. (1969a): Comments on the differential electrode technique, J. Acoust. Soc. Amer. *45*, 999-1007.
Dallos, P. (1969b): The combination tone $2f_1-f_h$ in microphonic potentials, J. Acoust. Soc. Amer. *46*, 1437-1444.
Dallos, P., Schoeny, Z. G., Worthington, D. W., and Cheatham, M. A. (1969a): Some problems in the measurement of cochlear distortion, J. Acoust. Soc. Amer. *46*, 356-361.
Dallos, P., Schoeny, Z. G., Worthington, D. W., and Cheatham, M. A. (1969b): Cochlear distortion: the effect of DC polarization, Science *164*, 449-451.
Dallos, P., and Sweetman, R. H. (1969): Distribution pattern of cochlear harmonics, J. Acoust. Soc. Amer. *45*, 37-46.
Goldstein, J. L. (1967): Auditory nonlinearity, J. Acoust. Soc. Amer. *41*, 676-689.
Goldstein, J. L., and Kiang, N. Y.-S. (1968): Neural correlates of the aural combination tone $2f_1-f_2$, Proc. IEEE *56*, 981-992.
Legouix, J. P., and Chocholle, R. (1957): Modification de la distortion du potential microphonique par le polarization de l'organ de Corti, Soc. de Biologie, Paris Compt. Rend. Hebd. *151*, 1851-1854.
Plomp, R. (1965): Detectability threshold for combination tones, J. Acoust. Soc. Amer. *37*, 1110-1123.
Sweetman, R. H., and Dallos, P. (1969): Distribution pattern of cochlear combination tones, J. Acoust. Soc. Amer. *45*, 58-71.
Tasaki, I., and Fernandez, C. (1952): Modification of cochlear microphonics and action potentials by KCl solution and by direct current, J. Neurophysiol. *15*, 497-512.
Tonndorf, J. (1958): Localization of aural harmonics along the basilar membrane of guinea pigs, J. Acoust. Soc. Amer. *30*, 938-943.
Wever, E. G., and Lawrence, M. (1954): Physiological Acoustics (Princeton University Press, Princeton, New Jersey), pp. 117-176.
Worthington, D. W. (1969): Spatial Patterns of Cochlear Difference Tones, unpublished Doctoral Dissertation, Northwestern University, Evanston, Ill., pp. 1-227.
Zwicker, E. (1955): Der ungewöhnliche Amplitudengang der nichtlinearen Verzerrungen des Ohres, Acustica *5*, 67-74.

DISCUSSION

Smoorenburg: Dr. Dallos, you did not find a CM according to the psychophysical CT's. Do you believe this to imply that the CT is not present at the, let me say, mechanical stage of the processing? If it was there it should give a CM, should not it?

Dallos: You are forcing me to speculate. What I imply by this is that, whatever mechanical process is required to generate the microphonic might not be the same process that is important for stimulating the auditory nerve. For example, von Békésy (1951) has shown that the radial shear between hair cells and tectorial membrane creates the microphonic, but this does not mean that this shear is important for creating the action potentials. This we infer generally but, actually, we don't know.

Goldstein: I know of no unequivocal proof that *mechanical* vibrations corresponding to CT's exist. We have measured neural combination tone phenomena, which in terms of their temporal properties and spectral selectivity are equivalent to tones in the stimulus; however, it is another matter to prove that corresponding mechanical vibrations exist (Goldstein and Kiang, 1968).

Johnstone: May I defend the CM to a small extent? You say that the difference tone $f_2 - f_1$ you have at low levels in the CM, which occurs roughly speaking in the same place as the fundamental, has no psychophysical correlate. When one thinks about how the system works, I would like to know how one can ever come to such a conclusion. As this difference tone is always present, how can we decide that there is no psychophysical correlate? Perhaps this correlate is also always there. What possible experiment can we do to detect it?

Dallos: All I am saying is that on the basis of audibility curves for the animal and of isopotential curves of CM we can very well calculate the amount of microphonic at the threshold of audibility. It appears, however, that in cases where one would expect, on the basis of such a calculation, that the CT should be audible, it is not. There is a discrepancy here, that is all I am saying.

Johnstone: It may be that the combination tone cannot be detected by trying beats or introducing matching tones, but it might be that this particular phenomenon shows up perhaps only as a light change in timbre.

Dallos: When you introduce a signal at a level above its presumed threshold, you should be able to detect it. Or do you mean to say that we have learned to ignore it?

Plomp: You mean that the combination tone is not present as a pure tone separated from the primary tones f_1 and f_2. If this is the case, then it would not be possible to detect the tone in psychophysical experiments.

Johnstone: That is just what I am questioning. If the combination tone stimulates the nerve fibres at one particular place, it may be that it does not beat with a probe tone stimulating a different place along the cochlear partition.

Schwartzkopff: I wonder if there is any chance that the cat is psychoacoustically different from man, which could explain the discrepancy.

Dallos: Yes, there is a very good chance but we should not ignore that the effect we are talking about is not a marginal one. These microphonic potentials are rather large in the animal. Of course, nobody has measured them in man, but if we compare the pure tone audibility curve of a guinea pig or a cat with the isopotential or microphonic curve, the differences are not very great between man and these animals. So, we can make these inferences. Obviously, it is a rather dangerous thing to do, I agree. I want to point out, however, that at least as far as the properties of $2f_1 - f_2$ are concerned, fairly good agreement was shown between psychoacoustical and 8th nerve data (Goldstein and Kiang, 1968). In spite of this, the corresponding CM data show no resemblance to the psychoacoustical data, thus what I really question is the direct causal relationship between CM and neural responses.

Kohllöffel: I do go along with Dr. Dallos in warning about extrapolation from nonlinearities in the CM for psychoacoustics, because there is another experiment that one can do. Psychoacoustically, a low-frequency tone masks a high-frequency tone more than the reverse. Now, when you look how the presence of a low tone affects the CM distribution of a high tone and the reverse, you will find just the opposite of the psychoacoustical situation. We found in our experiments that, for instance, the amplitude distribution of an 8 kHz tone is much more influenced by a simultaneous 13 kHz tone than the reverse (Fig. 1).

We may attempt an interpretation of Fig. 1 by assuming a nonlinearity in the mechano-electric transduction of the type suggested by Engebretson and Eldredge (1968). It is a feature of their model that in the case of two simultane-

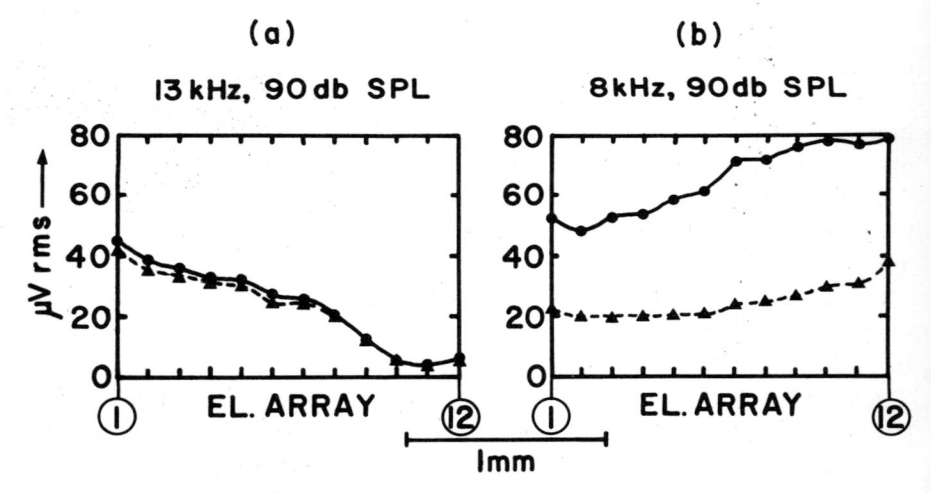

Fig. 1 (Kohllöffel). CM amplitude distributions obtained from a guinea pig. The mutual interference of two stimuli (13 kHz, 90 dB SPL and 8 kHz, 90 dB SPL) was measured. There is only slight reduction in the 13 kHz pattern while the 8 kHz pattern is strongly affected.
(a) circles   : CM amplitude at 13 kHz (90 dB).
    triangles: CM amplitude at 13 kHz (90 dB) in the presence of 8 kHz (90 dB).
(b) circles   : CM amplitude at 8 kHz (90 dB).
    triangles: CM amplitude at 8 kHz (90 dB) in the presence of 13 kHz (90 dB).
The electrode array is located in the basal scala tympani (see Kohllöffel, p. 110 of this volume, Fig. 1).

ous stimuli of unequal strength the output due to the stronger stimulus is much less affected than that of the weaker stimulus. Since the 13 kHz tone, as compared to the 8 kHz tone, causes relatively stronger mechanical disturbance in the basal part of the basilar membrane we would, indeed, expect the result shown in Fig. 1. (It should be noted that the amplitude of the local mechanical stimulus is *not directly* reflected in the CM amplitude; thus the CM amplitude at 13 kHz is smaller than that at 8 kHz. It is necessary to take into account the spatial filter effect.)

REFERENCES

Békésy, G. von (1951): Microphonics produced by touching the cochlear partition with a vibrating electrode, J. Acoust. Soc. Amer. *23*, 29-35.

Engebretson, A. M., and Eldredge, D. H. (1968): Model for the nonlinear characteristics of cochlear potentials, J. Acoust. Soc. Amer. *44*, 548-554.

Goldstein, J. L., and Kiang, N. Y.-S. (1968): Neural correlates of the aural combination tone $2f_1-f_2$, Proc. IEEE *56*, 981-992.

AURAL COMBINATION TONES

JULIUS L. GOLDSTEIN

Research Laboratory of Electronics
Massachusetts Institute of Technology
Cambridge, Mass.
and
Eaton-Peabody Laboratory of Auditory Physiology
Massachusetts Eye and Ear Infirmary
Boston, Mass., U.S.A.

INTRODUCTION

Tartini in 1714 discovered the psychophysical phenomenon of combination tones, in which human subjects perceive pitch sensations of simple tones at frequencies different from those in a complex acoustic stimulus (Jones, 1935). In its quarter millennium history the theories of the physical basis of this phenomenon have developed along with physical acoustics, mathematics of signal analysis and auditory physiology, and throughout this period the phenomenon was often taken as key evidence in favor of either a temporal or spectral theory of pitch. When the theory of acoustic vibrations was developed during the 17th and 18th centuries starting with Mersenne (1636), musical pitch became closely associated with vibratory rate. Combination tones fitted naturally into this view as being a simple consequence of the temporal interaction between any two simple sinusoidal vibrations (Lagrange, 1759; Young, 1800; Chladni, 1809).

Early in the 19th century Fourier (1822) generalized the mathematical concept that complex periodic signals can be represented as the sum of harmonically related simple sinusoidal signals (*e.g.* see Young's (1800) review). Ohm (1843) saw the relation between Fourier's theorem and the psychophysical fact that humans can perceive the pitch sensations of several overtones within complex musical sounds (Mersenne, 1636); he postulated that the total sound perception was compounded out of elemental pitch perceptions produced by each Fourier component in a periodic sound and thereby implicitly rejected the earlier temporal theory of combination tones.

Von Helmholtz (1863, Chs. 6 and 7) articulated Ohm's law by introducing the concepts that physical systems have limited frequency resolution and operate nonlinearly at large signal amplitudes. Limited frequency resolution gave time a rational place in auditory theory; von Helmholtz, however, downgraded the temporal effects of interactions among Fourier components as the cause only of undesirable dissonances in musical sounds. Musical sounds, he thought, were designed to evoke a group of spatially separated simple sinusoidal vibrations in the cochlea (the frequency-place hypothesis) and pitch was determined only by the locations of the vibrating members (the place-pitch hypothesis). Combination tones were fitted formally into this view as being distortion tones generated by middle ear waveform distortion in response to overloading by large amplitude sounds (von Helmholtz, 1863, Ch. 7 and Appendix 12).

The psychophysical facts on combination tones, according to Hermann (1891), did not support von Helmholtz's formal theory, because the perception of a pitch corresponding to the combination frequency $2f_1 - f_2$ $(1 < f_2/f_1 < 2)$ did not require intense two-tone sounds. The failure of von Helmholtz's spectral theory of combination tones, argued Hermann, meant that the phenomenon must be accounted for with a temporal theory and this made the place-pitch hypothesis questionable in every case. Rayleigh (1896) acknowledged Hermann's criticism by emphasizing the need for better data on the effects of sound intensities, although he judged that available facts on the balance supported von Helmholtz's theory of combination tones and therefore his theory of pitch. Rayleigh suggested that cancellation measurements, in which an externally generated tone is superimposed upon and cancels the presumed internal combination tone, provide a critical means for studying combination tones.

Contributions toward resolving the issues raised on combination tones in the 19th century came recently in psychophysical studies by Zwicker (1955), Plomp (1965), and Goldstein (1967), in which the former and latter supplied the measurements called for by Rayleigh. For primary acoustic stimuli consisting of two simple tones of frequencies f_1 and f_2 $(1 < f_2/f_1 < 2)$, measurements at the frequency of the most prominent combination tone $2f_1 - f_2$ contradicted von Helmholtz's theory of overloading distortion and instead were in accord with Hermann's qualitative observations, while measurements at $f_2 - f_1$ were in reasonable accord with von Helmholtz. In particular, the relative amplitude of the cancellation tone at $2f_1 - f_2$ was nearly independent of overall stimulus amplitude, but decreased typically at 100 dB per octave with increasing frequency interval, being 15-20% at $f_2/f_1 = 1.10$ (Goldstein, 1967). These measurements provided no basis for rejecting von Helmholtz's hypothesis that mechanical nonlinearity in the ear mediates combination tone phenomena; it was only necessary to reject the *nature* of the mechanical nonlinearity hypothesized.

However, in the absence of a clear physiological basis for introducing mechanical nonlinearity, no *a priori* reason exists for rejecting other theories of combination tones such as those based on temporal theories of pitch. Goldstein and Kiang (1968) provided relevant physiological evidence by investigating the responses to two-tone acoustic stimuli in single auditory-nerve fibers of anesthetized cats; they found that the auditory nerve responds as if the two-tone stimuli also contain a tone with frequency $2f_1-f_2$ and an amplitude very similar to that found in the psychophysical studies. However, since the auditory nerve responds to simple tones with both time and place information (Kiang et al., 1965, Chs. 6 and 7), the higher auditory system may proceed to use either or both as mediators of pitch. Hence the combination tone phenomena studied do not provide evidence directly relevant to theories of pitch (Schouten et al., 1962), but rather they are relevant to the physics of the transformations performed by the cochlea. The purpose of this paper is to present in integrated form the complementary portions of the psychophysical and physiological studies that were previously reported separately by Goldstein (1967) and Goldstein and Kiang (1968). The details of methodology will not be repeated here.

THE PSYCHOPHYSICAL PHENOMENON OF COMBINATION TONES

In psychophysical experiments on human perception of acoustic stimuli consisting of two simple tones of frequencies f_1 and f_2, the most frequently reported combination tones are of frequencies $2f_1-f_2$ and f_2-f_1, where $1<f_2/f_1<2$ (Zwicker, 1955; Plomp, 1965; Goldstein, 1967). A measure of the comparative SPL's of these and other combination tones is given by the results of the extensive psychophysical pitch cancellation experiment shown in Fig. 1. The two primary tones were presented monaurally and the existence of a pitch perception corresponding to a given combination frequency was ascertained. A cancellation tone at the given combination frequency was then added and its amplitude and phase were null adjusted to completely remove the pitch of the combination tone. These data appear to be consistent with results from other subjects and reports by other contemporary investigators of combination tones. The SPL of the cancellation tone is given relative to the equal levels of the stimulus tones $(L-L_{CT})$ and the level of the stimulus tones is referenced to quiet threshold at $f_1(SL_1)$. For the three combination frequencies of the form $f(n)=f_1-n(f_2-f_1)$ the level of each of the two tones in the primary stimulus was 50 dB SL_1 (approximately 50 dB SPL in this case). At the difference frequency (f_2-f_1) a higher stimulus level of 70 dB SL_1 was required, because the pitch at this combination frequency was inaudible at 50 dB. The relative level of f_2-f_1 for a 50 dB stimulus is suggested by the dashed line, which is an extrapolation based upon studies of the amplitude growth of this cancellation tone with increasing stimulus amplitude (Goldstein, 1967, Fig. 19).

Several features of the cancellation data are noteworthy:
(1) The relative amplitude at combination frequencies of the form $f(n)$ decreases

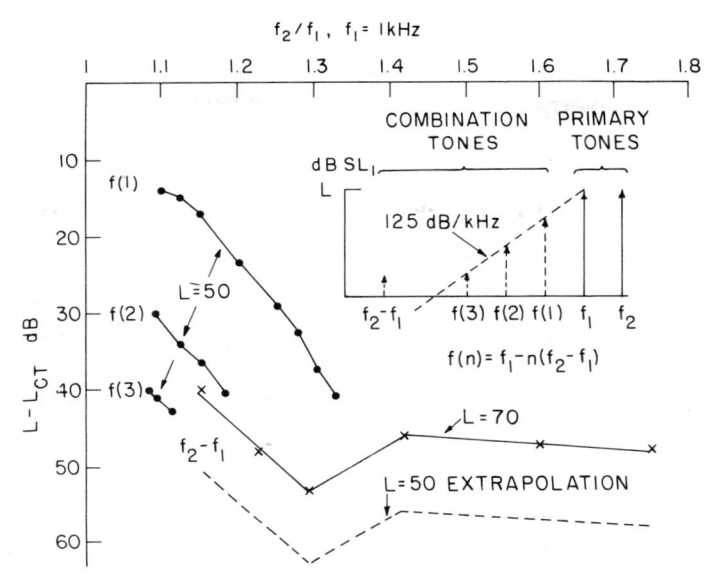

Fig. 1. Relative SPL's of pitch cancellation tones for several combination tones (subject JLG).

sharply with increasing stimulus frequency interval, while the amplitude at the difference frequency displays no clear trend. Measurements of $f(n)$ for f_2/f_1 less than about 1.10 were not possible, because the pitch of these frequencies could no longer be perceived and therefore pitch cancellation was precluded.

(2) The relative level at $2f_1 - f_2$ is generally much greater than that at $f_2 - f_1$; the former is -14 dB (20%) at $f_2/f_1 = 1.10$ and falls off at about 100 dB per octave, while the latter hovers about -55 dB (0.2%).

(3) The cancellation tone amplitudes at $f(n)$ are roughly related by a geometcir progression, which can be compactly described by the relation sketched in the insert of Fig. 1:

$$L - L_{CT}(n) \approx 125(1 - f(n)/f_1) \text{ dB.}$$

(The relation provides an approximate rule for predicting when a combination tone of the form $f(n)$ can be heard, by assuming that the CT level must be above quiet threshold and that its frequency separation from neighboring CT's must be greater than the limit of introspective frequency analysis of about 10%. Experimental data discussed later (Figs. 5 and 7) for acoustic stimuli with amplitudes and frequencies different from those in Fig. 1 suggest that the form of the above relation is not *greatly* dependent upon f_1 and L.)

THE NEUROPHYSIOLOGICAL PHENOMENON OF COMBINATION TONES

Goldstein and Kiang (1968) investigated whether peripheral auditory processing of two-tone acoustic stimuli causes single auditory-nerve fibers in anesthetized cats to respond as though the stimulus contained energy at combination frequencies. The aspects of the neural responses to the complex stimuli that were investigated were those which have been shown by Kiang et al. (1965, Chs. 6 and 7) to be characteristic of the responses to single tones; namely that all single fibers in the auditory nerve are highly frequency selective and their spike responses are time-locked to individual cycles of stimulus tones with frequencies below 4 to 5 kHz.

Fig. 2 shows the result of an experiment in which the two frequencies of the stimulus ($1 < f_2/f_1 < 2$) were chosen so that their combination frequency $2f_1 - f_2$ approximated the fiber's most sensitive frequency (called characteristic frequency or CF). Here a periodic stimulus was used and histograms of the average neural response per stimulus period, $1/(f_2 - f_1)$, are shown for one minute records of neural responses to the two-tone stimulus as well as to either tone alone. (We shall refer to these averaged response histograms as PZT histograms, an abbreviation for post, positive-going, zero-crossing time histograms. The zero-crossings of the synchronizing signal determines the zero time and the time base, or width, of the histogram. A time base of several cycles of the periodic synchronizing signal may be used for convenience. The data record length for each histogram in this paper is one minute. The full scale time is

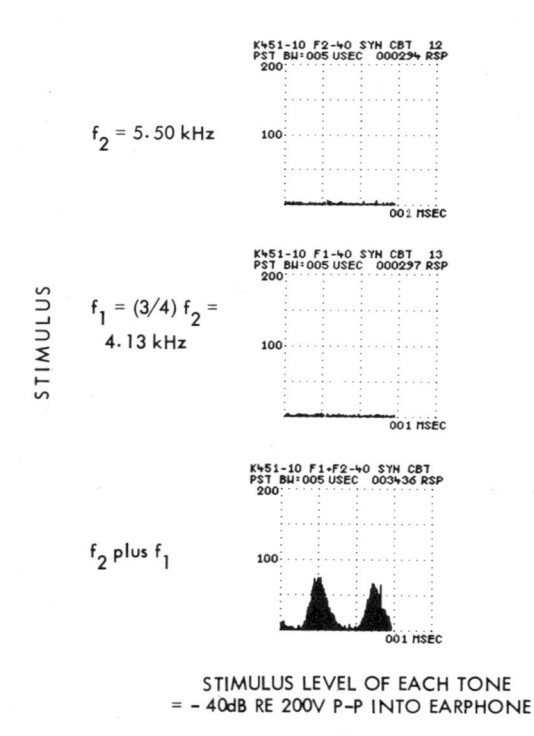

Fig. 2. PZT histograms of spike responses in single auditory-nerve fibers for single-tone and two-tone stimuli (unit K451-10, CF = 2.69 kHz). All histograms are synchronized to the periodic frequency of the two-tone harmonic stimulus $(f_2 - f_1)$.

given at the bottom of each histogram and the total number of spikes processed is given at the top — second line, right. Zero dB stimulus level into the earphone corresponds approximately to 120 dB SPL.) The fiber responded to the two-tone stimulus as though it were presented with a single tone of frequency $2f_1-f_2$; while its response to each tone alone differs insignificantly from spontaneous. The level of each tone is approximately 80 dB SPL; similar results were obtained with this stimulus at much lower and higher levels (Goldstein and Kiang, 1968, Figs. 5, 7 and 11).

When the frequency separation between stimulus tones is sufficiently small, a fiber tuned to $2f_1-f_2$ will show clear evidence of responses to the stimulus tones as well as to the combination tones. To examine a single neural record for time coherence to each of these tones, the frequencies in the two-tone stimulus must be inharmonically related. An example of this analysis is given in Fig. 3 for three levels of the two-tone stimulus. The time base of all histograms is three cycles of the indicated synchronizing tone, so that these PZT histograms reveal the average time coherence between neural responses in the one-minute records and a stimulus-locked synchronizing tone. The phase relations among the four synchronizing tones in Fig. 3 were fixed but arbitrary.

A tentative interpretation of the physiological data in Figs. 2 and 3 in terms of the model given in Fig. 4 is that the separate channels of the cochlear fre-

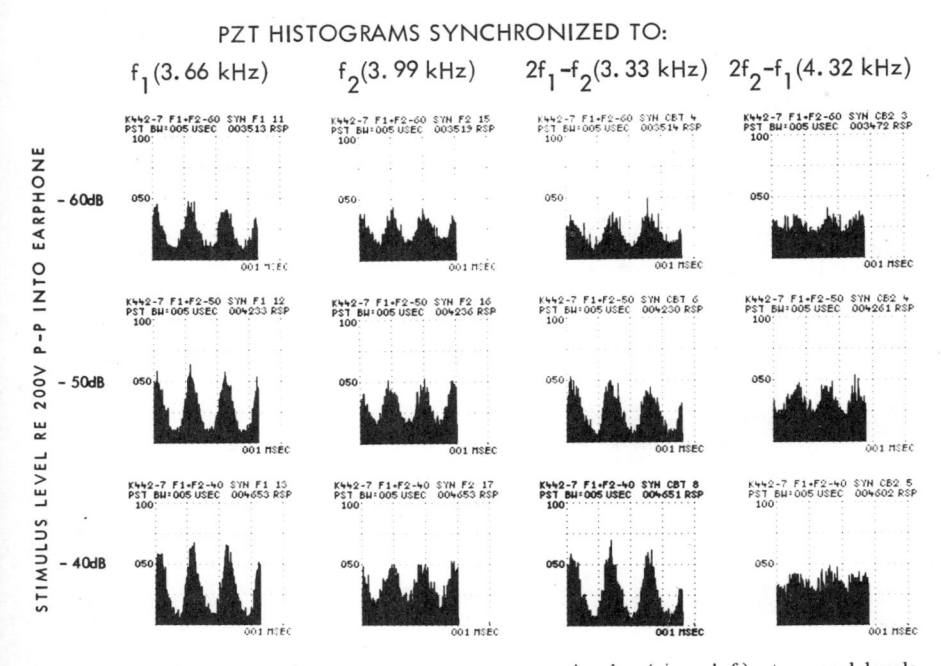

Fig. 3. PZT histograms of responses to a two-tone stimulus (f_1 and f_2) at several levels (unit K442-7, CF = 3.20 kHz).

quency analyzer respond to distortion tones arising from nonlinear waveform distortion of the stimulus that occurs *prior* to some frequency analysis. Thus the filtering process acts upon the distortion components so that a particular channel can respond selectively to a combination tone of frequency $f(n)$ only when the stimulus is sufficiently wide-band, while the analyzing filters cannot separately resolve combination tones $f(n)$ produced by a narrow-band stimulus. This view is compatible with von Helmholtz's theory of combination tones, in which he hypothesized that combination tone phenomena are generated by mechanical overloading distortion in the middle ear (von Helmholtz, 1863).

An alternative hypothesis we have considered is that combination tone phenomena are generated by the nonlinear transformation of the transducer system R-N in Fig. 4, and that the band-pass filters do not process combination tones as separate spectral components. Specifically, if the rectifier is taken as a memoryless nonlinear transformation and the probabilistic transformation as a generator of brief spikes with an instantaneous probability of occurrence that is proportional at any instant to the rectifier output, then time coherence to combination tones could be associated with the form of the rectifier nonlinearity. Thus coextensive time coherences to f_1, f_2 and $2f_1 - f_2$ as in Fig. 3 can be attributed to an odd-order nonlinearity in the rectifier transformation. An even-order nonlinearity, as in a linear halfwave rectifier, would generate coherence only to even-order combination frequencies, such as $f_2 \pm f_1$. It is evident, however, that a simple transducer mechanism cannot account for combination tone data that manifest selective tuning to spectral components at combination frequencies, as in Fig. 2 (see also Figs. 3-7 and 10 in Goldstein and Kiang, 1968). In contrast, all the combination tone data are describable as effects equivalent to a tone of frequency $2f_1 - f_2$ in the stimulus, in accord with the waveform distortion hypothesis.

BASIC EFFECTS OF STIMULUS AMPLITUDE

The failure of von Helmholtz's hypothesis of overloading distortion to account for the level dependence of the most significant combination tone

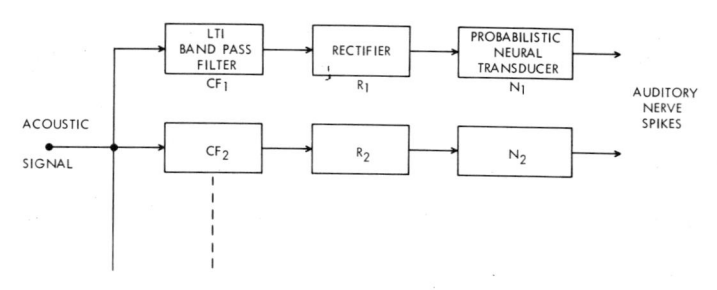

Fig. 4. Simple black-box model of auditory-nerve response.

($2f_1-f_2$) was shown by psychophysical pitch cancellation experiments in which the amplitude of the cancellation tone was found to grow almost linearly with stimulus amplitude. Fig. 5 shows this result in which the relative amplitude of the cancellation tone ranges between about 4 to 8% over a 40 dB range of the stimulus levels. In contrast, a simple cubic overloading distortion mechanism would yield a combination tone $2f_1-f_2$ of amplitude that grows in proportion to the stimulus amplitude cubed; so that the relative cancellation tone level would be expected to vary by 80 dB in this case.

If the psychophysical and physiological combination tone phenomena are mediated by the same nonlinear transformation, then the auditory nerve should respond to two-tone acoustic stimuli as though the stimuli also contain the combination tone $2f_1-f_2$ with an amplitude that is approximately proportional to the stimulus amplitude. Evidence consistent with this behavior was obtained by examining the effects of stimulus amplitude upon the average spike discharge rate of single auditory-nerve fibers for both two-tone and single-tone stimuli. Fig. 6 shows that the form of the rate functions for a given fiber is very similar for a wide variety of stimuli. For both fibers the range of stimulus levels over which the rate increases is approximately 30 dB. This dynamic range is typical for auditory-nerve fibers in response to simple tones (Kiang et al., 1965, Ch. 6). If overloading distortion were responsible for the combination tones evidenced by the neural responses, one would anticipate that the neural discharge rate should depend differently upon stimulus level depending upon (1) whether the acoustic stimulus is a single tone (curve A) or two tones (curves B-F) and (2) whether the neural channel is evincing responses to combination tones pre-

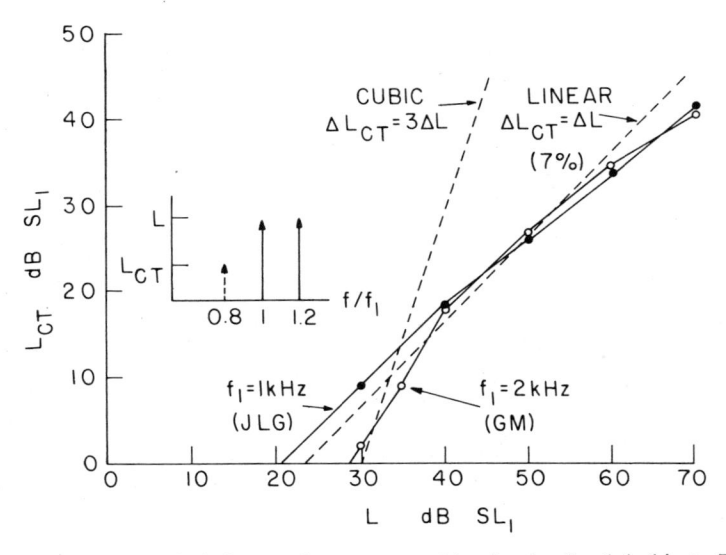

Fig. 5. Growth of level of pitch cancellation tone with stimulus level (subjects JLG and GM).

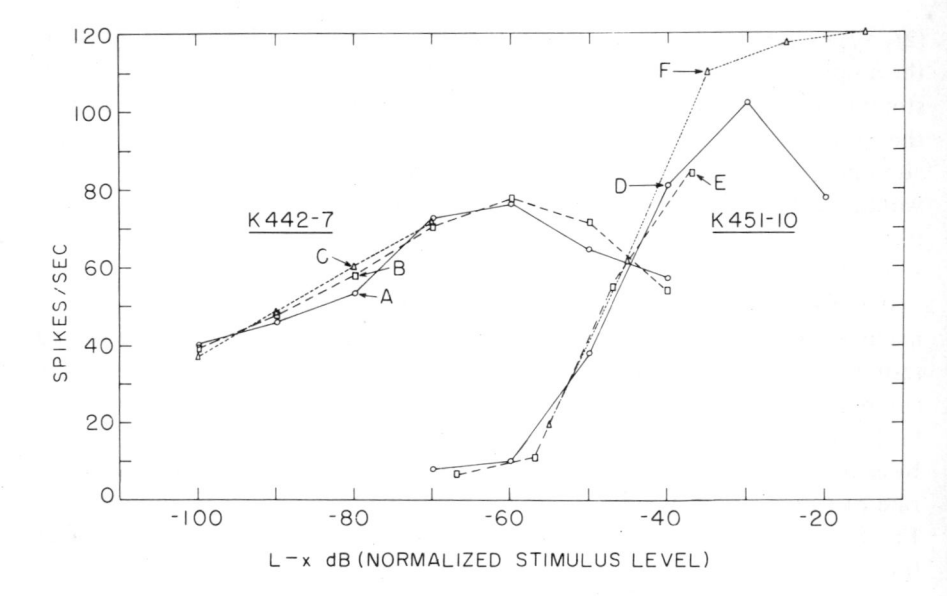

Fig. 6. Comparison of rate functions for different stimuli (units K442-7 and K451-10). L = stimulus level in dB *re* 200V P-P into earphone. X = normalizing factor given below.

Curve	f_1 (kHz)	f_2 (kHz)	X (dB)	Curve	f_1 (kHz)	f_2 (kHz)	X (dB)
A	3.33	0	0	D	3.75	4.70	0
B	3.66	3.99	20	E	4.13	5.50	7
C	4.00	4.67	30	F	1.58	2.15	−25

dominantly (curves D-E), significantly (curves B-C) or insignificantly (curves A and F). (The basis for these categories is provided by the PZT histograms shown in Goldstein and Kiang, 1968.)

BASIC EFFECTS OF STIMULUS FREQUENCIES

The data appear to be consistent with waveform distortion prior to the cochlear spectrum analysis (Fig. 4), but the distortion cannot be characterized as overloading because the relative amplitude of the distortion tone of frequency $2f_1 - f_2$ depends little upon stimulus amplitude. It might now appear that the data require a revision of only the form of the distortion mechanism that von Helmholtz hypothesized in the middle ear. However, direct studies of the middle ear (Guinan and Peake, 1967) provide no evidence for amplitude insensitive nonlinearities, and the evidence found for the low-pass behavior of the middle ear appears to be incompatible with the frequency sensitivity of the combination tone of frequency $2f_1 - f_2$.

The frequency sensitivity of the amplitude of the cancellation tone in psychophysical pitch cancellation experiments was shown previously in Fig. 1. Pitch cancellation measurements for $2f_1-f_2$ were made throughout the main portion of the audible spectrum and were found to be similarly dependent upon relative frequency interval f_2/f_1 (see Fig. 7). The rate of fall-off of the cancellation tone level is roughly 100 dB per octave in the middle frequency region (0.5 kHz $\leq f_1 \leq$ 2 kHz), while it can be considerably greater or less outside of this frequency range.

Analogous physiological cancellation experiments were performed in anesthetized cats (Goldstein and Kiang, 1968). The stimulus paradigm was the same as in the psychophysical studies, but the criterion for adjusting the cancellation tone was the removal of time coherence between the neural discharge of a single auditory nerve fiber and the combination tone. (Explicit examples will be given later in Fig. 9.) The data for this technically difficult experiment are currently limited to 20 cancellation measurements; in every case attempted it was possible to execute a cancellation measurement when sufficient time was available to adjust the cancellation tone. All the data taken over a range of stimulus levels and frequencies are *quantitatively* similar to both the relative levels and rates of frequency sensitivity of the cancellation tone for $2f_1-f_2$ found in psychophysical pitch cancellation experiments (see Fig. 8).

It is almost certain that both psychophysical and physiological experiments are measuring the effects of the same mechanism of nonlinear waveform processing within the peripheral auditory system. The site of this nonlinear mechanism is almost certainly within the inner ear, because the frequency dependence of the amplitude of the combination tone ($2f_1-f_2$) is compatible with the frequency selectivity of the inner ear (Kiang et al., 1965, Ch. 7) but not with that of the middle ear (Guinan and Peake, 1967).

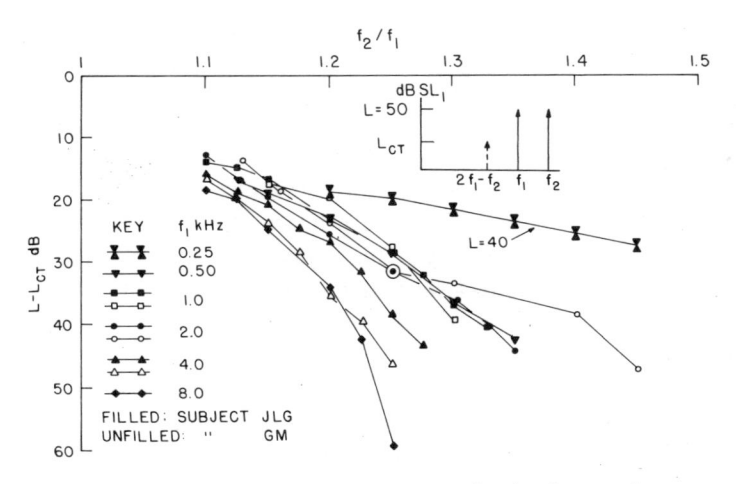

Fig. 7. Level of pitch cancellation tone *vs.* stimulus frequencies.

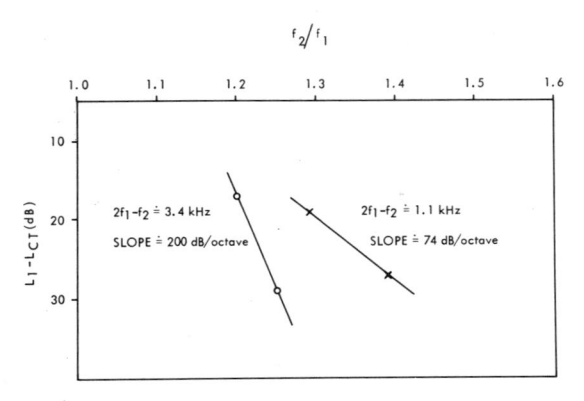

Fig. 8. Level of coherence cancellation tone vs. stimulus frequencies (units K504-14, 15, 30).

BASIC EFFECTS OF RESPONSE CHANNELS

The quantitative similarity of the behavioral and physiological data is indeed surprising in view of the facts that (1) the mechanisms responsible for combination tones are almost certainly within the distributed inner ear system, and that (2) the behavioral responses involved the whole organism whereas the physiological responses were taken from single auditory-nerve fibers. Could it be that the human subjects obtain all the information leading to the percept "pitch of $2f_1-f_2$" only from those auditory-nerve fibers with characteristic frequencies approximating $2f_1-f_2$, i.e., in accord with a place-pitch transformation?

To answer this question physiological cancellation experiments were performed in which recordings from different single auditory-nerve fibers were obtained for the same primary stimulus (f_1 and f_2). The main results of this experiment are shown in Fig. 9. The time base of all the PZT histograms is two cycles of the cancellation tone of frequency $2f_1-f_2$, so that these histograms reveal the average time coherence between the neural response in one-minute records and the externally generated tone of frequency $2f_1-f_2$. The histograms in the left column are recordings from a fiber with CF of 1.582 kHz; it was the first unit in the sequence of recordings. The top histogram in this column displays the strong time coherence to $2f_1-f_2$ (1.48 kHz) of the responses to the two-tone stimulus. The third histogram from the top shows that this time coherence is essentially removed by adding to the primary stimulus the cancellation tone with appropriate phase and level. This adjustment is a null measurement that allows discrimination of 1 dB level changes and 5° phase changes in the region of the "flat histogram" for one-minute records. The second histogram from the top shows the response to the cancellation tone without the primary stimulus. As is expected, the time pattern of the cancellation tone response is 180° out of phase with respect to the internally generated combination tone response. Another interesting point is that the rate of neural discharge in response to the two-tone stimulus is typically less than the response rate of the cancellation

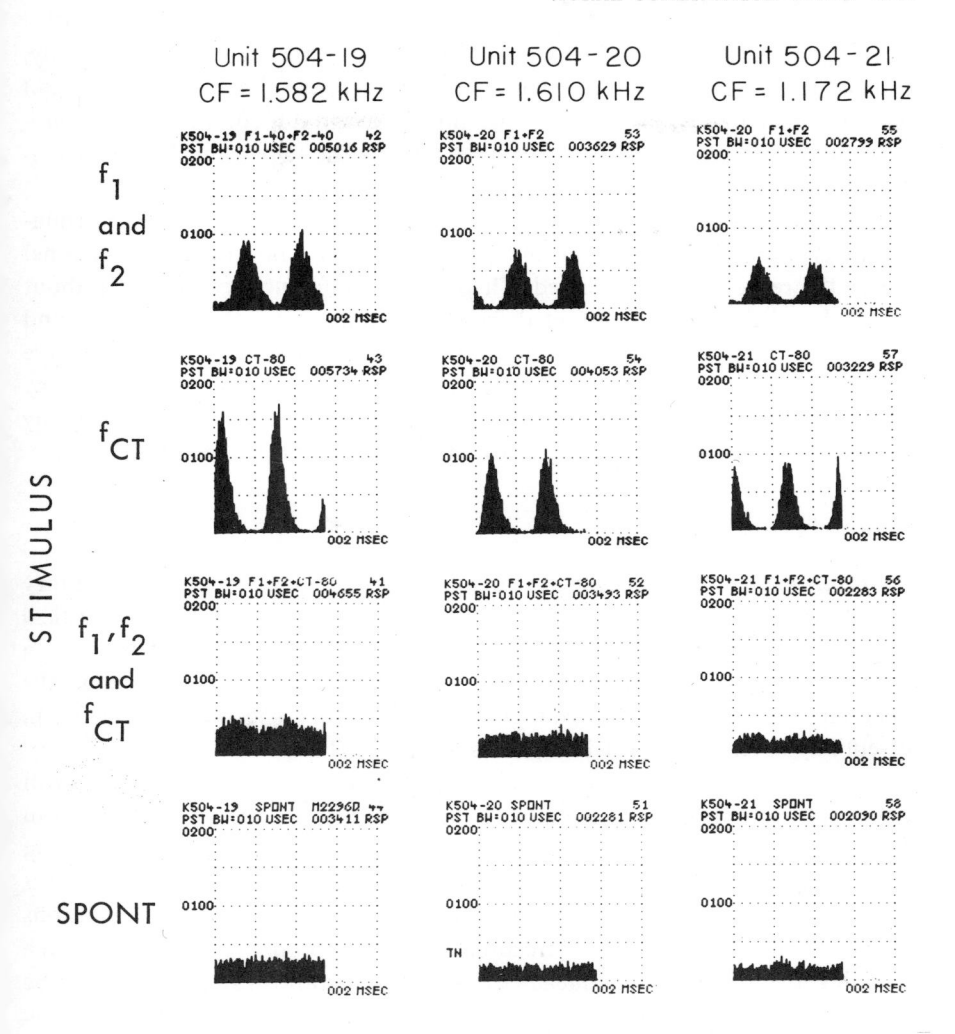

Fig. 9. PZT histograms of responses of different fibers to the same stimulus. All histograms are synchronized to $2f_1-f_2$. Stimulus frequencies: $f_1 = 2.10$ kHz; $f_2 = 2.72$ kHz; $f_{CT} = 2f_1-f_2 = 1.48$ kHz. Stimulus levels *re* 200 V P-P into earphone: $L_1 = -54$ dB; $L_2 = -54$ dB; $L_{CT} = -74$ dB.

tone alone. This effect might be attributable to a partial suppression by the primary tones of the response to the internally generated combination tone (Sachs and Kiang, 1968). The bottom histogram shows the spontaneous response of the fiber. The rate of discharge with the combination tone cancelled is between the spontaneous rate and the rate in response to the primary stimulus.

Recordings from the next fiber in the sequence are analyzed in the middle column. The four stimulus conditions are as before except that the phase of the cancellation tone was changed by 4° to improve the cancellation slightly. Finally in the right column recordings are analyzed for a fiber with a CF about half an octave lower than the previous unit. The stimulus conditions are identical in all respects with the previous case. Clearly the same cancellation tone is effective for this fiber.

It is clear that the combination tone measured in the auditory nerve is primarily determined by the stimulus and is not dependent upon the neural channel where the responses are observed. Therefore, no inference can be made about the place where pitch is extracted from the similarity of the behavioral and physiological data. Indeed, not only single fibers, but the whole auditory nerve responds to two-tone stimuli as if the stimuli also contain a tone with frequency $2f_1 - f_2$. As is the case for responses to simple tones, the auditory system may proceed to use time and/or place information as mediators of pitch.

COMMENTS ON FURTHER WORK

Spectrum analysis is a key feature of all modern functional models of auditory signal processing that have been used to provide quantitative theoretical descriptions for psychophysical and physiological phenomena. Combination tones are pertinent to these models, because their properties challenge the generality of the classical assumption that auditory spectral filtering is an essentially linear, time-invariant process.

An extreme example of the spectral interactions that could not occur with linear spectrum analysis is given by the psychophysical experiment outlined in Fig. 10. The top graph describes the parameters of a two-tone stimulus along with the frequencies corresponding to audible combination tones; the cancellation data for this stimulus were given in Fig. 1. The middle and bottom graphs in Fig. 10 show that complex perceptual effects are produced by adding to the stimulus a third tone of frequency $2f_2 - f_1$. The middle graph shows that the pitch corresponding to the combination frequency $2f_1 - f_2$ was eliminated by the third tone for a particular phase setting and for levels equal to or greater than 35 dB (the highest level tested was 55 dB). In contrast with the preceding "saturation" adjustment, the bottom graph shows that the combination tone $3f_1 - 2f_2$ was eliminated by a null adjustment of the third stimulus tone.

Besides the need for a more complete functional model of the cochlea, there is the important and related need for an understanding of the structural or physiological basis of the combination tone phenomena. The physiological study has reasonably established that the auditory nerve responds to complex stimuli as though the cochlear spectrum analyzer induces a distortion upon the stimulus and then proceeds to analyze that distorted waveform.

Some tentative speculations have been offered that the physical basis for combination tones is a nonlinear mechanical motion of the cochlear partition

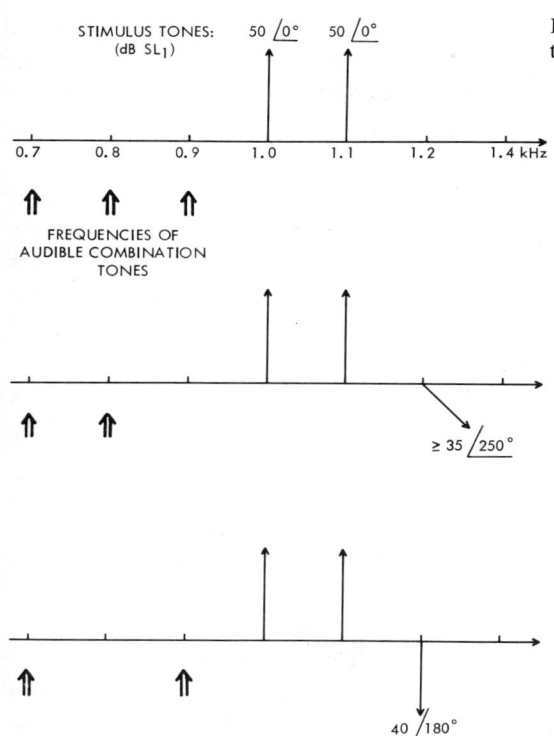

Fig. 10. Perception of complex interaction among tones.

that is produced either by hydrodynamic nonlinearities in the cochlear fluids or by nonlinear mechanical coupling between the basilar and tectorial membranes (Goldstein, 1967; Crane, 1968; Tonndorf, 1969). The information necessary to describe the physics of these phenomena is yet incomplete and therefore gross assumptions are necessary to obtain quantitative comparisons between theory and data. Some physically meaningful model computations were executed as outlined elsewhere (Goldstein, 1967). The linear mechanical band-pass response of each place along the cochlear partition was assumed to be distorted by a normalized cubic function whose form was suggested by the empirical data. The distortions arising at different places in the distributed system were then weighted by a linear function equivalent to the response of each place and then added to the effective stimulus. In the original computations, the phase characteristic of the linear response was ignored. Since the phase characteristic could greatly influence the relationship between the net distortion and the stimulus parameters (Whitfield and Ross, 1965), machine computations were recently executed using various *minimum phase* response and weighting functions.

In one computation, the minimum phase response function was a two-piece Bode approximation to the basilar membrane response reported by von Békésy

(Flanagan and Bird, 1962) and the weighting function was a constant. The amplitudes of the various distortion tones ($2f_1 \pm f_2$, $2f_2 \pm f_1$, $3f_1$ and $3f_2$) did not in the remotest way bear the same relation to the frequency ratio f_2/f_1 as in the empirical data. Similar computations were made with various high-Q minimum phase weighting function was added to the high-Q model, did the computations and again the computations did not model the data. Only when the minimum phase weighting function was added to the high-Q model, did the computations roughly model the data. These computations do not encourage the view that the combination tone phenomena can be accounted for simply with an odd-order amplitude normalized distortion of the broadly tuned displacement of the cochlear partition.

Further physiological studies of cochlear processes are clearly necessary for a detailed physiological understanding of combination tones. A possible source of pertinent new psychophysical information is suggested by the study of monaural diplacusis reported by Ward (1955) in which a pathological auditory state produces behavioral responses to simple tones that mimic the combination tones generated by normal cochlear responses to two-tone stimuli.

SUMMARY

Cochlear sound processing throughout its full operating range incorporates a nonlinear mechanism that generates perceptual and neural responses corresponding to distortion products of neighboring spectral components in the stimulus. Evidence for these distortion products was first obtained from psychophysical studies in man of aural combination tones heard with stimuli consisting of two simple tones. The most prominent combination tone has the frequency $2f_1 - f_2$, where $1 < f_2/f_1 < 2$; its relative amplitude is nearly independent of overall stimulus amplitude, but decreases at typically 100 dB per octave with increasing frequency interval, being 15-20% at $f_2/f_1 = 1.10$. Neural responses from single auditory-nerve fibers in anesthetized cats were next studied to examine the physiological validity of the psychophysical conception that perception of combination tones is mediated by cochlear distortion products. The auditory nerve responds to two-tone acoustic stimuli as if the stimuli also contain a tone with frequency $2f_1 - f_2$ and with an amplitude as found in the psychophysical studies.

ACKNOWLEDGEMENT

I am grateful to Dr. Nelson Y. S. Kiang for his major contribution to the reported research; he performed all of the physiological experiments. The continuing stimulation and criticism of my colleagues Dr. Nelson Kiang in particular and Professors William Peake, William Siebert and Tom Weiss contributed to the development of this research.

This work was supported in part by the National Institutes of Health (Grant 5 P01 GM14940-03), the Joint Services Electronics Program; and in part by Grant NB-01344, National Institute of Neurological Diseases and Blindness, NIH, PHS.

REFERENCES

Chladni, E. F. F. (1809): Traité d'acoustique (Courcier, Paris), pp. 251-254.
Crane, H. D. (1968): Mechanical impact and fatigue in relation to nonlinear combination tones in the cochlea, Stanford Research Institute, unpublished report.
Flanagan, J. L., and Bird, C. M. (1962): Minimum phase responses for the basilar membrane, J. Acoust. Soc. Amer. *34*, 114-118.
Fourier, J. (1822): The Analytical Theory of Heat (Dover Reprint, 1955, New York).
Goldstein, J. L. (1967): Auditory nonlinearity, J. Acoust. Soc. Amer. *41*, 676-689.
Goldstein, J. L., and Kiang, N. Y. S. (1968): Neural correlates of the aural combination tone $2f_1-f_2$, Proc. IEEE *56*, 981-992.
Guinan, J. J., and Peake, W. T. (1967): Middle-ear characteristics of anesthetized cats, J. Acoust. Soc. Amer. *41*, 1237-1261.
Helmholtz, H. L. F. von (1863): Sensations of Tone (Dover Reprint, 1954, New York).
Hermann, L. (1891): Zur Theorie der Combinationstöne, Pflügers Arch. ges. Physiol. *49*, 499-518.
Jones, A. T. (1935): The discovery of difference tones, Am. Phys. Teacher *3*, 49-51.
Kiang, N. Y. S., et al. (1965): Discharge Patterns of Single Fibers in the Cat's Auditory Nerve (Research Monograph No. 35, M.I.T. Press, Cambridge, Mass.).
Lagrange, J. L. (1759): Recherches sur la nature et la propagation du son, in: Oeuvres de Lagrange, Vol. 1, compiled by J. A. Serret (Gauthier-Villars, 1867, Paris), pp. 39-148.
Mersenne (1636): as discussed in Rayleigh, *op. cit.*, Vol. 1.
Ohm, G. S. (1843): Ueber die Definition des Tones, Ann. Phys. Chem. *59*, 513-565.
Plomp, R. (1965): Detectability threshold for combination tones, J. Acoust. Soc. Amer. *37*, 1110-1123.
Rayleigh, J. W. S. (1896): The Theory of Sound (Dover Reprint, 1945, New York), Vol. 2, Ch. 23.
Sachs, M. B., and Kiang, N. Y. S. (1968): Two-tone inhibition in auditory-nerve fibers, J. Acoust. Soc. Amer. *43*, 1120-1128.
Schouten, J. F., Ritsma, R. J., and Cardozo, B. L. (1962): Pitch of the residue, J. Acoust. Soc. Amer. *34*, 1418-1424.
Tonndorf, J. (1969): Nonlinearities in cochlear hydrodynamics, J. Acoust. Soc. Amer. *45*, 304-305(A).
Ward, W. D. (1955): Tonal monaural diplacusis, J. Acoust. Soc. Amer. *27*, 365-372.
Whitfield, I. C., and Ross, H. F. (1965): Cochlear microphonic and summating potentials and the outputs of individual hair-cell generators, J. Acoust. Soc. Amer. *38*, 126-131.
Young, T. (1800): Outlines of experiments and inquiries respecting sound and light, Phil. Trans. Roy. Soc. London, Pt. 1, 106-150.
Zwicker, E. (1955): Der ungewöhnliche Amplitudengang der nichtlinearen Verzerrungen des Ohres, Acustica *5*, 67-74.

DISCUSSION

De Boer: I want to compliment Dr. Goldstein for his beautiful presentation. I think it is important that two types of descriptions are possible, yours and the one developed in my paper. I agree with your point of view that there are cases which suggest that a very specific nonlinearity precedes the linear filtering. My description can always be applied but appears to be insufficient for all data. The puzzling questions of a distortion relatively independent of SPL and, for example, the audibility of $2f_1-f_2$ while $2f_1+f_2$ is not heard, still have not been solved.

Goldstein: The main point of my talk today is that all the physiological data on combination tones are describable as effects equivalent to a tone of frequency $2f_1-f_2$ in the stimulus. For some of the data, when the fibre is responsive to both tones in the stimulus, we have no basis yet for distinguishing between transducer nonlinearity as you propose or spectral nonlinearity as I propose. I make a conceptual distinction between these two types of nonlinearity for convenience; ultimately I believe that after we do more experiments and our understanding grows we will find that the nonlinear processes are coupled and that the transducer process is more complex than you propose. Indeed the *whole* puzzle of combination tones in psychophysics and the auditory nerve has not been solved; however, I think some progress has been made.

Smoorenburg: Do you think that a certain phase locking of nerve spikes to an externally generated frequency f implies that adequate information for a sensation corresponding to this frequency f is present in the auditory system? Notwithstanding the phase locking, an interval histogram of the same train of spikes does not necessarily show intervals corresponding to integral multiples of $1/f$ as you can deduce from the reasoning in de Boer's paper (pp. 208-212).

Goldstein: I do not believe that phase locking between the neural spikes and an external reference tone is adequate to provide a sensation of that tone. The most important evidence for this is that neural phase locking to the difference frequency (f_2-f_1) can be found for all sound levels, while the pitch corresponding to the difference frequency is perceived only at high sound levels. Conversely, no neural phase locking can be found above 5 kHz, while psychophysical measurements of $2f_1-f_2$ extend up to 7.2 kHz and there is no clear evidence for any upper frequency limit on the perception of this combination tone (Goldstein, 1967). Yes, phase locking to a combination tone only implies that the stimulus underwent a nonlinear transformation. Other data are required, as we found (Goldstein and Kiang, 1968, Figs. 3-7 and 10), to determine whether the nonlinear transformation is imbedded in or followed by a system with memory, *e.g.* a filtering process.

Helle: We did some experiments with cancellation tones (Helle, 1969/70). We varied the levels of f_1 and f_2 (L_1 and L_2) simultaneously, starting at a low level, while $L_2 = L_1 - 10$ dB. With f_2 about equal to $1.3 f_1$ and primary levels of about 50 dB, we found that the level of the cancellation tone did not increase monotonically with increasing L_1 and L_2; there was a discontinuity and even a region where the level of the cancellation tone decreased with increasing L_1 and

L_2. This discontinuity was accompanied by a large phase shift. Did you find similar relations in your experiments?

Goldstein: I have no systematic psychophysical cancellation data for the condition you mentioned. For all the conditions that I studied, the cancellation level increased nearly linearly with overall stimulus level. While my finding is the interesting result to me, research on the cancellation behaviour should not stop with that finding. The most complete physiological combination tone data on the effects of level that we have appears in Figs. 2, 3, 7, 11 and 12 of the paper I coauthored with Nelson Kiang (Goldstein and Kiang, 1968). We have no data inconsistent with the published data.

REFERENCES

Goldstein, J. L. (1967): Auditory nonlinearity, J. Acoust. Soc. Amer. *41*, 676-689.
Goldstein, J. L., and Kiang, N. Y.-S. (1968): Neural correlates of the aural combination tone $2f_1-f_2$, Proc. IEEE *56*, 981-992.
Helle, R. (1969/70): Amplitude und Phase des im Gehör gebildeten Differenztones dritter Ordnung, Acustica *22*, 74-87.

Section 6

PITCH PERCEPTION

Any theory of hearing has to deal with the question of whether pitch perception is based upon frequency analysis or/and upon periodicity detection. This question is reconsidered critically in the following papers. The experiments reported are on a variety of topics, including both complex and simple tones, periodic and nonperiodic sounds, and monaural and binaural presentations. In several cases the results can be explained quite nicely in terms of periodicity models, in others alternative explanations are proposed.

PERIODICITY DETECTION

ROELOF J. RITSMA

Institute of Audiology
University Hospital
Groningen, The Netherlands

INTRODUCTION

Nowadays von Helmholtz's concept (1863) that there is a one-to-one relation between the frequency of the acoustic signal and a specific pitch is generally interpreted as composed of two hypotheses in hearing theory. The first hypothesis states the one-to-one relation between the frequency of the acoustic signal and the place of maximal stimulation on the basilar membrane. The validity of this relation has been experimentally confirmed by the work of von Békésy (1960), among others. This implies that the ear is capable of carrying out a Fourier analysis of any complex sound, although with only a limited power of resolution. A trained observer is able to perceive the lower harmonics of a periodic pulse individually.

Along with this analyzing faculty, we are able to hear a complex as one single percept. This percept may have a decidedly low pitch, though no corresponding low frequency need be physically present. This phenomenon contradicts von Helmholtz's second hypothesis, that there is a one-to-one relation between the place of maximal stimulation on the basilar membrane and a specific pitch, in such a way that pitch decreases gradually from the apical to the basal end of the membrane. However, this phenomenon may fulfil an alternative hypothesis, formulated by Wundt (1880), that tones give rise to synchronous nerve impulses whose rate determines pitch. This hypothesis implies that the ear can determine the periodicity in a sound.

A closer study of pitch of complex sounds is of interest, because there are situations where average frequency and periodicity are widely different and may be varied fairly independently.

COMPLEX SOUNDS WITH BROAD BANDWIDTH

In the past many scientists reported pitch measurements of complex sounds with broad bandwidth. Dependent on the sounds used, they referred to these pitches as residue pitch, periodicity pitch, time separation pitch, repetition pitch, time difference tone, reflection tone, and sweep tone.

The question arises whether these pitch effects are really different or whether they are essentially of the same nature. Let us consider the examples given in Fig. 1.

The first signal, A, is an unfiltered periodic unipolar pulse train. The listener will hear a definite pitch, corresponding to the fundamental. This pitch often does not change even though the fundamental and a number of lower harmonics are eliminated. The only perceivable change is one of timbre. This joint perception of a number of neighbouring higher harmonics of a periodic signal Schouten (1940) called the residue. The general review of Schouten (pp. 41-58 of this volume) gives a wealth of information in this field.

This pitch phenomenon can be explained rather simply both in the frequency domain and in the time domain. Problems crop up if the even-numbered pulses are phase shifted over 180°. In that case one obtains a periodic pulse train of alternating polarity. For repetition rates above 300 pps a pitch is heard corresponding to the fundamental, but for repetition rates below 300 pps a low pitch is heard roughly corresponding to the number of pulses per second. As was reported by de Boer (1956), Flanagan and Guttman (1960), Guttman and Flanagan (1964), and Rosenberg (1965), who performed accurate listening experiments, two pitch values are found which have a significant departure from the pulse rate. One pitch value is somewhat higher, the other somewhat lower than the pulse rate. These pitch values cannot be explained in terms of either frequency analysis or of time analysis alone.

A similar situation occurs when taking signal B: pulse pairs with a constant

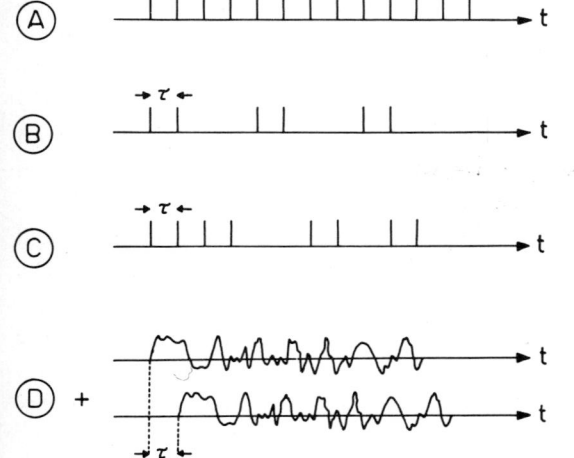

Fig. 1. Examples of sounds used in pitch experiments.

time distance τ presented periodically, or signal C: pulse pairs presented at random. The spectral envelopes of both types of signals are identical, but signal B has a discrete frequency spectrum whereas signal C has a continuous frequency spectrum. These signals evoke a pitch referred to as time separation pitch and time difference tone in the literature. If both pulses of a pulse pair have the same polarity, the pitch is the same as the pitch of a pure tone with a frequency of $1/\tau$ (Small and McClellan, 1963; McClellan and Small, 1963, 1965a, 1965b, 1966, 1967). They used different kinds of signal: a periodic sequence of pulse pairs, a random sequence of pulse pairs, periodic sequences of pairs of different or identical noise bursts, and single pulse pairs. Because they gave the original sound and the repetition always the same polarity, they could not decide whether spectral cues or timing cues are responsible for the perception of this pitch. One thing, however, appeared to be evident, namely that time separation between successive sound events alone is insufficient for the evocation of pitch. Obviously, successive sound events must be highly correlated. Small and McClellan further concluded that for the test signals listed above the amount of information available per unit time is not particularly important, because the distribution of pitch matchings appeared to be about the same for the different test signals. Reversing the following pulse, that is, shifting it 180° in phase, a pitch jump upward or downward is perceived (Thurlow, 1958). As Nordmark (1963) and Bilsen (1966) found, the antiphasic condition has two pitches, one a little higher, the other a little lower than for the cophasic condition. For the antiphasic condition no explanation can be given in terms of frequency analysis exclusively or of time analysis exclusively.

Finally, consider signal D, in which noise and its delayed repetition are added. The frequency spectrum of signal D is the same as that of signal C. The main difference between them is that for signal C the pulse pairs are given successively whereas for signal D pulse pairs are intertwined. Signal D evokes a tone, the repetition pitch or reflection tone, with a pitch that corresponds to $1/\tau$. According to Fourcin (1965) and Wilson (1966), if the delayed repetition is shifted 180°, the signal will evoke a pitch sensation corresponding to $7/(8\tau)$. Again, the antiphasic condition is ambiguous, one pitch is a little higher, the other a little lower than for the cophasic condition (Bilsen, 1966). These pitches cannot be explained by an autocorrelation function, or by the spacing of the spectral peaks, or by the position of the first of the series of spectral peaks.

If we are to hypothesize the same pitch mechanism both for the case of equal polarity pulses and for the case of alternating polarity pulses, the mechanisms proposed in the literature for the case of equal polarity are not acceptable. Thus, the problem of formulating a pitch mechanism that is valid independently of the signal applied is still open.

In approaching this problem we will consider the pitch perception of complex sounds with narrow bandwidth.

COMPLEX SOUNDS WITH NARROW BANDWIDTH

Narrow-band signals are, for example, the signals A, B, C, and D of Fig. 1 to which a 1/3-octave band-pass filter is applied, and amplitude modulated sinusoids. Experiments with models of the basilar membrane indicated that the response to wide-band signals may depend on an analyzing or filtering mechanism (Schouten, 1940; Flanagan, 1962, 1965), because different waveforms appeared at different places along the membrane (Fig. 6). Narrow-band signals, however, produce a "ready made" envelope at the place on the basilar membrane corresponding to the frequency region of the signal, and so we can be sure that the displacement waveform at that place is about the same as the acoustic waveform.

Many experiments with narrow-band signals have been reported in the literature. It was found that stimulation of a particular area of the basilar membrane may give rise to widely different sensations of pitch ranging from the highest for stimulation with a pure tone down to about one-twentieth of that value (Ritsma, 1962; Walliser, 1969a, 1969b). The explanation of residue pitch and of time separation pitch in terms of nonlinear behaviour of the mechanical part of the ear has been disproved by a number of arguments, including the occurrence of both pitches at moderate loudness, and the behaviour of the residue with respect to beats and masking (Schouten, 1940; Licklider, 1954; Thurlow and Small, 1955). Hence, the pitch must be evoked at the stimulated area exclusively.

Licklider (1951) assumes the *temporal envelope* to be the most important clue for residue pitch. This possibility has been disproved by the following experiment. A simple signal, yet exhibiting the residue effect, is a sinusoid of frequency $f=2000$ Hz, modulated in amplitude by a sinusoid of frequency $g=200$ Hz. The spectrum consists of three components with the frequencies 1800, 2000, and 2200 Hz. The sound is heard to have a pitch corresponding to a pure tone of 200 Hz. Within certain limits the pitch does not vary when the centre frequency f is a different multiple of 200 Hz. The pitch of such a complex changes when the frequency of each of the components is increased by the same amount. For example, if the components are raised by 50 Hz, the pitch jumps from 200 Hz to about 205 Hz.

This phenomenon is known in the literature as the first effect of pitch shift (Schouten, 1940; de Boer, 1956). It follows that, in this situation, the envelope of the time function cannot be a relevant parameter in determining pitch, since the frequency of the envelope remains 200 Hz, irrespective of the frequency shift. Moreover, the pitch cannot be determined on the basis of *frequency differences*, as the spacing between the components remains the same.

In the case of continuous noise with its repetition (signal D of Fig. 1) pitch cannot possibly result from a process of detection of a temporal envelope.

A second possibility has been suggested by Schroeder (1966) and by Walliser (1969a): pitch is based on *subharmonics* of one dominant frequency component in the signal spectrum. As Walliser pointed out, the lowest frequency component of the frequency spectrum appears to be the dominant frequency component. For an AM signal with frequency components $(f-g)$, f, and $(f+g)$ the pitch corresponds to the pitch of a pure tone with frequency P:

$$P = (f-g)/(n-1) \tag{1}$$

with $n = f_o/g$ if f_o is the nearest centre frequency giving a harmonic complex.

This possibility has been disproved by experiments on the pitch behaviour of a quasi-FM signal, derived from an AM signal by shifting the phase of the centre frequency 90°. In the AM situation, the sound is heard to have a distinct pitch and a sharp timbre; in the quasi-FM situation, the sound has a much weaker, low, pitch with a fuller timbre. It is not clear whether the low pitch of the quasi-FM signal corresponds to the frequency g or $2g$.

Ritsma and Engel (1964) made histograms of the distribution of pitch judgments of two subjects both in the region of g and of $2g$. As the form of these histograms did not in the first instance differ from each other, they join the four histograms into one histogram. Fig. 2 gives the combined histogram of pitch judgments for a quasi-FM signal with a ratio $n = f/g = 12$, a modulation index $m = 2.55$, and a centre frequency $f = 2000$ Hz. For simplicity, the frequency values found in the region of the frequency g have been multiplied by a factor of 2. The dash-dotted line represents the Gaussian distribution of the judgments. For the AM signal with the same parameters, both subjects heard a pitch corresponding to the modulation frequency $g = 167$ Hz.

From this experiment one must conclude that, in themselves, the frequency components constituting the complex do not contribute to the resulting sensation of pitch. This conclusion fits in with the pitch behaviour of the signals of type C and D of Fig. 1 which do not possess a dominant frequency component

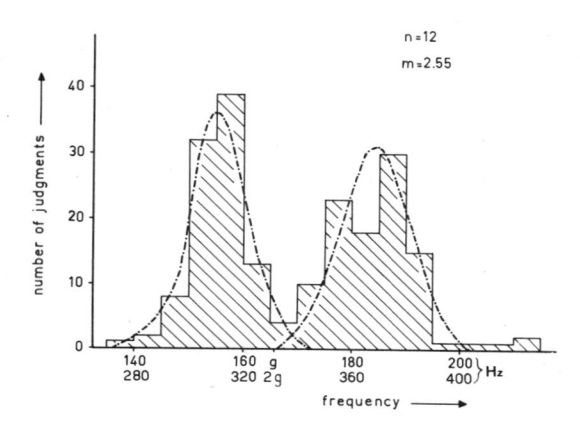

Fig. 2. Histogram of 200 pitch judgments by two subjects for a quasi-FM signal with a ratio $n = f/g = 12$, a modulation index $m = 2.55$, and a centre frequency $f = 2000$ Hz. The dash-dotted line represents the Gaussian distribution of the judgments.

with a determinable subharmonic and with the sweep-tone effect that has been observed in beats of mistuned consonances (Plomp, 1967b).

Because of the dependence on phase one must conclude that the time function of the stimulation in a particular area on the basilar membrane is the decisive factor in bringing about the pitch perception.

A remaining possibility is that the pitch of narrow-band signals is determined by the time interval between two positive peaks in the *fine structure* of the waveform in or near two successive crests of the envelope of this waveform (de Boer, 1956; Schouten et al., 1962). This is indicated in Fig. 3a. The solid line represents the waveform of a sinusoid of frequency f modulated 100 per cent by a sinusoid of the frequency g. Mathematically this signal has a period $1/g$ if and only if g is an integral divisor of f.

The pitch of this signal corresponds closely to the time interval $1/P$. Therefore, the time interval $1/P$, an integral multiple of the distance $1/f$, serves as a "period". The pitch values found in the experiment with quasi-FM signals have confirmed the concept of "fine structure" detection. The tops of the histogram of Fig. 2 are seen to be in reasonable agreement with the time intervals $1/P_1$ and $1/P_2$ between the peaks in the fine structure near the crests of the envelope of the time function of the quasi-FM signal (Fig. 3b).

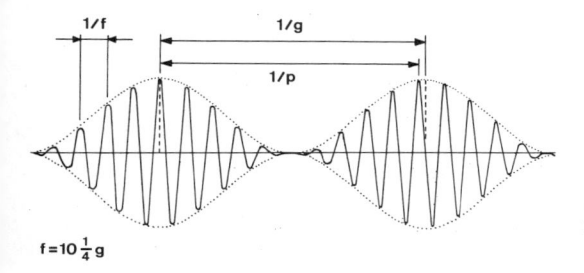

Fig. 3a. Displacement waveform of an amplitude modulated sinusoid.

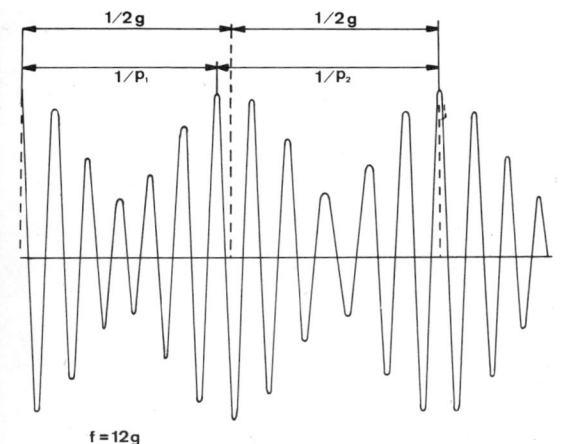

Fig. 3b. Displacement waveform of a quasi-FM signal.

The pitch behaviour of inharmonic complexes cannot be described precisely by

$$P = f/n \tag{2}$$

or by the concept of subharmonics of a dominant frequency component predicting a pitch P according to Eq. (1). This is shown in Fig. 4 giving the residue pitch for an AM signal as a function of centre frequency f for constant modulation frequency g.

The residue pitch P can be described by the empirical formula

$$P = (f - cg)/(n - c). \tag{3}$$

The empirical constant c as a function of the ratio $n = f/g$ is given in Fig. 5. According to Eq. (3) the residue pitch of an AM signal with constant centre frequency decreases for small increases in modulation frequency g. The deviation of the real pitch value from the theoretical pitch value according to Eq. (2) is mentioned the second effect of pitch shift.

From experiments it is known that the constant c increases for higher sensation levels. For a sensation level of about 15 dB c is zero, irrespective of the value of n.

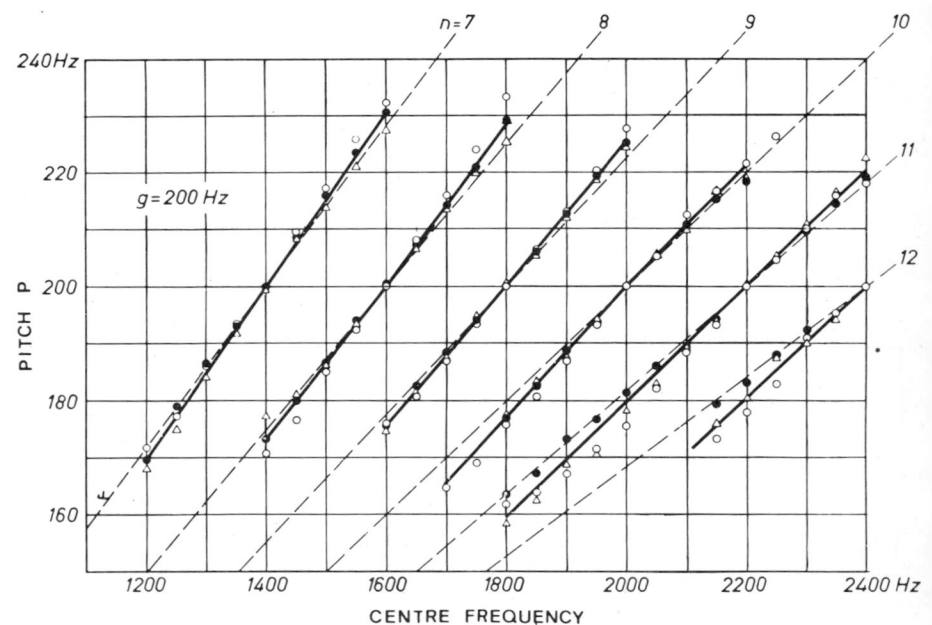

Fig. 4. Pitch as a function of the centre frequency for a three-component complex. The modulation frequency g is 200 Hz. The circles, triangles and dots represent the means of twelve matchings made by three subjects, respectively. The sensation level is about 35 dB. Solid lines are drawn to best fit the experimental points. Dashed lines represent the first effect of pitch shift given by Eq. (2) (after Schouten et al., 1962).

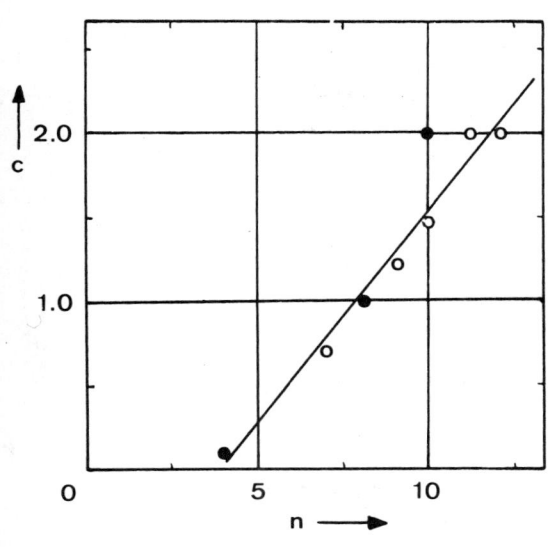

Fig. 5. The empirical constant c of Eq. (3) as a function of $n = f/g$. The open circles represent the results given in Fig. 4; the closed circles have been measured separately. Sensation level is 35 dB.

THE CONCEPT OF DOMINANCE

We now return to the problem of pitch in broad-band signals. These signals give displacement waveforms shaped by the specific properties of the membrane at different places along the membrane. Dependent on the signals used, these membrane waveforms will differ strongly at different places. Fig. 6 gives an outline of various membrane responses to broad-band periodic pulses of alternating polarity (*cf.* Flanagan, 1965).

The displacement waveform of the basilar membrane at the place that is maximally sensitive to the frequency f_o is similar to the waveform of the acoustic signal passing through an adequate band-pass filter with centre frequency f_o and bandwidth Δf.

The response of an ideal band-pass filter with centre frequency f_o and bandwidth Δf to a Dirac pulse can be shown to be equal to

$$\delta_o(t) = \frac{2}{\pi t} \sin \pi \Delta f t \cos 2\pi f_o t. \qquad (4)$$

In the same way we have, for the response to a 90° phase shifted Dirac pulse,

$$\delta_{90}(t) = \frac{-2}{\pi t} \sin \pi \Delta f t \sin 2\pi f_o t. \qquad (5)$$

For $t = 0$ the envelope $\frac{2}{\pi t} \sin \pi \Delta f t$ is at maximum. Therefore, in this vicinity the fine structure has its greatest peaks. In particular, $\delta_o(t)$ has its major positive peak for $t = 0$; $\delta_{90}(t)$ for $t = -1/(4f_o)$; $\delta_{270}(t)$ for $t = 1/(4f_o)$; and $\delta_{180}(t)$ has two major peaks, one for $t = 1/(2f_o)$ and the other for $t = -1/(2f_o)$.

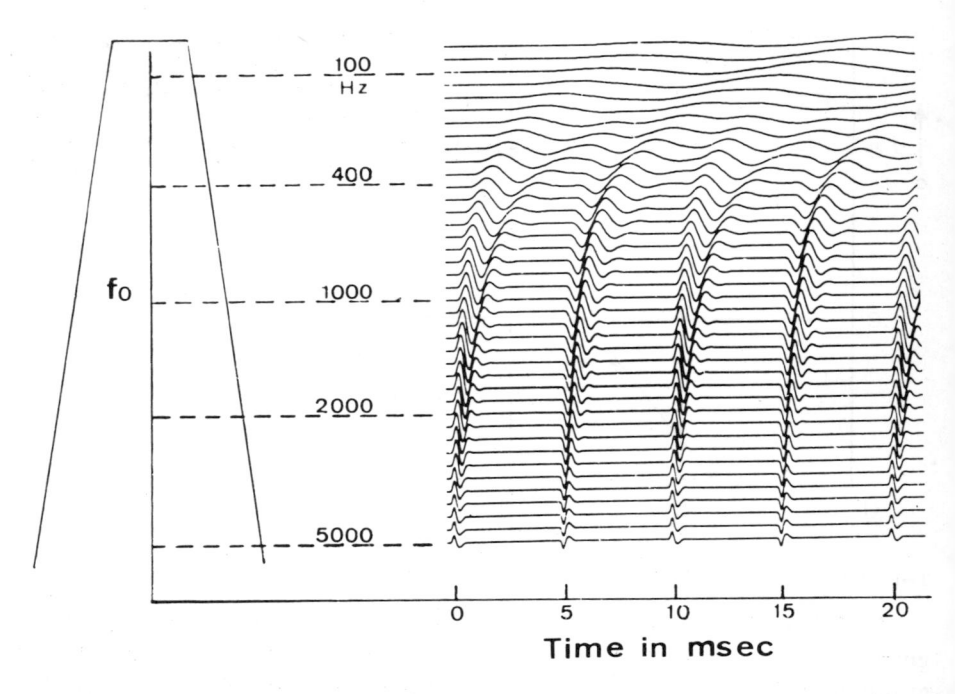

Fig. 6. Basilar membrane responses at various points to broad-band periodic pulses of alternating polarity (after Flanagan, 1965).

Now, if pitch P_φ due to the interaction of a Dirac pulse and a $\varphi°$ phase shifted Dirac pulse (pulse interval τ) in perception, is indeed correlated with the reciprocal value of the time distance between the major positive peaks of the fine structure, the pitch behaviour for the place on the basilar membrane that corresponds to f_o must satisfy the following relations:

$$P_0 = 1/\tau \qquad P_{90} = 1/\{\tau - 1/(4f_o)\}$$
$$P_{180} = 1/\{\tau \pm 1/(2f_o)\} \qquad P_{270} = 1/\{\tau + 1/(4f_o)\}. \qquad (6)$$

Theoretically, broad-band periodic pulses of alternating polarity do not evoke a definite pitch as the displacement waveforms are different in fine structure over the frequency range (P_{180} is a function of f_o), but empirically this signal evokes a definite pitch. This problem can be solved by introducing the concept of dominance. That is: *if pitch information is available along a large part of the basilar membrane the ear uses only the information from a narrow band. This band is positioned at 3-5 times the pitch value.* Its precise position depends somewhat on the subject.

By means of this concept of dominance the pitch value can be calculated if we replace f_o by 3-5 times the pitch value. The pitches evoked by the signals A, B, C, and D of Fig. 1 for the antiphasic condition should possess values corresponding to

$$P_{180}=\frac{2x-1}{2x\tau} \text{ and } \frac{2x+1}{2x\tau} \text{ (Eq. (6) with } f_o=x\cdot P_{180}).$$

In taking $x=4$, $P_{180}=7/(8\tau)$ and $9/(8\tau)$. These calculated pitch values are in accord with the experimental results (Ritsma, 1967; Bilsen and Ritsma, 1967).

To check that the concept of dominance is not merely a "rule of thumb" for calculating the pitch values, two types of experiments have been carried out. In a first experiment the broad-band signal was mixed with either high-pass or low-pass masking noise. The cut-off frequency of the masking noise could be varied by the experimenter. Subjects made pitch matchings for various values of the cut-off frequency of the masking noise. The results for one subject in the case of a broad-band periodic alternating-polarity pulse train (100 pps) mixed with high-pass masking noise are represented in Fig. 7.

The low and the high pitches did not change in their values if the frequency range above 600 Hz was completely masked by the noise. But for masker cut-off frequencies below 600 Hz, both pitches changed; the low pitch tended to be lower, the high pitch to be higher. Thus, the temporal behaviour of the basilar membrane at places tuned to characteristic frequencies greater than 600 Hz does not influence the overall pitch perception. Similar results were found in the case of low-pass masking noise. The low and the high pitches did not change in their values if the frequency region below 400 Hz was masked by the noise. For cut-off frequencies of the low-pass masking noise above 400 Hz, the low pitch tended to be higher, the high pitch to be lower. From this result it can be concluded that the frequency region below 400 Hz does not influence the overall

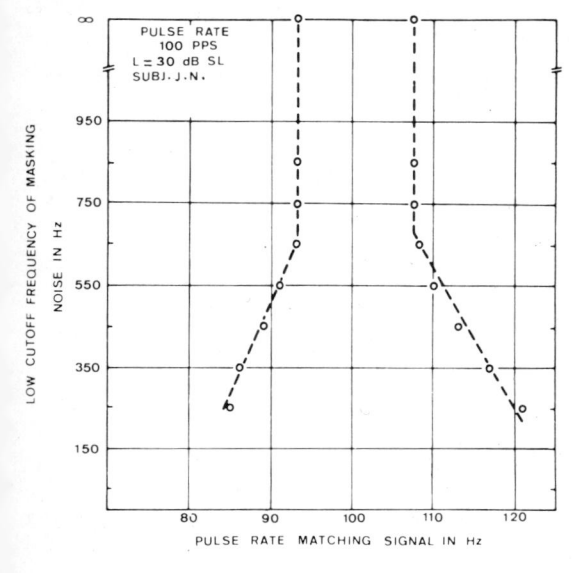

Fig. 7. Pitch of broad-band periodic alternating-polarity pulses influenced by high-pass masking noise with various cut-off frequencies.

pitch perception. The results of both experiments together confirm the view that for a pulse rate of 100 pps only the frequency region between 400 and 600 Hz is important with respect to pitch.

In a second type of experiment the concept of dominance was checked in the following way. Subjects listened to a pair of stimuli which are presented schematically in Fig. 8. The first stimulus consists of the sum of a low-frequency band and a high-frequency band of a signal with pitch corresponding to $1/\tau$ and a centre frequency band of the same kind of signal with pitch corresponding to $1/(\tau+\Delta\tau)$. In the second stimulus the pitch values were interchanged. Listening to both stimuli in succession either a pitch rise or a pitch fall is heard. In the case of a pitch fall the low and (or) the high frequency band must be dominant; in the case of a pitch rise the centre frequency band must be dominant. It was found that when the centre band corresponded to 4 times the pitch value, this band tends to dominate the pitch sensation as long as it is more than about 10 dB above threshold, irrespective of the level of the other frequency bands in the signal.

This experiment has been done both with a periodic unipolar pulse train (Ritsma, 1967; Plomp, 1967a) and with a signal built up by noise added to its repetition after a delay τ (Ritsma et al., 1967).

From these experiments we conclude that the concept of dominance is generally valid for all types of signals shown in Fig. 1.

DISCUSSION

It is concluded that there is no difference in pitch perception for the types of sounds mentioned in Fig. 1. For all these signals the pitch mechanism works on the same principles: *The sound is subjected to a spectral analysis on the basilar membrane. Because of the limited resolving power of the membrane on each place*

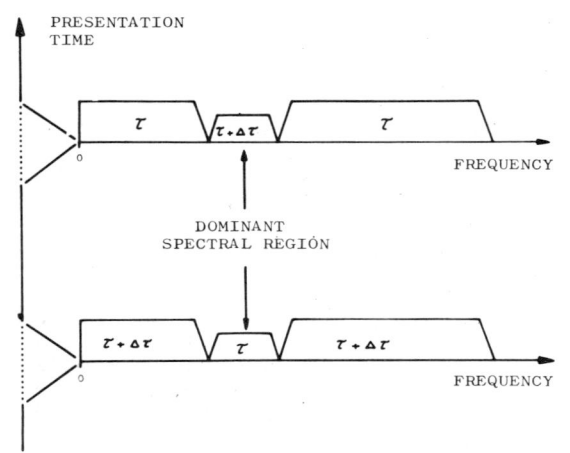

Fig. 8. Schematic representation of the test signal.

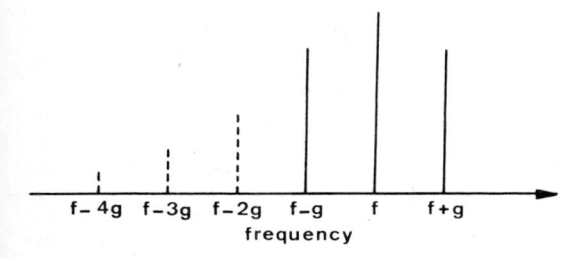

Fig. 9. Frequency spectrum of the displacement waveform on the basilar membrane for an AM signal. Solid lines represent the stimulus components; dashed lines represent the combination tones $(f-kg)$.

of the membrane a waveform is generated. According to the concept of dominance only one region on the basilar membrane is dominant with respect to the perception of pitch. This region is roughly 4 times the pitch value. On the waveform generated in this dominant region the ear performs an auto-correlation process determining the time interval between two pronounced positive peaks in the fine structure.

Within this framework the second effect of pitch shift remains as an artefact. As has been stressed by Schroeder (1966) and Fischler (1967), this second effect may be accounted for by deviations of the displacement waveform on the basilar membrane from the original acoustic waveform. Both authors assume that AM signals presented to the ear undergo a certain amount of phase modulation, synchronous with their amplitude modulation, in the mechanical and neural processing. Fischler states that this phase modulation is due to asymmetry introduced in the sideband energy of the acoustic signal as a result of the mechanical filtering by the inner ear.

A second possibility arises from the experiments on combination tones as done by Goldstein (1967): for an AM signal one must expect activities on the basilar membrane for a number of discrete frequency values $(f-kg)$ below the stimulus frequencies (Fig. 9).

These activities increase alinearly with decreasing modulation frequency g and with increasing sensation level. From the experiments on dominance it is known that for a complex sound with certain bandwidth the frequency region nearest the dominant frequency region will be dominant if the activity in that region exceeds a certain threshold. Thus, for higher sensation levels, the dominant region will shift to lower frequency values. This trend agrees with the experimental fact that the constant c of Eq. (3) which is a measure for the second effect of pitch shift, increases with increasing sensation levels. According to the concept of dominance this second effect of pitch shift must be zero for $n \approx 4$ irrespective of the sensation level. Experimental results as given in Fig. 5 confirm this statement.

REFERENCES

Békésy, G. von (1960): Experiments in Hearing (Mc Graw-Hill, New York).

Bilsen, F. A. (1966): Repetition Pitch: monaural interaction of a sound with the repetition of the same, but phase shifted sound, Acustica *17*, 295-300.

Bilsen, F. A., and Ritsma, R. J. (1967): Repetition pitch mediated by temporal fine structure at dominant spectral regions, Acustica *19*, 114-116.

Boer, E. de (1956): On the "Residue" in Hearing, Doctoral Dissertation, University of Amsterdam.

Fischler, H. (1967): Model of the "secondary" residue effect in the perception of complex tones, J. Acoust. Soc. Amer. *42*, 759-767.

Flanagan, J. L. (1962): Models for approximating basilar membrane displacement; Part II: effects of middle-ear transmission and some relations between subjective and physiological behaviour, Bell System Techn. J. *41*, 959-1009.

Flanagan, J. L. (1965): Speech Analysis, Synthesis, and Perception (Springer, New York).

Flanagan, J. L., and Guttman, N. (1960): Pitch of periodic pulses without fundamental component, J. Acoust. Soc. Amer. *32*, 1319-1328.

Fourcin, A. J. (1965): The pitch of noise with periodic spectral peaks, in: Rapports 5e Congrès International d'Acoustique, Liège, Vol. Ia, B 42.

Goldstein, J. L. (1967): Auditory nonlinearity, J. Acoust. Soc. Amer. *41*, 676-689.

Guttman, N., and Flanagan, J. L. (1964): Pitch of high-pass filtered pulse trains, J. Acoust. Soc. Amer. *36*, 757-765.

Helmholtz, H. L. F. von (1863): Die Lehre von den Tonempfindungen als physiologische Grundlage für die Theorie der Musik (F. Vieweg & Sohn, Braunschweig), 1st Ed.

Licklider, J. C. R. (1951): The duplex theory of pitch perception, Experientia *7*, 128-137.

Licklider, J. C. R. (1954): 'Periodicity' pitch and 'place' pitch, J. Acoust. Soc. Amer. *26*, 945.

McClellan, M. E., and Small, A. M. (1963): Pitch perception of randomly triggered pulse pairs, J. Acoust. Soc. Amer. *35*, 1881-1882.

McClellan, M. E., and Small, A. M. (1965a): Time-separation pitch associated with correlated noise bursts, J. Acoust. Soc. Amer. *38*, 142-143.

McClellan, M. E., and Small, A. M. (1965b): Time-separation pitch associated with correlated noise pulses, J. Acoust. Soc. Amer. *38*, 939.

McClellan, M. E., and Small, A. M. (1966): Time-separation pitch associated with noise pulses, J. Acoust. Soc. Amer. *40*, 570-582.

McClellan, M. E., and Small, A. M. (1967): Pitch perception of pulse pairs with random repetition rate, J. Acoust. Soc. Amer. *41*, 690-699.

Nordmark, J. (1963): Some analogy between pitch and lateralization phenomena, J. Acoust. Soc. Amer. *35*, 1544-1547.

Plomp, R. (1967a): Pitch of complex tones, J. Acoust. Soc. Amer. *41*, 1526-1533.

Plomp, R. (1967b): Beats of mistuned consonances, J. Acoust. Soc. Amer. *42*, 462-474.

Ritsma, R. J. (1962): Existence region of the tonal residue. I, J. Acoust. Soc. Amer. *34*, 1224-1229.

Ritsma, R. J. (1967): Frequencies dominant in the perception of pitch of complex sounds, J. Acoust. Soc. Amer. *42*, 191-199.

Ritsma, R. J., and Engel, F. L. (1964): Pitch of frequency modulated signals, J. Acoust. Soc. Amer. *36*, 1637-1644.

Ritsma, R. J., Bilsen, F. A., de Jong, Th. A., and Verkooyen, C. J. (1967): Spectral regions dominant in the perception of repetition pitch, IPO Annual Progress Report *2*, 24-30.

Rosenberg, A. E. (1965): Effect of masking on the pitch of periodic pulses, J. Acoust. Soc. Amer. *38*, 747-758.

Schouten, J. F. (1940): The perception of pitch, Philips Techn. Rev. *5*, 286-294.

Schouten, J. F., Ritsma, R. J., and Cardozo, B. L. (1962): Pitch of the residue, J. Acoust. Soc. Amer. *34*, 1418-1424.

Schroeder, M. R. (1966): Residue pitch: a remaining paradox and a possible explanation, J. Acoust. Soc. Amer. *40*, 79-82.

Small, A. M., and McClellan, M. E. (1963): Pitch associated with time delay between two pulse trains, J. Acoust. Soc. Amer. *35*, 1246-1255.

Thurlow, W. R. (1958): Some theoretical implications of the pitch of double-pulse trains, Amer. J. Psychol. *71*, 448-458.

Thurlow, W. R., and Small, A. M. (1955): Pitch perception for certain periodic auditory stimuli, J. Acoust. Soc. Amer. 27, 132-137.
Walliser, K. (1969a): Zusammenhänge zwischen dem Schallreiz und der Periodentonhöhe, Acustica 21, 319-329.
Walliser, K. (1969b): Zur Unterschiedsschwelle der Periodentonhöhe, Acustica 21, 329-336.
Wilson, J. P. (1966): Psychoacoustics of obstacle detection using ambient or self-generated noise, in: Les systèmes sonars animaux, R. G. Busnel, Ed. (Frascati), pp. 89-114.
Wundt, W. (1880): Grundzüge der physiologischen Psychologie, Vol. I (Verlag W. Engelmann, Leipzig), 2nd Ed., 315-317; as quoted by R. Plomp (1966): Experiments on Tone Perception, Doctoral Dissertation, University of Utrecht.

DISCUSSION

Schouten: If we are here to create a better understanding between psychophysicists and physiologists, I would like to make a remark on the role of fine structure in pitch detection. Our most convenient way of interpreting pitch and ambiguities of pitch is in terms of time intervals between peaks of the carrier frequency that are located at the crests of the envelope (Fig. 3 of the paper). But we have no psychophysical evidence that we should take time intervals between peaks at the crests; pitch might just as well be related to time intervals between peaks at some other part of the envelope or to time intervals between zero crossings of the fine structure. Perhaps a collaboration with the neurophysiologists may give the solution of this problem.

Ritsma: It is indeed rather difficult to answer this question by means of psychophysical experiments. There is some evidence, however, that pitch related to time intervals between peaks or zero crossings at the minima of the envelope is an idea somewhat in contradiction with the pitches found for quasi-FM signals.

Smoorenburg: By applying narrow-band signals you try to minimize the influence of frequency analysis of the basilar membrane which allows you to identify, practically, the acoustic time pattern with the cochlear one. You use the term "ready made envelope". I wonder whether this identification can be made because we know that essential nonlinearities giving rise to combination tones are involved in these narrow-band signals. Moreover, I showed in my paper (pp. 267-277) that they play a part in the second effect of pitch shift. Don't you think it is questionable to relate pitch to time intervals in the acoustic time pattern, in particular with respect to the important implications you ascribe to the quasi-FM experiment?

Ritsma: We did measure an effect of phase on pitch. Moreover, the pitch values found for the quasi-FM signals are in good agreement with those calculated from the hypothesis that the distance between two positive peaks in the fine structure near two adjacent crests of the envelope of the signal constitutes a measure of the pitch. So there must have been a true quasi-FM signal within the cochlea. We used sensation levels of about 30 dB where the second effect of pitch shift and thus the influence of combination tones is rather small.

Piazza: How do you measure the pitch? Do you prefer a pitch scale based on simple tones or on complex tones?

Ritsma: We always used a pitch matching procedure. It appears to be difficult for untrained subjects to match pitches of tones with different timbre. They have to learn to distinguish between pitch and timbre, and to adjust for equal pitches irrespective of timbre. This is especially difficult if the pitch of a complex test tone is to be matched with the pitch of a simple tone. Matchings with a simple tone are about three times less accurate than matchings with complex tones of about the same composition as the complex test tones. For AM signals in the pitch range of 200 to 400 Hz we found an accuracy of 0.3% independent of the constituent harmonics of the matching signal. We prefer to match with complex tones in another frequency band than the test tone to avoid a comparison of coinciding frequency components (Ritsma, 1965).

De Boer: I like to compliment Dr. Ritsma on his beautiful endeavour to synthesize many kinds of pitches that occur in the literature. I like to express, however, some doubts about the theoretical aspects. One of the main reasons for ascribing a temporal mechanism to pitch detection has been the phase effect, which you illustrated by an example in which you compared the pitches of AM and quasi-FM signals. But now we are confronted with a frequency region dominant in pitch perception that is located at a place where we do *not* find a phase effect. Next, you talk about the second effect of pitch shift and you invoke a contribution of combination tones to explain this. Then I fail to see how you can explain in an easy way this second effect of pitch shift on the basis of temporal resolution. I would suggest that there may be another principle at work like, for example, an extended type of a place principle. This implies that when the ear can resolve some components it will try to ascribe a pitch to the whole complex by looking for, let us say, some kind of a common denominator. And, whenever this mechanism fails, the temporal resolution mechanism sets in and supplies the missing information. (Not the missing fundamental!)

Ritsma: I think it is unlikely that in the dominant frequency region the ear resorts to a pitch mechanism different from the mechanism active in the region where the phase effect is observed. The accuracy of the pitch matching remains constant. Besides, I do not think that the ear utilizes temporal information only when it cannot resolve components. In fact, I think that the whole residue pitch stops if the components cannot be resolved separately. I do not mean "separately" in the sense that the components can be distinguished as separate quantities but I mean that the existence region of the tonal residue is limited by the disappearance of any frequency resolution.

Terhardt: I think that we should not assign too much significance to the discrepancy between theory and experimental results in the second effect of pitch shift. You found yourself, for instance, that this discrepancy was very small at the low sensation levels applied. Moreover, the discrepancy might be explained by some artefact arising from the matching signal that you used. The results of Walliser (1969), who matched with simple tones, do not show this discrepancy between theory and experiment. Therefore, I think, it is not certain that this discrepancy really exists and that combination tones, which are almost not audible, would play a role.

Secondly, I have a question about the dominant region for pitch perception. You have stated that one cannot speak of a certain dominant frequency region independent of pitch but that one merely should speak of a dominant region consisting of the third to the fifth harmonics. Looking to Fig. 7, I notice that for low pulse rates the sixth to seventh harmonics may dominate in pitch perception. So, what is the difference between a dominant frequency band from 500 to 1500 Hz and a dominant frequency region consisting of the third to the fifth or perhaps seventh harmonics?

Ritsma: With respect to your first question, I think the best answer will be given by the paper of Mr. Smoorenburg (pp. 267-277). On your second question, you have to determine the dominant region for every subject separately. And for individual subjects you will find that the dominant frequencies shift proportionally to the pitch. Accidentally, the subject of Fig. 7 has a rather high dominant frequency region.

Whitfield: As I have stated in my paper, I find it difficult to see how the nervous system might perform such a fine-structure temporal analysis. Moreover, we have shown that the periodicity as recorded in the cochlear nucleus, remains the same if one shifts equidistant components up in frequency. Why does everybody want to equate pitch with time structure? Pitch seems to be an attribute of a lot of sounds of which the simple tone happens to be one. And my understanding is that the pitch of a simple tone is rather less secure than the pitch of a complex tone in that it can change with intensity. So, why a temporal relationship to pitch? I am with Dr. de Boer here, that it is not without significance that the pitch of complex tones is controlled from a frequency region where you can resolve harmonics and where the ear is most sensitive. When you present the ear with one of these extraordinary signals, then the situation is that the nervous system says in effect: "What on earth can be the likely source of this?" and endeavours to assign a pitch to it on the basis of its experience of real sources with somewhat similar, though not identical, properties.

Ritsma: For all pitch phenomena mentioned in the introduction, like the pitches arising from random pulse pairs or from pairs of noise bursts, a tight correlation is required for a pitch sensation. And that means that the actual mechanism may be something like an autocorrelation process. I do not think that all pitch phenomena can be described simply by a mechanism operating on the spectral information.

Scharf: It seems that we are getting into semantic problems with pitch as we did with inhibition. There seems to be some problem about matching the pitch of a residue to the pitch of a simple tone and this may be due to the large difference in timbre. It is also possible that there are two different kinds of pitch. And, this leads me to wonder about all the other kinds of pitch you mentioned. Do you have any information on the perceptual similarity between the various pitches?

Ritsma: All these pitches are perceptually quite similar, but they are coupled with different timbres so that matching may be difficult.

Wilson: I was very surprised when I first listened to periodically gated noise. These signals have a very weak pitch while we get a very good neural following. Should not temporal theories expect a strong pitch effect in this case?

Ritsma: There is a marked perceptual difference between periodically interrupted noise and periodically interrupted periodic noise. Bilsen (1968) observed that interrupted noise resembled noise with a flutter superimposed, but he could not hear any tonality. On the contrary, the interrupted periodic noise was clearly tonal, had a definite pitch. McClellan and Small (1965, 1966) reported that they could not perceive Time-Separation Pitch due to two successive uncorrelated noise bursts. This proved to be possible only if the bursts were correlated. Therefore, I do not think that neural following alone is sufficient in generating pitch. Pitch is perceived only, when sound events are sufficiently correlated.

REFERENCES

Bilsen, F. A. (1968): On the interaction of a sound with its repetitions, Doctoral Dissertation, University of Delft.
McClellan, M. E., and Small, A. M. (1965): Time-separation pitch associated with correlated noise bursts, J. Acoust. Soc. Amer. *38*, 142-143.
McClellan, M. E., and Small, A. M. (1966): Time-separation pitch associated with noise pulses, J. Acoust. Soc. Amer. *40*, 570-582.
Ritsma, R. J. (1965): Pitch discrimination and frequency discrimination, in: Rapports 5e Congrès International d'Acoustique, Liège, Vol. Ia, B 22.
Walliser, K. (1969): Zusammenhänge zwischen dem Schallreiz und der Periodentonhöhe, Acustica *21*, 319-329.

PITCH OF TWO-TONE COMPLEXES

GUIDO F. SMOORENBURG

Institute for Perception RVO-TNO, Soesterberg
and
Physics Laboratory, Department of Medical and Physiological Physics
Utrecht University
Utrecht, The Netherlands

1. INTRODUCTION

It is now about thirty years since Schouten (1938, 1940) published the results of his experiments on the perception of pitch. The generally accepted idea that the fundamental frequency component of a complex tone determines the pitch was shown to be untenable. Listening to synthesized complex tones, such as pulse trains with many harmonics, Schouten noticed that, besides the separately audible lower harmonics, there is another perceptive element with a sharp sound and a pitch corresponding to the fundamental frequency. He concluded that this pitch must arise from the combined action of higher harmonics; their time pattern giving the relevant information. He called the collective sensation of such a group of harmonics "the residue". This conclusion was supported by de Boer's (1956) investigations of inharmonic complexes. In his opinion, the smallest number of frequency components necessary to obtain a stable and distinct residue was five. Ritsma (1962), however, showed in extensive studies with sinusoidally amplitude modulated (SAM) signals that their three frequency components are already sufficient to achieve a tonal residue. Shifting the phase of the carrier frequency component of such a SAM signal over 90° gives a quasi frequency modulated signal that produces a different pitch whereas the stimulus frequencies remain the same. Ritsma and Engel (1964) demonstrated that this pitch can be related to the time structure similarly as for SAM signals.

Pursuing the reduction of the number of frequency components, it is of interest to know whether a signal consisting of two frequency components (the simplest case with the fewest parameters) also has a pitch corresponding to the fundamental frequency. There are three major reasons to suppose that indeed

such a pitch can be expected.
(1) The time pattern of two frequency components summated with equal amplitudes is rather similar to the time pattern of a 100%-SAM signal which gives rise to a clear pitch of the complex tone.
(2) The number of stimulus frequency components itself does not seem to be important. The marked difference for the signals under (1) — two vs. three frequency components — is not recovered perceptually because of the introduction of combination tones such as $2f_1 - f_2$ in the ear.
(3) The results of experiments on the spectral region dominant in the perception of pitch of complex tones (Ritsma, 1967a, 1967b; Bilsen, 1968) suggest that two low harmonics can be sufficient to generate a pitch of the complex tone. This is in agreement with the idea that the pitch of the strike note of bells is determined by the second and third harmonics (Schouten and 't Hart, 1965).

2. EXPERIMENTS WITH TWO FREQUENCY COMPONENTS OF EQUAL SPL

There appeared to be an important disparity among subjects in pitch perception of two-tone stimuli. Subjects perceived the complex tone as a whole or pitches of individual partials. This was investigated more closely by presenting 42 subjects with two 160 msec tone bursts consisting of the frequencies $f_1, f_2 =$ 1750,2000 Hz and 1800,2000 Hz, successively, through headphones at a 40 dB sensation level. Subjects who perceive the complex tone as a whole will hear a pitch drop corresponding to the successive fundamental frequencies of 250 Hz and 200 Hz. Possible difference tones corresponding to these frequencies were masked sufficiently with an octave band of noise around 250 Hz added to the stimulus. On the other hand, subjects perceiving pitches of individual partials will hear an increasing pitch corresponding to the changing lower frequency component. The results of this experiment showed that about half the number of subjects perceived consistently the complex tone as a whole, while the others perceived consistently pitches of individual partials. A repeat of this experiment after a month gave for each subject the same result.

The pitch of the complex tone as a whole was further investigated for inharmonic stimuli where the frequencies are no integral multiples of the frequency difference. The SPL of each of the two components was 40 dB above hearing threshold. At this level it is not likely that the difference tone $f_2 - f_1$ which might be introduced in the ear by nonlinear distortion will play a role (Plomp, 1965). Moreover, the results for stimuli with a constant frequency difference (200 Hz) showed pitch variations which indicated that the difference tone could not be responsible for the pitch reported.

The matching signal had the same SPL as the test signal and consisted of a pair of adjacent harmonics. For the matching signal the harmonic numbers of the test signal were used and also numbers one or two differing to be able to exclude the artifact of adjustments simply based upon coinciding frequency components. Inharmonic test signals and harmonic matching signals, both producing the same pitch, will have one or two different frequency components. Sometimes, this brought about that during the pitch adjustments observers were hindered by jumping pitches of partials attracting the attention, especially in case of an almost completed match when the pitch of the complex tone does not jump anymore.

Two trained observers who perceived the complex tone as a whole in the introductory experiment on disparity among subjects matched the pitches over a wide range of stimulus frequencies for $f_2 - f_1 = 200$ Hz. The pitch, expressed in the (absent) fundamental frequency P of the harmonic matching signal, appeared to be independent of the various matching signals, so that average values could be given in Fig. 1. Also, 90%-SAM signals were used as matching signals, with identical results. A pitch corresponding to 200 Hz was found whenever the frequencies f_1 and f_2 were integral multiples of the difference frequency. A certain shift of the frequencies away from these central harmonic situations resulted in a clear pitch shift with a magnitude depending on the initial harmonics. The solid lines in Fig. 1 represent predicted values to be discussed in the next section.

3. DISCUSSION

The periodicity of the stimulus corresponds exactly to 200 Hz if the frequencies f_1 and f_2 are integral multiples of the difference frequency. Thus, the equal pitches in these cases are easily understood on the basis of a periodicity theory of pitch. The stimulus is described by

$$S(t) = A \cos 2\pi f_1 t + A \cos 2\pi f_2 t = 2A \cos 2\pi \frac{f_2 - f_1}{2} t \cos 2\pi \frac{f_2 + f_1}{2} t.$$

Herewith, a certain phase relation between the two components is chosen in agreement with the actual harmonic matching signal. We can interpret this signal as a sine wave of frequency $\frac{1}{2}(f_1 + f_2)$, of amplitude $|2A \cos 2\pi \frac{1}{2}(f_2 - f_1)t|$, and with a phase jumping 180° every time the amplitude is zero. If the frequencies are integral multiples of a fundamental frequency g, for instance $f_1 = ng$ and $f_2 = (n+1)g$, the stimulus can be written as

$$S(t) = 2A \cos \tfrac{1}{2} \cdot 2\pi g t \cos (n + \tfrac{1}{2}) \cdot 2\pi g t.$$

In Fig. 2 this signal is depicted for $n = 7$. It can be seen that the periodicity is indeed g, because in the time interval $\tau = 1/g$ the fine structure shows $7\frac{1}{2}$ oscillations and the phase has once jumped 180° at zero amplitude.

The periodicity g is lost if both frequencies are shifted by Δf ($\Delta f < g$) away

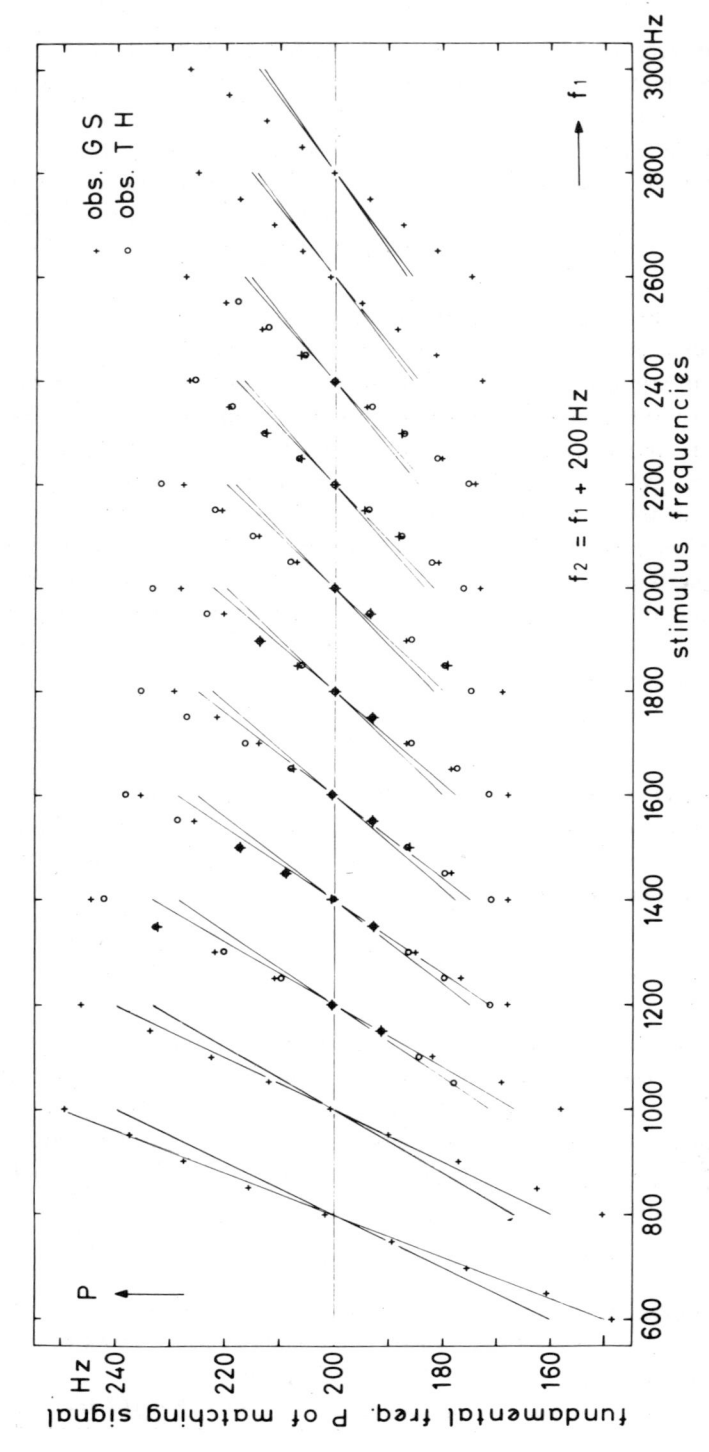

Fig. 1. Pitch of two-tone complexes with a constant frequency difference of 200 Hz expressed in the fundamental frequency P of a harmonic matching signal. The solid lines border the area in which the pitch is predicted.

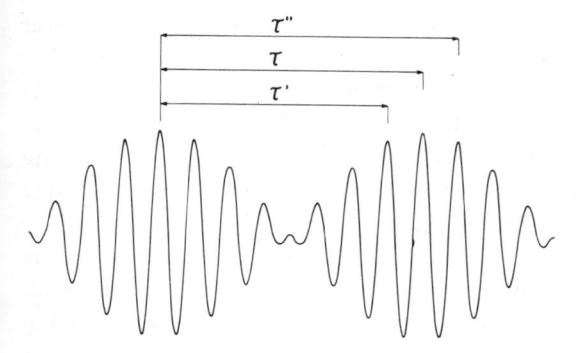

Fig. 2. The waveform of a two-tone stimulus consisting of the 7th and 8th harmonics with equal amplitudes. Between different peaks there are indicated several time intervals τ, τ', and τ'' which can be related to pitches produced by this signal.

from the central harmonic situation $f_1 = n.g$ and $f_2 = (n+1).g$. A pitch prediction can be made with the aid of a simple model proposed by de Boer (1956) and by Schouten, Ritsma, and Cardozo (1962). The pitch is presumed to be related to the time interval between, for example, peaks of the waveform. Different time intervals (τ, τ', and τ'' for the harmonic signal in Fig. 2) may be obtained by taking various peaks. This suggests an ambiguity of pitch which is indeed found experimentally, as can be seen in Fig. 1. The pitch corresponding to the time interval τ (Fig. 2) is determined by $(n+\frac{1}{2})$ oscillations of a frequency $(n+\frac{1}{2}).g$ in the fine structure. After a frequency shift of Δf the time interval is determined by $(n+\frac{1}{2})$ oscillations of a frequency $(n+\frac{1}{2}).g + \Delta f$. Thus a pitch shift corresponding to $\Delta f/(n+\frac{1}{2})$ may be expected. This holds only for components of equal amplitude. However, the frequency components may not have equal amplitudes at the place of pitch detection. In extreme cases the pitch might practically be determined by only one of the frequency components. This results in pitch shifts corresponding to $\Delta f/n$ or $\Delta f/(n+1)$, respectively. So, theoretically, pitch shifts are expected to lie within these limits represented in Fig. 1 by the solid lines.

The standard deviation of the pitch adjustments typically amounts to 1.2 Hz and is somewhat dependent on the frequency shift $|\Delta f|$ (values between 0.75 and 2.5 Hz). Each point in Fig. 1 is determined by 9 adjustments, so the standard deviation of the mean is typically 0.4 Hz. Apparently, the measured pitches deviate significantly from the predicted values. This phenomenon is comparable to the "second effect of pitch shift" (Schouten *et al.*, 1962).

The time pattern has been presumed to give the relevant information for pitch. The empirical deviations invite a closer look at this statement with an experiment in which the time pattern is transformed simply by changing the amplitude ratio of the two components.

4. PITCH FOR A VARIABLE SPL OF THE HIGHER FREQUENCY COMPONENT

Only frequency shifts away from the central harmonic situation $f_1 = 1600$ Hz and $f_2 = 1800$ Hz are considered in this section. The level (L_1) of the lower frequency component was fixed at 40 dB above threshold. For the higher frequency component the levels (L_2) 25 dB, 35 dB, and 45 dB were used. At these combinations the pitch is about equally pronounced. (The matching signal in this experiment was a harmonic 90%-SAM signal.) Again, matching signals with different sets of harmonics were chosen which gave the same fundamental frequencies for matched pitches.

The influence of a change in SPL of the higher frequency component on the pitch adjustments turns out to be insignificant over the range investigated. Within the small experimental error the pitch settings in these three cases and in the case of equal SPL of Sec. 2 were similar; this in spite of the variations in time pattern.

It is clear that thinking in terms of the acoustic waveform is insufficient, even in the case of frequency components closely spaced with respect to the critical bandwidth. Surprisingly, the data rather suggest that pitch is based upon lower frequency components than actually presented because the experimental slopes in Fig. 1 are steeper than expected. Therefore, it is of interest to consider carefully whether secondary tones, introduced in the ear, have played a part.

5. PITCH IN RELATION TO COMBINATION TONES

Recent publications (Zwicker, 1955, 1968; Plomp, 1965; Goldstein, 1967) show clearly that for the applied stimuli combination tones are introduced by an essential nonlinearity of the ear. A striking property of the combination tones is found in a restricted existence region below f_1 at moderate SPL. In this region the combination tones described by $f_1 - k(f_2 - f_1)$ with k as high as 5 or 6 can be heard. Both the frequency relation and the existence region support the idea that in explaining the results of the pitch adjustments reported above, the combination tones should be taken into account.

We may think of the pitches shown in Fig. 1 as to be determined by an *effective* frequency. Such an effective frequency can be defined by $\eta = \Delta f / \Delta P$ and then it can be derived from the data of Fig. 1 by calculating the slopes of best linear fits in the sense of least squares. Properly, the effective frequency has been expressed relative to the constant frequency difference of the two components. This gives a good overall picture. The definition of the effective frequency can be understood easily from the speculations in the time domain of Sec. 3. The relation is also understandable from pitch theories based upon frequency-pattern recognition (de Boer, 1956) or upon subharmonics (Terhardt, pp. 281-286 of this volume). The effective frequency is not necessarily restricted to (integral) harmonic numbers. For example, in case of equal amplitudes an effective frequency of $(n + \frac{1}{2})$ would be expected in Sec. 3.

The effective frequency is depicted in Fig. 3 (filled circles connected by solid

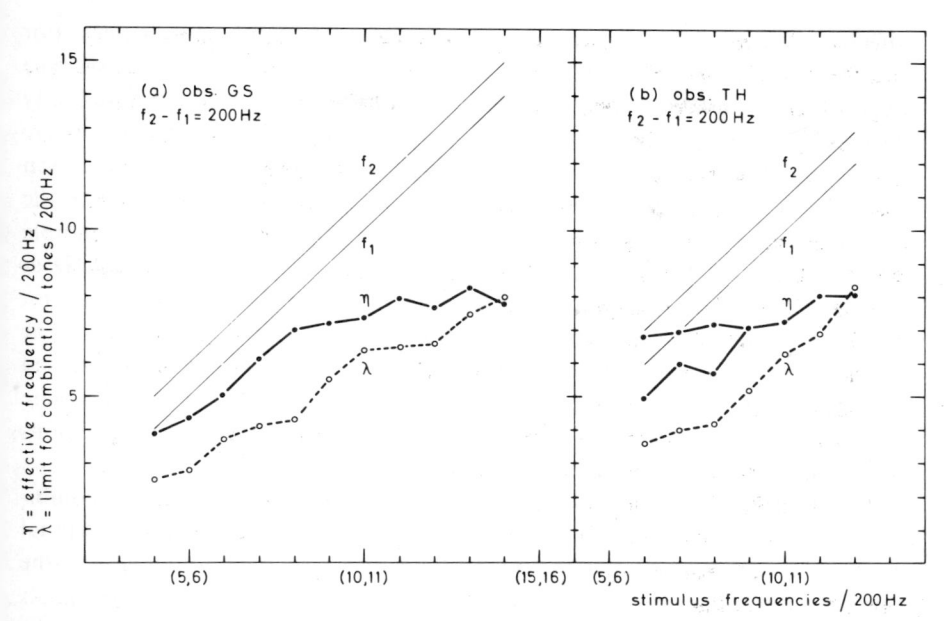

Fig. 3. The effective frequency η determining the pitch of a two-tone stimulus in relation to the lower limit λ of the existence region for combination tones for two observers. The frequencies η and λ are expressed relative to the frequency difference $g = 200$ Hz. In case of bifurcation of the effective frequency for observer TH the upper line holds for $P < 200$ Hz and the lower one for $P > 200$ Hz.

lines). It appeared that η did not exceed a value of about 8 for both observers, notwithstanding an interpersonal difference in the highest frequencies at which a pitch still could be heard. Interpersonal differences were also found for low harmonics. For each observer pitch adjustments at lower harmonics than presented in Fig. 1 could not be made, because in those situations the observers were not able to perceive the complex tone as a whole; the observers were diverted by the prominent individual partials. Above this lower limit one straight line did not fit the adjustments by observer TH. Two different values of η were found; the lower one for $P > 200$ Hz and the higher one for $P < 200$ Hz. This may be explained by a more pronounced higher pitch that dominated over the lower one in the ambiguous pitch situations. The lower pitch may not have been recognized and partial adjustments may have been made inadvertently. The pitch settings by observer TH for $(n,n+1)=(6,7)$ and $(n,n+1)=(7,8)$ support this view. The splitting up of the effective frequency for $(n,n+1)=(8,9)$ may be explained by a combination tone becoming effective above a certain Δf (if combination tones are involved in pitch detection of these complex tones). All other data did not deviate significantly from the linear relationship $\Delta f = \eta . \Delta P$; $\eta = $ const.

If the pitch originates from combination tones, it is necesssary that the effective frequency lies within the existence region for combination tones. For the SPL used in the pitch matching experiment the lower limit of the existence region for combination tones was determined as a function of k by gradually decreasing f_2, starting from $2.f_1$, and observing the appearance of the successive combination tones. The lowest combination tone present in the condition $f_2-f_1=200$ Hz is of particular interest in this experiment. Yet, the limit of the existence region should preferably not be expressed in the lowest audible combination tone because then this limit will show discrete steps of 200 Hz at frequencies subject to experimental variability. In Fig. 3 we preferred to plot the lower limit of the existence region continuously as a function of f_1. The appropriate value of the lower limit for the situation $f_2-f_1=200$ Hz is obtained by interpolating the experimental limit as a function of k. The observers were able to hear the combination tones for k up to 5 or 6. Thus no extrapolations were necessary as can be deduced from Fig. 3.

For the sake of completeness it may be remarked that the presence of combination tones introduces the theoretical possibility that the pitches of individual partials (combination tones included) instead of the overall pitches of the complex tones have been matched, even in case of matching and test signals without common frequency components. Yet, this is improbable because the possibility of coincidence of partials around the effective frequency depends on the choice of the harmonic numbers of the matching signal so that partial comparisons would give an effective frequency depending on the harmonic numbers. This has not been found. Moreover, adjustments with simple tones gave similar results.

The results show that the effective frequency determining the pitch lies within the existence region for combination tones. Further, the limit of tonality for high stimulus frequencies seems to correspond to a situation where the lowest just-audible combination tone does not exceed the eighth harmonic.

6. CONCLUSIONS

The acoustic waveform appeared to be insufficient to account for the pitch of a two-tone stimulus, even in case of close frequency components. This is understandable if one realizes that for these stimuli a nonlinearity generating combination tones plays a part. An improved model is difficult to propose because we do not yet know how these combination tones are generated. But the deviations from the expected pitch indicated that these combination tones may play a part in pitch detection and it appeared, indeed, that appropriate combination tones were always present. Preliminary experiments in which the combination tones are masked with noise showed an increase of the effective frequency which strongly suggested a pitch detection in the frequency region of the combination tones. Further results on this subject will be published (Smoorenburg, 1970).

This research was supported by the Netherlands Organization for the Advancement of Pure Research (ZWO).

REFERENCES

Bilsen, F. A. (1968): On the Interaction of a Sound with its Repetitions, Doctoral Dissertation, Technological University of Delft.
Boer, E. de (1956): On the Residue in Hearing, Doctoral Dissertation, University of Amsterdam.
Goldstein, J. L. (1967): Auditory nonlinearity, J. Acoust. Soc. Amer. *41*, 676-689.
Plomp, R. (1965): Detectability threshold for combination tones, J. Acoust. Soc. Amer. *37*, 1110-1123.
Ritsma, R. J. (1962): Existence region of the tonal residue. I, J. Acoust. Soc. Amer. *34*, 1224-1229.
Ritsma, R. J., and Engel, F. L. (1964): Pitch of frequency modulated signals, J. Acoust, Soc. Amer. *36*, 1637-1644.
Ritsma, R. J. (1967a): Frequencies dominant in the perception of pitch of complex sounds, J. Acoust. Soc. Amer. *42*, 191-198.
Ritsma, R. J. (1967b): Frequencies dominant in pitch perception of periodic pulses of alternating polarity, IPO Annual Progress Report *2*, 14-24.
Schouten, J. F. (1938): The perception of subjective tones, Proc. Kon. Nederl. Akad. Wetensch. *41*, 1086-1093.
Schouten, J. F. (1940): The residue, a new component in subjective sound analysis, Proc. Kon. Nederl. Akad. Wetensch. *43*, 356-365.
Schouten, J. F., Ritsma, R. J., and Lopes Cardozo, B. (1962): Pitch of the residue, J. Acoust. Soc. Amer. *34*, 1418-1424.
Schouten, J. F., and 't Hart, J. (1965): De slagtoon van klokken, in: Publication No. 7 of the Nederl. Akoest. Genootschap (Stieltjesweg 1, Delft), pp. 20-49.
Smoorenburg, G. F. (1970): Pitch perception of two-frequency stimuli, J. Acoust. Soc. Amer. *48*, 924-942.
Zwicker, E. (1955): Der ungewöhnliche Amplitudengang der nichtlinearen Verzerrungen des Ohres, Acustica *5*, 67-74.
Zwicker, E. (1968): Der kubische Differenzton und die Erregung des Gehörs, Acustica *20*, 206-209.

DISCUSSION

Goldstein: You invoked combination tones as a possible means for changing the time pattern that one would expect in response to the two-tone stimulus. From a psychophysical or descriptive point of view I agree that on the basis of current knowledge you have made the most sensible choice. Nevertheless I would like to add a word of caution lest this description be taken more literally than may be correct. From electrophysiological data it is quite clear to me that all the temporal effects of interaction between two simple tones probably should not be explained exclusively on the basis of combination tones which are effectively added to the stimulus. To be specific, Sachs and Kiang (1968) and Sachs (1969) have worked on two-tone suppression (see also p. 202 of this volume) and the interactions that they have published deal with the overall firing rate in response to two-tone stimuli. Privately, Sachs and Kiang have discussed with me the temporal interactions which occur when the inhibitory tone is in a frequency region where one gets time locking to simple tones. It seems as though the inhibitory tone can chop the response to the, let us make it simple, excitatory tone in a manner that follows the time pattern of the inhibitory tone. Here then you are provided with another mechanism which changes the time pattern of a response in a manner which is rather different from what one might expect from linear interaction prior to a simple transducer. In my opinion, therefore, in the absence of complete physiological knowledge a reasonable precondition for invoking combination tones as an explanation for the larger pitch shift than expected is that there be behavioural responses to the combination tones themselves. For example, you could ask: can you hear a combination tone in the sense that if you cancel at the frequency of the combination tone something disappears which you can identify with the pitch of a simple tone at that frequency?

Smoorenburg: I have asserted that combination tones may play a part in pitch detection and that an improved model is hard to propose because we do not even know how combination tones are generated. The larger than expected pitch shift may be explained by changes in the time pattern, but pitch detection based on time pattern is a hypothesis. The model based on time structure, which I described in the paper, appears to be insufficient to explain the experimental results. I prefer to leave the question of the pitch mechanism open. But the results of pitch measurements with the combination tones masked by the noise suggest strongly that combination tones are indeed involved in pitch detection.

As to your question about hearing the combination tones, I determined the lower limit of the existence region for combination tones by gradually lowering f_2 from $2f_1$ to f_1. Then the combination tones become successively audible. They can be followed for quite a while as they rise in pitch. Their pitch can easily be matched with a simple tone. With this procedure you do not need a cancelling technique. During the pitch matchings I am not aware of the presence of combination tones. The possibility that individual partials or combination tones of the matching signal and the test signal were compared, was examined very carefully. It appeared to have happened only in situations with low stimulus frequencies for observer TH.

Zwislocki: I very much like the experiment on disparity among subjects. Does not this experiment indicate rather clearly that pitch is not a unitary sensation, that there are at least two criteria, and that some people use one and some the other? I think it may be very interesting to know whether it is possible to talk people out of one group and into the other.

Smoorenburg: Yes, I made an effort in that direction. I tried to bias someone judging pitch on the basis of the complex tone as a whole by presenting at first the frequency components of 1750 Hz and 1800 Hz solely. We might then expect that the listener recognizes the pitch jump corresponding to these frequencies after the addition of 2000 Hz to both stimuli. This did not happen. The observer judged the pitch of the complex tone 1750/2000 Hz to be higher than the pitch of the complex tone 1800/2000 Hz. Each listener seems to have a stable criterion. It is not just a matter of rivalry between criteria.

Terhardt: We should not forget that the whole phenomenon of perceiving the residual pitch of only two harmonics is very, very weak. In our Institute we constructed a melody with signals consisting of two harmonics and no subject heard any melody corresponding to the expected periodicity pitch. I like to remark that we should be careful in discussing a phenomenon like the second effect of pitch shift by means of a phenomenon which itself is so weak.

Smoorenburg: Well, you have seen that there is an important disparity among subjects. The first pitch impressions of 16 subjects out of a total of 42 are in complete disagreement with your finding.

Plomp: I personally hear only the pitch jump of individual partials but I fully believe that people hearing it the other way were correct. Among them are also listeners quite familiar with psychoacoustical experiments. Many of the subjects are surprised that others hear it the other way around.

Goldstein: I would like to say that our psychophysical findings are not in agreement with what Dr. Terhardt mentioned. We have played melodies with computer-controlled signals consisting of two successive harmonics. From note to note the harmonic numbers were randomly chosen and yet, everybody hears the melody. Problems arise only if the harmonic numbers and absolute frequencies are to high (Houtsma, 1970).

Zwicker: It cannot be denied that there is also a melody. But, I myself found the melody in such a situation by listening to the rhythm.

Schügerl: In the work of Wellek (1939), two types of listeners are distinguished. The one is the "polar", the other the "linear" type. The "polar" listener is a ratio estimator with good performances in evaluating the harmonic quality of a complex. He perceives the sound as a whole, in a synthetic manner, and does not note the distance between the components. On the other hand, the "linear" listener hears analytically, he is well able to pick out the components of a complex (as in listening to polyphonic music) and gives good estimates of the distance, but only poor estimates of the quality. I see a strong relation between these types and the grouping of observers you touched upon.

REFERENCES

Houtsma, A. J. M. (1970): Perception of musical pitch, J. Acoust. Soc. Amer. *48*, 88 (A).
Sachs, M. B. (1969): Stimulus-response relation for auditory-nerve fibers: two-tone stimuli, J. Acoust. Soc. Amer. *45*, 1025-1036.
Sachs, M. B., and Kiang, N. Y.-S. (1968): Two-tone inhibition in auditory-nerve fibers, J. Acoust. Soc. Amer. *43*, 1120-1128.
Wellek, A. (1939): Typologie der Musikbegabung im deutschen Volke (Beck'sche Verlagsbuchhandlung, München).

FREQUENCY ANALYSIS AND PERIODICITY DETECTION IN THE SENSATIONS OF ROUGHNESS AND PERIODICITY PITCH

E. TERHARDT

Institut für Elektroakustik
Technische Hochschule München
München, Germany

INTRODUCTION

An acoustical sensation that clearly seems to be a matter of time structure detection by the ear is the sensation of roughness. In the first section some basic findings about this sensation will be described.

Pitch sensations like "periodicity pitch", "pitch of the residue", "time separation pitch", "repetition pitch" in many cases are interpreted as the result of a time structure detection process, too. In the second section some phenomena reported by Walliser (1968) will be described which lead to a more complex interpretation. This new interpretation will be discussed in the third section.

1. ROUGHNESS

Though the sensation of acoustic roughness as a result of interference of two or more tones in the ear is very easy to observe, there were only few systematic investigations of roughness in the past. The important data from von Helmholtz (1863), Wever (1929), von Békésy (1935) and Mathes and Miller (1947) were not complete enough to allow conclusions about the functions of the ear. Therefore we made many psychoacoustical measurements to find basic facts about the perception of roughness. These investigations were made with amplitude modulated (AM) tones as well as with two beating sinusoids (Terhardt, 1968a, 1968b). The most important results are the following.

1.1. *Roughness and degree of amplitude modulation*

The roughness of an AM tone with given carrier and modulation frequency depends practically only on the *relative* amplitude fluctuation $m = \Delta p/p$ of the sound pressure. A tone with $m=1$ and a sound pressure level (SPL) of 80 dB has almost the same roughness as a tone of 60 dB and the same relative amplitude fluctuation $m=1$, although the absolute amplitude fluctuations differ by a factor of 10.

Measuring the roughness function $r = F(m)$ with the fractionation method of Stevens (1957), i.e. by asking the subjects for double or half roughness settings, showed consistent and reproduceable results. The roughness function can be simply described by the power function

$$r = \text{const}.m^2. \tag{1}$$

The exponent 2 was consistently found for different SPL's as well as for different carrier and modulation frequencies.

1.2. Roughness as a function of modulation and carrier frequency

Since the modulation degree m was found to be a very useful measure of roughness, the influence of carrier and modulation frequency (f_{car}, f_{mod}) could be measured easily. Fig. 1 shows the roughness of an AM tone of given SPL and m-value as a function of modulation frequency f_{mod} and carrier frequency f_{car}. The diagram shows that the values of f_{mod} for maximum roughness as well as for vanishing roughness depend on f_{car} if $f_{car} < 2$ kHz. For $f_{car} > 2$ kHz the modulation frequencies for maximum roughness as well as for vanishing roughness are constant. Fig. 2 shows the maximum modulation frequency (in case of AM tone) and the maximum beat frequency f_b (in case of two beating tones) up to which roughness could be observed, as a function of carrier

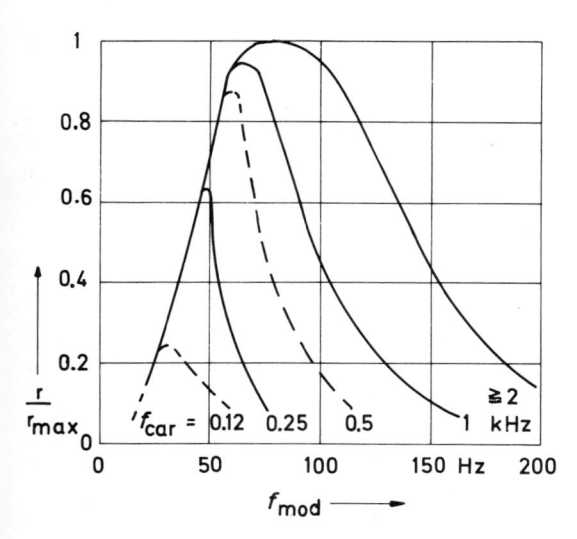

Fig. 1. Relative roughness r as a function of modulation frequency f_{mod} (abscissa) and carrier frequency f_{car}. The dashed curves are not experimental results but interpolated functions.

frequency. In the carrier frequency range below 1 kHz this boundary for the existence of roughness agrees very well with the critical bandwidth (Zwicker, 1961). For carrier frequencies above 2 kHz the modulation (beat) frequency where roughness disappears remains constant as already seen in Fig. 1.

The results described and some other results not mentioned here (see Terhardt, 1968c) lead to an interpretation which shows some already known frequency analyzing properties of the ear and some less known inertial influences on the perception of amplitude fluctuations.

1.3. *Interpretation of the results*

Because of the spatial frequency selectivity of the inner ear, amplitude fluctuations of different carrier frequencies produce fluctuating stimulations at different places of the organ of Corti. From the results described in 1.1, it follows that each place along the cochlear partition detects fluctuations of the stimulation in the same way, namely according to Eq. (1), regardless of the frequency of the fluctuations and of the SPL.

The results represented in Figs. 1 and 2 show that the amplitude fluctuations of the excitating stimulus in a certain region along the cochlear partition are influenced by the frequency analyzing mechanism working in the inner ear: if two tones differ in frequency by more than one critical bandwidth there seems to remain only a relatively small region of the cochlear partition wherein both frequencies interfere, so that no roughness is perceived. This confirms the critical bandwidth as a good first approximation for the frequency analyzing power of the inner ear. However, Figs. 1 and 2 show that for high carrier frequencies — this means for stimulations at the upper half of the cochlear partition — the roughness depends no longer on the critical bandwidth. In this region the roughness vanishes when the fluctuation frequency is greater than about 250 Hz. This shows that the ear cannot follow very fast amplitude fluctuations even if the spatial frequency selectivity of the inner ear allows strong interference of the partials. So the upper part of the curve "$f_{car} \geq 2$ kHz" in Fig. 1 seems to be characteristic for the "low-pass" character of the neural transmission of fluctuating excitations. It seems possible that any limit of the

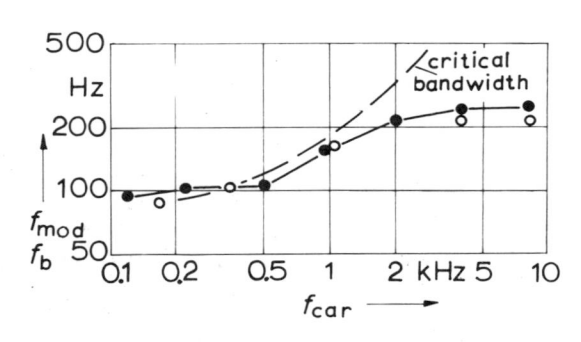

Fig. 2. Modulation frequency f_{mod} and beat frequency f_b above which roughness of AM tones or beating tones vanishes. Points: results for AM tones; open circles: results for two beating pure tones with middle frequency f_{car}.

ear's periodicity detection ability has something to do with this limit for the perception of roughness.

2. PERIODICITY PITCH

This section deals with some recent investigations by Walliser (1968) on the pitch of complex sounds (called periodicity pitch, PP). These investigations may be separated into two parts.

In the first part Walliser (1969a) completed former work of de Boer (1956) and Schouten et al. (1962). It could be pointed out that the PP of harmonic or inharmonic sounds (for the definition of harmonicity see section 2.1) may be described with good approximation by means of frequency analysis as well as time analysis principles. The simple rule for the determination of the PP of any sound with equidistant partials where the lower harmonics are removed is: the frequency f_{PP} corresponding to the perceived PP is that subharmonic of the lowest present partial f_u which lies nearest to the beat frequency f_b between two neighbouring partials. This means

$$f_{PP} = \frac{f_u}{n}, \qquad (2)$$

with n an integer, so chosen that $|f_{PP} - f_b|$ is minimal.

In the second part some new experimental phenomena were found. These phenomena concern quantitatively only small effects, but they are important because they lead to an interpretation of the PP phenomenon that is different from other published interpretations. Some of the most important results will be described in this section. Walliser (1969a) derived from these data a second rule for determining the PP which is an even better approximation than Eq. (2). This second rule will be reported too.

2.1. *Even for harmonic sounds PP does not correspond to the beat frequency*

The frequency reciprocal to the envelope period of the investigated sounds (or of the stimulation in the inner ear) shall be called beat frequency f_b. It is known from investigations of de Boer (1956) and Schouten et al. (1962) that the PP differs from the pitch corresponding to the beat frequency f_b in those cases where the sound is "inharmonic" (this means that f_v/f_b is not an integer, where f_v are the frequencies of the partials present).

By using the frequency of a pure tone with equal pitch as a measure for the PP, Walliser (1969b) showed that even in the *harmonic* case the PP does not exactly correspond to the beat frequency. As an example for this phenomenon Fig. 3 shows the frequency f_{PP} of a tone having the same pitch as the test sound, as a function of the beat frequency. Although in this experiment the test sounds

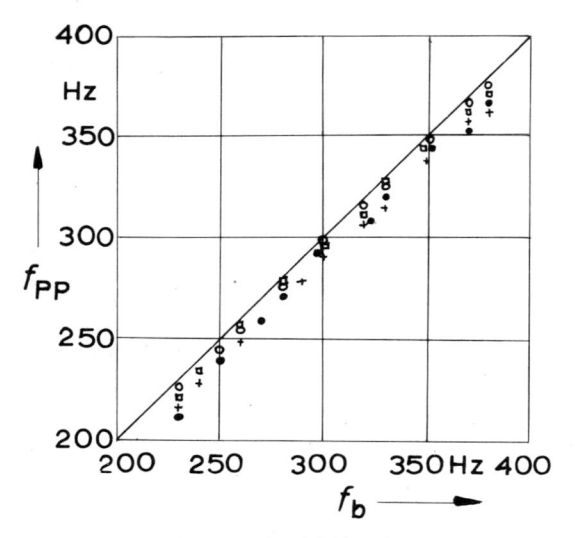

Fig. 3. Frequency f_{PP} of a pure tone with the same pitch as produced by a test signal consisting of short pulses with fundamental frequency (identical with beat frequency) f_b which are filtered through an octave band from 1.4 to 2.8 kHz. SPL of pure tone 70 dB, of complex tone 50 dB. Different symbols correspond to median values of different subjects.

were always harmonic (f_v/f_b = integer), the median frequency f_{PP} for equal pitch from all subjects is significantly lower than f_b.

2.2. PP shifts when a noise is added to the test sound

If a white noise is added to the test sound, the PP of this sound rises with growing noise intensity. This means that the beat frequency f_b of a complex sound presented alone has to be made higher than the beat frequency f_{bN} of the same sound presented together with a noise for equal pitches of both sounds. Fig. 4 shows the relative difference $\dfrac{\Delta f_b}{f_b} = \dfrac{f_b - f_{bN}}{f_b}$ as a function of the white noise SPL L_{WN} (curve marked "PP complex sound"). In this diagram the reference sound pressure defining the level L_{WN} is the white noise SPL in the case where the white noise just masks the sound completely, so $L_{WN} = 0$ dB is equivalent to the masked threshold of the test sound.

The curves marked "pure tone f_{PP}" and "pure tone f_u" in Fig. 4 show two other results from the same subject. Apparently, the influence of the added noise on the pitch of a single pure tone with the frequency of the lowest present partial f_u of the former experiment is very similar to the noise's influence on the PP of the complex sound (compare curves "PP complex sound" and "pure tone f_u"). Also it becomes clear from Fig. 4 that the noise's influence on the pitch of a single pure tone with frequency f_{PP} is different from the two phenomena just mentioned.

2.3. PP shifts when the sound's SPL is changed

It is known that the pitch of a simple tone depends on SPL (Stevens, 1935; Walliser, 1969c) although this phenomenon quantitatively depends on the subjects participating in the experiments. Walliser (1969b) compared the SPL's

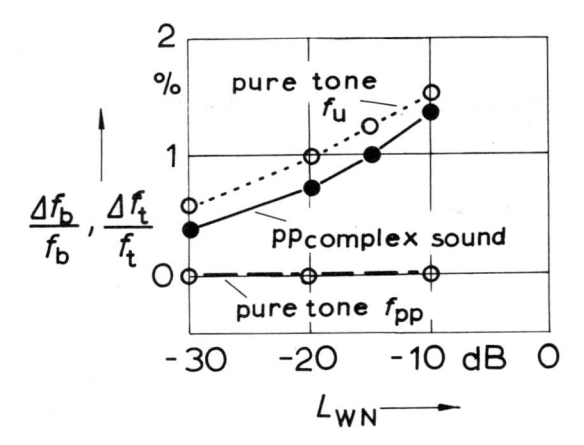

Fig. 4. Pitch shift of a harmonic sound and of pure tones with frequencies f_{PP} and f_u, respectively, caused by masking white noise. Test sound: octave band 2.8–5.6 kHz filtered out of the spectrum of short pulses with $f_b = 400$ Hz. $\Delta f_b/f_b$ is the relative difference between the beat frequencies of the unmasked and the masked test sound for equal pitch. $\Delta f_t/f_t$ is the equivalent value for pure tones with the frequencies $f_{PP} = 400$ Hz and $f_u = 2.8$ kHz. L_{WN} is the SPL of the white noise *re* masked threshold. SPL of the test sounds was 50 dB (from Walliser, 1969b).

influence on the PP of a complex sound with the SPL's influence on the pitch of single tones with the frequencies f_{PP} and f_u, respectively. Fig. 5 shows such a comparison for one subject. The curve "PP complex sound" shows the relative difference between the beat frequency f_b of a complex sound with variable SPL and the frequency f_{b50} of the same sound with constant SPL of 50 dB, for equal PP's and as a function of L. The curves "pure tone f_{PP}" and "pure tone f_u" show equivalent results with single tones. It becomes clear from Fig. 5 that the influence of SPL on the PP of a complex sound is similar to the SPL's influence on a single tone with frequency f_u, but clearly different from the influence on a single tone with frequency f_{PP}.

2.4. *The "second rule" for PP*

The results described in 2.1 to 2.3 are a few examples from many experimental findings. These findings suggest that the PP of a complex tone without lower partials seems to be derived from the pitch of the lowest partial present, together with the beat frequency f_b. The beat frequency seems to determine the approximative PP and its exact value seems to be derived from the pitch of the lowest partial. The pitch extracting mechanism for the PP seems to be clearly different from the mechanism involved in the pitch of low pure tones.

If one has to predict the PP with some approximation, one could use arbitrarily a time analysis method or the rule Eq. (2). However, consideration of the described small but significant phenomena strongly suggests a new rule for the

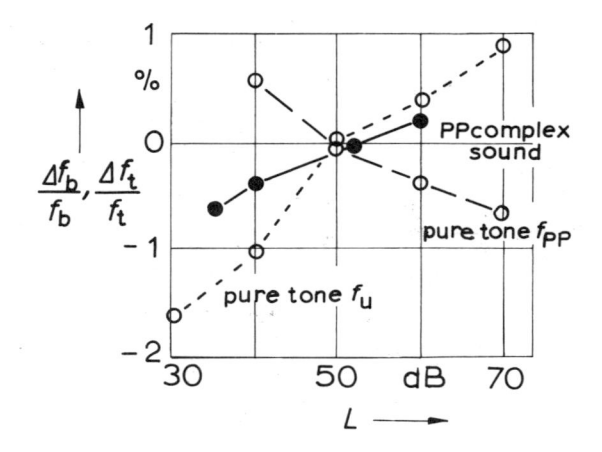

Fig. 5. Pitch shift of a harmonic sound and of pure tones with frequencies f_{PP} and f_u, respectively, caused by changes in SPL. Test sound: octave band 2.8–5.6 kHz filtered out of the spectrum of short pulses with $f_b = 300$ Hz. $\Delta f_b/f_b$ is the relative difference between the beat frequencies of the test sound with variable SPL L and the test sound with fixed SPL of 50 dB for equal pitches. $\Delta f_t/f_t$ is the equivalent value for pure tones with the frequencies $f_{PP} = 300$ Hz and $f_u = 2.8$ kHz for equal pitches.

determination of PP. This "second rule" is similar to the "first rule" Eq. (2). There is only one modification of the first rule necessary: instead of deriving the PP from the frequency f_u of the lowest present partial, by dividing this frequency through an integral number, we have to use the *pitch* corresponding to f_u as a cue for PP. Then the "second rule" for the determination of PP is:

PP is that *subjective* subharmonic of the lowest present partial's *pitch* which lies nearest to the pitch equivalent to the beat frequency f_b.

By this rule even the discrepancy between f_{PP} and f_b in the harmonic case (see Fig. 3) is explained, because *subjective* harmonics do not correspond exactly to *objective* harmonics: Ward (1954) showed that subjective octave intervals between pure tones always correspond to frequency ratios a little greater than 2. Walliser (1969d) confirmed these results. From these investigations there could be derived a scale of subjective harmonics as a function of objective harmonics. It was found that the discrepancy between the subjective and the objective harmonics derived from Ward's and Walliser's experiments with pure tones has the same size and direction as the discrepancy between f_{PP} and f_b in the harmonic case (Fig. 3). This is the reason why the term "subjective subharmonics" in the "second rule" explains the phenomenon of Fig. 3.

3. DISCUSSION

Up to now there is no theory that explains all experimental findings concerning PP phenomena, especially the new results of Walliser; thus we are looking for an interpretation or a hypothesis which at least makes these phenomena easier to understand from a common viewpoint.

For the generation of PP Walliser's "second rule" requires at least one pitch belonging to one partial of the test sound and, additionally, some information about its beat frequency (or period of envelope). Assuming that the information about the beat frequency is represented in the roughness sensation, we should expect that the existence of the PP is connected with the existence of roughness as well as with the existence of separate pitches. Fig. 6 gives a comparison of these existence regions. The "roughness boundary" curve shows the maximal beat frequency f_b between partials for the perception of roughness, as a function of the middle frequency of the partials. The "pitch separation" curve shows the minimal frequency difference f_b between two simple tones for the perception of two separate pitches (Terhardt, 1968a; Plomp, 1964). The shaded area represents the existence region of PP or "tonal residue" (Ritsma, 1962, 1963; Walliser, 1968). Comparison leads to the following conclusions.

(a) The lower boundary of the PP existence region is almost identical with the "pitch separation" curve. Apparently, beats between two neighbouring partials contribute to the perception of PP only, if there is a certain degree of pitch separation.

(b) Apparently, the existence of roughness is no strong condition for the existence of PP. PP may exist although there is no longer any roughness (upper boundary of PP existence region).

Regarding Walliser's results together with these conclusions, there is some evidence for the following hypothesis, which permits the general interpretation of the PP phenomenon we were looking for: "Periodicity pitch, the residue, pitch of the missing fundamental, *etc.* are secondary sensations, derived on a higher level of sensory perception from primary sensations like pitch (of partials) and roughness".

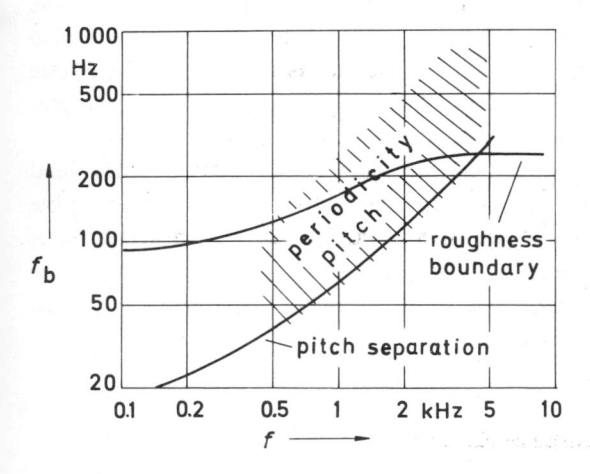

Fig. 6. "Pitch separation": minimal frequency difference f_b for pitch separation of two simultaneously sounding tones. "Roughness boundary": maximal frequency difference (beat frequency) f_b of two beating tones for the existence of roughness.
"Periodicity pitch": existence region of PP; a few partials of a harmonic sound generate PP only if the combination of beat frequency f_b and middle frequency f lies in the shaded area.

This hypothesis, which is similar to the "mediation theory" published by Thurlow (1963), helps us understand some phenomena more easily. Here are four examples:

1. *The results of Walliser's investigations.* The hypothesis was suggested by his data. The information which Walliser needs in his "second rule" is available in terms of the primary sensations pitch of partials (and pitch differences) and/or roughness.

2. *The very existence of a lower boundary of PP existence region.* This lower boundary would otherwise be difficult to explain.

3. *The uncertainty involved in the perception of PP under extreme conditions (i.e., when only a few partials are presented).* Some observers perceive PP easily, others only with the aid of pitch changes, others not at all. An observer may one day perceive PP very easily, the next day he may have great difficulties. Octave (and other) errors in pitch matching experiments are usual. These phenomena are easier to understand if one considers PP a "secondary sensation" which is not automatically established by the stimulated ear, but is derived from "primary sensations" by a perception process of much greater complexity than usually assumed.

4. *The "dominance" of the frequency range of approximately 500 Hz to 1500 Hz in the generation of PP* (Plomp, 1967; Ritsma, 1967). In the light of our hypothesis this dominance could be explained as follows. The human hearing mechanism is used to "regenerate" the fundamental pitch of a sound (*i.e.*, of the human voice) out of this dominant frequency range even if the fundamental is present, because the conditions for perception (hearing threshold, frequency discrimination of the "primary perception mechanism") at very low and very high frequencies are worse than in the medium frequency range.

Summarizing we can state:

(a) The results of the roughness investigations support the hypothesis that the frequency analyzing power of the inner ear is represented roughly by the critical bandwidth (Frequenzgruppe; Zwicker, 1961). These investigations also show some inertial properties involved in the perception of fast amplitude fluctuations; fluctuations faster than about 250/sec produce no roughness (where the definition of roughness is given by the experiments; see Terhardt, 1968a).

(b) Walliser's PP investigations together with the results about roughness lead to a new interpretation of the "residue" phenomenon which could be a suitable basis for further experiments. However, in connecting the PP with the pitches of the partials, no decision is made about the question whether those "pure tone pitches" are established and separated by a "frequency analyzing" or by a "periodicity detecting" mechanism.

REFERENCES

Békésy, G. von (1935): Ueber akustische Rauhigkeit, Z. Techn. Phys. *16*, 276-282.

Boer, E. de (1956): On the "Residue" in Hearing, Doctoral Dissertation, University of Amsterdam.
Helmholtz, H. L. F. von (1863): Die Lehre von den Tonempfindungen als physiologische Grundlage für die Theorie der Musik (F. Vieweg & Sohn, Braunschweig, 6th Ed. 1913).
Mathes, R. C., and Miller, R. L. (1947): Phase effects in monaural perception, J. Acoust. Soc. Amer. *19*, 780-797.
Plomp, R. (1964): The ear as a frequency analyzer, J. Acoust. Soc. Amer. *36*, 1628-1636.
Plomp, R. (1967): Pitch of complex tones, J. Acoust. Soc. Amer. *41*, 1526-1533.
Ritsma, R. (1962): Existence region of the tonal residue. I, J. Acoust. Soc. Amer. *34*, 1224-1229.
Ritsma, R. (1963): Existence region of the tonal residue. II, J. Acoust. Soc. Amer. *35*, 1241-1245.
Ritsma, R. (1967): Frequencies dominant in the perception of the pitch of complex sounds, J. Acoust. Soc. Amer. *42*, 191-198.
Schouten, J. F., Ritsma, R. J., and Cardozo, B. L. (1962): Pitch of the residue, J. Acoust. Soc. Amer. *34*, 1418-1424.
Stevens, S. S. (1935): The relation of pitch to intensity, J. Acoust. Soc. Amer. *6*, 150-154.
Stevens, S. S. (1957): On the psychophysical law, Psychol. Rev. *64*, 153-181.
Terhardt, E. (1968a): Ueber die durch amplitudenmodulierte Sinustöne hervorgerufene Hörempfindung, Acustica *20*, 210-214.
Terhardt, E. (1968b): Ueber akustische Rauhigkeit und Schwankungsstärke, Acustica *20*, 215-224.
Terhardt, E. (1968c): Beitrag zur Ermittlung der informationstragenden Merkmale von Schallen mit Hilfe der Hörempfindungen, Doctoral Dissertation, University of Stuttgart.
Thurlow, W. R. (1963): Perception of low auditory pitch: A multicue, mediation theory, Psychol. Rev. *70*, 461-470.
Walliser, K. (1968): Zusammenwirken von Hüllkurvenperiode und Tonheit bei der Bildung der Periodentonhöhe, Doctoral Dissertation, Technische Hochschule München.
Walliser, K. (1969a): Ueber ein Funktionsschema für die Bildung der Periodentonhöhe aus dem Schallreiz, Kybernetik *6*, 65-72.
Walliser, K. (1969b): Zusammenhänge zwischen dem Schallreiz und der Periodentonhöhe, Acustica *21*, 319-329.
Walliser, K. (1969c): Ueber die Abhängigkeiten der Tonhöhenempfindung von Sinustönen vom Schallpegel, von überlagertem drosselndem Störschall und von der Darbietungsdauer, Acustica *21*, 211-221.
Walliser, K. (1969d): Ueber die Spreizung von empfundenen gegenüber mathematisch harmonischen Intervallen bei Sinustönen, Frequenz *23*, 139-143.
Ward, W. D. (1954): Subjective musical pitch, J. Acoust. Soc. Amer. *26*, 369-380.
Wever, E. G. (1929): Beats and related phenomena resulting from the simultaneous sounding of two tones, Psychol. Rev. *36*, 402-418.
Zwicker, E. (1961): Subdivision of the audible frequency range into critical bands (Frequenzgruppen), J. Acoust. Soc. Amer. *33*, 248.

DISCUSSION

Plomp: I have two points related to your data on the roughness of two simultaneous tones. The first one is that I am quite amazed that you found such a sharp boundary between frequency distances that give roughness and larger distances that do not give roughness. In our own experiments (Plomp and Steeneken, 1968), we found a rather broad transition range. Perhaps you used a better criterion.

My second point concerns your finding that roughness diminishes so rapidly when the frequency separation exceeds critical bandwidth (Fig. 1 of the paper). Because of the considerable overlap of the stimulation patterns along the basilar membrane, I would have expected a much more gradual course.

Terhardt: I believe that the discrepancy in transition range in your and our roughness measurements is mainly due to differences in the instructions of the subjects and in the method. Your subjects had to change the frequency of one of the two oscillators until no roughness was heard anymore. In my experiment the stimulus was presented successively with random frequency separation and the subject had to say only whether there was roughness or not. Moreover, we found that the roughness sensation diminishes slightly for longer presentation times of, say, 2 sec, so we have an effect comparable with adaptation. It might be that this effect had also some influence on your experiments, as the subjects were listening during a rather long period while adjusting the oscillator.

Your second question can be explained by considering that roughness appears to be proportional to the square of the relative amplitude fluctuations, so it depends very strongly on this relative measure, not on the absolute amplitude fluctuation. And, indeed, we find only a small region within the overlap of the stimulation patterns of the simultaneous tones where the amplitudes of both patterns are about equal. Therefore, it is not astonishing that roughness is rather small for frequency distances beyond the critical bandwidth.

Zwicker: Roughness can be understood as some kind of integral over the whole range with only a small part of overlap really contributing.

Smoorenburg: I think it is dangerous to take the modulation depth as the only criterion for roughness because that implies that phase as such does not play a part. I can show you some stimuli with the same envelope, but with rather different roughness.

Terhardt: Using modulation depth as a measure for roughness does not imply that phase is irrelevant. We compared amplitude-modulated (AM) and frequency-modulated (FM) signals with equal power spectrum and found the roughness of the FM tone to be very much smaller than of the AM tone as long as the spectrum did not exceed the critical bandwidth. For wider spectra this difference in roughness disappeared. So there is indeed a strong influence of phase within the critical band.

Smoorenburg: Sure, but changing from AM to FM implies also a change in envelope. What I mean is not a difference in roughness by changing the envelope but by changing only the phase itself. Do you agree that such a change may result in a difference in roughness?

Zwicker: What do you mean by phase itself?

Smoorenburg: By an effect of phase itself I mean a difference in roughness for two signals with the same envelopes, the same frequency components, but different phase relations. For example, shifting the carrier frequency of an AM signal 20° in one direction or 20° in the other give two signals differing in roughness, whereas the envelopes are just the same.

Zwicker: Is the envelope the same? How can you then change the phase?

Terhardt: I don't have any result referring to such a situation.

Smoorenburg: I would like to put forward a second question. Fig. 6 of your paper suggests that the boundary of the existence region of periodicity pitch agrees rather well with the minimum frequency difference required to hear two frequency components separately. I checked the numbers, but it seems to be somewhat different. For instance, for a carrier frequency of 200 Hz, Ritsma's (1962) existence region goes up to about 4000 Hz, whereas Plomp's (1964) data indicate that two tones with a frequency difference of 200 Hz can be distinguished up to about 2000 Hz which value is still lower for more complex stimuli. What I would conclude from this is that, apparently, there are stimuli of which the frequency components cannot be heard individually, whereas we do hear periodicity pitch.

Terhardt: I think, neither the boundary of the existence region of periodicity pitch nor the curve representing the minimum frequency difference required to hear two frequency components separately are known so precisely that you may draw such a conclusion.

Scharf: I'd like to return to the question of roughness. It seems to me that the range of frequency separations of the components of a complex stimulus over which roughness is heard ought to depend strongly on the SPL because if you increase the SPL, the excitation patterns of the components will become wider also and, therefore, the overlap region would become more important. Do you have any data on the effect of SPL?

Terhardt: It is not sufficient to have overlap, you need additionally similar amplitudes. The region over which the amplitudes are similar does not change very much as a function of SPL.

Zwicker: The dependence on SPL has been investigated. Dr. Terhardt found that there is no dependence and therefore he can explain roughness in the way he did.

Schouten: I would like to go back to Mr. Smoorenburg's remarks. Dr. Terhardt, you have stated specifically that the tonal residue can be heard only if the pitches of the single partials can be separated by the ear and, later, that the limit of pitch separation is almost identical to the lower boundary of the existence region of periodicity pitch. Now this boundary coincides roughly with the 20th harmonic, which would mean that you can hear the individual partials up to the 20th. The unaided ear, however, is able to hear about the 8th harmonic, and with great difficulty the 10th. But even then you have still a factor of 2 unexplained, which is extremely important in this case. We should be careful in our conclusions and I think that your reasoning does not hold.

Terhardt: May I refer to what Dr. de Boer said in his paper about the domi-

290   DISCUSSION

nance of certain frequency regions?

De Boer: Those were much lower harmonics.

Plomp: In my opinion, it is somewhat dangerous to use the pitch separation curve based on only two frequency components (Fig. 6) in explaining the pitch of a multicomponent signal. Actually, you should use the critical bandwidth curve.

Terhardt: The only conclusion I make, regarding Fig. 6 of my paper, is that the existence of periodicity pitch seems to depend on the ear's ability of resolving the partials of the "dominant region". Even if one believes in this principle one can of course not expect that both psychoacoustical results (lower boundary of existence region and pitch separation of two simple tones) would coincide exactly.

Schügerl: In my opinion, roughness depends not solely on frequency distance, but, to a less degree, on frequency ratio too. When, starting from unison, the frequency ratio is increased, you will hear roughness; then, on further increasing the frequency distance beyond critical bandwidth, the roughness disappears at first but, when the frequency ratio approaches simple frequency ratios, where we get beats of mistuned consonances, a faint roughness will be heard again. Hence, it is not quite correct to say that roughness diminishes monotonically with increasing frequency difference.

Terhardt: Well, even for a frequency ratio of 2:1, there is some overlap which may cause roughness. Therefore, I cannot see any discrepancy.

REFERENCES

Plomp, R. (1964): The ear as a frequency analyzer, J. Acoust. Soc. Amer. *36*, 1628-1636.
Plomp, R., and Steeneken, H. J. M. (1968): Interference between two simple tones, J. Acoust. Soc. Amer. *43*, 883-884.
Ritsma, R. (1962): Existence region of the tonal residue. I, J. Acoust. Soc. Amer. *34*, 1224-1229.

REPETITION PITCH; ITS IMPLICATION FOR HEARING THEORY AND ROOM ACOUSTICS

FRANS A. BILSEN

Applied Physics Department
Delft University of Technology
Delft, The Netherlands

INTRODUCTION

"Repetition pitch" is a pitch sensation that occurs when a sound reaches the listener's ear together with the (delayed) repetition of the sound.

In 1693, Huygens described such an observation, made at the castle at Chantilly in France. Standing at the foot of a big stony staircase leading to the garden, he noticed that the noisy sound coming from a fountain near the staircase was producing a certain pitch. He concluded that the pitch had to be caused by the (periodic) reflections of the sound of the fountain against the successive steps of the staircase. In 1944, Supa *et al.* showed that blind men use reflected sound to locate objects. It turned out that an observer approaching a flat, sound-reflecting, surface in the presence of certain sounds perceives a pitch varying inversely with distance from the surface.

When the phenomenon occurs in rooms, studios (Fig. 1), or concert halls, it has an influence, appreciated or not, on the perceived quality of speech and

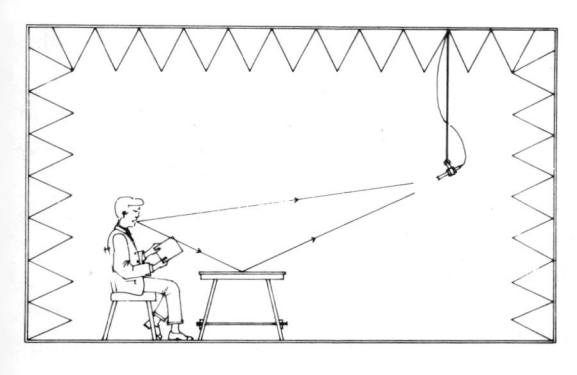

Fig. 1. Illustration of the pitch effect due to the interference, at the microphone, of the speaker's direct sound with the reflection from the table.

music (Somerville et al., 1966).

In psychoacoustics, a similar pitch effect was investigated by Thurlow and Small (1955). Shifting a sequence of band-pass filtered pulses (a, b, c, ...) continuously in time with respect to a second identical sequence (a', b', c', ...) a "sweep pitch" was perceived that, at any given instant, appeared to be related to the time separation a-a' or a'-b, depending on which was shorter. An extensive series of experiments dealing with matchings of this "time separation pitch" has been reported by Small and McClellan (1962-1967; references to these authors, and a more complete historical review are given by Bilsen, 1968). They used different kinds of signals: a periodic or random sequence of pulse pairs, a periodic sequence of pairs of (un)identical noise bursts, and single pulse pairs.

Experiments with digital-noise signals were reported by Fourcin (1965). He noticed that the monaural presentation of wide-band noise together with the same noise delayed by means of a shift register (delay time τ) gave rise to a pitch corresponding to $1/\tau$.

An inspection of these pitch effects revealed that all are, in fact, one and the same pitch effect. A necessary condition for the pitch to exist is that the original and the repeated sound are sufficiently correlated; in other words: a "time separation" alone is insufficient and unnecessary, there must be a repetition in the true sense of the word. Therefore, we called this pitch "repetition pitch" (RP) (Bilsen, 1966).

"FREQUENCY" PITCH OR "TIME" PITCH?

In the following some essential characteristics of RP are summed up, the knowledge of which is of fundamental importance for an understanding of RP.
(1) By filtering out, or masking, the lower frequency range of the signal, the place of the basilar membrane corresponding to the frequency of the perceived pitch can be made inactive. However, the pitch continues to exist even at the lowest sensation levels. Thus, RP is not due to a distortion product equal in frequency to the pitch observed.
(2) Presenting the original sound to the left ear and the repetition to the right ear, no pitch is evoked (Fourcin, 1965).
(3) No pitch is perceived when the original sound is fed through a high- (or low-) pass filter and the repetition through a low- (or high-) pass filter in the absence of a common pass band (Bilsen, 1967/68). Obviously, there must be detection of the original and repeated sound in a common region of the basilar membrane and the peripheral nervous system (see also 2).
(4) It is important to note that, in the case of continuous noise with its repetition, RP cannot possibly result from a process of detection of a temporal envelope because this is, essentially, missing. (Several authors assume the temporal

envelope to be the most important pitch clue for periodicity pitch (Walliser, 1969).)

(5) Reversing the polarity of the repetition (addition with negative sign) an upward as well as a downward pitch jump (ambiguity) can be observed (Thurlow, 1958; Fourcin, 1965). The results of accurate matchings of RP to a pure tone are represented in Fig. 2. In this figure, the pitch values $RP_{\varphi_2}^{\varphi_1}$ are given in their dependence on the delay time τ and the phase shift φ_1 and φ_2 of all the frequency components of the original and the repeated sound, respectively (φ_1, $\varphi_2 = 0°$, 90°, 180°, or 270°). The schematized signals, together with their frequency (amplitude) spectra, are also pictured in this figure. An inspection of Fig. 2 reveals that there is no point in looking for a simple explanation of RP as a "time" pitch or "frequency" pitch (corresponding to the lowest "cosine mountain" in the spectrum, or corresponding to a difference tone).

(6) Band-pass filtering of the signal until only about one spectral cosine mountain remains, does not make the pitch to disappear (Bilsen and Ritsma, 1970). Thus, only one spectral maximum (of a particular shape!) offers enough information for the pitch to be evoked. At first sight only, this finding seems to be in contradiction with the fact that at least two or three spectral maxima (components) of a complex periodic signal, separated from each other by an amount about equal to the "residue" pitch, are needed for this pitch to exist (Smoorenburg, pp. 267-277 of this volume).

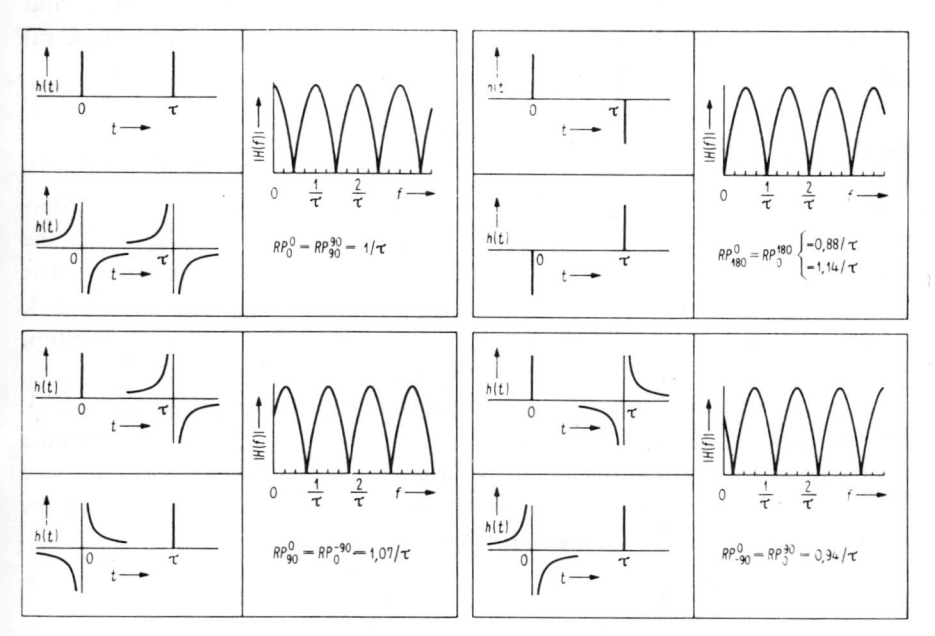

Fig. 2. Different configurations of 0°–, 90°–, 180°–, or 270°–pulses forming pulse pairs, together with their spectrum and corresponding RP (Bilsen, 1966).

THREE ASSUMPTIONS

Although RP cannot be explained simply in terms of time- or frequency analysis, an explanation is possible if one makes the following assumptions (Bilsen and Ritsma, 1967/68, 1969):

(1) *Spectral analysis.* A signal with a wide-band spectrum undergoes a spectral analysis on the basilar membrane (von Békésy, 1928). Due to the limited resolving power of the membrane, however, the signal is not completely analysed into its frequency components. Therefore, on each place of the basilar membrane a temporal displacement waveform is present which is due to the interaction of several neighbouring frequency components.

(2) *Temporal fine-structure detection.* The time interval between two prominent, unidirectional, peaks in the fine structure of the temporal displacement waveform present at a place on the basilar membrane is an adequate parameter for pitch (de Boer, 1956; Schouten et al., 1962).

(3) *Spectral dominance.* Signals with a wide-band spectrum possess a so-called dominant spectral region (Ritsma, 1967; Ritsma et al., 1967). This implies that the information in a specific spectral region is dominant in the perception of pitch over the information in other spectral regions. It has been found experimentally that the dominant spectral region has a centre frequency of about four times the pitch value.

Taking together these assumptions we conclude that the pitch of wide-band signals is determined by the temporal fine structure in the dominant spectral region.

EXPERIMENTS

In order to check the concept of temporal fine-structure detection we undertook a series of listening tests with 1/3-octave filtered pulse pairs of which the second pulse was given a phase shift of 90°, 180°, or 270°, respectively. The actual responses of the filter used in the experiments are reproduced in Fig. 3. It can be shown (Bilsen, 1968) that these responses, in particular the position of the major positive peaks, are in accord with the theoretical expectation.

If, due to the interaction of two successive filtered pulses (pulse distance τ), RP is indeed correlated with the reciprocal value of the time distance between the major, positive, peaks of the fine structure, the pitch behaviour for that place on the basilar membrane, which corresponds to the frequency f_o, must fulfil the following relations (see the length of the arrows, Fig. 3):

$$[RP_0^0]_{f_o} = 1/\tau, \qquad [RP_{90}^0]_{f_o} = 1/(\tau - 1/4f_o),$$
$$[RP_{180}^0]_{f_o} = 1/(\tau \pm 1/2f_o), \quad [RP_{270}^0]_{f_o} = 1/(\tau + 1/4f_o).$$

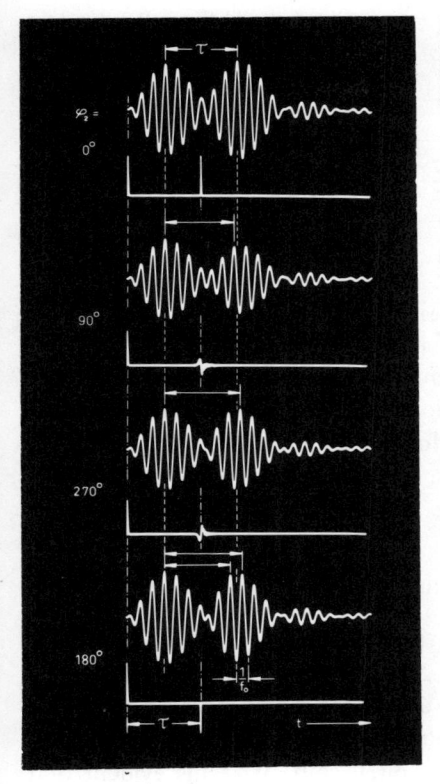

Fig. 3. The responses of the 1/3-octave band-pass filter with centre frequency f_o (each upper trace) to pulse pairs consisting of a 0°- and a 0°-, 90°-, 180°-, or 270°-pulse, respectively.

These relations are represented in Fig. 4 by the solid lines, for $\tau = 2$ msec. The theory is confirmed by the results of the listening tests, represented by the measured points in Fig. 4. Each measured point is the average of ten matchings of the low pitch heard by two subjects. The test signal and the matching signal were of the same character: the test signal built up of a 0°-pulse and a 90°-, 180°-, or 270°-pulse, respectively, the matching signal built up of two 0°-pulses. (The way of repeating the pulse pair, periodically or at random, is not of significant influence on the perceived pitch.)

In Fig. 4 the wide-band RP-values from Fig. 2 are denoted by the dashed lines. These lines intersect with the curves at a centre frequency f_o of about 2000 Hz, this value being 4 times the values of RP_0^0 ($= 500$ Hz). The value "4" is in full accordance with the results of the listening tests concerning spectral dominance (Ritsma et al., 1967).

CONCLUSION I

Repetition pitch, and likewise the pitch of periodic signals, corresponds to the reciprocal value of the time interval between two prominent peaks in the

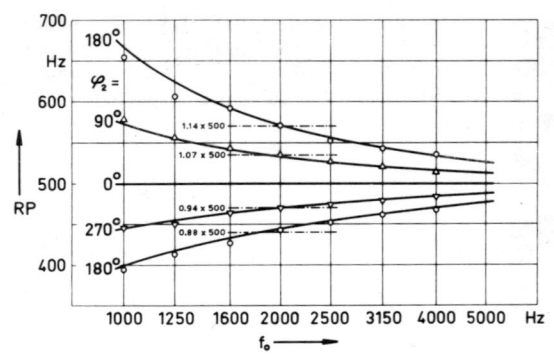

Fig. 4. Solid lines: RP due to a 1/3-octave filtered 0°- and φ_2-pulse as a function of the filter centre frequency f_o; $\tau = 2$ msec. Dashed lines: wide-band RP-values (Bilsen, 1968).

temporal fine structure of the displacement waveform present on the basilar membrane at a dominant frequency region. Thereby, this dominant region appears to have a centre frequency about four times higher than the frequency value of the pitch in question. RP, thus, is the result of limited frequency analysis, followed by time analysis (see Fig. 5).

A NOTE ON THE SECOND EFFECT OF PITCH SHIFT

The pitch of three-component signals (Schouten et al., 1962) and two-component signals (Smoorenburg, pp. 267-277 of this volume) can be explained with the concept of temporal fine-structure detection. In general, however, there appears to be a systematic deviation of the real pitch value from the theoretical pitch value, known as the second effect of pitch shift. It seems evident that

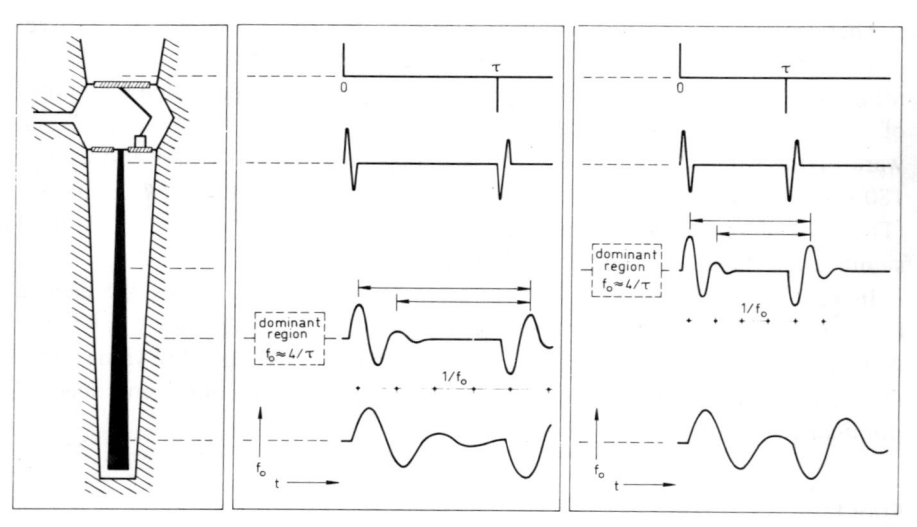

Fig. 5. Illustration of how pitch corresponds to the reciprocal value of the time distance(s) between prominent, unidirectional, peaks in the displacement waveform of the basilar membrane at the dominant frequency region.

combination tones play a role in pitch perception and may be the cause of this second effect of pitch shift (Ritsma, pp. 261-266; Smoorenburg, pp. 272-277 of this volume).

We checked the fine-structure theory once more for 1/3-octave filtered pulse pairs. We choose $\tau = 5$ msec and $f_o = 2000$ Hz. As regards the temporal fine structure, this signal can be compared with a three-component signal with carrier frequency $f_o = 2000$ Hz and modulation frequency $g = 1/\tau = 200$ Hz. Further, making a negligible error, we can compare this signal also with a two-component signal consisting of $f_1 = 2000$ Hz and $f_2 = f_1 + 200$ Hz.

Histograms of pitch matchings by three subjects are given in Fig. 6. Measured pitch values belonging to the three-component signal (Ritsma, p. 256 of this volume, Fig. 4) and the two-component signal (Smoorenburg, p. 270 of this volume, Fig. 1) are also represented in that figure. It turns out that, contrary to the pitch values of the three- and two-component signals, the measured RP-values do not deviate significantly from the theoretical values.

Assuming with Ritsma and Smoorenburg that combination tones are the cause of the second effect of pitch shift, we may explain the absence of a deviation in the case of RP by stating that pulse pairs with their particular, continuous, spectrum are not able to generate combination tones producing RP.

PERCEPTIBILITY OF PITCH AND TIMBRE (COLOURATION)

It is self-evident that the pitch sensation will fade away when the repetition level is gradually attenuated with respect to the original-sound level. At some

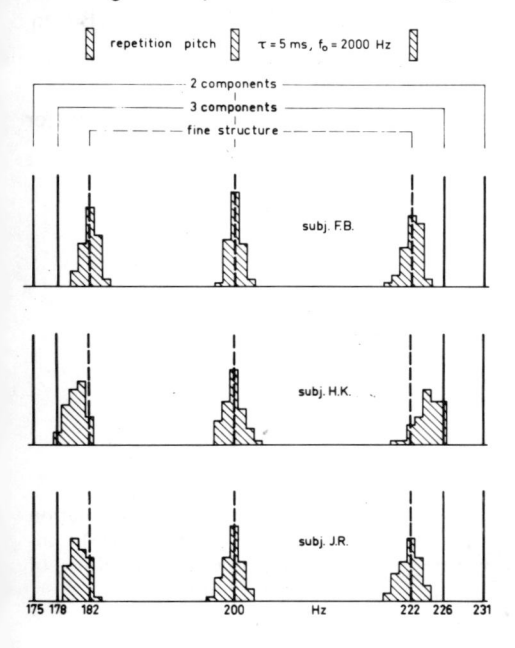

Fig. 6. The second effect of pitch shift: the deviation of the measured pitch values from the theoretical pitch values for a two-component signal, a three-component signal, and an RP-signal.

critical level difference, pitch and timbre will become undetectable. Some basic information on the build-up of pitch in the time- and frequency domain can be obtained from the behaviour of this critical level difference, further called "the threshold of perceptibility of pitch and timbre", as a function of time and frequency parameters (Bilsen, 1967/68; Bilsen, 1968; Bilsen and Ritsma, 1970).

For room acoustics, the threshold of perceptibility of pitch and timbre is also a threshold of perceptibility of colouration (of an original sound due to the addition of reflections), since no other qualities than pitch and timbre form part of the notion colouration.

A parameter having a great influence on the threshold of perceptibility of pitch and timbre is the pitch value itself. Using experimentally determined threshold values, and an autocorrelation model for the perceptibility of pitch and timbre (Bilsen, 1968), it was possible to derive a general expression for the perceptibility of pitch and timbre: pitch and timbre are perceptible if

$$\varphi(\tau)/\varphi(0) > 0.063/\rho(\tau), \text{ with } \varphi(\tau) = \int_{-\infty}^{+\infty} f(t) \cdot f(t+\tau) \, dt.$$

In this expression, $\varphi(\tau)$ is the autocorrelation function of the signal $f(t)$. Thus, $\varphi(\tau)/\varphi(0)$ is a well-suited physical measure for the quantity of timing information present in a signal. $\rho(\tau)$ is a subjective weighting function, which indicates that the perceptibility is dependent on τ due to the (partly unknown) properties of the hearing organ.

The general validity of the expression has been shown and the experimental function $\rho(\tau)$ (see Fig. 7) has been determined for the following signals (Bilsen, 1968):

(1) white noise with its attenuated repetition;
(2) white noise with an infinite number of periodic repetitions forming a flutter echo; this type of signal occurs naturally in room acoustics;

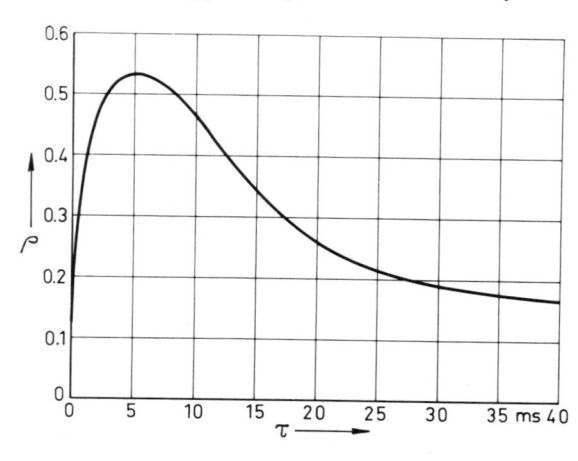

Fig. 7. The autocorrelation weighting function $\rho(\tau)$, indicating the perceptibility of pitch and timbre.

(3) "periodic white noise" masked by normal white noise. Note that periodic white noise has a line spectrum equal to that of a periodic pulse.

CONCLUSION II

The perceptual similarity between a sound with its repetition and a pure periodic signal can be represented in a general model for the perceptibility of pitch and timbre. Like the explanation of repetition pitch, time separation pitch, and periodicity pitch, the model finds its description in the time domain.

The model is of practical importance too, because it can be applied in electroacoustics and room acoustics in order to predict whether colouration (pitch and timbre) will be perceptible or not in a particular situation.

REFERENCES

Békésy, G. von (1928): Zur Theorie des Hörens; Die Schwingungsform der Basilarmembran, Phys. Z. 29, 793-810.
Bilsen, F. A. (1966): Repetition pitch: monaural interaction of a sound with the repetition of the same, but phase shifted, sound, Acustica 17, 295-300.
Bilsen, F. A. (1967/68): Thresholds of perception of repetition pitch. Conclusions concerning colouration in room acoustics and correlation in the hearing organ, Acustica 19, 27-32.
Bilsen, F. A. (1968): On the Interaction of a Sound with its Repetitions, Doctoral Dissertation, Delft University of Technology.
Bilsen, F. A., and Ritsma, R. J. (1967/68): Repetition pitch mediated by temporal fine structure at dominant spectral regions, Acustica 19, 114-115.
Bilsen, F. A., and Ritsma, R. J. (1969): Repetition pitch and its implication for hearing theory, Acustica 22, 63-73.
Bilsen, F. A., and Ritsma, R. J. (1970): Some parameters influencing the perceptibility of pitch, J. Acoust. Soc. Amer. 47, 469-475.
Boer, E. de (1956): On the Residue in Hearing, Doctoral Dissertation, University of Amsterdam.
Fourcin, A. J. (1965): The pitch of noise with periodic spectral peaks, in: Rapports 5e Congrès International d'Acoustique, Liège, Vol. Ia, B 42.
Huygens, C. (1693): En envoiant le problème d'Alhazen en France ..., in: Oeuvres Complètes, Vol. 10, 570-571.
Ritsma, R. J. (1967): Frequencies dominant in the perception of the pitch of complex sounds, J. Acoust. Soc. Amer. 42, 191-198.
Ritsma, R. J., Bilsen, F. A., Jong, Th. A. de, and Verkooyen, C. J. (1967): Spectral regions dominant in the perception of repetition pitch, I.P.O. Annual Progress Report 2, 24-30.
Schouten, J. F., Ritsma, R. J., and Cardozo, B. L. (1962): Pitch of the residue, J. Acoust. Soc. Amer. 34, 1418-1424.
Somerville, T., Gilford, C. L. S., Spring, N. F., and Negus, R. D. M. (1966): Recent work on the effects of reflectors in concert halls and music studios, J. Sound Vib. 3, 127-134.
Supa, M., Cotzin, M., and Dallenbach, K. M. (1944): Facial vision; the perception of obstacles by the blind, Amer. J. Psychol. 57, 133-183.
Thurlow, W. R. (1958): Some theoretical implications of the pitch of double pulse trains, Amer. J. Psychol. 71, 448-450.
Thurlow, W. R., and Small, A. M. (1955): Pitch perception for certain periodic auditory stimuli, J. Acoust. Soc. Amer. 27, 132-137.
Walliser, K. (1969): Zusammenhänge zwischen dem Schallreiz und der Periodentonhöhe, Acustica 21, 319-329.

DISCUSSION

Gruber: We did some neurophysiological experiments on this repetition pitch and recorded the responses of single fibres of the cat's auditory nerve under stimulation with noise with its delayed repetition added or substracted (0-, π-repetition noise, respectively). Fig. 1a shows how the spectral density at a particular frequency f_c, equal to the fibre's characteristic frequency, varies over the range $0.3/f_c < \tau < 1.7/f_c$. In Fig. 1b the discharge rate, in spikes per sec, is reproduced for the same range of τ-values. The similarity between these graphs indicates that the discharge rate is mainly determined by the spectral density at its characteristic frequency. In Fig. 2 some interval histograms are represented, both for 0- and π-repetition noise. We see again that the discharge rate is greatest for τ-values for which the spectral density has a maximum at f_c. Moreover, we see that $\tau = 1.78$ msec gives the same periodicity as $\tau = 1.18$ msec $= 1/f_c$, indicating that this periodicity is due to the characteristic frequency rather than the delay time of the stimulus. So one can conclude that in single fibres of the auditory nerve a neural correlate of repetition pitch is not to be found.

Schwartzkopff: I don't think that such a general conclusion is justified on the negative findings of some units investigated.

Bilsen: I think I am not surprised that no periodicity corresponding to the

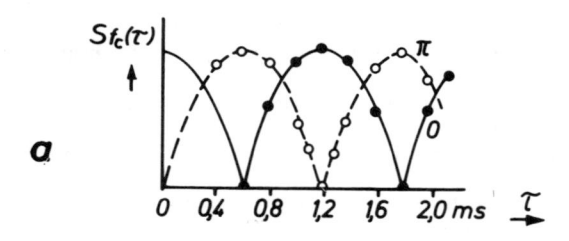

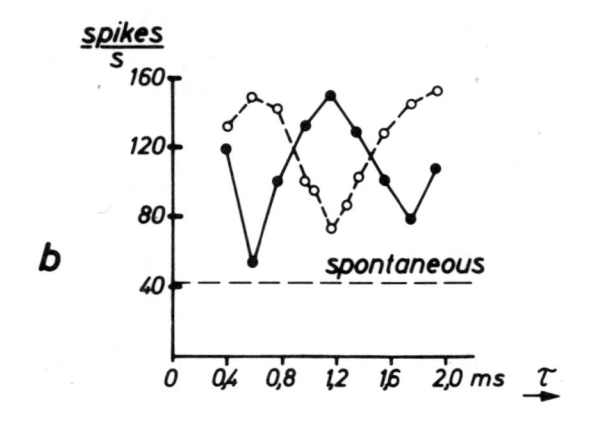

Fig. 1 (Gruber). (a) Spectral density as a function of τ. (b) Discharge rate as a function of τ.

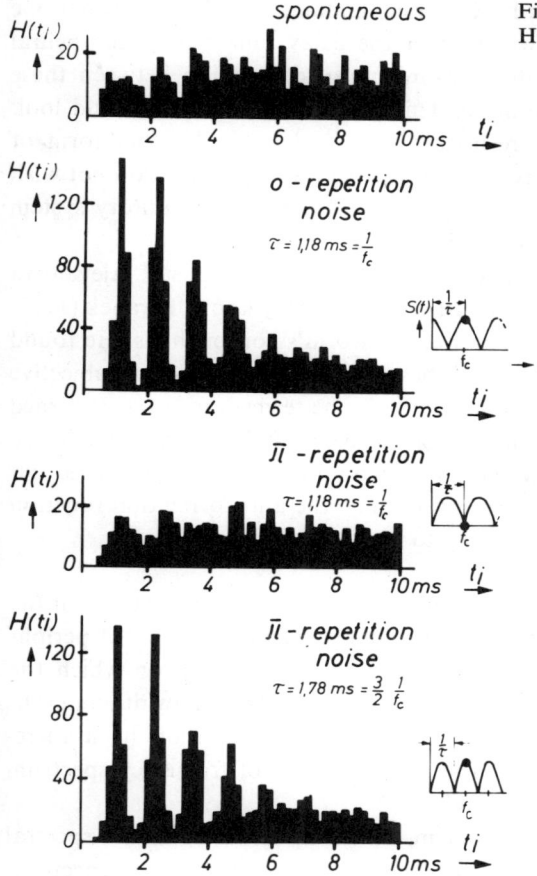

Fig. 2 (Gruber). Unit 16-16, $f_c = 850$ Hz, interval histograms.

spontaneous

o-repetition noise
$\tau = 1.18$ ms $= \frac{1}{f_c}$

π-repetition noise
$\tau = 1.18$ ms $= \frac{1}{f_c}$

π-repetition noise
$\tau = 1.78$ ms $= \frac{3}{2}\frac{1}{f_c}$

delay time was found in these experiments, because the fibre's characteristic frequency f_c was chosen within the range $0.3/\tau < f_c < 1.7/\tau$. Listening to narrow-band signals with spectral energy in a region below $2/\tau$ we never perceived repetition pitch (compare my Fig. 4). This fact is in accordance with our finding that only the timing information in the spectral region above $2/\tau$, particularly at $4/\tau$ (thus at the 3rd, 4th and 5th spectral peak: the dominant spectral region), is effective in evoking repetition pitch.

Gruber: We made also interval histograms with the characteristic frequency coinciding with the 3rd, 4th and 5th spectral peak, respectively, with the same negative results.

Goldstein: It is not a negative result at all that the interval histograms of the responses to repetition noise do not show a periodicity corresponding to the delay time. These histograms do not contain all information about the neural response. On the basis of what de Boer has shown (de Boer, 1967) and on the basis of our own confirmation of his empirical finding, I would expect that one form of signal processing that will recover the delay time τ from the neural responses to the echo noise is the first-order crosscorrelation between these responses and the original white noise. This correlation function should look like a sequence of two impulse responses separated by τ. Another form of processing would be the autocorrelation of the neural responses. I do not wish to claim, however, that I can now support the idea that the auditory system performs this processing (Licklider, 1959).

Carterette: I have a remark somewhat remotely related but still relevant to repetition pitch as related to room acoustics. As you may know, Beranek (1962) did considerable work on the acoustics of the world's concert halls. He found that their musical-acoustic qualities can be given well by a scale on subjective categories. One of the most important categories he termed *presence*. It turned out that presence depends on the time difference between the direct and reflected arrival of a sound. This difference should be in a range between a few msec and less than 20 msec, precisely the range over which the autocorrelation function of Fig. 7 (see p. 298) runs before falling to asymptote.

Bilsen: Yes, indeed.

Fourcin: In principle, it seems to be impossible at present to be sure that the pitch of these interacting noise signals does not result from some sort of peripheral frequency analysis. The presence of temporal information on which the pitch perception could be based requires a sufficiently large bandwidth and thus it is always necessarily accompanied in the monaural situation by a corresponding set of spectral attributes. The use of this set of frequency spectrum features has not been shown to be unlikely.

Bilsen: When using a 1/3-octave filter in, for example, the dominant spectral region and attenuating the level of the delayed sound, the power spectrum becomes more and more flat and much alike the pass-band characteristic of the filter. Even in that case, repetition pitch can be heard. Then, the signal is not accompanied by a set of spectral attributes, but only by one flat spectral peak. All facts on repetition pitch, known until now, can be explained from the model I have given. Although a certain event in the time domain always has a related effect in the frequency domain, a description in terms of frequency alone cannot be given.

REFERENCES

Beranek, L. L. (1962): Music, Acoustics, Architecture (John Wiley & Sons, New York).
Boer, E. de (1967): Correlation studies applied to the frequency resolution of the cochlea, J. Audit. Research 7, 209-217.
Licklider, J. C. R. (1959): Three auditory theories, in: Psychology: A Study of a Science, S. Koch, Ed. (McGraw-Hill, New York), Vol. 1, pp. 41-144.

AN AUDITORY AFTER-IMAGE

J. P. WILSON

Department of Communication
University of Keele
Keele, Staffordshire, England

In general, auditory stimulation does not lead to clear after-effects (Bishop, 1921). After prolonged stimulation at high SPL's, however, tinnitus frequently occurs and the effects of temporary threshold shift become obvious. Tinnitus, however, is not related in any obvious way with the stimulation frequency. Rosenblith *et al.* (1947) described a "jangly" after-effect impressed upon certain sounds, apparently unrelated in quality to the pulse train necessary to evoke it. More recently Zwicker (1964) reported that stimulation by a broad-band noise with a half-octave stop band led to a tonal after-image corresponding in pitch to the missing region. This effect occurs only within a limited SPL range. Lummis *et al.* (1966) rendered this effect more audible by repetition. Wilson (1966) briefly reported that a frequency spectrum with periodic peaks and dips can lead to a negative after-image heard against a background of white noise immediately following the initial stimulus. Although this may be related to Zwicker's after-image, the presence of this "after-field" of white noise allows the effect to be observed over a much wider range of stimulus conditions.

The purpose of this report is to describe these experiments more fully; to describe some recent experiments using a different method; to compare the results; and to consider what implications these findings may have for theories of hearing.

The type of signal used in these experiments is illustrated in Fig. 1. It is derived from white noise by taking a direct version, $N(t)$, and adding to it a delayed version, $mN(t+\Delta t)$, where Δt is the delay, and m the relative amplitude of the delayed signal, referred to later as the spectral modulation depth because this determines the peak-to-valley ratio of the spectrum. This is referred to as the NORMAL signal and the first peak in its spectrum, f_1, lies at $1/\Delta t$, the second at $2/\Delta t$, etc. The INVERTED signal, $N(t)-mN(t+\Delta t)$, is obtained by inverting the phase of either the direct or delayed noise prior to addition, and the spectral

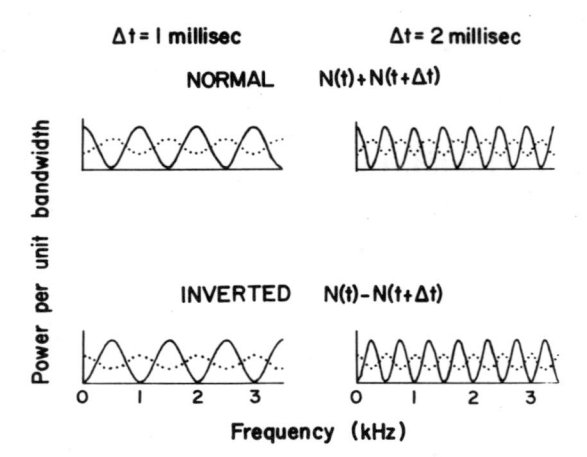

Fig. 1. Frequency spectra of noise added to delayed noise for NORMAL stimuli (in-phase components) and INVERTED stimuli (out-of-phase components). The dotted lines represent the equivalent spectra of the after-images, being inverted and having reduced modulation depths.

peaks lie at $1/2\Delta t$, $3/2\Delta t$, $5/2\Delta t$, etc.; the inverse of the NORMAL signal spectrum. The important feature of these signals for the investigation of after-images is that they are complementary and give rise to characteristically different sensations.

Such a signal is of interest on theoretical grounds because it can be analysed by a mechanism operating either in the frequency domain (Fig. 1) or in the time domain because one of its constituents is a delayed repetition of the other. The demonstration that a particular phenomenon depends on one type of mechanism and not the other would lead to a better understanding of hearing. Certain characteristics of the after-image indicate that it may be wholly a place/frequency phenomenon. For example, it might result from localised adaptation of the hair cells at positions that correspond to the spectral peaks, so that subsequent stimulation with a uniform noise spectrum produces relatively greater activity between these positions. The transient after-pattern of activity central to this adaptation would then be indistinguishable from that produced by the inverse spectrum. Such an hypothesis requires that the original spectral pattern should be resolved by the cochlea, since the pattern could not be recovered at higher levels if it were "blurred" out at this stage.

EXPERIMENTS

Four methods have been used, illustrated in Fig. 2. A diagram of the apparatus used to generate the stimulus sequences and to record the results is given in Fig. 3. The way in which this apparatus is inter-connected depends on the experiment concerned and will be described later. The time delay, Δt, between the two identical noise signals, adjustable through zero, is obtained on 2 tracks of a tape recorder with a micrometer screw on one of the heads.

Sound pressure levels were measured using a condenser microphone (B & K

AN AUDITORY AFTER-IMAGE 305

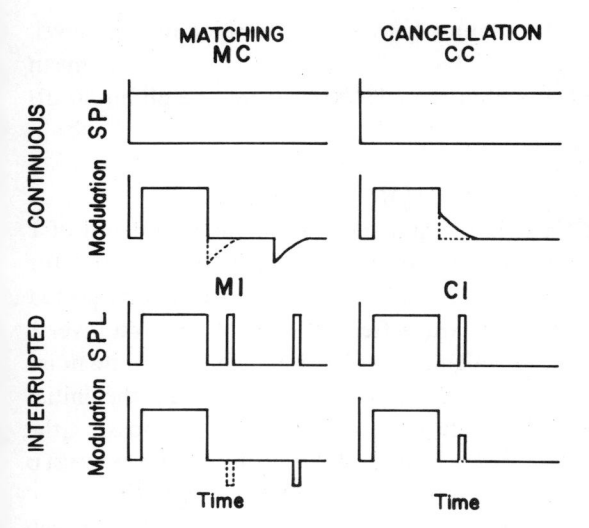

Fig. 2. Four different stimulus sequences for measuring the parameters of the after-image. The upper graphs in each case give the SPL and the lower graphs give the spectral modulation depths of the stimuli and test signals. The dotted lines indicate the subjective impressions during the course of the physical stimulus, solid lines.

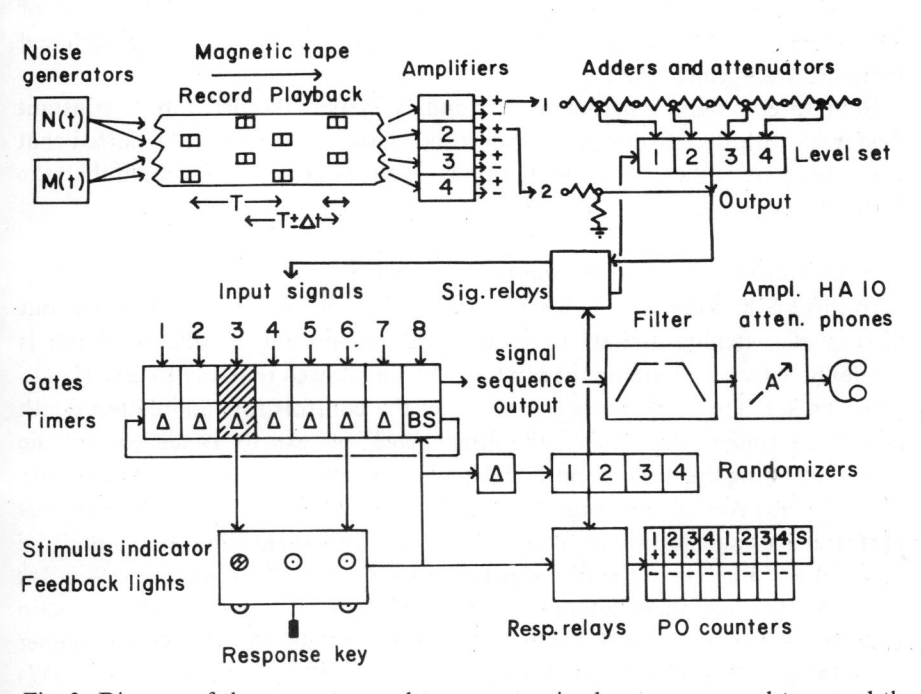

Fig. 3. Diagram of the apparatus used to generate stimulus sequences and to record the responses of the subject. Noise from the tape recorder is added in the appropriate phase and at four alternative preset modulation depths. One of these levels is chosen and the signal sequence set by relays controlled by randomisers; parallel relays channel the response to the appropriate counter. Monostable timers and gates pattern the input signals into the appropriate signal sequence; a bistable awaits the response of the subject to restart the cycle and reset the randomisers.

4133) with a flat plate coupler. The figures quoted refer to the unfiltered level. As the phones (Sharpe HA 10) have a bandwidth of about 10 kHz the mean spectrum level, or power per Hz, within the pass band would be about 40 dB below the figures given.

1. *Matching method using a continuous signal (MC)*

In this method (Fig. 2, MC) a signal with a spectral modulation depth of 1 (Fig. 1) is presented for 5 sec, then switched to white noise of the same SPL for 5 sec. Immediately after switch-over, a brief after-image, represented by dotted lines in Figs. 1 and 2 (MC), could be heard. After the 5 sec of noise an inverse spectral modulation was introduced which decayed exponentially to white noise again. The task of the subject was to adjust on successive trials the initial modulation depth and the rate of decay until it matched as nearly as possible the after-image immediately preceding it. This was done for NORMAL and INVERTED spectra, for delays of 5.7, 1.43, and 0.37 msec, and for 30, 60, 90 dB SPL.

Five subjects were used giving similar results ($m = 0.37 \pm 0.04$ and t between 0.7 and 1.5 sec). There were no significant differences as a function of Δt or SPL except for the 5.7 msec condition ($m = 0.26 \pm 0.05$) which was excluded from the averaged result.

Some preliminary experiments also indicated that the decay time constant decreases with decreasing stimulation time. Two of the subjects reported that the initial part of the decay is more rapid and the later part is less rapid than the exponential decay.

2. *Cancellation method using a continuous signal (CC)*

In this method the stimulation is the same as in the previous experiment but instead of switching directly to white noise the spectral modulation depth is suddenly reduced (retaining the same sign) then allowed to decay to zero (*i.e.* to white noise). The subject's task was to set the controls for modulation depth and decay time constant until the after-image was exactly cancelled and the epoch following the stimulation sounded like white noise without any impression of either the inverted spectrum or continuing stimulation. This experiment was performed using the two types of signal, the same three time delays and the same three SPL's as in the first experiment. As before there was no significant difference between these conditions except when $\Delta t = 5.7$ msec which was again slightly weaker at $m = 0.28 \pm 0.10$. The average values were somewhat higher ($m = 0.44 \pm 0.10$ and $t = 1.2 \pm 0.2$ sec, JPW; $m = 0.7 \pm 0.2$, $t = 1.0 \pm 0.2$ sec, CAW) than for the MC condition, possibly resulting from continuation of the cancelling signal.

As the impression of the two subjects in this experiment was again that the decay was not quite exponential, experiments were devised which could measure the decay function without any prior assumptions about its form. Two possible experiments are illustrated in Fig. 2 (MI and CI) and will be described in detail below. As both of these involve presenting short test pulses of noise, it was first necessary to decide what duration to make them. Long segments would allow decay of the after-image during the presentation, and would not give sufficient temporal resolution. Very short segments would not allow differences between a positive or negative image to be perceived.

3. *Threshold spectral modulation depth as a function of duration*

In this experiment a two-alternative forced-choice (2AFC) method was used. The subject's task was to decide whether interval A or interval B contained the stimulus which was assigned to one of these at random and without bias. Interval A was presented 1 sec after the subject pressed the response key; interval B occurred 1 sec after interval A was complete. One interval contained the stimulus at one of four possible spectral modulation depths around threshold, while the other contained uniform noise at the same overall SPL (60 dB). A and

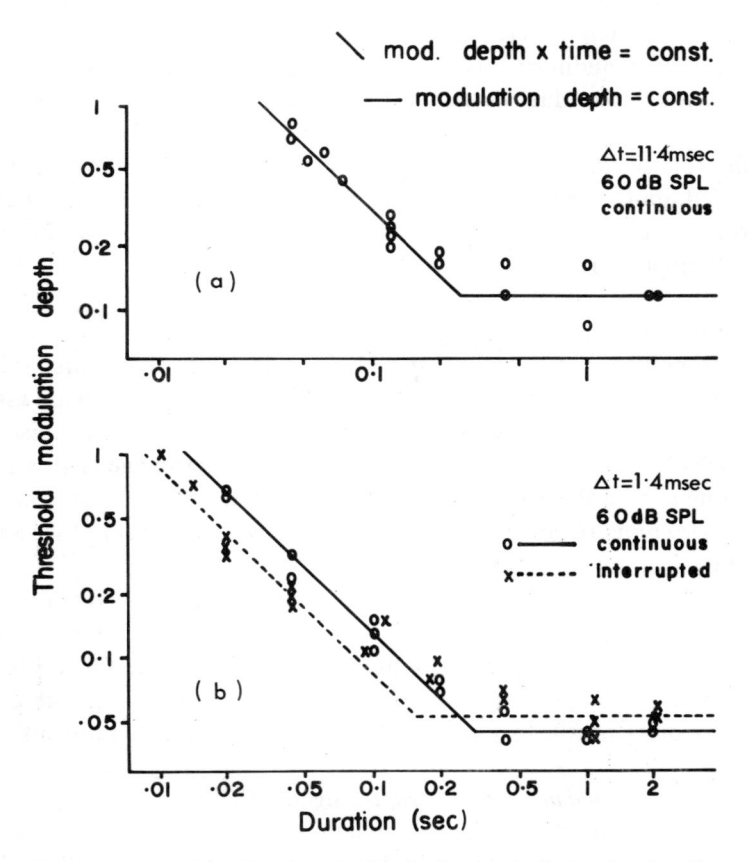

Fig. 4. Threshold modulation depth as a function of stimulus duration.

B were always equal in duration which was varied from 10 msec to 2 sec. The subject was given indication of correct sequence after each response. The experiment was performed at two values of time delay, 11.4 msec and 1.43 msec, and for two conditions of stimulus sequence: (1) in the interrupted condition (dotted line in Fig. 4b) the periods between the intervals A and B were silent, whereas in (2), the continuous condition, these periods were filled in with white noise of equal SPL (the intervals were defined by neon indicator lights). Each point was calculated graphically from 100 responses and represents 76% correct response (*i.e.* $d' = 1$ for 2AFC).

It will be seen from the results shown in Fig. 4 that, although the overall threshold values differ by a ratio of about 2.5:1 between the two time delays, the turnover points are similar at about 0.25 sec. Above this the threshold remains constant whereas below there is a trading relation between modulation depth and duration: the system appears to be integrating the information available.

The interrupted condition (dotted line in Fig. 4b) corresponds approximately to the continuous condition from 0.1 sec upwards whereas below 0.1 sec its threshold is definitely lower.

The value for auditory integration time derived in this experiment falls within the range of values determined for other types of signal by numerous other workers.

A compromise value of 0.1 sec for the test signal in the following experiments was decided upon as this gives a reasonably low threshold. Although 0.2 sec or more would give slightly greater sensitivity, this advantage would be outweighed by loss of temporal resolution of the initial part of the decay.

4. *Matching method using an interrupted signal (MI)*

In this technique (Fig. 2, MI) the stimulating signal is presented first, followed by a variable silent interval, during which the effects of the stimulus are assumed to be decaying unheard, then follows the 0.1 sec test signal of white noise which allows the subject to perceive the magnitude of the after-image at that time; a further long silence then follows to allow the after-image to decay completely and a 0.1 sec comparison signal of low spectral modulation depth follows to be matched against the test signal.

It was found in preliminary experiments, however, that it was very difficult to compare these short intervals in this situation. Furthermore, the time necessary to collect data is approximately doubled compared with the CI method discussed below. The method was therefore not used although it is possible that it might prove more satisfactory with longer test and comparison intervals.

5. *Cancellation method using an interrupted signal (CI)*

With this method the standard stimulation period (which was reduced to 4 sec to make the sequence faster and less tedious) was again followed by a variable silent interval, then a 0.1 sec test interval as in the MI method. The test signal in this case, however, contains spectral modulation of the same sign as the stimulus in just sufficient proportion to cancel out the after-image. It should therefore sound neutral like white noise. For the purpose of the experiment, the modulation depth was set to one of four values centred around the neutral value (determined roughly first or by extrapolation). The task of the subject was to categorize it as similar to the adapting stimulus or inverse to it. The data were analysed as if it were a 2AFC experiment (although it is recognized that there are possible objections to this). Again each point represents 100 decisions. Obviously it was not possible to provide the subject with feedback of the "correct" response in this situation. The presentation of NORMAL and INVERTED stimuli was randomized. It was necessary for the subject to have some practice before the experiment started in order that the two types of stimulus were readily recognized and distinguished.

Four parameters: (a) Δt, (b) spectral region, (c) SPL, and (d) stimulus duration were investigated in this experiment. The results of (a), (b), and (c) are presented together on Fig. 5. Three spectral regions were selected using a band-pass filter with slopes of 30 dB per octave (Allison 2AB): (1) 150-900 Hz with $\Delta t = 5$ msec, (2) 1.2-2.4 kHz with $\Delta t = 2.5$ msec, and (3) 2.4-9.6 kHz with $\Delta t = 0.625$ msec. Δt was set to the values given so that the reflection tone could be heard clearly and so that NORMAL and INVERTED spectra could readily be distinguished.

The data from two subjects can be seen to agree for the same experimental

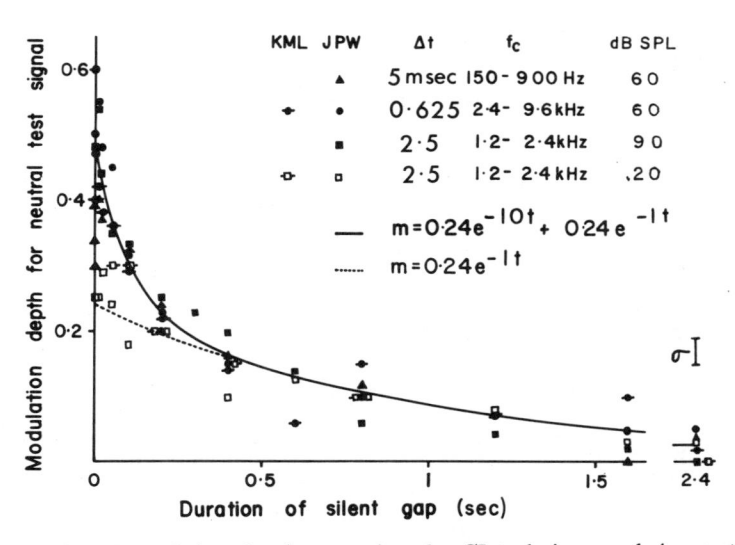

Fig. 5. Decay function of the after-image using the CI technique and 4 sec stimulus duration. Apart from the 20 dB SPL condition the data for different delays (Δt), band-pass filter settings, and SPL's are fitted by the sum of two exponentials.

conditions. There is no consistent difference between low and high parts of the frequency spectrum (dots and triangles) nor between medium and high SPL's although the points for low levels (20 dB, open squares) do lie below the others for the initial part of the decay. It is obvious that the data do not fit a single exponential decay function though they do fit a curve for the sum of two exponentials as shown (containing four constants). It will be seen, however, that the same data also fit a log time function (with only two constants) as shown by the crosses in Fig. 6. These points represent the averaged data of Fig. 5 (excluding the 20 dB SPL) and were also used in deriving the previous equation.

Fig. 6 also gives the data from one subject, JPW, for shorter stimulation periods of 1 sec and 0.25 sec. These experiments were performed exactly as the previous one using the 2.4-9.6 kHz band with a delay of 0.625 msec and at 60 dB SPL. It will be seen that these conditions also give straight line plots parallel with the 4 sec condition. The best lines through these data are separated by slightly less than the square root of the stimulation time relationship described by Taylor (1966) for various perceptual tasks although the difference is probably not significant.

A log time decay function appears inherently unlikely because it eventually passes through zero and becomes negative. Although some points are shown below zero, this may be experimental error as the judgement becomes more difficult and less reliable for the later part of the decay. Such a relationship has, however, been reported in another study on auditory decay (Plomp, 1964). In the present case, although the spectral modulation depth is plotted on a linear

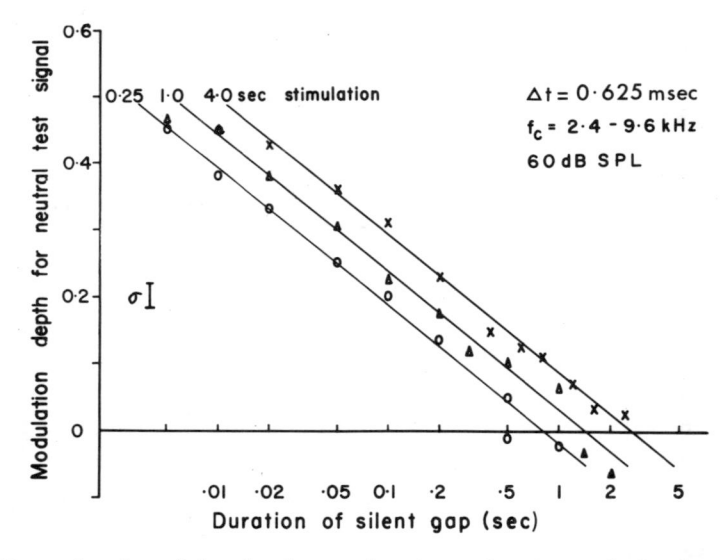

Fig. 6. Decay function of the after-image plotted on a log time scale for three stimulus durations. The 4.0 sec condition is derived from the same data as Fig. 5.

scale, this is proportional to the peak-to-valley ratio in dB's within the range used and thus corresponds with the scales used in TTS and auditory decay. One surprising feature of these results is that the graph (Fig. 6) continues to rise as it approaches the axis on the left in spite of the fact that the gap then becomes very much shorter than the test signal. Some experiments were therefore performed with test signals of 0.05 and 0.2 sec duration and each gave results similar to the 0.1 sec test signal for the same silent interval length. A possible objection to this experiment is that if the effects of stimulation last for a long time, the test signal will be affected also by the preceding stimulus and, possibly, by still earlier ones. As the sequence is random, however, and the final result is an average for many presentations, it would appear that this effect would average out and merely increase the variability of the results.

DISCUSSION

Comparing the results obtained by the three methods it is clear that they lead to very similar results. If one allows for the slightly shorter stimulation period used in the CI method and compares this with the optimal single exponential decay of the MC experiment, the curves agree within experimental error from about 0.1 sec upwards. The results from the CC method are also comparable although somewhat higher and longer lasting as might be expected for this method. Nevertheless, it is likely that the CI method gives the most accurate picture of the decay function, which is of the form $m = a - b\log t$. It would appear from these comparisons that the after-effects of stimulation decay regardless of whether any further sound is present (*i.e.* silence appears not to preserve the situation nor, alternatively, noise to accelerate the decay).

As the pitch and timbre of the after-image invariably correspond to the pitch and timbre of the inverse to the stimulating spectrum (Wilson, 1966), a simple model based on local adaptation suggests itself as outlined in the introduction. Whether the adaptation takes place at the hair cells or at some higher neural level is an open question. The hair cell mechanism was suggested here merely to illustrate a possible place/frequency basis for the negative after-image. The significant feature of this model is that it implies that the spectral pattern is resolved by the mechanical action of the cochlea. In temporal terms, one could also imagine local adaptation taking place at specific points along the parallel output pattern of an autocorrelator. The difficulty with such a suggestion is that phase information in the auditory nerve is only preserved for the lower frequencies. The after-image, on the other hand, shows exactly the same magnitude and decay properties from 150-900 Hz as it does from 2.4-9.6 kHz so that one would be led to the rather unlikely conclusion that there were two distinct mechanisms for the after-image: an autocorrelation mechanism for low frequencies and a place/frequency mechanism for high frequencies, both with the same constants. (Admittedly 2.4 kHz is not well above the upper limit of phase preservation, but the effects would nevertheless be much weaker whereas it has been confirmed qualitatively that the after-image still occurs for frequencies above 7 kHz.)

The corollary to a place/frequency basis for the after-image is that the basilar membrane is capable of quite reasonable resolution of the frequency components of a signal. It is possible then, that the *pitch* perceived for the NORMAL and INVERTED spectra, and of course for the after-image, may result from place analysis. Some experiments on spectral dominance briefly reported below are consistent with this view.

THE PITCH OF SIGNALS WITH MULTIPLE SPECTRAL PEAKS

The ear has a frequency resolving power which is approximately uniform on a logarithmic scale. A signal with uniformly spaced peaks on a linear scale will therefore be analysed differently in different parts of the spectrum. The low frequency peaks of Fig. 1 would be broad on a logarithmic scale and could not be expected to have a sharply defined pitch. At the very high frequencies the peaks would be too closely spaced to be separately resolved. In between these two conditions peaks would be sharp enough to have a clear pitch and yet be far enough apart to be resolved by the ear. This region might be expected to be dominant in perception. If one particular peak is dominant, it then becomes obvious why the pitch shifts to say 7/8 of the NORMAL value when the INVERTED signal is presented (Fourcin, 1965). In this case the dominant region would lie between 3.5 and 4 so that, for the NORMAL spectrum, peak number 4 would be dominant whereas for the INVERTED spectrum the nearest peak to the dominant region would be $7/2\Delta t$. Furthermore in the corresponding 90° phase shifted condition the dominant peak would lie at $15/4\Delta t$ leading to the ratio of 1.07 observed by Bilsen (1968) for this condition. As Wilson (1966) had found that the pitch difference between NORMAL and INVERTED signals was not constant but depended on Δt, it is of interest to know whether the dominant peak number is also a function of Δt. As peak number 4 is two octaves above the first peak and one octave above the second peak, it is conceivable that there are 'musical' constraints that cause peak number 4 to be dominant, when *pitch* is used as criterion. In order to eliminate this possibility, colouration rather than pitch was judged in this experiment. It is assumed that the region of a complex signal having the lowest relative threshold will be dominant at higher levels.

The dominant region for colouration was determined from the threshold curves obtained using high-pass and low-pass filtered multiple peak signals added to low-pass and high-pass noise signals as illustrated in the insets of Fig. 7. The threshold spectral modulation depth is plotted as a function of the frequency at which the cross-over from peaks and dips to a flat spectrum

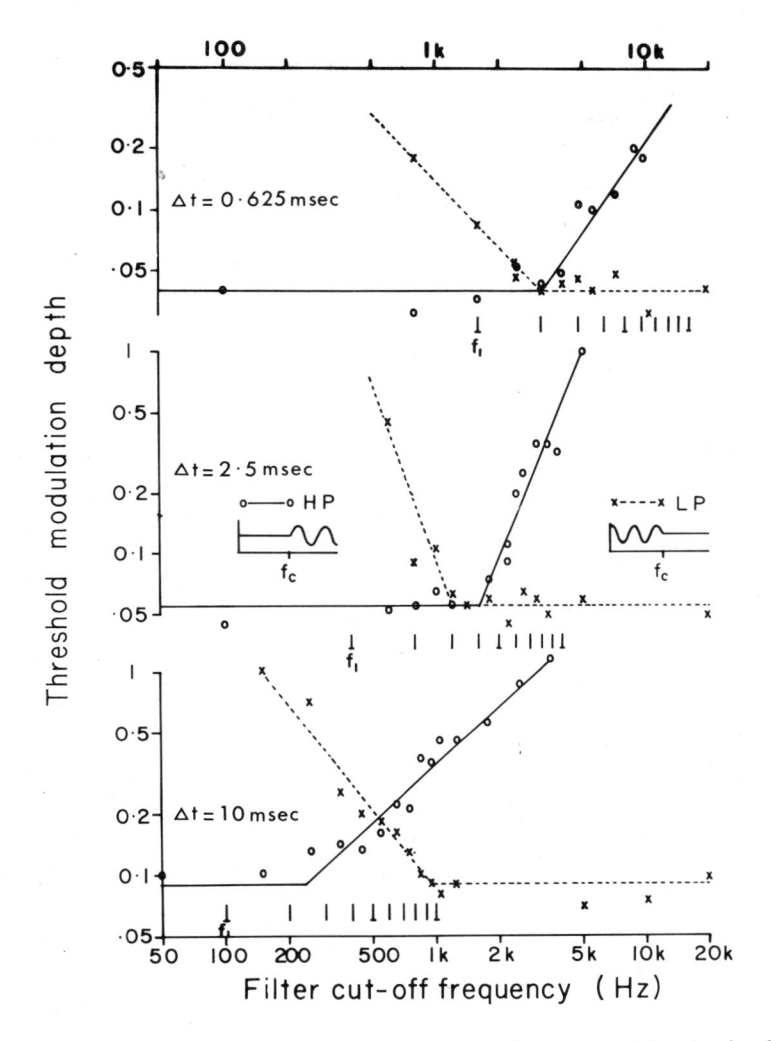

Fig. 7. Threshold modulation depth for high-pass and low-pass filtered stimuli as a function of cut-off frequency, f_c. The regions beyond f_c are filled in with uniform noise at the same mean spectral density as indicated in the inset spectra. The dominant spectral region for colouration is considered to be that where the high-pass and low-pass thresholds overlap. The vertical bars under each graph indicate the positions of the first ten spectral peaks in the stimulus where $f_1 = 1/\Delta t$.

occurs for three values of Δt. The dominant region is defined as the area of overlap between the low-pass and high-pass curves: the presence of peaks above or below this region does not contribute to a lowering of the threshold. The dominant spectral regions calculated from many data like Fig. 7 are given in Fig. 8. The curve with its axis on the left expresses the data in terms of absolute spectral position as a function of $1/\Delta t$ while the other curve (open

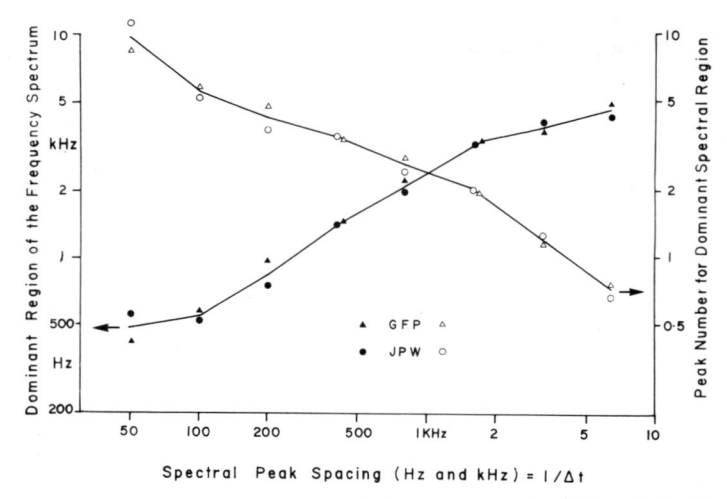

Fig. 8. Dominance for colouration expressed either as spectral position (left axis and filled data points) or as dominant spectral peak number (right axis and open data points) plotted as a function of spectral peak spacing ($=f_1=1/\Delta t$). Each point is derived from curves like those illustrated in Fig. 7.

points) expresses the same data in terms of dominant peak number. It can be observed that the dominant region of the spectrum for colouration extends from 500 Hz to 4 kHz and the dominant peak number ranges from about 10 to less than 1. Over the range studied by Ritsma (1967) and Bilsen and Ritsma (1968) in their pitch dominance experiments there is good agreement with the present results. Although perhaps tautological it appears that pitch may be a specific attribute of colouration. Furthermore the pitch ratios calculated from this colouration dominance experiment agree quantitatively with the pitch ratios given by Wilson (1966):

when $\Delta t = 25$ msec, $N_{dom.} = 10$ ∴ calculated ratio $10:10\frac{1}{2} = 1.05$
(observed ratio $= 1.06$)

when $\Delta t = 1$ msec, $N_{dom.} = 2\frac{1}{2}$ ∴ calculated ratio $2\frac{1}{2}:3 = 1.2$
(observed ratio $= 1.2$)

Although it is simpler to consider a single dominant peak, it may be more realistic to consider the whole spectral *pattern* over the dominant region: what the ear appears to do with one of these somewhat anomalous signals is to match the peaks within the dominant region to the nearest pattern based on a harmonic series and to assign to it the pitch that this would have.

It is therefore possible to explain the pitch of the signal and certain other similar signals without invoking neural timing mechanisms where previously it had been thought to be necessary (Bilsen, 1968; and pp. 291-302 of this volume).

SUMMARY

1. Stimulation with a sound containing multiple spectral peaks leads to a transient perception of a complementary spectrum on switching to white noise.
2. This effect occurs for a wide range of peak separations and for high and low regions of the frequency spectrum.
3. The effect can be observed at all SPL's although it might possibly be weaker below 20 dB SPL.
4. Three separate methods for its measurement give comparable results.
5. The simplest decay equation is of the form $m = a - b\log t$.
6. A place/frequency theory can account for the after-image in a single mechanism whereas an additional mechanism with identical constants would have to be invoked for a timing or autocorrelation theory.
7. A place/frequency theory can account for the pitch of the NORMAL and INVERTED spectra and for the pitch of the after-images and the 90° phase shifted condition.
8. Such a mechanism is based on matching the dominant peak (or peaks) with the nearest spectrum based on a harmonic series.

REFERENCES

Bilsen, F. A. (1968): On the Interaction of a Sound with its Repetitions, Doctoral Dissertation, Delft University of Technology.

Bilsen, F. A., and Ritsma, R. J. (1968): Repetition pitch mediated by temporal fine structure at dominant spectral regions, Acustica *19*, 114-115.

Bishop, H. G. (1921): An experimental investigation of the positive after-image in audition, Amer. J. Psychol. *32*, 305-325.

Fourcin, A. J. (1965): The pitch of noise with periodic spectral peaks, in: Rapports 5e Congrès International d'Acoustique, Liège, Vol. Ia, B 42.

Lummis, R. C., Guttman, N., and Bock, D. E. (1966): Auditory after-images in gated noise, J. Acoust. Soc. Amer. *40*, 1240 (A).

Plomp, R. (1964): Rate of decay of auditory sensation, J. Acoust. Soc. Amer. *36*, 277-282.

Ritsma, R. J. (1967): Frequencies dominant in the perception of the pitch of complex sounds, J. Acoust. Soc. Amer. *42*, 191-198.

Rosenblith, W. A., Miller, G. A., Egan, J. P., Hirsh, I. J., and Thomas, G. J. (1947): An auditory after-image?, Science *106*, 333-334.

Taylor, M. M. (1966): The effect of the square root of time on continuing perceptual tasks, Perception and Psychophysics *1*, 113-119.

Wilson, J. P. (1966): Psychoacoustics of obstacle detection using ambient or self-generated noise, in: Animal Sonar Systems, R. G. Busnel, Ed. (Frascati Symposium, 1966; Jouy-en-Josas, 1967), pp. 89-114.

Zwicker, E. (1964): Negative after-image in hearing, J. Acoust. Soc. Amer. *36*, 2413-2415.

DISCUSSION

Fourcin: It seems to me that one of the important findings Dr. Wilson has described concerns the uniformity of the after-image results over a wide frequency range. I agree that this makes it rather unreasonable to explain the results by temporal mediation since time structure is not preserved at high frequencies. Further, it occurs to me that the subjects in the experiments were asked to comment on the spectral colouration, and not to make pitch judgments. In consequence, they were not applying the same criteria as used by Ritsma (1967) and Bilsen and Ritsma (1968); so, as far as I can see, it follows that your objection to their use of temporal mediation no longer holds.

Wilson: Although the spectral dominance experiments to which you refer are based on thresholds for *colouration*, I believe they probably measure the same property of the auditory system as Ritsma (1967) and Bilsen and Ritsma's (1968) experiments because the results are similar. Furthermore, the quantitative predictions for the *pitch* of the INVERTED signal with high and low values of Δt and for the 90° phase shifted condition also turn out to be correct although again based on colouration thresholds. I do not think this experiment indicates a preference for either a place or a temporal theory: in fact both theories appear to rely on similar assumptions such as spectral dominance. However, as you say, the findings of the after-image experiments favour a place/frequency theory.

Fourcin: There are large differences between these signals in colouration and timbre, but not in pitch.

Wilson: For signals with small delays there is little impression of pitch, and judgements must be based on colouration. Within the 2-10 msec range, however, pitch is clear even for the weaker after-image. Although the difference in pitch between stimulus and after-image is not large, it is still readily appreciated over this range. All subjects have confirmed that the pitch and colouration of the after-image invariably correspond to those of the inverse of the stimulus.

Bilsen: I agree with Dr. Fourcin. I think we don't need to make a choice between time or frequency in your case, because both factors are working at the same time. It seems to me that you have investigated the spectral aspect since, for example, you found the after-image to be weaker below about 5.7 msec; in that region the spectral peaks come within the critical band, so they are not resolved completely and you will not find an optimal after-image. On the other hand, at 1600 Hz and 2700 Hz there is no repetition pitch (= reflection tone) whereas you found the same sort of after-effect having about the same magnitude as for $1/\Delta t$ values of 200-700 Hz. This also supports the view that you have studied the frequency aspect.

Wilson: I agree; although I would like to add that the pitch of the after-image is most striking in the range considered optimal for periodicity pitch.

Johnstone: In the graph showing the decay of the after-image (Fig. 6) you are using a log time scale. It is very noticeable that almost only in auditory work this log time scale ever shows up. I get the impression that some people want to explain this logarithmic behaviour by some obscure reactions in higher nuclei. Hood's (1950) work, for instance, shows threshold shifts agreeing extremely well with straight lines on a log time plot. Now today we heard a lot about theories but very little about mechanisms, of which the physical basis must be terribly obscure for such a nature. Could it be possible that this log time scale has to do with the summating potential which is the only physiological phenomenon, as far as I know, where you find logarithmic time relationships?

Wilson: Quite frankly, this log time relationship worries me somewhat. I think you were one of the first to point out to me the possibility of using a log time scale based on your findings on the summating potential. However, if the properties of the summating potential account for the after-image, this should be observable at the level of the cochlear nerve. My initial impression is that adaptation and after-effects are somewhat too rapid at this level but further experiments are clearly required to substantiate this.

Ward: I just want to point out that although we plot a lot of temporary threshold shift (TTS) curves on a log time basis (Reid first did this in 1946), it does not commit us to any theoretical substructure. Recently Botsford (1968) and Keeler (1968) have shown that you can fit the same data with ordinary exponentials which are a lot easier to understand on a physiological level. Furthermore, I plotted many years ago all the TTS data after Crozier's (1937) hypothesis and found straight lines, too. In my opinion, a good fit may be found for many relationships, except for data near the extremes (*i.e.* near zero time and as the TTS approaches zero), so you must make very precise measurements at the extremes in order to discriminate among these relationships.

Bosher: It seems to me that Dr. Wilson is presumedly analyzing the effect of fatigue on the hair cells and the nerve fibre system. I am surprised he obtained the same results in his continuous noise as in his interrupted noise experiments, since in the first case he is measuring adaptation or perstimulatory fatigue, and in the second case the recovery from this. From Hood's (1950) extensive studies on these two phenomena it is clear that the time course is quite different in the two instances and I would have consequently expected the results to have been different also. I was, incidentally, interested enough to replot his data on a log time scale and found much the same angles as Dr. Wilson has.

Wilson: It is important to realize that the ordinate scales are different. The modulation depth, m, (the relative amplitude coefficient of the delayed signal) is at first sight a linear measure. However, the peak-to-valley ratio in dB is very nearly proportional to m for the values measured, *i.e.* below 0.5. The type of relationship to which you refer is therefore most interesting.

Zwicker: I wish to raise the question whether "after-image" is the right word for what you were describing. When you have white noise and you suddenly add to it a 1/3-octave band of noise which is only about 5 dB higher in level than the white noise, also measured in 1/3-octave bands, then you hear the added noise band only for a short while. Then it disappears and you only hear the white noise again. Is not this similar to what you are doing by switching from your peaked signal to white noise? In doing this you add, in fact, the inverted signal. So, you should not be surprised to get a short decaying sensation according to this inverted signal.

Wilson: I am not quite sure what you mean but I believe the closest analogy in the situation you describe would be what one hears just after removing the extra 1/3-octave band of noise. The adaptation during stimulation is a very interesting phenomenon and is obviously related to the after-image, although with multiple peaked spectrum I do not think it disappears as completely as you suggest. I see no reason in this for rejecting the term "after-image". These and other features have many parallels in visual after-images.

REFERENCES

Bilsen, F. A., and Ritsma, R. J. (1968): Repetition pitch mediated by temporal fine structure at dominant spectral regions, Acustica *19*, 114-115.
Botsford, J. H. (1968): Theory of TTS, J. Acoust. Soc. Amer. *44*, 352.
Crozier, W. J. (1937): Strength-duration curves and the theory of electrical excitation, Proc. Nat. Acad. Sci. *23*, 71-78.
Hood, J. D. (1950): Studies in auditory fatigue and adaptation, Acta Oto-Laryngol., Suppl. 92.
Keeler, J. S. (1968): Compatible exposure and recovery functions for temporary threshold shift—mechanical and electrical models, J. Sound Vib. *7*, 220-235.
Reid, G. (1946): Further observations on temporary deafness following exposure to gunfire, J. Laryngol. Otol. *61*, 609-633.
Ritsma, R. J. (1967): Frequencies dominant in the perception of the pitch of complex sounds, J. Acoust. Soc. Amer. *42*, 191-198.

CENTRAL PITCH AND AUDITORY LATERALIZATION

A. J. FOURCIN

University College London
London, England

AIM

The sensation of pitch appears normally to arise from the operation of a multiplicity of mechanisms. It is not readily possible to distinguish their relative contributions and it would be an advantage to be able to study an effect which was dependent on the operation of only one type of mediation. The aim of the present experiments was to arrive at and to examine a situation in which only a neural mechanism contributed to the sensation.

Auditory lateralization can be made to depend only on the temporal difference between the stimuli applied to the two ears. When this condition is obtained the lateralization sensation is due only to the operation of a neural mechanism, since peripheral analysis cannot account for the effects observed. By examining the sensation of pitch which can be associated with this type of lateralization, it seems possible to arrange for a pitch sensation to be similarly independent of the action of peripheral analysis.

LATERALIZATION AND PITCH

In the absence of peripheral clues such as amplitude or spectral difference between the two ears, lateralization can occur for a temporal difference which may be as low as 10 μsec (Mills, 1958), or as high as 10 msec (Blodgett *et al.*, 1956). Although the point has not been explicitly investigated, the possibility of a physical, bone conduction, rather than a neural interaction being basic to these temporal lateralization phenomena is easy to exclude in informal listening and, because of the low sensation level (SL) at which the effects can occur, to dismiss experimentally. Fig. 1 outlines the basis of an experimental arrangement for which lateralization results only from the temporal difference, τ, in time of arrival at the two ears of otherwise identical stimuli. As τ is increased, the

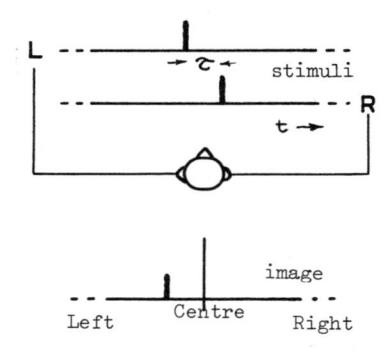

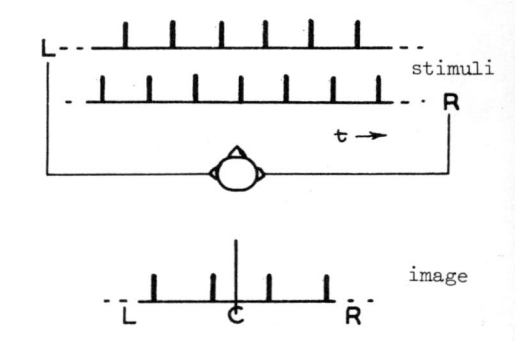

Fig. 1. Lateralization image representation for a left advanced click.

Fig. 2. Lateralization image representation for a left advanced click train.

auditory image moves further to the left until the delay is of the order of 1 msec. Beyond this limit, which corresponds roughly to the time needed for airborne sound to cross a distance comparable to the width of the listener's head, there is little practical utility in being able to process time differences for the purposes of lateralization. This ability, however, persists.

Since the mechanism exists for lateralizing a single click, the binaural presentation of a click train could result in an array of similar lateralization images. When this click train is periodic, as for the arrangement shown in Fig. 2, a sensation of pitch can be produced. This pitch is of course not dependent on the binaural nature of the presentation but in normal listening it would always be associated with the array's regular spacing. Two classes of array can be distinguished. The first is associated with an adjacent image spacing which could arise from two causes. Either from the lateralization of adjacent spatially separated independent acoustic sources or from the binaural reponse to a regular stimulus whose period is less than the maximum possible interaural delay. The second class of array can only be associated with the binaural response to a regular stimulus, one whose period is greater than the maximum possible interaural delay and less than about 10 msec.

Both of these classes of array could be basic to a central process of analysis akin to that postulated for the first in Licklider's Triplex Hypothesis (1962). The second class can have no lateralization function and if employed at all could only contribute to a process of signal analysis, one aspect of which, as a result of the array's regular spacing, might be related to the perception of pitch. This, however, is mere conjecture since an association between these arrays and pitch is as difficult to disentangle for ordinary stimuli as that between frequency and periodicity pitch. It would nevertheless be possible to test the principle, if a lateralization array could be set up using stimuli which were devoid of periodicity.

SYNTHESIS OF A LATERALIZATION ARRAY

The single lateralization image of Fig. 1 can be obtained for a click source or for noise with the delay line arrangement of Fig. 3. When noise is used the image is somewhat more diffuse but still well defined. The spectral envelope of the stimulus at each ear is entirely determined by the source, and on a long term basis can be completely free of peaks and troughs. The same technique can also be employed in order to build up a multiple image array. Fig. 4 shows how a three-lobe lateralization array can be constructed (the word "lobe" is employed here to denote the physical correlate of a possible subjective lateralization image). The use of a separate noise source for each lobe of the array, as for Fig. 3, produces an acoustic stimulus which has a smooth spectral envelope and a pressure-time waveform quite devoid of the periodic peaks which would ordinarily accompany a structured lateralization array of this type.

If, as in Fig. 4, the delays employed are of equal value, the pattern corresponds to the three central images that one might expect to get from an in-phase binaural periodic pulse train stimulation. Maximum congruence occurs for the delay, τ, equal to the pulse train period.

FREE-SPACE EXPERIMENT

A particularly simple way of producing the single lobe of Fig. 3 is shown in Fig. 5a where the sound source and the subject are in an anechoic room. For small θ, small interaural delays are obtained but $\theta = \frac{1}{2}\pi$ gives only a moderate delay, corresponding to the head width. This disadvantage is partly overcome with the arrangement of Fig. 5b by using microphones which separately drive sound excluding earphones and which are set on a head arm at a distance from

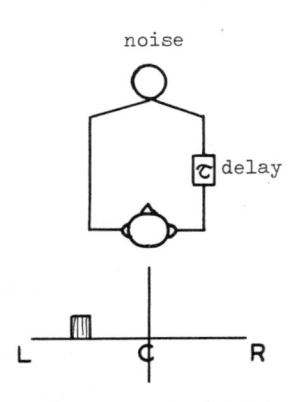

Fig. 3. Single lateralization lobe.

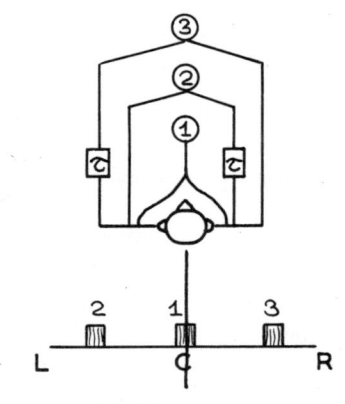

Fig. 4. Three lateralization lobes from three independent sources.

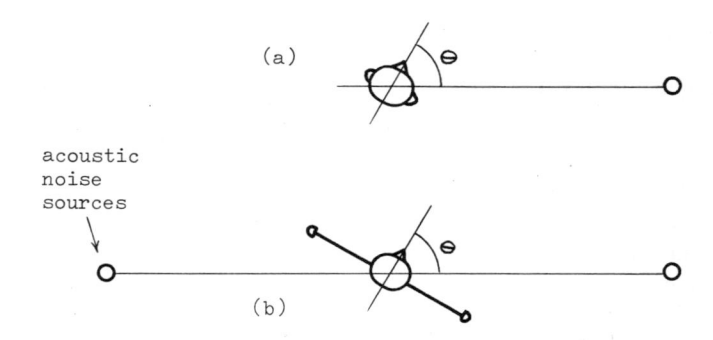

Fig. 5. (a) variable single lobe with small τ range; (b) variable double lobe with moderate τ range.

each other. The use of an additional independent sound source makes it possible to form two lobes, 2 and 3 of Fig. 4.

This experimental arrangement, Fig. 5b, was set up in a large anechoic room (10 m high and 14×9 m²) with a maximum horizontal diagonal spacing between the sound sources. The maximum interaural delay, τ, was 2 msec, using a symmetrically supported microphone arm and microphones of uniform polar response. With both sound sources operating there was a distinct pitch in the noise which varied with θ. As θ approached zero from $\frac{1}{2}\pi$, the pitch at first increased and then disappeared as the subject's head turned to face the source. The process was repeated cyclically for the other three quadrants with the pitch being at a minimum for $\theta = \pm\frac{1}{2}\pi$. It was not possible to hear this pitch if attention was directed to lateralization. With only one sound source operating, rotation of the head merely produced a slight diffuse change in the auditory quality of the stimulus as the lobe was displaced from the centre. With both sound sources but only one microphone operating, both pitch and lateralization disappeared leaving a uniform auditory sensation which was independent of rotation. It is worth noting that the pitch was most clearly heard when the head was actually turning, and that the pitch lagged on θ. Even for large τ the sensation of pitch was centrally located.

DELAY-LINE EXPERIMENT

The two most important results of the free-space experiment are that: (1) it is possible to derive a sensation of pitch from stimulus configurations capable of producing a lateralization, (2) the lateralization must have at least two lobes.

Another, simpler, way of providing two lobes is by removing noise source 3 from the arrangement of Fig. 4. This leaves an array of the type shown in Fig. 6a, which requires only one continuously variable delay line, and in-phase stimula-

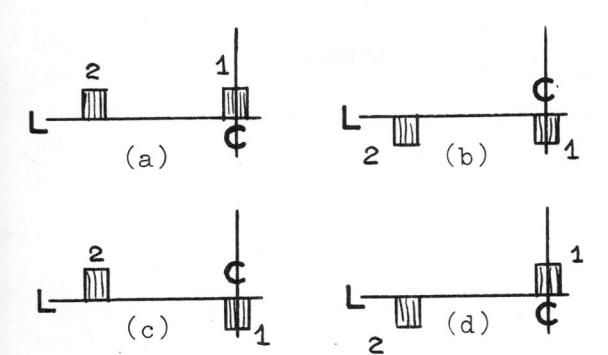

Fig. 6. Two-lobe lateralization arrays with in-phase and opposite-phase lobe stimulus connections.

tion. This was set up using 3 kHz low-pass clipped noise, a forty section shift register delay line, with a stepping rate continuously variable between 10 kHz and 100 kHz, circumaural headphones independently fed from 2 kHz low-pass filters and an SL of 25 dB. The stimuli for this lateralization pattern can also give rise to a sensation of pitch having the same characteristics as for Fig. 5b; and, as before, if one lateralization lobe is removed, in whatever fashion, the pitch disappears. It is important that, given the left lateralization, the pitch is produced by the addition of what would normally be regarded as in-phase masking noise to the two ears from another source.

The reversal of the connections to either headphone reverses the polarity of both lobes and this is represented in Fig. 6b. If the phase of noise source 1 is reversed at only one ear, Fig. 6c represents the lobe pattern and if a stimulus from noise source 2 is reversed at one ear the pattern of Fig. 6d results. All four of these configurations give rise to a sensation of pitch, and, in every case, if one of the four stimulus connections to the earphones is removed, the pitch sensation is completely eliminated. Not all listeners are sensitive to the pitch (although all players of stringed instruments respond) but most can hear it and the same characteristics apply as for the free-space experiment. It is, however, more noticeable with this arrangement that the pitch follows rapid changes of delay fairly slowly. Lateralization and pitch do not exist together and even for the lowest pitch the sensation appears to be centrally located rather than to belong to one ear or the other.

PITCH MATCHING

(1) *for a two-lobe central array*

When a pure tone pitch matching situation is arranged, with an SL of 25 dB and a pure tone range of 250 Hz to 1 kHz and a range of τ of 4 to 0.4 msec, the central-pitch producing patterns of Fig. 6 can be divided into two groups. For Figs. 6a and 6b the matches lie along the line $1/\tau = \frac{4}{3} f_t$; where f_t is the frequency of the pure tone given to the subject and to the pitch of which he must adjust the noise by varying τ. For Figs. 6c and 6d the matches lie along the line $1/\tau = f_t$, and their dispersion is somewhat less. In both cases octave confusions may be made. Once more an alteration of any part of the array will alter the response to the whole. The use of a right handed array, 1 and 3 of Fig. 4, instead of the present left handed configuration, makes no difference to the type or accuracy of pitch matching.

(2) for a three-lobe central array

If a right hand lobe is added to the previous array, by means of a third independent noise generator, as in Fig. 4, the use of symmetrical delays and phase changing results in the family of lateralization arrays shown in Fig. 7. Each one of these arrays has a pitch associated with it. Pitch matching as before and again with a 25 dB SL produces exactly the same division as before. Figs. 7a and 7b correspond to Figs. 6a and 6b and Figs. 7c and 7d correspond to Figs. 6c and 6d. Qualitatively the pitch sensation associated with this three-lobe array is just the same as for the two-lobe presentations. A phase-asymmetric three-lobe array has an essentially indistinguishable pitch. The use of extra symmetrical lobes to make a five element array does not improve the clarity of the pitch.

(3) for a two-lobe asymmetric array

When the fixed central lobe of any of the arrays of Fig. 6 is displaced to one side, by an amount T, the pitch associated with a change of τ for the variable lateralization lobe becomes quite different. The array of Fig. 8, for which $\tau < T$, gives a pitch which increases as τ is increased. When $|T-\tau| \leq 1$ msec the pitch disappears and re-appears, if τ is increased beyond T, when the variable lobe is symmetrically on the other side of the fixed lobe. It follows that the pitch is not dependent on the absolute delays employed but can be interpreted in terms of the relative lateralizations of the lobes. Pitch matching with $T = 3$ msec gave the same general results as before for this type of stimulus when the relative delays were plotted, $1/(T-\tau) = f_t$ (but f_t was only in the range 300 Hz to 700 Hz).

(4) for a three-lobe asymmetric array

In Fig. 9, τ has been re-defined to give the variable lobe delays relative to

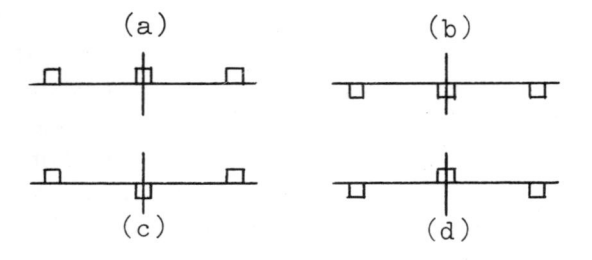

Fig. 7. Centre-symmetric three-lobe lateralization arrays.

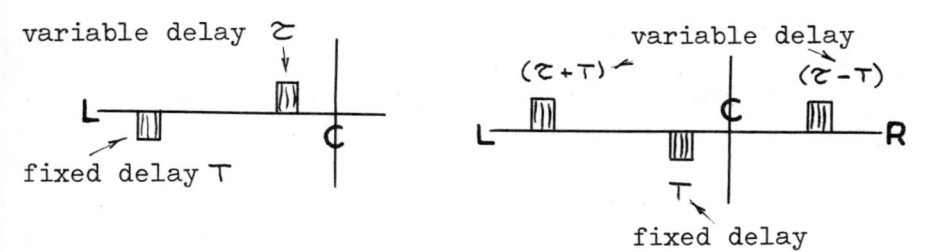

Fig. 8. Asymmetric two-lobe array.

Fig. 9. Centre-asymmetric three-lobe array.

those for the fixed lobe. When $T=0.3$ msec for this fixed lobe, pitch matching is essentially identical with that for the lobe pattern of Fig. 7c. When $T=1.5$ msec matching is more difficult but the basic relation between τ and f_t is unaltered.

POSSIBILITY OF AN ARTEFACT

The central pitch which has been described is only of interest if it requires a neural mechanism for its explanation. This need could be dispensed with by showing that the pitch effects described could result from a peripheral analysis. If there is even a small spectral peak in the acoustic stimulus to either or both of the ears, this peak might be resolved in an orthodox way. There are two basic ways in which such a peak might occur. First by an apparatus malfunction such that it is present explicitly in the original stimulus. Second by a physical interaction, cross talk, between the stimuli which takes place only when the subject is wearing the headphones.

(1) *Spurious stimulus pitch*. Long term spectral analyses (Fourcin, 1964) of the stimuli have been made which are capable of resolving peaks of less than 0.5 dB but no trace of a peak has been found. More convincing evidence is contained in the nature of the experiments themselves, however. When, for any one of the patterns of Fig. 6, the central noise is removed, to one ear or to both, the pitch effect disappears. Now it is easy to provide this noise from a variety of external sources, each of which is quite innocent of spectral peaks and independent of the τ-generating apparatus, but with any of them the pitch is restored. In consequence the pitch is a function of this independent noise which, if the pitch were an artefact, would ordinarily be expected to be a source of masking. This only leaves the lateralized lobe, and the results of the free-space experiment show that the pitch is not generated by the particular type of apparatus involved in the production of delay.

(2) *Cross talk*. Over the range 50 Hz to 10 kHz the interaural attenuation for the subjects and headphones employed was greater than 40 dB. With each lobe stimulus at 25 dB SL the result of interaction is superficially negligible. It is always possible however that an unknown facilitating effect is present in the

multiple-lobe presentation which could enhance the effect of cross talk. To investigate this possibility the nature of the pitch which cross talk would produce if it were present (echo pitch, due to noise ± itself delayed) has been separately investigated. Subjects matched echo pitch significantly differently from central pitch (Fourcin, 1965); in consequence the results reported here could not be due to a simple physical interaction between the stimuli.

The most convincing result in favour of the separate existence of a central neurally mediated pitch comes however from the observations made with the lateralization array of Fig. 8. Here the perceived pitch at first increases and then decreases with a steady increase in the delay producing the variable lateralization lobe. This result is quite consistent with the other central pitch observations. It is not possible to reconcile it with a frequency based pitch mediation.

CONCLUSION

The pitch observations which have been reported here can be neatly classified in terms of lateralization patterns but in no way related to peripheral single ear frequency analysis. In consequence central pitch owes its mediation to a neural mechanism capable of operating on the temporal structure of acoustic stimuli.

REFERENCES

Blodgett, H. C., Wilbanks, W. A., and Jeffress, L. A. (1956): Effects of large interaural time differences upon the judgment of sidedness, J. Acoust. Soc. Amer. 28, 639-643.

Fourcin, A. J. (1964): A note on the spectral analysis of unvoiced sounds, in: Proceedings of the Fifth International Congress of Phonetic Sciences, E. Zwirner and W. Bethge, Eds. (S. Karger, Basel), pp. 287-291.

Fourcin, A. J. (1965): The pitch of noise with periodic spectral peaks, in: Rapports 5ᵉ Congrès International d'Acoustique, Liège, Vol. Ia, B 52.

Licklider, J. C. R. (1962): Periodicity pitch and related auditory process models, Internat. Audiol. 1, 11-36.

Mills, A. W. (1958): On the minimum audible angle, J. Acoust. Soc. Amer. 30, 237-246.

DISCUSSION

De Boer: If you have two noise sources and you connect a different band-pass filter to each output such that the two sounds do not overlap in frequency, do you get this central pitch sensation and does the region of overlap have anything to do with the pitch?

Fourcin: The situation you describe could be associated with the two-lobe arrays of Figs. 6 and 8. It was, however, investigated using the arrangement shown in Fig. 5b. If there is no frequency overlap for the two associated noise sources there is no pitch. If the overlap occurs above about 2 kHz there is no pitch. If the overlap is small below 2 kHz then the sensation level and the range of the central pitch are small.

Piazza: Do you think that there is a relation between this effect and binaural beats?

Fourcin: Yes. I think both phenomena have a common neural origin but that the effects I have described depend on more processing.

Plomp: Do I understand your paper correctly that the time delays involved in this central pitch are significantly larger than in direct binaural listening with the consequence that we cannot hear this pitch normally?

Fourcin: The maximum interaural delay normally encountered in ordinary binaural listening corresponds to the minimum inter-lobe delay which can lead to a sensation of the present pitch. The maximum interlobe- delay associated with a central pitch is about 10 to 15 msec. This also corresponds to the maximum interaural delays for which judgments of sidedness can occur. It is interesting that both this lateralization effect and the central pitch described here cannot occur in normal listening. It seems unlikely that the neural mechanisms which lead to their existence have evolved uselessly. Just as for the sensation of binaural beats, these effects may be the slight by-products of essential components of our auditory processing. I believe that the function of the particular component concerned with central pitch is one of signal analysis — as opposed to detection and localization. The Huggins effect (Cramer and Huggins, 1958) is similar to those described here and I have found it also to be invariant with the type of translation shown in Fig. 8, but although its pitch can be equated with that of the stimuli of this paper its quality is quite different. This quality distinction can only be made as the result of quite elaborate processing in the nervous system. I think it likely that this processing is available in normal listening; not only binaurally but also monaurally and that it mediates the sensations of echo pitch and the really important residue pitch family.

Wilson: I was very impressed by the way in which the data points in your slides are closely grouped around the lines you drew. How does the spread for the pitch matchings in your experiments compare with the spread for the pitch of gated noise and for the residue pitch?

Fourcin: I have measured the dispersion of the pitch matches made with my stimuli but it is difficult to make an exact comparison with the results which have been obtained using gated noise (Harris, 1963) and with residue pitch stimuli (Cardozo and Ritsma, 1965). This difficulty arises because I have only employed pure tones as reference stimuli, in consequence my subjects have had a more than usually difficult pitch matching situation. The relative standard deviation of their responses however is only a little greater than that associated with residue pitch. In consequence it is much smaller than that for gated noise and significantly greater than that for pure tone matched against pure tone.

Goldstein: You said that the lines you drew are theoretical lines. Do you mean you have a theory of central pitch?

Secondly, you stated that your central pitch effect requires very careful listening with both ears. Could you elaborate on that?

Fourcin: The lines defined in the paper and shown in the slides are good fits to the experimental matches but they are derived from the idea that the effects are a side-product of a central process of signal-in-noise analysis. I would, however, prefer not to discuss a theoretical explanation of the whole phenomenon at present.

When listening to these stimuli one can either attend to the lateralization images with which they may be associated or to their pitch. In my experience, subjects cannot hear both percepts at once, the one precludes the other. The consequent direction of attention required to hear the pitch is initially not always easily obtained.

REFERENCES

Cramer, E. M., and Huggins, W. H. (1958): Creation of pitch through binaural interaction, J. Acoust. Soc. Amer. *30*, 413-417.

Harris, G. G. (1963): Periodicity perception by using gated noise, J. Acoust. Soc. Amer. *35*, 1229-1233.

Cardozo, B. L., and Ritsma, R. J. (1965): The pitch of noise with periodic spectral peaks, in: Rapports 5e Congrès International d'Acoustique, Liège, Vol. Ia, B 37.

JITTER DETECTION FOR REPEATED AUDITORY PULSE PATTERNS

IRWIN POLLACK

Mental Health Research Institute
University of Michigan
Ann Arbor, Michigan, U.S.A.

Two extreme cases of the detection of temporal irregularities within repeated pulse patterns can be identified. If temporal irregularities are introduced into a sequence of interpulse intervals, and the sequence is not repeated, we have the conditions for a 'jitter' detection test (Cardozo et al., 1966; Pollack, 1968a). If a temporal irregularity is introduced into a sequence of only two interpulse intervals, and the sequence of the two intervals is successively repeated, we have the conditions for a two-interval discrimination test (Small and McClellan, 1963). In the jitter detection test, temporal irregularities are perceived in terms of roughness, harshness, etc., especially at high pulse frequencies. In the two-interval test, the repeated temporal irregularity may be detected and perceived in terms of pitch changes. This paper seeks to explore the contribution by periodicity information to the detection of temporal irregularities within repeated auditory pulse trains.

APPROACH

A reference pulse train is defined in Fig. 1 in terms of successive interpulse intervals (IPI) produced by a clock. In the 'jitter' detection test, successive intervals are modified by adding plus or minus J to the center interpulse interval IPI_c. The jitter detection threshold is defined by the magnitude of J which must be introduced in order to attain a fixed proportion of correct responses.

Three further modifications of the jitter detection test are also shown in Fig. 1. The top pair of Fig. 1 illustrates the case where successive clock pulses are assigned at random to the separate ears; the middle pair illustrates the case where the signal at only one of the two ears within the top pair is presented to

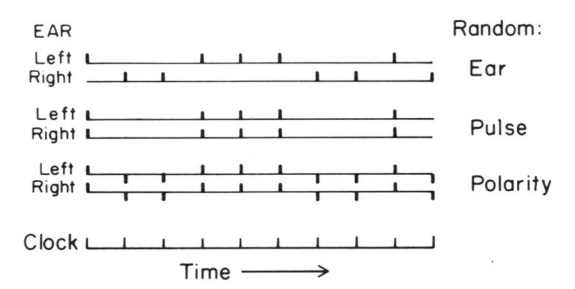

Fig. 1. Schematic representation of non-jittered, non-repeated pulse patterns based upon a periodic clock (reference procedure) and three modifications from the reference procedure.

both ears; and the bottom pair illustrates the case where the signal at only one of the two ears within the top pair is inverted in polarity and the resultant signal is presented to both ears. We shall refer to the three procedures as the *random ear*, the *random pulse*, and the *random polarity* procedures. In each case, jitter, J, is introduced upon successive clock pulses independently of the assignment to the separate ears or polarities.

The random ear and the random pulse procedures permit the testing of extreme bounds of how well the binaural system preserves temporal information antecedent to the perception of jitter. If the binaural system perfectly melds temporal information at the separate ears prior to the analysis of jitter, we would expect that jitter thresholds would be the same under the random ear and the binaural reference, or clock, procedures. If jitter thresholds are based entirely upon the temporal pattern at the separate ears, and there is no melding of temporal information across the head, we would expect that jitter thresholds would be the same under the random ear and random pulse procedures, and both might be expected to be poorer than under the reference procedure (*cf.* Pollack, 1968b).

The random polarity procedure preserves the temporal information at each ear, but varies the spectral information. If jitter detection is based primarily on temporal information, *i.e.* on the time-of-occurrence of successive pulses, jitter thresholds should be the same under the random polarity and the reference procedures. If jitter detection is based primarily upon the spectral information at the separate ears, jitter thresholds could differ substantially between the random polarity and the reference procedures.

METHOD

Pulse trains were constructed by a PDP-8 computer. The program varied J, the magnitude of jitter; $n/2$ of the interpulse intervals were increased by J; $n/2$ of the intervals were decreased by J; and the resulting n intervals were randomly ordered. An adaptive stimulus programming procedure (Taylor and Creelman, 1967) varied the magnitude of J to converge upon 50% correct responses in a 4-interval forced-choice test. The unit of temporal control was 0.375 μsec with extrapolation to one-half the temporal unit. Listeners, experienced in listening to pulse trains, wore earphones (Koss PRO-4) which served to stretch out the brief (10 μsec) pulses generated by the computer. Each point represents the average of at least two thresholds by at least 15, 14, 12, and 11 listeners under the reference, random ear, random pulse, and random polarity procedures, respectively. Listening was carried out at a comfortable listening level.

Fig. 1 illustrates the generation of a non-repeated, non-jittered pulse train of 9 intervals. The number of interpulse intervals is identified as n, and the number of repeated presentations as P. The pulse trains of Fig. 1 are identified as $n=9$; $P=1$. For a given total number of interpulse intervals, n times P, there is a tradeoff between n and P. Thus, for a train of 64 intervals, some possible conditions are: $n=2$, $P=32$; ..;$n=8$, $P=8$; ..;$n=64$, $P=1$. Jitter was imposed upon a pattern of n clock intervals, and the entire pulse pattern was repeated exactly. The last pulse of the $(P-1)$-th presentation was the first pulse of the P-th presentation.

RESULTS

Fig. 2 represents jitter thresholds obtained under the reference procedure as a function of the total number of interpulse intervals, n times P. The jitter thresholds, plotted on the ordinate, are expressed relative to the center interpulse interval, IPI_c. The points are represented by the clock code shown in the insert, which reflects n, the number of intervals within the pulse pattern. The separate sections represent the several center intervals. The thin dashed lines connect points representing the smallest size pattern (usually $n=2$); the thin solid lines connect points representing the largest n employed for the indicated $n \cdot P$ product. The distance between the thin lines represents the extent of the $n \cdot P$ tradeoff between the shortest and longest patterns in determining jitter thresholds.

In Fig. 2, and all successive graphs, there is no shift on the ordinate scale across conditions, but there are shifts along the abscissa scale to aid in the visual presentation of the results. It is noted in Fig. 2 that:

(1) An extremely wide range of jitter threshold sensitivity is obtained as a function of the total number of intervals, especially at high pulse frequencies, i.e. at short interpulse intervals, IPI_c.

(2) Relative to the wide range of thresholds as a function of the total number of intervals at high pulse frequencies, the effect of the tradeoff between n and P is small for the reference procedure. At most, thresholds change as a factor of 1.9 from the shortest to the longest pattern, but may differ by 100-fold as a function of the number of intervals at the highest pulse frequency.

(3) The effect of the length of the pulse pattern, n, as represented by the distance between the thin curves, may change over pulse frequency.

Figs. 3, 4, and 5 present jitter thresholds for the three experimental modifica-

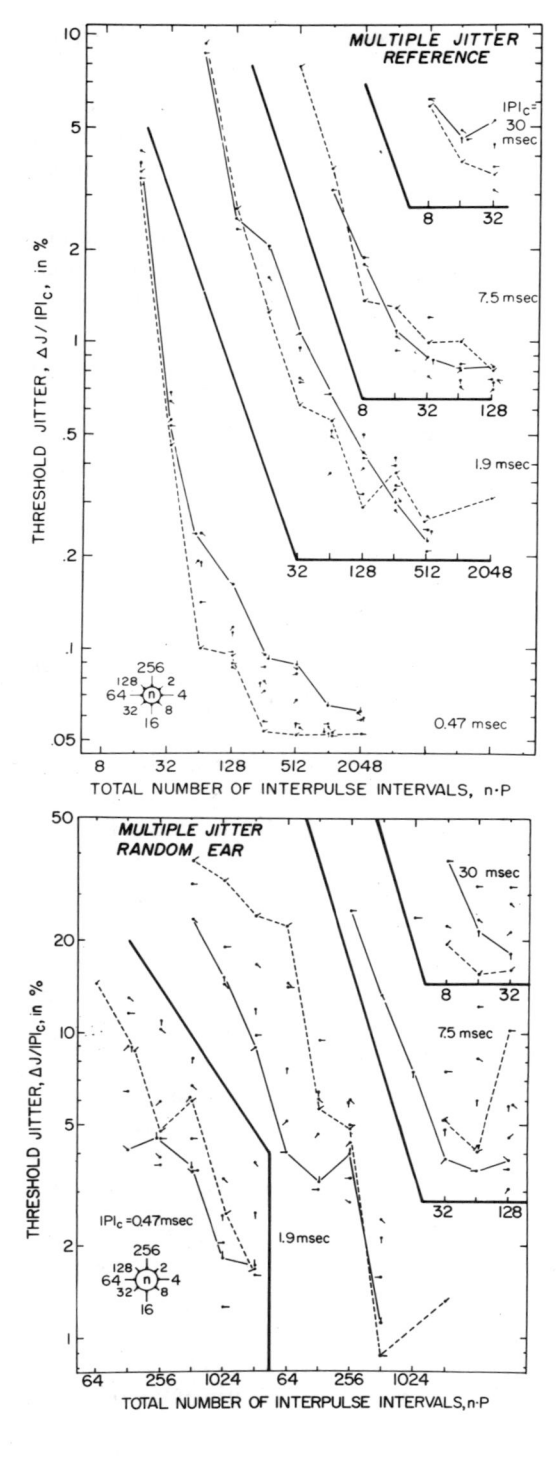

Fig. 2. Jitter detection thresholds for repeated auditory pulse patterns under the reference procedure. In Figs. 2-4, the ordinate represents the relative level of jitter, J, which was added to or subtracted from the center interpulse interval, IPI_c, to converge upon 50% correct trials in a four-interval forced-choice test. The abscissa is the total number of interpulse intervals, consisting of a pattern of n interpulse intervals presented P times. The direction of the tic marks indicates n, as shown in the insert. The four sections represent four IPI's. The thin dashed lines connect points of smallest n; the thin solid lines connect points of largest n for the specific $n \cdot P$ product.

Fig. 3. Jitter detection thresholds for repeated auditory pulse patterns under the random ear procedure.

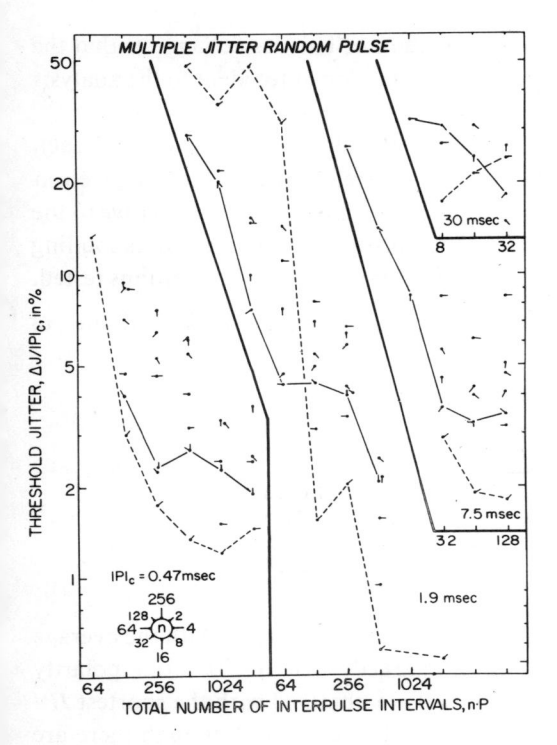

Fig. 4. Jitter detection thresholds for repeated auditory pulse patterns under the random pulse procedure.

tions of the reference procedure.

Relative to Fig. 2, it is noted that the fine jitter detection thresholds (below 0.5%) achieved with the reference procedure at high pulse frequencies are absent with the experimental modifications. With respect to the role of periodicity, however, there is a substantial advantage for repeated short patterns (dashed lines) over longer patterns, especially for the random pulse and random polarity modifications. It is also noted that the random pulse procedure yields

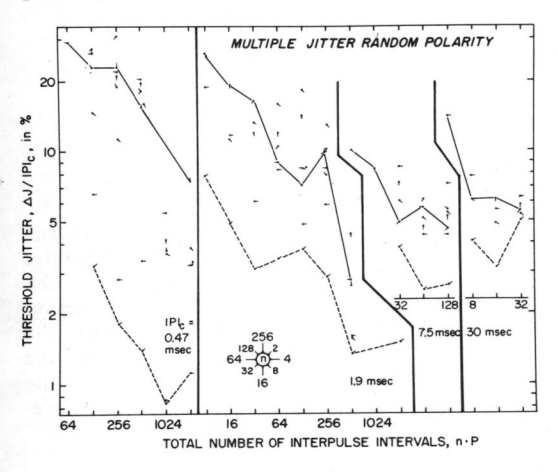

Fig. 5. Jitter detection thresholds for repeated auditory pulse patterns under the random polarity procedure.

equivalent thresholds to the random ear procedure. This result suggests that the binaural system cannot preserve temporal information antecedent to the analysis of jitter.

A direct comparison between the jitter thresholds obtained with each modification and that obtained with the reference procedure, is given in Figs. 6-8. The ordinate is the ratio of the jitter threshold obtained, relative to the jitter threshold obtained with the reference procedure for the corresponding condition. Table I presents the median ratios, averaged over all conditions tested.

Table I. Median jitter threshold ratio for three experimental modifications, relative to jitter thresholds for reference procedure.

Interpulse interval msec	Experimental modification		
	random ear	random pulse	random polarity
0.46	63.10	48.88	130.72
1.88	10.65	9.68	17.65
7.5	5.52	5.00	6.75
30.0	4.59	4.65	1.24

In Figs. 6, 7, and 8, we note that the threshold ratios decrease with the average interpulse interval. The change is most marked with the random polarity procedure where threshold ratios of near 500 are obtained with the shortest *IPI* and ratios of near unity are obtained with the longest *IPI*. Although there are striking exceptions for particular combinations of experimental modification, *IPI*, and pattern-length (*e.g.* random pulse procedure, 1.9 msec, short patterns), the threshold ratio is not extremely sensitive to the total number of intervals. The role of periodicity is, in some cases, greater in terms of the ratio measure because, relative to the reference procedure, the effectiveness of short and long patterns may be opposite to that of the reference procedure.

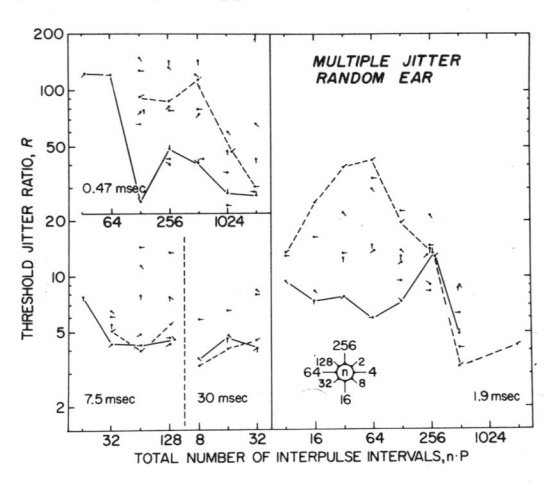

Fig. 6. Threshold jitter ratios for repeated auditory pulse patterns under the random ear procedure. In Figs. 6-8, the ordinate is the ratio of the jitter detection threshold obtained under the indicated procedure over the threshold obtained under the reference procedure for corresponding conditions.

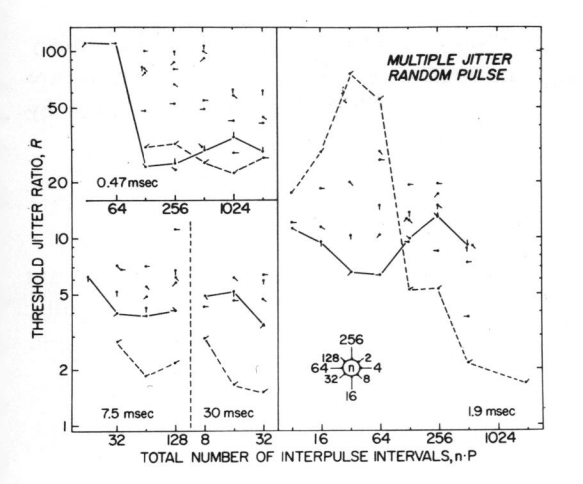

Fig. 7. Threshold jitter ratios for repeated auditory pulse patterns under the random pulse procedure.

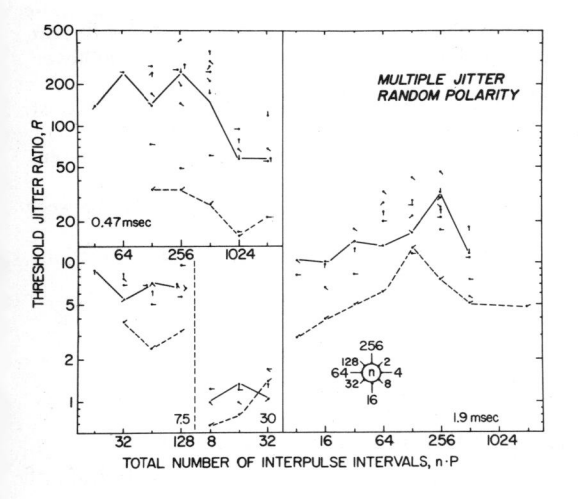

Fig. 8. Threshold jitter ratios for repeated auditory pulse patterns under the random polarity procedure.

DISCUSSION AND SUMMARY

The near-unity threshold ratios for long interpulse intervals and the extremely high threshold ratios for short interpulse intervals of Figs. 6-8 suggest that the auditory system 'solves' the problem of detection of temporal perturbations differently at low and high pulse frequencies. At long interpulse intervals, all experimental operations yield nearly equivalent results, perhaps pointing to the importance of preserving temporal, rather than spectral, information. At short interpulse intervals, all experimental operations are equally ineffective although temporal information is preserved in the random polarity procedure. Presumably, jitter discrimination is based upon the spectral information at the separate ears, since there is substantial loss in sensitivity when temporal information is distributed across the ears. These results confirm those obtained previously (Pollack, 1968b). The additional important finding is that the major results are not substantially modified by the introduction of periodicity information by means of repeated presentations of short interval patterns.

ACKNOWLEDGMENT

The research was supported in part by National Sciences Foundation grant GB 6148. The writer wishes to acknowledge the assistance of Mr. Peter Headly, who wrote the computer program, and of Mrs. Dorothy LaBarr, who supervised the experimental tests.

REFERENCES

Cardozo, B. L., Ritsma, R. J., Domburg, G., and Neelen, J. J. M. (1966): Unipolar pulse trains with perturbed intervals; perceptibility of jitter, IPO Annual Progress Report, No. 1, 17-27 (Instituut voor Perceptie Onderzoek, Insulindelaan 2, Eindhoven, Holland).
Pollack, I. (1968a): Detection and relative discrimination of auditory jitter, J. Acoust. Soc. Amer. *43*, 308-315.
Pollack, I. (1968b): Can the binaural system preserve temporal information for jitter?, J. Acoust. Soc. Amer. *44*, 968-972.
Small, A. M., Jr., and McClellan, M. E. (1963): Pitch associated with time delay between two pulse trains, J. Acoust. Soc. Amer. *35*, 1246-1255.
Taylor, M., and Creelman, C. D. (1967): PEST: Efficient estimates on probability functions, J. Acoust. Soc. Amer. *41*, 782-787.

DISCUSSION

Rose: How large is the jitter detection threshold, expressed in μsec?

Pollack: It is almost arbitrarily fine. Subjects were showing sensitivities of the order of 0.1 % or even finer. So, if we are willing to deal with high-frequency pulse trains then, at 10,000 pps, we find values below 0.1 μsec. But one might say that these high-frequency pulse trains should be disallowed because in the auditory system they are probably not represented in the time domain. Still there are precisions of about 1 μsec at 1000 pps and 2 or 3 μsec at 500 pps.

Rose: Your figures do not amaze me. You can have any time 5 to 10 μsec on the basis of physiological data.

Pollack: Remember that the ear is showing jitter discrimination thresholds of the order of 0.1 % whereas the individual units are showing a jitter of the order of 30 % (standard deviation divided by the mean).

Rose: When you say individual unit, do you mean neurons?

Pollack: Yes, I am thinking of the data of Geisler and Goldberg (1966) and all those other data which show a very wide, a shockingly wide, variability.

(These studies are reviewed in Pollack, 1968.)

Rose: I can assure you that a good lateralization neuron can recognize a time difference of about 10 μsec at least.

Pollack: I think we have to be careful. As far as binaural experiments with lateralization neurons are concerned, I agree with you. But I am talking about the monaural case in which one investigates the standard deviation of the time delay between the presentation of a click and the occurrence of the nerve spike. For such experiments the data in the literature suggest that the jitter is substantially of the order of about 30-50%.

Goldstein: Why does one have to assume that the mechanism uses but one neuron? Surely, it has a large number. And, secondly, why does one have to assume that we are measuring one interspike interval? Why cannot the auditory system look between several intervals which might give a much better precision? There are many ways in which you can get the required precision from the time domain (Colburn, 1969; Siebert, 1970).

Pollack: I am in complete agreement with your statement. If people show this kind of very fine temporal resolution then we have to conclude that there must be some overall ensemble properties of which the system takes advantages. Even if we assume that individual units are statistically independent—which they are obviously not—and that the ensemble variability is inversely proportional to the square root of the number of units in the ensemble, we are hard-pressed to find sufficient units. A reduction from 50% variability to 0.1% jitter implies $(500)^2$ or 250,000 neural units. This far exceeds the estimates of the number of auditory neural units. Only if we also assume that responses to successive pulses are also statistically independent—which they are probably not—would we gain an additional factor of 10 in a 100-pulse train. The right ballpark has been reached, but the extreme assumptions cast doubt on the calculation exercise.

Zwicker: I think that we have touched upon a very important matter and like to make a somewhat more general remark. It seems to me that there is a deep gap between what we know from psychoacoustical measurements and what we expected to hear from physiologists. To all of us, the frequency-to-space relation is almost clear. On the other hand, there is, at least to me, a big difference between what physiologists mean by time and what psychoacousticians do. To me, periodicity pitch and interspike histograms are just different things. I cannot see any agreement between these two things. As soon as we compare psychological data with single unit responses and find that there seems to be no relation, there follows a comment like Dr. Goldstein's that we should think of what is going on in many units. But the big problem is that we know nothing about the whole complex. So, I would like to suggest that when the psychoacousticians talk about time domain they should try to build some model so that we can discuss something realistic. We should avoid a situation in which the answer to a problem shifts to "maybe it is one of the many possibilities".

Terhardt: Dr. Pollack, you said that your subjects were asked to give responses on the basis of their temporal sensations. But they cannot judge the temporal irregularities at repetition frequencies of, for example, 2000 Hz. How is the sensation in that condition?

Pollack: All our tests are carried out with forced-choice procedures. Subjects have to pick out the sound that is different from the others. I know very little about what they actually heard except in so far as I served as a subject myself in all my experiments. With large degrees of jitter the signal sounds rough, hoarse. Near the threshold of jitter perception it is very difficult to describe how such a sound differs from a sound without jitter.

Terhardt: Would it not be better to formulate the question whether there is temporal jitter perceptible additionally on the side of sensations and not only in terms of whether temporal irregularities themselves are perceptible? For you can describe the temporal patterns with the same exactness in the frequency (or space) domain.

Pollack: I defined jitter operationally in terms of the stimulus. I did not use a psychological variable like you did with roughness (p. 278 of this volume).

Zwislocki: There is another point if I understood Dr. Terhardt correctly. Why do you want to explain the detection in the time domain? I think you have a spectral variation too and your results could be very well explained in that domain. I do not see that you have to look for a very precise timing in the nervous system.

Pollack: Yes, I agree. For the presentation I was attempting to take the extreme position that jitter was associated only with the temporal properties of the neural message. And from that extreme position, I was asking what temporal requirements must be met by the neural system. For the extremely precise jitter thresholds at high pulse frequencies mediation through spectral coding seems far more attractive than through temporal coding. For lower pulse frequencies, however, I prefer an explanation in terms of the time domain. And even at very low pulse rates the neurophysiological data that I have looked at show variabilities, that are, on a pulse-by-pulse basis, at least one order of magnitude greater than the threshold of jitter discrimination.

REFERENCES

Colburn, H. S. (1969): Some Physiological Limitations on Binaural Performance, Doctoral Dissertation, Massachusetts Institute of Technology, Cambridge, Mass.
Geisler, C. D., and Goldberg, J. M. (1966): A stochastic model of the repetitive activity of neurons, Biophysics J. *6*, 53-69.
Pollack, I. (1968): Detection and relative discrimination of auditory jitter, J. Acoust. Soc. Amer. *43*, 308-315.
Siebert, W. M. (1970): Frequency discrimination in the auditory system: place or periodicity mechanisms?, Proc. IEEE, in press.

THE PERCEPTION OF JITTERED PULSE TRAINS

BEN L. CARDOZO

Institute for Perception Research
Eindhoven, The Netherlands

INTRODUCTION

One of the central points in hearing theory is the acuteness of frequency discrimination by the human ear. In place theories, there is a need of some sharpening mechanism, for instance lateral inhibition along the cochlear partition, in order to bridge the gap between the low-Q travelling wave pattern in the cochlear partition and the very high frequency discrimination. But also in periodicity theory there is a problem, because the pitch matching accuracies attained in psychoacoustical experiments correspond to timing accuracies which seem to go beyond anything known from physiological evidence. In Ritsma's (1965) pitch matching experiments subjects had to match the pitch of an amplitude-modulated sinusoid centered in one frequency region to the pitch of a similar stimulus in another frequency region. This was done with a standard deviation of 0.5% for a pitch corresponding to 100 Hz and with a standard deviation of 0.3% for a pitch corresponding to 400 Hz. If one accepts that these results are to be interpreted in terms of an internal representation of the time pattern of the stimuli, then one must conclude that the time intervals are represented with high precision. We know that 3 to 10 pitch periods are sufficient for obtaining this high precision (Cardozo and Ritsma, 1965), so that we estimate the standard deviation of the internal time interval to be of the order of 100 μsec for the 100 Hz tone and 25 μsec for the 400 Hz tone. It may be remarked, that these timing accuracies are not in conflict with data from binaural lateralization experiments which seem to indicate at least as acute timing of nerve pulses (even at a non-peripheral level).

Concerning physiological evidence there are two points to be made.
(1) It is difficult to assess what the absolute minimal time resolution in the nervous system would be likely to be. The leading edge of spikes seems to be very steep indeed, but it is not clear whether this would be sufficient for

a timing accuracy in the microseconds range.
(2) Timing accuracies are often found to be more or less proportional to the inter-spike intervals. That is, the relative standard deviation tends to be constant in a particular neuron (Pollack, 1968).

The aim of this paper is to try and estimate what further evidence on the timing accuracy of the hypothetical internal representation can be obtained by means of psychoacoustical experiments with jittered pulse trains. There are various ways of introducing jitter into a periodic sound (*cf.* Rosenberg, 1966; Cardozo *et al.*, 1966; Pollack, 1968, pp. 329-338 of this volume), but we shall confine ourselves to Gaussian jitter, that is, the pulse intervals form a Gaussian distribution around an average T with variance τ^2. τ will be called jitter, it has the dimension time. The relative jitter will be denoted by $J = \tau/T$.

We will concentrate furthermore on pulse trains within the domain of existence of the residue (Ritsma, 1962), that is with repetition rates of 50 to 400 Hz. We know from many experiments that, as a rule, the pitch in these stimuli is not derived from the fundamental component in the spectrum. This does not necessarily imply, however, that jitter is detected by the same mechanism which is responsible for the perception of pitch.

EXPERIMENT I

It is instructive to consider a simple experiment which demonstrates once more that in complex sounds a pitch can be heard which is not based on the perception of the fundamental nor on the perception of any other isolated Fourier component. A train of perfectly periodic pulses with a rate of 100 Hz and a pulse width of 100 μsec is filtered in the band 0-4 kHz. The amplitudes of its components are measured with a wave analyser to be A_n, with n the number of the harmonic. The signal is presented binaurally with earphones PDR 8 at a sensation level (SL) of 30 dB. We now add noise to the pulse train. The level and spectral density of the noise are so adjusted, that it is able to mask every single component. This is checked with a sinusoid which is given the amplitudes A_n at frequencies n times 100 Hz. When listening to the pulse train with the noise added, it is not difficult to hear the pulse train with its characteristic residue pitch corresponding to 100 Hz. This can be easily explained because in the higher frequency range, a number of harmonics fall in the same critical band. Their energies add. The situation may be described by saying that certain clusters of harmonics remain unresolved and form together a band-filtered version of the original periodic pulse (Schouten, 1940). In a band, in which *e.g.* 4 components are combined in this way, the signal-to-noise ratio is raised by 6 dB. It is therefore quite natural that the residue or periodicity pitch is heard, without any of the Fourier components being audible. The band-filtered version exhibits the original periodicity of the pulse train, be it perturbed to a certain extent by the noise.

This experiment is described here only qualitatively. It may be regarded as another evidence of pitch perception on the basis of an internal representation of the time structure of the stimulus.

EXPERIMENT II

It is interesting to investigate what happens when, in the situation described under experiment I, the periodic pulse train itself is perturbed. When a large amount of jitter is applied to it such that $J=0.1$ to 0.2, then the pulse train is inaudible, that is, one cannot discriminate between noise alone and noise plus jittered pulse train. This is compatible with previous experiments (Cardozo and Ritsma, 1968) in which it was found that subjects heard no pitch in pulse trains with a relative jitter of 10 to 20 percent and were unable to perform pitch matchings with acceptable accuracy. When a very small amount of jitter is introduced, one does not hear the difference with a perfectly periodic pulse train, but with moderate relative jitter one can tell the periodic pulse train from the jittered one. This point is considered proof that the ear is able to detect jitter in sounds (with a repetition rate that is in concordance with the domain of existence of the residue) without the spectral properties of the sound being available.

The formal experiment II was conducted in the following way. In a binary choice experiment the subject is presented with two stimuli of 0.25 sec duration each, separated by an interval of 0.8 sec. The level of the stimuli is 30 dB SL. Stimulus A consists of a train of pulses with interval T and $J=0$. Stimulus B consists of a train of pulses with average interval T and relative jitter $J \neq 0$. The order of A and B is randomized. Both A and B are embedded in a white noise which is audible continuously. In responding, the subject has to press one of two buttons according as he observes jitter in the first or in the second stimulus. A sequential up-and-down technique (cf. Cardozo, 1966) determines semi-automatically the value of J for which the subject produces about 75 percent correct responses. This threshold of relative jitter will be designated by J'. Parameters in the experiment are the signal-to-noise ratio and the repetition rate of the pulses.

The signal-to-noise ratio S/N was determined for each subject separately in the following way. After having found the 30 dB SL for the repetition rate of the pulses, the amplitude A_n of the 1 kHz component was measured and noise added such that it would just mask this component. The ratio of the power of the pulse train over the noise was then arbitrarily called 1, that is 0 dB. This value of S/N corresponded, when averaged over the subjects and over the pulse rates 50, 100, 200 and 400 Hz to -18 dB as measured with the voltmeter in the band 0-4 kHz. Two experienced subjects performed the experiment. Fig. 1 presents

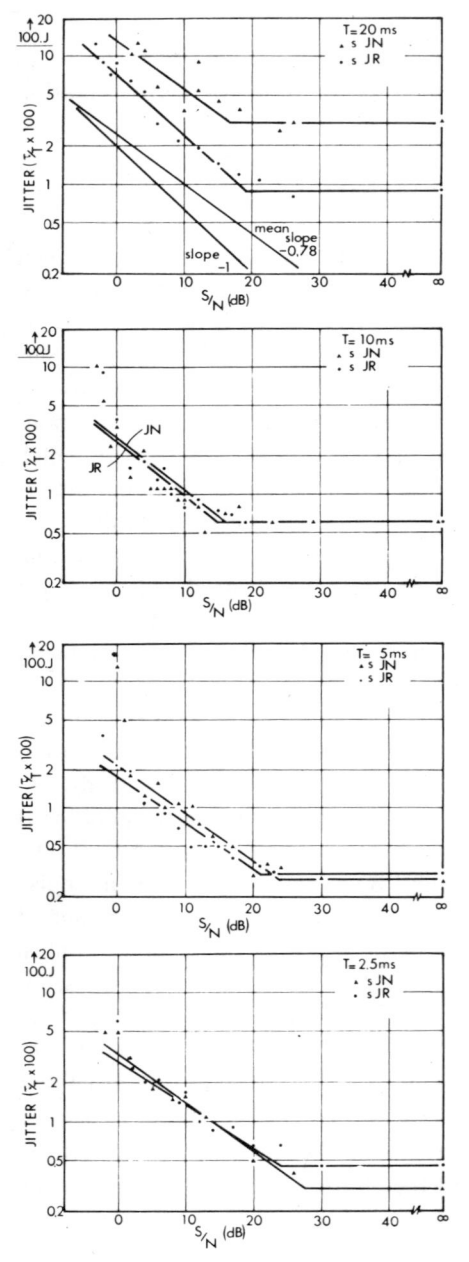

Fig. 1. Just noticeable jitter percentage as a function of the signal-to-noise ratio. J' was measured by two subjects (JN and JR) at 4 pulse rates (50 Hz, 100 Hz, 200 Hz and 400 Hz). $S/N=0$ dB corresponds to the situation in which the individual Fourier components of the pulse train were just masked by the noise. In the region between $S/N=0$ dB and $S/N=20$ dB, a slanting line provides a best fit to the data points. The average of the slope over subjects and repetition rates is -0.78.

the results. Plotted points represent, as a rule, the average of two threshold determinations. The points have been fitted with a pair of straight lines which may be regarded as asymptotes of some curve, not drawn, which provides a

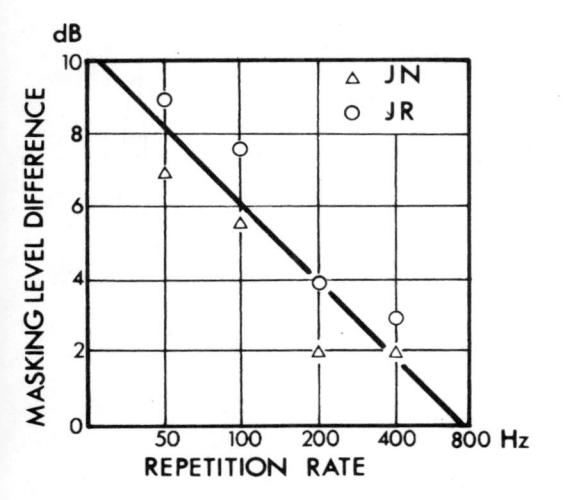

Fig. 2. Difference between the signal-to-noise ratio at which the individual Fourier components of a pulse train are just masked by the noise and the signal-to-noise ratio at which the pulse train itself is just masked. Both signal and noise are low-pass filtered at 4 kHz.

best fit to the data.

In the region of high S/N, the horizontal line indicates that the detection of jitter is not affected by the addition of a small amount of white noise to the pulse train. Then, at $S/N =$ about 20, the effect of white noise is an increase of J'. The average slope is -0.8 and differs significantly from -1.0. As a first approximation it seems justified, however, to conclude: the jitter threshold rises in proportion to the noise *amplitude*. There are a few data points at negative S/N values, but in general the possibility to detect jitter becomes very small indeed. Fig. 2 presents the difference in S/N between the situation in which the periodic pulse train as a whole is just masked and the situation in which the separate components are just masked by the noise. It is evident that the present experiment must break down at a pulse rate around 800 Hz.

DISCUSSION

We have established that at a signal-to-noise ratio 1 in the arbitrary scale, the perception of jitter must be due to the detection of the perturbation of the time structure. We have developed the notion that this time structure must be taken from a band-filtered version of the original pulse train. Now, if we decrease the noise level whilst keeping the signal level constant, the noise-induced perturbation of the time structure, which we shall term $\tau_n = j_n T$ (the "noise jitter") must decrease in proportion to the rms noise amplitude N.

This can be derived from Fig. 3. The damped sinusoid is a hypothetical waveform at a certain point along the basilar membrane. Superimposed on it is noise, which must have passed through the same band filter as the pulses. This noise will be able to shift the time position of the principal maximum of the sinusoid. In this maximum the time derivative of the combination of sinusoid

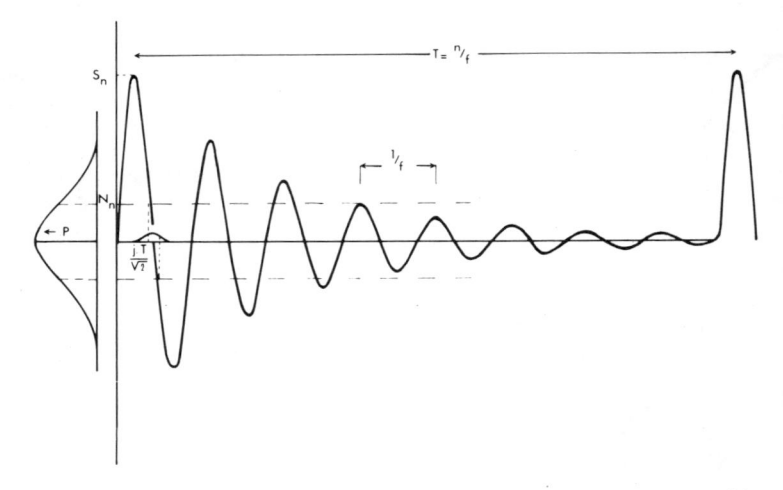

Fig. 3. Hypothesized displacement waveform of a point at the cochlear partition corresponding to a frequency $f=n/T$, T being the pulse period and n the number of the harmonic. The principal maximum has an ordinate S_n. Additive noise in the displacement waveform is indicated by the dashed lines at $\pm N_n$, this being the rms amplitude. The Gaussian probability function P of this noise is drawn to the left of the ordinate. The noise produces random time shifts of the zero crossings of the pulse response. The probability function of a zero is also drawn. By definition, its standard deviation equals the noise induced jitter $j_n T/\sqrt{2}$.

plus noise will be zero. Now, if we differentiate this combined signal, the effect is, in first approximation, a 90° phase shift. In stead of looking for the statistical distribution of principal maxima we may, therefore, look as well at the distribution of negative-going zeros of the first wave. In the first place, it may be remarked that multiple zeros are very unlikely to occur because the period of the band-filtered noise is more or less equal to the period of the sinusoid. However, the noise may shift the zero position along the time axis. The probability distribution is, to a first approximation, a Gaussian one with mean shift zero and with an rms value which must, by definition, be equal to $\tau_n/\sqrt{2}$. It follows from Fig. 3, that

$$\tau_n = \frac{\sqrt{2}}{2\pi} \cdot \frac{T}{n} \cdot \frac{N_n}{S_n}, \quad \text{provided } \sin\left(\frac{N_n}{S_n}\right) = \frac{N_n}{S_n},$$

with n the number of the harmonic which is characteristic for the specific place along the basilar membrane, N_n the rms amplitude of the band-filtered noise at this specific place, and S_n the amplitude of the *maximal* top of the signal at the specific place along the membrane. As N_n is proportional to N, the noise jitter τ_n must be proportional to the rms noise amplitude, provided $N_n/S_n \ll 1$.

Returning to Fig. 1, it is clear that the threshold of relative jitter J' decreases almost linearly with N up to $S/N \approx 20$ dB. It seems hardly plausible that this behaviour would be the result of a gradual coming into play of another jitter detecting mechanism, that is, a detection of the spectral properties of the signal. The very fact, that the decrease of J' is, in fact, slightly slower than would be predicted by the lowering of τ_n would be an additional obstacle to such a hypothesis: in this case the two mechanisms together would perform worse than one alone. On the contrary, it is natural to suppose that the leveling off of the J' curve is caused by internal jitter $J_i = \tau_i/T$ gradually taking over the role of the noise jitter.

It is interesting to estimate the magnitude of the internal jitter. There are two methods.

The first method is a rather direct one, and consists of the application of a variant of statistical decision theory to the listener *in the noise-free situation*. He is then presented with two series of pulse intervals. The B series is characterized by $\tau = J'/T$. In the internal representation, the A intervals will have a standard deviation τ_i around T and with the B intervals the standard deviation will be $\sqrt{\tau^2 + \tau_i^2}$ around the same T. We assume that the internal jitter may be treated as Gaussian and that it is not correlated with the external jitter.

Now the subject has to decide which of the two stimuli A or B has the greater perturbation. He may compare standard deviations or he may search for the maximal perturbation occurring or he may use some other strategy. As long as we do not know which strategy is used, we shall assume that he bases his decisions on the standard deviation, which is, statistically speaking, the best thing to do, although the loss in efficiency when looking for the greatest perturbation occurring would be comparatively small. Both τ_i and $\sqrt{\tau^2 + \tau_i^2}$ are to be regarded as average values, which apply to a great number of stimulus presentations. In fact, there are bound to be cases, when the A stimulus happens to be contaminated with a large amount of internal jitter while in the B stimulus the internal and the external jitter happen to cancel partially. In such cases, the subject will be likely to make an incorrect response. In fact the subject produces 25% incorrect responses when $J = J'$. The ratios $(\tau_i^2 + \tau^2)/\tau_i^2$ which produce 25% incorrect decisions are tabulated in the F-test. These threshold ratios also depend upon the number of intervals in the stimulus, which contribute to the detection. This corresponds to the number of degrees of freedom in the F-test. This number can be obtained by finding the duration of the stimuli, necessary for optimal jitter discrimination. In experiment II, the duration of the stimuli was fixed to 250 msec, but in a separate experiment along similar lines, these minimal durations were measured. The results are shown in Table I. It must be remarked, that these minimal durations were measured rather inaccurately. From Pollack's (1968) Figs. 1, 2, and 3 one would derive values which tend to be about two times as large. Also, previous experiments in our Institute (Cardozo et al., 1966) seem to indicate a larger value of m. However, as the effect of m in the computation of τ_i is comparatively small, we will not worry about this.

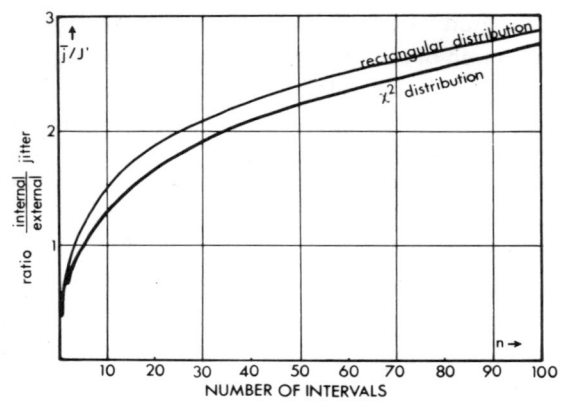

Fig. 4. Ratio of internal jitter over external jitter as a function of the number of pulse intervals in the pulse train. The heavy line corresponds the approach discussed in the text.

Table I

repetition rate	50 Hz	100 Hz	200 Hz	400 Hz
J' (average 2 subj.)	3%	0.6%	0.3%	0.3%
number of intervals m	12	9	14	28
estimated internal relative J_i	4%	0.8%	0.5%	0.6%
estimated internal abs. τ_i	800 μs	80 μs	25 μs	15 μs

Because $F_m^m = (\tau_i^2 + \tau^2)/\tau_i^2$, we find $\tau_i/\tau = 1/\sqrt{F_m^m - 1}$. This factor is depicted in Fig. 4 as a function of m. The resulting τ_i values are represented in Table I.

The second way of estimating the internal jitter is based on the cross-over from noise jitter to internal jitter around $S/N = 20$ dB in the arbitrary scale in Fig. 1. This allows the internal jitter to be expressed in terms of the bandwidth of the internal filter and the harmonic number n, characteristic of the place along the basilar membrane from which the timing signal is taken. The computation will not be given here. Introducing reasonable values for the two parameters mentioned above, it is possible to arrive at the same values for the internal jitter as presented in Table I. This gives some indication of an internal consistency of the theory.

We may end this discussion by establishing that internal jitter in the range of 100 to 400 Hz is in good agreement with the figures given in the introduction. Attention is called to the contrastingly large value of the internal jitter for 50 Hz.

CONCLUSION

We have discussed at some length psychoacoustical experiments with noise-embedded jittered pulse trains. The notion that jitter detection takes place with pulse trains in the range 50 to 400 Hz, on the basis of spectral properties, was refuted. This does not preclude the possibility that, for high values of relative jitter, spectral properties do become perceptible, like in harsh, raucous sounds. Also, the present experiments must not be generalized or extrapolated to higher pulse rates falling outside the domain of existence of the residue.

A linear extrapolation from the noise-immersed situation to the noise-free situation fits with the observed behaviour of the jitter threshold and leads to an interpretation of the jitter threshold in the noise-free situation in terms of internal jitter. This internal jitter proves to be a more or less constant percentage rather than a constant amount of time and may be taken to be responsible for the difference limens of periodicity pitch.

ACKNOWLEDGEMENT

The author feels deeply indebted to J. F. F. N. Rijckaert and J. J. M. Neelen for carrying out the experiments.

REFERENCES

Cardozo, B. L. (1966): A sequential up-and-down method, I.P.O. Annual Progress Report *1*, 110-114.
Cardozo, B. L., and Ritsma, R. J. (1965): Short-time characteristics of periodicity pitch, in: Rapports 5e Congrès International d'Acoustique, Liège, Vol. Ia, B 37.
Cardozo, B. L., and Ritsma, R. J. (1968): On the perception of imperfect periodicity, IEEE Trans. on Audio and Electro-acoustics, Au-*16*, 159-164.
Cardozo, B. L., Ritsma, R. J., Domburg, G., and Neelen, J. J. M. (1966): Unipolar pulse trains with perturbed intervals, I.P.O. Annual Progress Report *1*, 17-27.
Pollack, I. (1968): Detection and relative discrimination of auditory 'jitter', J. Acoust. Soc. Amer. *43*, 308-316.
Ritsma, R. J. (1962): The existence region of the tonal residue. I, J. Acoust. Soc. Amer. *34*, 1224-1230.
Ritsma, R. J. (1965): Pitch discrimination and frequency discrimination, Rapports 5e Congrès International d'Acoustique, Liège, Vol. Ia, B 22.
Rosenberg, A. E. (1966): Pitch discrimination of jittered pulse trains, J. Acoust. Soc. Amer. *39*, 920-928.
Schouten, J. F. (1940): The perception of pitch, Philips Techn. Rev. *5*, 286-294.

DISCUSSION

Zwislocki: You said you added so much noise to a pulse train that all the single spectral components were just masked. I imagine that you heard the pulses because the energy of the spectral components is integrated within a critical band. In this case it seems to me that if you jitter the pulse train, you change the width of the spectral components. Is that correct?

Cardozo: The peak level of a particular component will indeed decrease and the width of each individual line will increase, but the energy of the jittered components integrated in the critical band will not change substantially.

Kuyper: One of your parameters is the harmonic number n related to a place at the basilar membrane. If you filter the pulses, you are able to eliminate this parameter and you know exactly where you are working on the basilar membrane. However, by filtering you may introduce new unwanted features for detection; perhaps it is possible to filter the pulses first and to jitter next.

Cardozo: We have been working with filters but as soon as you have a steep filter slope, this will introduce a spurious amplitude modulation. We have also been working with noise with a gap at some frequency. In this way the artefact of amplitude modulation is avoided. But such experiments are very difficult to perform and the only conclusion I am able to draw from them is that the optimal region for jitter detection is somewhere between 500 and 2000 Hz.

Smoorenburg: I think I can say something about Kuyper's proposal. One may assume that the difference limen for periodicity pitch is related to internal temporal jitter. To exclude the possibility of judging on the basis of frequency components, the DL may be measured by determining the just-noticeable difference in repetition frequency of two pulse trains filtered by different 1/3-octave filters. I found that this just-noticeable difference depends on the repetition frequency rather than on the filter frequency. Similar results are found by Ritsma (1965). By jittering the pulses externally and measuring the increase of the just-noticeable difference it is possible to obtain an estimate of the internal jitter. And, in accordance with the preceding, I found that the internal jitter depends on the repetition frequency rather than on the filter frequency (70 μsec for 200 Hz and 35 μsec for 400 Hz). I did not find a significant contribution from peripheral internal jitter of, for example, the detection mechanism; a jitter which is expected to be related to the place of detection at the basilar membrane or to the filter frequency. However, electrophysiological data suggest that the peripheral jitter cannot be of minor importance because it is certainly not a magnitude smaller than the estimates of 70 μsec and 35 μsec which I obtained for the whole process.

Cardozo: I see your problem, but I do not see the solution.

Schouten: Even at our institute we have strong arguments *pro* and *con* the hypotheses whether detection is linked with time measurements in the nerve or with some event on the basilar membrane. One of the things which intrigues me is that you can interpret a pulse train equally spaced in time but with one pulse out of its place as a superposition of a regular pulse train and a double pulse, consisting of a negative and a positive pulse, which substitutes the irregularly placed pulse and cancels the regular one. In terms of the basilar membrane this means that we have a regular pattern and that at some moment there is a transient. And this is what one actually hears; if there are a few pulses somewhat out of place one hears the regular tone and from time to time some pips. The basilar membrane plays a part but whether this part is sufficient to explain the phenomenon is another question. I mentioned this to show that we are not always trying to describe our experiments in terms of time analysis.

Zwicker: What Dr. Schouten was saying, is very important I think, and maybe both place and time people can meet each other here. The ear neither performs a Fourier analysis in an exact manner, nor does it operate only in the time domain. The ear works somehow in between, like a short-term Fourier analysis. It acts as a filter set, the output of which will show important time patterns in the envelopes. I would say that we do best to look at the excitation pattern and how it changes in time. I mean quick changes with time constants of perhaps 3 msec that correspond to the bandwidths.

Cardozo: I quite agree. Even so, there still is the problem that the shorter you make the time base of the Fourier process, the wider the intrinsic line width becomes. If you accept a kind of short-term Fourier transform with a reasonably wide intrinsic line width, it will be a very hard job for the ear to detect the relatively small broadening of spectral components caused by the jitter.

Zwicker: The ear is doing a hard job, I think.

REFERENCES

Ritsma, R. J. (1965): Pitch discrimination and frequency discrimination, in: Rapports 5e Congrès International d'Acoustique, Liège, Vol. Ia, B 22.

A COMPARISON OF THE EFFECTS OF SIGNAL DURATION ON FREQUENCY AND AMPLITUDE DISCRIMINATION*

G. BRUCE HENNING

Defence Research Establishment Toronto
Downsview, Ontario, Canada

INTRODUCTION AND SUMMARY

We wish to determine whether the information in auditory stimuli used by the ear to produce the residue tone contributes to the basic detection and resolution capabilities of the ear. In particular, we wish to determine if there is any evidence to support the hypothesis that the frequency resolution of the ear at low frequencies is determined by timing information known to be present in primary auditory neurons. Although the data presented here do not permit firm conclusions, indirect evidence indicates that several models which postulate a single time-invariant mechanism as sufficient to describe both the frequency and amplitude resolution capabilities of the auditory system are wrong.

PROCEDURE

Amplitude discrimination

Two experienced observers[1], seated in a sound attenuating room, participated in a standard two-alternative forced-choice (2AFC) amplitude discrimination experiment. On each trial, a tone of a given frequency, duration, and amplitude was presented, followed after a pause by a signal of the same frequency and duration but different amplitude. The observers were required to indicate which of the two signals had been greater in amplitude. The probability of the higher amplitude tone being in the first interval was 0.5 on each trial. After each trial, the observers were informed which interval had in fact contained the higher

---

* D.R.E.T. Research Paper No. 758.
[1] Observer 1 was the author; observer 2, a former bugler in the Royal Canadian Artillery.

amplitude tone. The signals were presented in a background of continuous broadband Gaussian noise of 20 dB uniform average Spectrum Level. The signals, gated on and off with an essentially rectangular envelope at a zero axis crossing, were presented binaurally over TDH-39 earphones driven in phase.

The level of the higher amplitude tone was kept constant at 85 dB SPL while the lower amplitude signal was 2.0, 0.8, or 0.3 dB fainter. Signal duration was kept constant for 200 trials at each of the amplitude ratios to be discriminated. The duration of the signals was then changed to durations ranging from 1 to 250 msec depending on the frequency of the signal. Frequencies ranged from 250 to 8000 Hz.

Two aspects of this procedure require comment. First, the level of the signals to be discriminated remained constant in spite of differences in the duration of the signals. Thus, the energy in each signal varied with signal duration; on the other hand, the signal power remained constant. It has been shown for rectangularly gated 1000 Hz sinusoids in the standard 2AFC procedure used here that signal energy has no effect on amplitude discrimination unless the ratio of signal energy to noise-power density is less than about 20 dB (Henning and Psotka, 1969). This minimum ratio was exceeded by 15 dB at all durations and frequencies used. Assuming variations in signal energy to be equally ineffective at all frequencies once the ratio of signal energy to noise-power density exceeds 20 dB, the change in signal energy with duration should produce no effect. The alternative method — maintaining constant signal energy with changing duration — necessarily involves changes in the power of the signals to be discriminated. Changing the signal power has two disadvantages: (1) amplitude discrimination performance improves with increasing signal power (Corliss, 1967) and (2) it is difficult to achieve consistent relative amplitude settings at different levels of signal power. Fixing the signal amplitude both removes the effect of varying signal power and permits precise control of the relative amplitudes of the tones to be discriminated given the stability of a passive attenuation network and the ease of an initial null setting.

Second, the interval between the offset of the first signal and the onset of the second decreased as the signal duration increased. In fact, the interval between onset of the two signals was kept constant at 600 msec. Intersignal interval is known to have a small effect on both amplitude and frequency discrimination; but, the effect is small over the range of interstimulus intervals used in this experiment.

Frequency discrimination

The same observers also participated in a standard 2AFC frequency discrimination experiment. The procedure was similar to that used in the amplitude discrimination experiment. On each trial a signal was presented, followed by another of the same amplitude and duration but different frequency. The observers were required to indicate which of the signals had been higher in frequency and were informed, after each trial, which interval had in fact contained the higher frequency tone. The function relating the percentage of correct responses in 200 trials to frequency difference was determined by varying the frequency difference. The arithmetic mean frequencies of the tones to be discriminated were the same as the frequencies of the tones used in the amplitude discrimination experiment. As in the amplitude discrimination experiment, the signals were presented binaurally, in-phase, and in a background of continuous noise of 20 dB uniform average Spectrum Level. Data were obtained for all durations at a given mean frequency before the mean frequency of the signals was changed to another value. It should be noted that since the signals were gated on and off at zero axis crossings, the durations of two signals of slightly different frequency were not exactly the same. The lower frequency signal is, at most, one-half its period longer than the higher frequency signal. This maximum difference in duration is not discriminable, however, provided the duration of the signal is long with respect to its period (Creelman, 1962). For only the shortest duration of the 250 Hz signal does the difference in duration exceed the "just-noticeable difference" for duration discrimination above which observers may discriminate frequency differences on the basis of signal duration. Further, a difference exists between the total energies of the signals, this difference being proportional to the difference in the duration of the signals. The energy difference should not be discriminable until the signal duration is at least an order of magnitude less than values at which frequency discrimination begins to deteriorate. Finally, not only will the energy density spectrum of the lower frequency signal be centred at a lower frequency but it will also be somewhat narrower, with its major lobe narrower in bandwidth by at most the period of the signal. A slight improvement in frequency discrimination performance at short durations might in general be anticipated from this effect, but estimates of the magnitude of the effect will depend on the particular model used.

RESULTS

Figs. 1-6 show the results of both the amplitude and the frequency discrimination experiments. The ratios of "just-noticeable amplitude difference" to amplitude ($\Delta V/V$) and "just-noticeable frequency difference" to frequency ($\Delta F/F$) are plotted on logarithmic coordinates as functions of signal duration. Data are plotted separately for each observer at each of three different frequencies. The "just-noticeable difference" is taken as the difference corresponding to the 75% correct response level obtained by interpolation for both the amplitude and the frequency discrimination cases. (Functions similar in shape to those of Figs. 1-6 are obtained if either the 90% or the 60% correct levels are used.) It should be noted that the values of $\Delta F/F$ have been scaled to

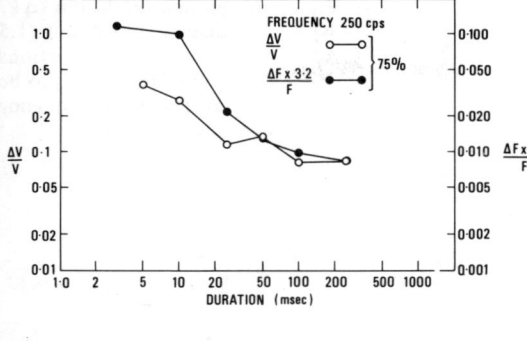

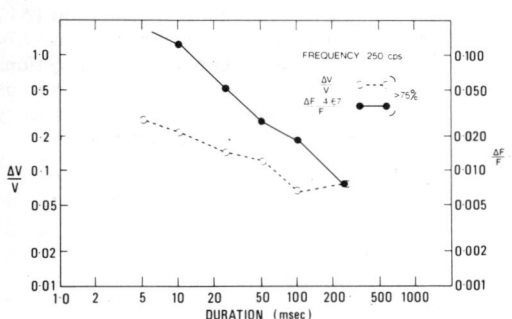

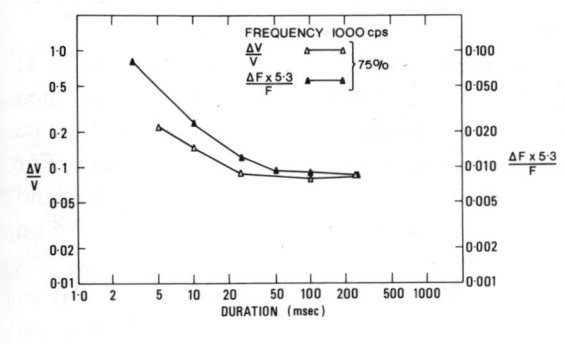

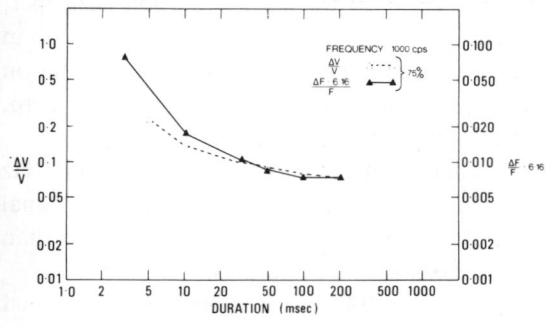

Fig. 1. Amplitude resolution ($\Delta V/V$) and frequency resolution ($3.2\Delta F/F$) for Observer 1 as functions of the duration of the signals to be discriminated. The coordinates in this and the following figures are logarithmic and the "just-noticeable differences" ΔF and ΔV are the frequency and amplitude differences corresponding to 75% correct responses in 200 trials of standard two-alternative forced-choice experiments. The mean frequency of the signals was 250 Hz.

Fig. 2. Amplitude resolution ($\Delta V/V$) and frequency resolution ($2.85\Delta F/F$) for Observer 2 as functions of the duration of the signals to be discriminated. The mean frequency of the signals was 250 Hz.

Fig. 3. Amplitude resolution ($\Delta V/V$) and frequency resolution ($5.3\Delta F/F$) for Observer 1 as functions of the duration of the signals to be discriminated. The mean frequency of the signals was 1000 Hz.

Fig. 4. Amplitude resolution ($\Delta V/V$) and frequency resolution ($6.16\Delta F/F$) for Observer 2 as functions of the duration of the signals to be discriminated. The mean frequency of the signals was 1000 Hz.

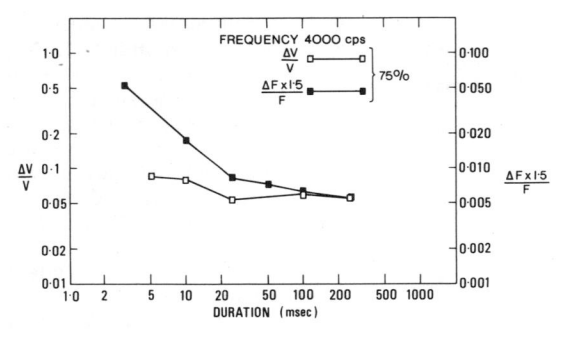

Fig. 5. Amplitude resolution ($\Delta V/V$) and frequency resolution (1.5 $\Delta F/F$) for Observer 1 as functions of the duration of the signals to be discriminated. The mean frequency of the signals was 4000 Hz.

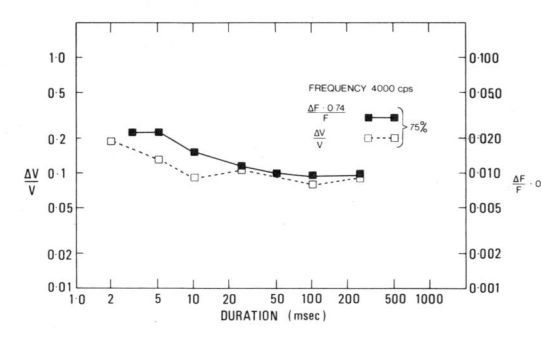

Fig. 6. Amplitude resolution ($\Delta V/V$) and frequency resolution (0.74 $\Delta F/F$) for Observer 2 as functions of the duration of the signals to be discriminated. The mean frequency of the signals was 4000 Hz.

permit representation and comparison of both amplitude and frequency discrimination data on the same graph.

Figs. 1 and 2 show the results obtained with each observer for signals at 250 Hz. Consider first the amplitude discrimination data (plotted as open circles). Amplitude discrimination performance at 250 Hz appears to be independent of signal duration for durations greater than about 100 msec. For durations shorter than 100 msec $\Delta V/V$ increases (performance deteriorates) and the data are described reasonably well by a straight line with a slope of -0.5 on the logarithmic coordinates used here.

The frequency discrimination data (plotted as solid circles) are similar in form and show decreasing frequency resolution with decreasing frequency. The slope relating log $\Delta F/F$ to the logarithm of the signal duration is steeper, however, having the value between -0.5 and -1.0 for durations less than about 50 msec. It appears possible to describe these frequency discrimination data on logarithmic coordinates by three straight lines with slopes of zero, -0.5 and -1.0, as Liang and Chistovich (1961) have done.

Figs. 3 and 4 show data of a similar form for each observer for 1000 Hz signals. At this frequency, amplitude discrimination is independent of signal duration for durations in excess of about 25 msec. For durations shorter than 25 msec, amplitude discrimination performance deteriorates.

Frequency discrimination performance at 1000 Hz is independent of signal duration for durations in excess of about 50 msec, and deteriorates more rapidly then amplitude discrimination performance as duration decreases below 50 msec.

At 4000 Hz, (Figs. 5 and 6) frequency discrimination deteriorates for signals below 25 msec duration whereas amplitude discrimination performance remains virtually independent of signal duration for durations in excess of 10 msec.

DISCUSSION

The results of both the frequency and amplitude discrimination experiments may be compared with those obtained in similar experiments. The relation between frequency resolution and signal duration seen in this experiment is similar to that reported by Turnbull (1944) and by Liang and Chistovich (1961). At all frequencies, frequency resolution decreases with decreasing duration of the signal once the signal duration falls below a given value. The duration below which an inverse square-root relation holds and above which resolution is independent of duration, or is a very slowly changing function of it, is frequency dependent at least between 128 and 4000 Hz, being shorter for higher frequency signals.

The results of the amplitude discrimination experiments reveal a similar relation between amplitude resolution and signal duration, with one major exception: at each frequency, the duration below which amplitude resolution becomes dependent on signal duration is less than the corresponding duration for frequency resolution. There appears to be a range of durations over which frequency resolution depends on duration but amplitude resolution does not. The range is small, and at each frequency the critical duration for amplitude resolution is about one-third that for frequency resolution.

While the differences in the effect of signal duration on frequency and amplitude resolution are of some theoretical interest, the emphasis given the result should be tempered by (a) the small range of durations involved in the effect and (b) differences among the results of the amplitude discrimination experiment reported here and by Henning and Psotka (1969), those of Chocholle and Krutel (1968) and those of Garner and Miller (1944). The data of Chocholle and Krutel indicate that amplitude resolution varies inversely with the square-root of signal duration for durations less than 400 msec. Further, their data show no dependence on the frequency of the signal. On the other hand, the results of Garner and Miller indicate that the function relating $\Delta V/V$ to signal duration is dependent on the intensity of the signals. At low intensities (40 dB SL) discrimination performance with 500 Hz signals deteriorates more rapidly with decreasing signal duration over the range 400 to 20 msec than the data in the present experiment would indicate. The change is shallower than an inverse square relation, however, and at higher intensities (70 dB SL) there is reasonable agreement between Garner and Miller's data and those of the present study.

It should be noted that in the experiments by Chocholle and Krutel the rise and fall time of the signals to be detected was 5 msec. Garner and Miller used 10 msec rise and fall time. Thus in both studies what amounts to band-pass filtered signals were used as opposed to the rectangularly gated signals used in the present study. While filtering the signals to be discriminated leads to some uncertainty as to the effective duration of the signals, it appears not to be the principal source of the discrepancies among the results of the various experiments; for, when amplitude discrimination performance was measured with shaped signals (10 msec rise and fall time) the shape of the functions relating amplitude and frequency resolution to signal duration did not appear to change. The results of this experiment are shown in Fig. 7 together with the comparable data for the same observer listening to rectangularly gated signals. While both frequency and amplitude resolution are markedly affected by shaping the signal, a substantial range over which frequency and amplitude resolution depend differently on signal duration remains.

If it is accepted that signal duration affects amplitude and frequency discrimination in the way suggested by the data of the present study, the implications for models of the auditory system are important.

For example, the model developed by Siebert (1968) in which certain properties of the peripheral auditory nervous system are incorporated into a mathemat-

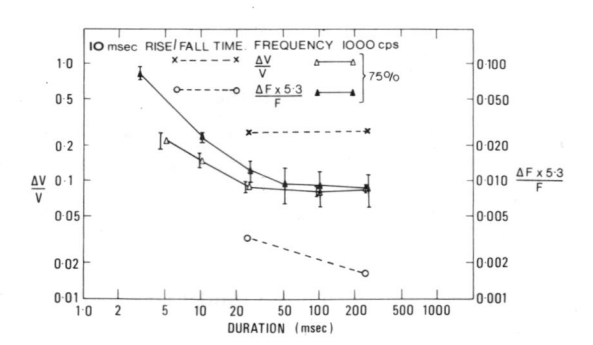

Fig. 7. Data of Fig. 3 together with frequency and amplitude discrimination data from the same observer when judging signals with 10 msec rise and fall times. The signal durations in the latter case were taken as the duration of a rectangularly gated signal with the equivalent envelope area. The vertical bars on the data points indicate a range one standard deviation above and one standard deviation below the mean values of $\Delta F/F$ and $\Delta V/V$, respectively. The estimates were obtained by assuming binomial variability in the performance measure from which the values of $\Delta F/F$ and $\Delta V/V$ corresponding to 75% correct responses were obtained and then reflecting the performance levels one standard deviation above and below the 75% correct level through the functions relating performance to ΔF and ΔV at each duration.

ical analogue of the ear with specific behavioural implications predicts, reasonably it would appear, that both frequency and amplitude resolution vary inversely with the square-root of the signal duration. Presumably, however, the range of durations over which this relation holds should be very nearly the same for both frequency and amplitude resolution, contrary to the findings of the present study. The prediction arises from the Poisson character of the neural noise assumed to be limiting discrimination performance; the ratio of mean to the standard deviation of the decision statistic on which the observer bases his decisions varies with the square-root of the duration of the signal.

Again, both the modified energy detection models of Green and Swets (1966) and of Henning (1967) predict that amplitude discrimination performance should be independent of duration provided only that the signal energy passing the initial band-pass filter proposed in these models exceeds the noise-power density by a ratio of about 20 dB. While the latter model makes reasonable predictions for the duration at which frequency discrimination should begin to deteriorate at each frequency, both models predict that amplitude resolution should not be affected by signal duration in the present experiment until the signal durations are on the order of tens of microseconds, quite contrary to the results obtained.

One modification of the energy detection models that would allow them to predict the effects of signal duration on amplitude discrimination found in the present study is the assumption (Scholl, 1962) of a marked increase in the bandwidth of the initial filter with decreasing duration.

Consider, now, the comprehensive model of Corliss (1967). In this model, the energy resolution limits of the auditory system are basic measures for which values for different frequency signals are determined empirically from amplitude discrimination data or deduced from the "absolute threshold" of hearing and the properties of a least-count limited receptor. Much of the classic psychophysical data is then very nicely predicted by assuming a set of resonant analysers specified simply in terms of their resonant frequency and Q.

Because of the relatively high values of Q required by the model, (60, 150, and 200 at 250, 1000, and 4000 Hz, respectively) it will be seen that quite long durations will be necessary for frequency resolution of the analysers postulated in the model to approximate their maximum resolving power. An analysis of the dynamic resolving power of resonant frequency analysers by Kharkevich (1960) shows that the signal durations below which analysers of the Q used in the Corliss model would show psychophysically just measurable differences in frequency resolution are only slightly larger than those measured in the present study.

While the Corliss model predicts the effects of signal duration on frequency resolution quite well, it is difficult to see how analysers with such high Q can predict the effect of signal duration on the detection of signals in noise (Green *et al.*, 1957). Indeed the dynamic response of a resonant analyser having the Q value used in the Corliss model at 1000 Hz suggests that the detectability of constant energy signals at 1000 Hz in continuous noise should be measurably poorer for signal durations less than about 140 msec. This is contrary to the data of Green *et al.* (1957).

The analogy between Kharkevich's analysis and Corliss' mechanistic model may be inappropriate in detail; nonetheless, it appears that there are discrepancies between the predictions of each model and the measured effects of signal duration on frequency resolution, amplitude resolution, or detection. While each of the models considered above relies, as von Helmholtz's model does, on some form of a "place" mechanism for frequency analysis, it is clearly unreasonable on the basis of the data presented here to conclude that "place" models are inappropriate and that a counting mechanism extracting frequency information from the time-locked response of the peripheral auditory system necessarily provides the representation of frequency for the observer.

It is instructive, nonetheless, to consider the effect of signal duration on frequency and amplitude discrimination at frequencies where time-locked response is not present in the peripheral system, and at which the residue phenomenon does not occur (Ritsma, 1962). Unfortunately, frequency resolution is difficult to measure at frequencies much above 4000 Hz (Henning, 1966) and at very low frequencies it is difficult to arrange different frequency tones of nearly identical duration. We have made one attempt, however, to measure the effect of signal duration on frequency and amplitude resolution at 8000 Hz. In the frequency discrimination experiment, the tones to be discriminated were set equal in amplitude at the tip of a probe microphone located in the ear canal under the earphone. The observer's head was fixed in position on a hardened dental wax impression of his teeth. If, at the end of a set of 100 trials, the measured amplitudes of the tones to be discriminated differed by 1/2 dB or more the results were discarded and the condition repeated. About 25% of the trials were lost using this technique and there is no assurance that temporary changes in amplitude occurring during the set of trials do not significantly affect the measures.

The data for Observer 1 are shown in Fig. 8. As in the experiments with signals ranging from 250 to 4000 Hz, amplitude discrimination is roughly independent of signal duration down to very short durations. This finding is consistent with the decrease in critical duration with increasing frequency noted at the lower frequencies. In the case of frequency discrimination, however, the data for signals centred at 8000 Hz differ from those obtained at lower frequencies. First, frequency resolution at 8000 Hz appears to deteriorate more rapidly with decreasing duration; the ratio $\Delta F/F$ varies inversely with the duration. Further, it appears that the duration below which frequency discrimination deteriorates with decreasing duration is at least 250 msec. The implication of

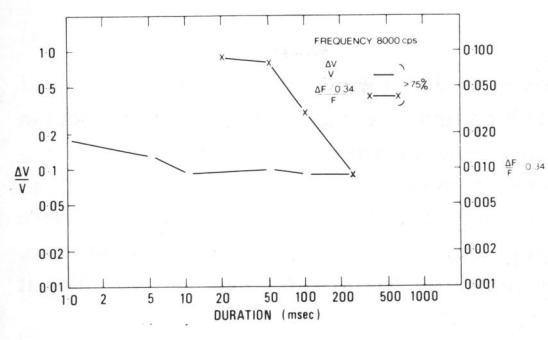

Fig. 8. Amplitude resolution ($\Delta V/V$) and frequency resolution (0.34 $\Delta F/F$) for Observer 1 as functions of the duration of the signals to be discriminated. The mean frequency of the signals was 8000 Hz. The signals were presented monaurally and, in the frequency discrimination experiment, the amplitudes of the signals were set equal in the external ear canal using a Muirhead probe microphone (Shaw, 1966).

the data taken at lower frequencies that the critical duration for frequency discrimination should decrease with increasing frequency is not confirmed at 8000 Hz. Whether this finding can be substantiated and a firm conclusion reached concerning the importance of timing information in the classic psychophysical measures remains to be seen.

REFERENCES

Chocholle, R., and Krutel, J. (1968): Les seuils auditif différentiels d'intensité en fonction de la durée des stimuli, C.R. Soc. Biol. (Paris) *162*, 848-851.
Corliss, E. L. R. (1967): Mechanistic aspects of hearing, J. Acoust. Soc. Amer. *41*, 1500-1516.
Creelman, C. D. (1962): Human discrimination of auditory duration, J. Acoust. Soc. Amer. *34*, 582-593.
Garner, W. R., and Miller, G. A. (1944): Differential sensitivity to intensity as a function of the duration of the comparison tone, J. Exp. Psychol. *34*, 450-463.
Green, D. M., Birdsall, T. C., and Tanner, W. P. (1957): Signal detection as a function of signal intensity and duration, J. Acoust. Soc. Amer. *29*, 523-531.
Green, D. M., and Swets, J. A. (1966): Signal Detection Theory and Psychophysics (Wiley, New York).
Henning, G. B. (1966): Frequency discrimination of random amplitude tones, J. Acoust. Soc. Amer. *39*, 336-339.
Henning, G. B. (1967): A model for auditory discrimination and detection, J. Acoust. Soc. Amer. *42*, 1325-1344.
Henning, G. B., and Psotka, J. (1969): Effect of duration on amplitude discrimination in noise, J. Acoust. Soc. Amer. *45*, 1008-1013.
Kharkevich, A. A. (1960): Spectra and Analysis (Consultants Bureau, New York).
Liang, C., and Chistovich, L. A. (1961): Frequency-difference limens as a function of tonal duration, Sov. Phys. Acoustics *6*, 75-80.
Ritsma, R. J. (1962): Existence region of the tonal residue. I, J. Acoust. Soc. Amer. *34*, 1224-1229.
Scholl, H. (1962): Das dynamisches Verhalten der Frequenzgruppen, Acustica *12*, 101-107.
Shaw, E. A. G. (1966): Earcanal pressure generated by a free sound field, J. Acoust. Soc. Amer. *39*, 465-470.
Siebert, W. M. (1968): Stimulus transformations in the peripheral auditory system, in: Recognizing Patterns, P. A. Kolers and M. Eden, Eds. (M.I.T. Press, Cambridge, Mass.), p. 68.
Turnbull, W. W. (1944): Pitch discrimination as a function of tonal duration, J. Exp. Psychol. *34*, 302-316.

DISCUSSION

De Boer: I like to make a comment about your mentioning the model of Mrs. Corliss. I think there is something fundamentally wrong with her model in that she assumes that there is a linear filter at work, and that this linear filter is of the first order. In that case there is indeed, as a function of duration, an intimate connection between the amount of energy change that you can perceive and the amount of frequency change that you perceive, as you have mentioned. But when you go to higher order filters, this relation no longer exists. I think it is worth-while to be warned against using models that are too simple.

Goldstein: For completeness I would like to note that Siebert has formulated *other models* of the processing of the auditory nerve data than the one you mentioned, for instance, one in which frequency discrimination is based upon temporal information in the neural data (Siebert, 1968, 1970). The temporal model predicts a difference limen, ΔF, that is inversely proportional to the three-halves power of duration. The "place" model you referred to predicts a ΔF inversely proportional to the square root of duration. I should also like to note that bandwidth *per se* of the analyzing filter is decidedly not a critical parameter in Siebert's place model; rather, it is the filter's rate of attenuation above and below its maximum response as measured along the cochlear partition.

Klinke: Perhaps I may mention some facts which should not be overlooked in psychophysiological experiments. Dr. Henning, you wrote in the paper that your experiments were performed in a soundproof room, and it is surely necessary to do so. But let me draw your attention to a paper published by Fruhstorfer and Bergström (1969) on potentials evoked by acoustic stimuli. They tried to check the influence of vigilance, that is, the degree of alertness of a subject, on auditory evoked potentials. The vigilance was judged from electro-encephalogram. They performed their experiments in a soundproof but not darkened room and found that within 10 to 20 minutes all subjects (7) fell into a very low state of vigilance, followed by a period in which their state of vigilance fluctuated very much. The subjects were not required to respond to any acoustic stimulus. I think the state of vigilance of the subjects should always be considered in psychophysical experiments.

Henning: I can only assure you that our observers remained sufficiently awake to make fine frequency discrimination judgements every three seconds.

Wilson: I think it is a common experience in psychophysical experiments with well practiced subjects in a forced-choice situation that threshold does not depend on the intensity of a *conscious* act of attention to the stimulus. On a number of occasions subjects have reported to me that I should not accept a particular experimental run because they were not "attending". Retesting with a considerable effort of attention has never shown a significant change. This of course has nothing to do with the effects of tiredness which can be considerable.

Klinke: During an experiment, the subject's state of vigilance is always fluctuating. So I think during psychophysical experiments the electroencephalogram ought to be recorded and only those responses should be evaluated together which occurred during a particular state of vigilance (see Fruhstorfer and Bergström (1969) for method). This would lead to more reliable research.

Bosher: May I just support Dr. Klinke. Working in a neurological hospital where the patients may be severely ill or sedated, I have found such patients tend to have variable pure tone thresholds, although they appear fully attentive. It has always seemed to me that this is attributable to some diminution of alertness and surely this danger cannot be ignored in psychoacoustical experiments demanding considerable mental attention.

Zwislocki: Well, I am quite aware that this discussion has nothing to do with Dr. Henning's paper, really, but I cannot leave the situation the way it is now. We have done quite a lot of measurements as a function of time under reasonably well controlled conditions, *e.g.* forced-choice procedures, and we found that even under such conditions the threshold of experienced subjects varies when one keeps them a long time in a booth. The threshold usually goes up by about 4 dB, and in some measurements that's a lot.

Zwicker: Can you explain the cross in Fig. 7 where the duration is 200 msec? Does this result mean that, for 200 msec duration, the subjects could not discriminate between the two signals unless the amplitude difference was more than 25%?

Henning: That is correct. Possibly because the observers were accustomed to listening to rectangularly gated signals, the observers performed poorly with shaped signals.

REFERENCES

Fruhstorfer, H., and Bergström, R. M. (1969): Human vigilance and auditory evoked responses, Electroenceph. Clin. Neurophysiol. *27*, 346-355.
Siebert, W. M. (1968): Stimulus transformations in the peripheral auditory system, in: Recognizing Patterns, P. A. Kolers and M. Eden, Eds. (M.I.T. Press, Cambridge Mass.), pp. 104-133.
Siebert, W. M. (1970): Frequency discrimination in the auditory system: place or periodicity mechanisms? Proc. IEEE, in press.

EXPERIMENTS ON BINAURAL DIPLACUSIS AND TONE PERCEPTION

G. VAN DEN BRINK

Department of Biological and Medical Physics
Medical Faculty Rotterdam
Rotterdam, The Netherlands

INTRODUCTION

In this paper a survey will be given of some aspects of a study on diplacusis and pitch perception in relation to pure tone thresholds and the shape of isophones.

Although extensive data have been collected during the preceding years, only a small part of it has been published (van den Brink, 1965). We hope, however, to accomplish a more detailed publication of these data and a literature survey in the near future. In this paper, however, I shall only present some of the results of our own experiments.

Diplacusis is the phenomenon that presentation of a single tone results in the perception of two pitches. If this occurs simultaneously in one ear, the phenomenon is called monaural diplacusis. This usually is a symptom of a hearing defect. There have been reported pathological cases of triplacusis monauralis, which, as far we can understand, must be due to central neurological defects.

If binaural presentation of the tone results in the perception of different pitches in the two ears, we deal with binaural diplacusis. In severe cases of binaural diplacusis the two pitches can be perceived simultaneously; it is not amazing that most of the described cases concern musicians. "Normal" subjects, however, do not notice it: the two pitches melt together and result in one pitch laying in between the pitches for the separate ears. Only when tone pulses are presented alternatively to the two ears, a pitch difference is noticed. Evidently the frequency-pitch relation is different for the left and the right ear, as is illustrated exaggeratedly in Fig. 1. In the upper part of this figure the frequency is plotted horizontally and some arbitrary measure of pitch vertically. If the two curves, regardless their shape, do not coincide, different pitches are being heard

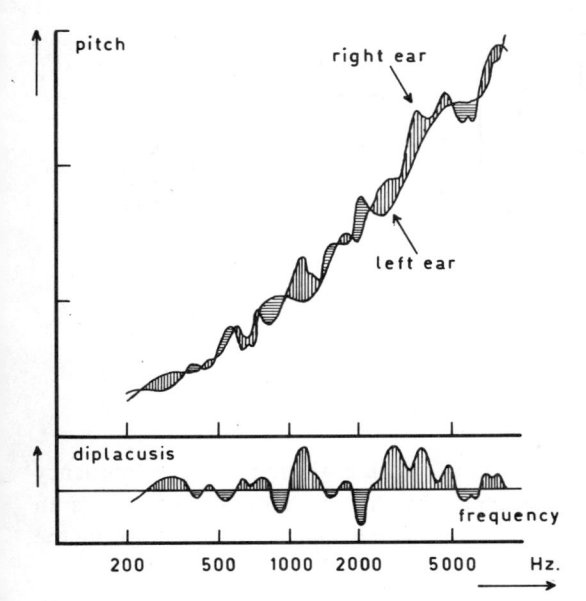

Fig. 1. Different frequency-pitch relations result in binaural diplacusis. Pitch scale is arbitrary.

in the two ears for equal frequencies, as shown in the lower part of the figure.

MEASURING PROCEDURE AND APPARATUS

Quantitative information of diplacusis can be obtained by presenting tone pulses alternatively to the two ears. The pulses are taken from independent oscillators. For a certain frequency of the tone in one ear, the frequency of the tone in the other ear is adjusted to have equal pitches for both ears. These matchings are repeated for different frequencies in the reference ear.

In our experiments the stimuli are presented in cycli with a duration of 3.2 sec: two pairs of tone pulses are presented (left-right-left-right), each pulse with a duration of 0.4 sec, followed by a silent interval of 1.6 sec. This interval turned out to be very useful because, especially at low loudness levels, a cue is necessary whether pulses were presented to the left or to the right ear. Such a cue is not present for continuous presentation of alternating tone pulses.

A block diagram of the apparatus is presented in Fig. 2. Separate channels provide tone pulses in the two ears *via* attenuators and a program generator. The program generator consists of a modified record player. The disk rotates once every 3.2 seconds. On the disk a vertical perspex cylinder is mounted. On this cylinder slips of black adhesive tape interrupt narrow light beams between light sources and photodiodes, according to the desired program. Mercury-wetted relays, controlled by the photodiodes, switch the stimuli on and off.

As a function of the frequency of the stimulus in the left ear, the pitch of the

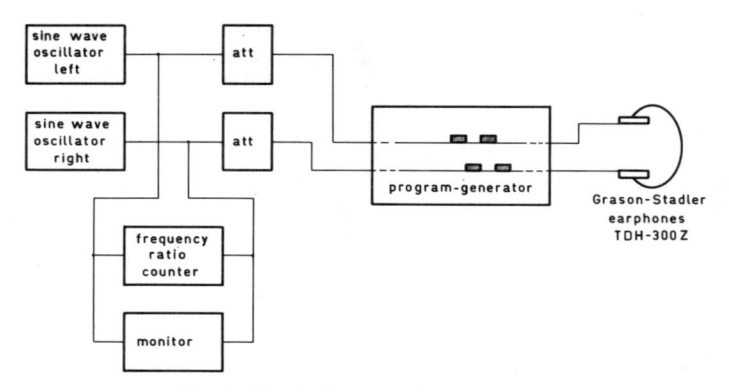

Fig. 2. Block diagram of the apparatus.

stimulus in the right ear is matched with the pitch in the left ear. The frequency ratio $\Delta v/v$ is presented in three decimals by the counter. The monitor is an additional visual aid, only used to check the signals, but switched off during the settings to avoid visual cues.

Experienced observers usually are able to reproduce their frequency settings within 0.1 to 0.2%.

DIPLACUSIS PATTERNS FOR DIFFERENT SUBJECTS

In Fig. 3 examples are given of diplacusis patterns of five subjects. For each of them the relative frequency difference $\Delta v/v$ is plotted against the frequency in the left ear. All curves show a rather irregular pattern with maxima and minima which usually do not exceed 1 to 2%. For different subjects the curves are entirely different. Below we shall discuss the possible origin of these strictly personal, fingerprint-like, patterns.

Repetition of an experiment within few days usually results in practically identic curves. There are, however, differences between results separated by a longer time. These differences concern primarily the height of the maxima, rather than the frequencies where they occur. Two maxima which have, for instance, about the same height on one day may have entirely different heights on another day. In the latter case, one maximum can be so pronounced, that the next maximum is only an irregularity along the slope of the first one; sometimes it is not detectable at all. However, if a maximum is present again, which usually is the case, it is located at a frequency that is remarkably constant. This is shown in Fig. 4, presenting four curves which are obtained over a period between January 1963 and April 1969. Maxima occur here at rather constant frequencies between 500 and 6000 Hz. Below 500 Hz the measurements do not reproduce, even not when they are repeated immediately; neither do they above 6000 Hz.

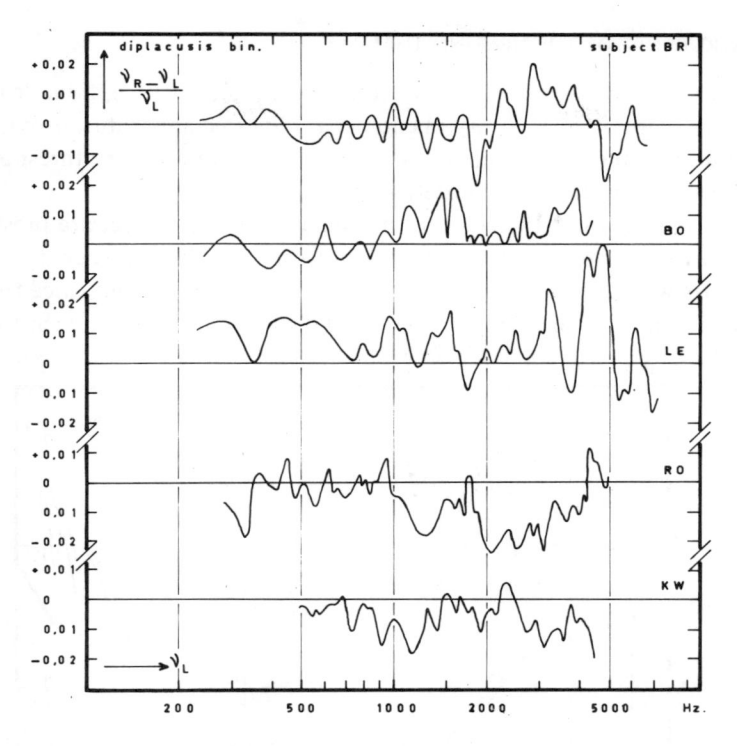

Fig. 3. Some examples of diplacusis patterns.

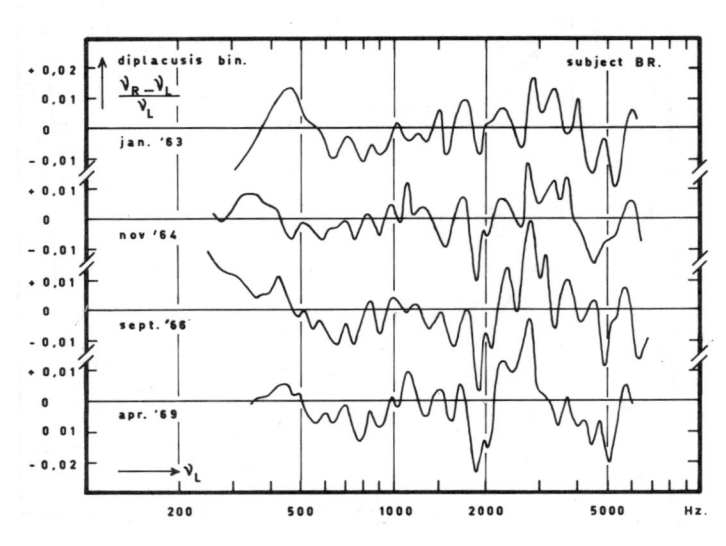

Fig. 4. Four diplacusis patterns of the same subject, measured with long intermissions in between.

FINE STRUCTURE OF THE HEARING THRESHOLD

In an effort to correlate the fine structure in pitch perception, leading to diplacusis, with other auditory phenomena, the same apparatus, provided with a continuous 10 dB attenuator, was used to measure auditory thresholds as a function of frequency as accurately as possible.

The results of these threshold measurements of one subject are presented in Fig. 5. In order to eliminate systematic errors, the measurements were done four times with different strategies, twice from low to high and twice from high to low: in one case of each pair the thresholds of both ears for each frequency

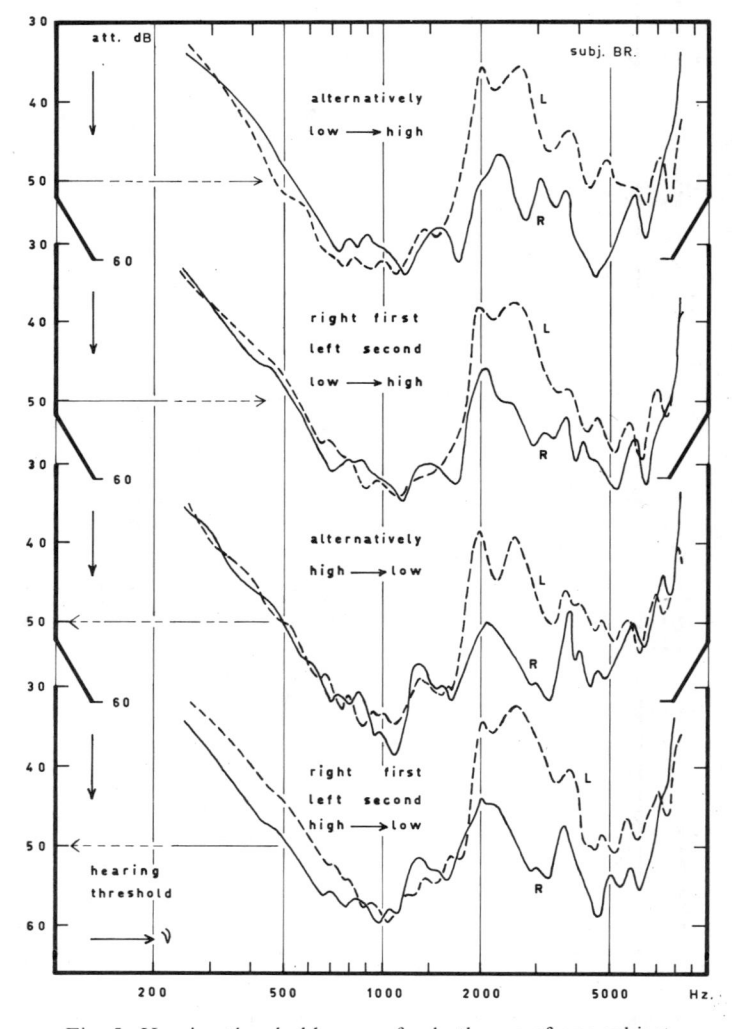

Fig. 5. Hearing threshold curves for both ears of one subject.

were determined immediately after each other; in the other case first the whole threshold curve for the right ear and then that for the left ear were determined.

There appears to be a satisfactory conformity between the four pairs of curves. Although the results do not reproduce as nicely as the pitch matchings in the diplacusis measurements, there is a good correspondence between the maxima and between the minima. Besides, there seems to be a certain correlation in the fine structures for the left and right ears.

Because the frequencies where maxima and minima occur are not very constant, conventional averaging of the data would blur out the fine structure. Therefore we determined for each of the maxima and minima its frequency and SPL. From these data we calculated the most probable frequency of occurrence

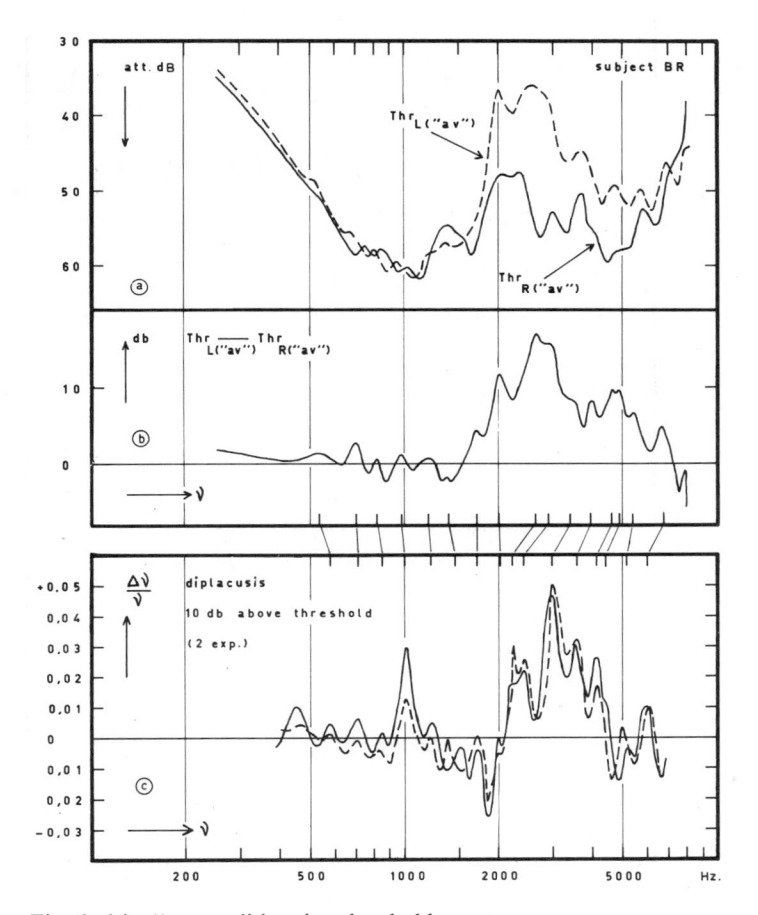

Fig. 6. (a): "average" hearing threshold curves.
(b): difference of the "averages".
(c): diplacusis at 10 dB above threshold (two measurements).

of such an extreme and its dB value. The most probable threshold curves obtained in this way are given in Fig. 6a, indicated with "av".

Since diplacusis is due to differences between the ears, we subtracted the values of the threshold in the right ear from those in the left. This difference is presented in Fig. 6b. Fig. 6c gives the results of a diplacusis measurement at levels which are 10 dB above the average threshold for every frequency.

On the lines between Fig. 6b and Fig. 6c the frequencies are indicated where maxima in the difference curve and maxima in the diplacusis curve occur. There seems to be a fair correspondence between these frequencies, indicated by thin

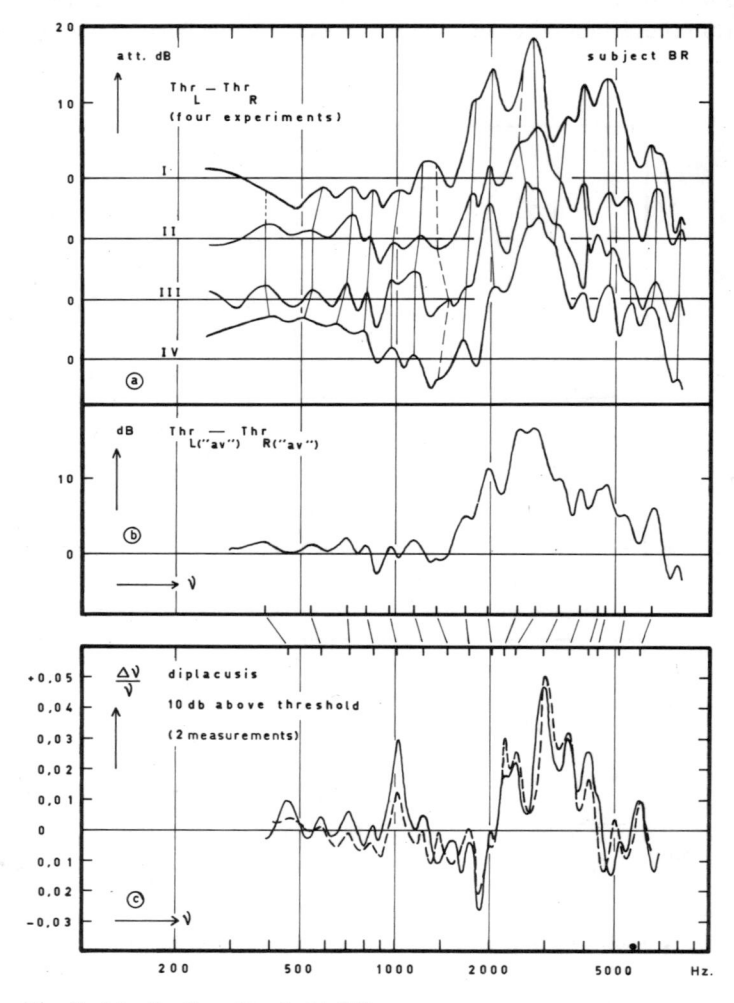

Fig. 7. (a): the four threshold difference curves.
(b): the "average" of the differences.
(c): diplacusis at 10 dB above threshold (two measurements).

lines, although there is a sudden jump just above 2000 Hz. The fact that this subject has a hearing loss, different in both ears and beginning at about 2000 Hz, may be the cause of this sudden jump; in the case of perception deafness the place of maximal excitation of the organ of Corti shifts for certain frequencies.

In Fig. 7 the same data are treated in a different way. Fig. 7a gives the four separate threshold difference curves and Fig. 7b the most probable difference. This procedure leads to a result which is nearly the same as the first procedure in which the differences were taken after averaging the four threshold curves for the left and the right ear.

Two other subjects repeated the experiment. Their results are presented in Figs. 8 and 9. Although it was not easy to identify the maxima in either of the curves, like we did for the first subject, there appears to be a correlation. For both subjects the number of maxima for each of the curves is much greater than for the first subject. One subject has 29 maxima between 500 and 5000 Hz in both curves and the other subject has 28 maxima in the threshold difference curve and 27 in the diplacusis curve over the same frequency range. For the first subject this number was 14.

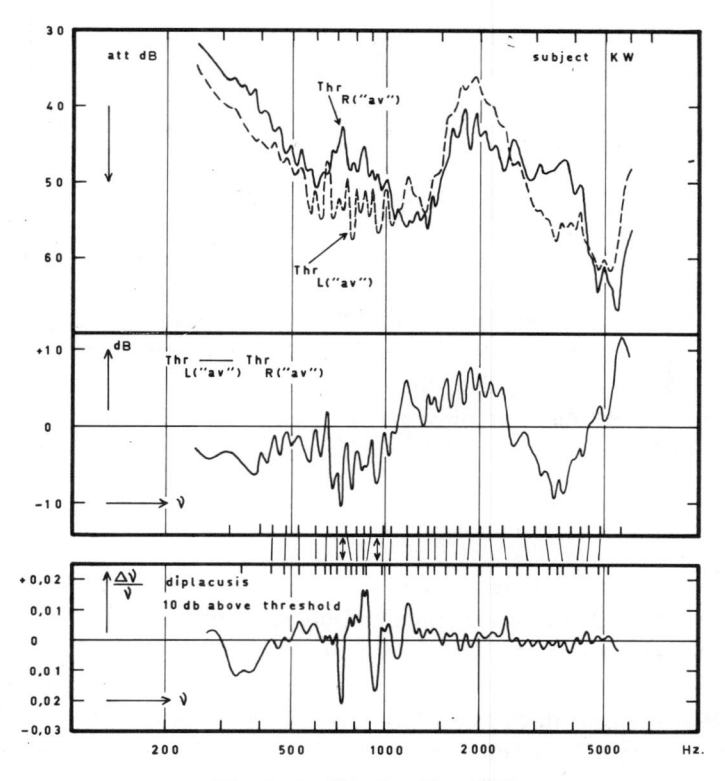

Fig. 8. As Fig. 6, subject KW.

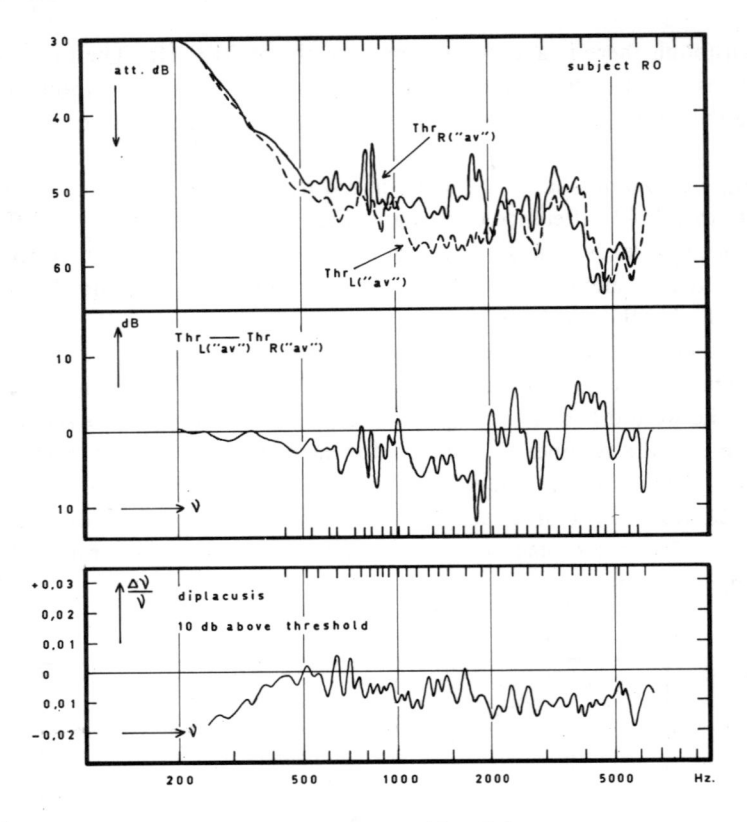

Fig. 9. As Fig. 6, subject RO.

Although we realize that these data are no proof, they are rather strong evidence that a correlation between the fine structure in the threshold difference curve and the diplacusis pattern does exist. Factors, other than the one that causes this fine structure, like, for example, perceptive hearing losses, may cause some deviations that prevent us from complete identification of the maxima in both curves.

FINE STRUCTURE IN THE ISOPHONES

The pattern of a diplacusis *vs.* frequency measurement appears to be nearly independent of the SPL of the stimuli. (Only with different SPL's in both ears evident changes occur, in such a way that the maxima are shifting along the frequency scale (van den Brink, 1966). This gives us more information concerning the relation between pitch and SPL; this relation appears to be much more complex than usually is supposed, because close frequencies sometimes show opposite "Broca-effects" with changing SPL.)

If, as we supposed initially, the fine structure in the hearing threshold curves represents small perceptive dips, recruitment should make them disappear at higher SPL's. However, since we became convinced that this pattern is related to the diplacusis pattern, either because one is the cause of the other or because both have the same cause, and furthermore diplacusis is approximately independent of the SPL, we must conclude that the fine structure in the threshold curve has nothing to do with perceptive deafness. Consequently, there should not be recruitment of the small dips in the threshold curves.

This has been checked by several subjects and an example of the results of these difficult measurements is shown in Fig. 10. The upper curve gives threshold as a function of frequency. The other curves are the results of equal loudness matchings with a 1000 Hz tone of 40, 60 and 80 dB SPL, respectively.

More data will be published in the future, but this figure shows already that, although there seems to be some recruitment at low SPL's for the extreme peaks, there is hardly any significant difference between the 40, 60 and 80 phone curves. They all show a comparable and highly correlated structure, not caused by small perceptive hearing losses, because there is no recruitment.

DISCUSSION

As far as pure tones are concerned we can conclude that binaural diplacusis

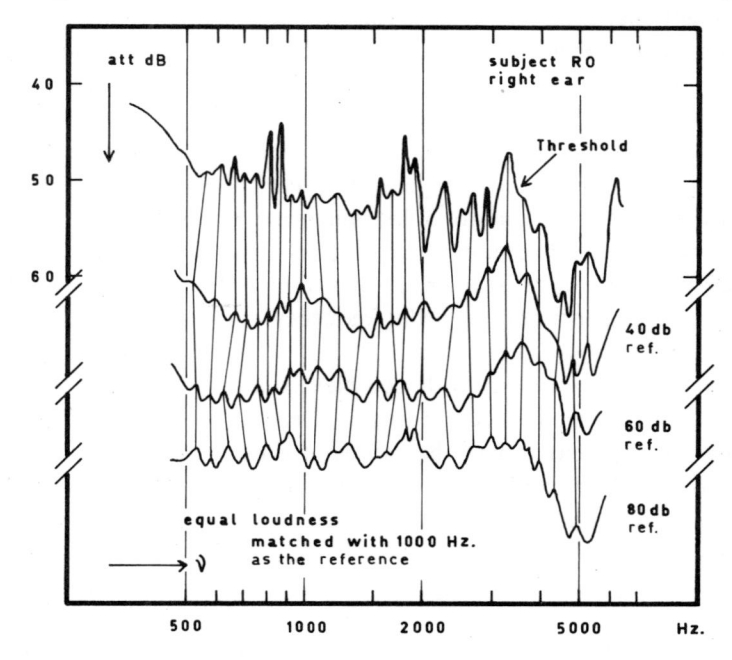

Fig. 10. Threshold and isophones obtained by matching with a 1000 Hz tone of 40, 60, and 80 dB SPL, respectively.

is due to different frequency-pitch relations in the two ears. Because there seems to be a correlation between the diplacusis patterns and the fine structure in hearing threshold and equal loudness curves, this fine structure cannot be caused by small perceptive hearing losses. The most obvious way to explain the phenomena is by supposing that their origin is located in the mechanical part of the middle or the inner ear. Our supposition, that the patterns are due to a fine structure in the mechanical properties of the basilar membrane is supported by a personal communication from Zwislocki, who built an electronic model of the cochlea and found that he needed 1 % tolerance components for a model that showed response irregularities of a few percents. This is of the same order as we actually found. It means that 1 % irregularities of the mechanical properties of the basilar membrane on top of the smooth course of its mechanical properties is enough to explain the fine structure of hearing threshold and equal loudness curves. Its origin cannot be explained yet. It is possible that the fine structure is the result of long lasting exposure to a large variety of sounds not loud enough to cause perceptive losses. The fact that there is, too, a correlation between the threshold patterns in the two ears (see Fig. 5) may support this supposition, although we realize that the patterns might be inborn as well, either centrally or peripherally. Experiments to verify this, by studying the influence of age and heredity, are planned.

If the phenomena are due to long lasting, though not trauma causing, sound causing slight changes of the mechanical properties of the tissue of the basilar membrane, one of the most likely sources of this sound might be our own voice. Efforts, however, to correlate the fine structure with the spectral composition of half an hour of speech did not yet give a definite answer, but will be continued.

Another aspect of these studies is pitch perception of complex sounds. A diplacusis pattern obtained with periodic pulses showed somewhat different results below 700 Hz, but did not differ much from the pure tone pattern above this frequency. An experiment in which the pitch of a residue signal was matched with a pure tone indicated that the pitch of a residue follows the fine structure of the pure tone diplacusis pattern in the frequency range of the components (van den Brink, 1965).

More detailed experiments with complex sounds are in preparation. If a confirmation of the mentioned experimental results with complex sounds will be found, a reconsideration of the causes of diplacusis, threshold and equal loudness irregularities will be necessary. In that case it will become more likely to think of more central processes.

REFERENCES

Brink, G. van den (1965): Pitch shift of the residue by masking, Internat. Audiol. *4*, 183-186.
Brink, G. van den (1966): Relation between intensity and pitch, Acta Physiol. Pharmacol. *14*, 50 (A).

DISCUSSION

Zwislocki: I have one little comment. Especially in the frequency range between 1000 and 3000 Hz, one would expect variations in the threshold of audibility because of the transmission of the middle ear. I suggest that your correlation between the diplacusis pattern and the threshold would have been even better if the middle ear did not contribute to these threshold variations.

Van den Brink: Well, this might be possible. With respect to this I can mention that von Békésy (personal communication) suggested that the whole fine structure in the diplacusis pattern as well as in the threshold might be caused by the transmission characteristic of the middle ear, but I am rather inclined to believe that it has its origin in the inner ear.

Zwislocki: I agree with you. My only point was that the middle ear plays some role.

Schügerl: What about diplacusis with complex tones? This would be very interesting in relation to the concept of periodicity pitch, since any irregularity in the ear can only affect the spatial distribution, but not the periodicity.

Van den Brink: I did some experiments on that. First, I did an experiment in which I used periodic pulses instead of pure tones. It was, after all, a rather silly experiment because pure tone as well as periodic pulse diplacusis measurements turn out to be reliable only for frequencies above 500 Hz, due to experimental errors. Periodic pulses gave the same result as pure tones. We did another experiment, however, (van den Brink, 1965) in which a carrier frequency was sinusoidally modulated; the carrier frequency was always a multiple of the modulation frequency so that we had a harmonic complex of three frequency components. Both frequencies were chosen in such a way that they coincided as much as possible with either three successive maxima or three successive minima of the diplacusis curve for pure tones. The residue pitch of the complex was matched with a pure tone in the other ear. Many different modulation frequency and carrier frequency settings showed that the diplacusis of the residue correlates with the diplacusis of the frequency components, not with the diplacusis of the modulation frequency. So, the diplacusis found in this experiment must be due to the residue signal itself. (Recent experiments, to be published in J. Acoust. Soc. Amer., confirmed these findings convincingly.)

Scharf: Did you also measure the diplacusis between a pure tone presented monaurally and a tone presented binaurally?

Van den Brink: I did not actually measure this. In this case the binaural pitch lies in between the pitches perceived when the signal is presented to the left and the right ear separately.

Scharf: Did you notice any difference in pitch quality when a pure tone is presented monaurally and binaurally? I have not much experience but my impression is that the pitch of a binaural tone is a little bit purer than that of a monaural tone. Did you find a similar difference?

Van den Brink: I agree that the tones sound differently but I am not able to describe the difference.

REFERENCES

Brink, G. van den (1965): Pitch shift of the residue by masking, Internat. Audiol. *4*, 183-186.

Section 7

FREQUENCY ANALYSIS AND MASKING

The frequency-analyzing power of the ear becomes manifest in our ability to distinguish among simultaneous tones. A measure of the selectivity of this analysis is the threshold elevation, or masking, of one tone brought about by the introduction of another tone. Above threshold the filtering process becomes apparent in the dependence of timbre upon frequency spectrum, the interference of close frequency components, and loudness summation. These are the topics discussed in the following papers.

MASKING AND PSYCHOLOGICAL EXCITATION AS CONSEQUENCES OF THE EAR'S FREQUENCY ANALYSIS

E. ZWICKER

Institut für Elektroakustik
Technische Hochschule München
München, Germany

INTRODUCTION

So far there is no possibility to pursue the transformation of a physically measurable sound by physiological measurements up to the final sensation. Only the first steps, the transformations to the ear drum, through the middle ear, and into the inner ear, are known physiologically. Even though the configuration in the inner ear became clear with the work of von Békésy (1942), Zwislocki (1948) and Ranke (1950) in terms of the displacement of the basilar membrane, the question of the actual stimulation value of the sensory cells remains unsolved. Some decades of scientific work may be necessary to understand the hearing process in terms of physiological equations and functions. In the meantime we have to content ourselves more or less with the results of psychoacoustical measurements, and it may be allowed to discuss these results a little more hypothetically than I am normally used to and somewhat beyond the point up to which we have a solid scientific foundation. This may stimulate the discussion, too.

THE APPROXIMATION OF THE FREQUENCY SELECTIVITY

Simplifying, von Helmholtz (1863) can be considered as the scientist who introduced the transformation of stimulus frequencies to specific places along the basilar membrane. In addition to that important relation von Békésy (1943) was able to show that a single tone produces displacements of the basilar membrane not only at a certain point or small range but over a wide area of the inner ear, with steeper slope toward the helicotrema and flatter slope toward the oval window. He also pointed out (von Békésy, 1953) that different parts of the organ of Corti show different directions of movement, suggesting the possibility of a somewhat smaller range of stimulation of the hair cells than the range of displacement of the basilar membrane. Nevertheless, the fact remains that a certain range in the cochlea responds to a single frequency, resulting in what is called the hydromechanical frequency selectivity of the ear. Whether and how much this selectivity is sharpened by neural processes, we don't know physiologically. Psychoacoustical data may be helpful in this situation.

Experiments on the hearing threshold of complex sounds by Gässler (1954), on two-tone masking (Zwicker, 1954) and on phase sensitivity (amplitude modulation *vs.* frequency modulation; Zwicker, 1952) at threshold of modulation have led to the first approximation of the frequency selectivity of the ear in a psychoacoustical measure: the critical band (Zwicker *et al.*, 1957; Zwicker, 1961). It may be understood as a band filter with infinitely steep slopes and a frequency-dependent width Δf_G (see Fig. 7). Adding the successive critical bands up to the frequency f, the relation between the physical frequency scale and one that corresponds more closely to psychological results can be calculated. It is the critical band scale, plotted in Fig. 1 as critical band rate z in Bark with linear and logarithmic frequency scales as abscissa. The just-noticeable difference (JND) in frequency, the location of the maximum of the displacement along the basilar membrane and the mel scale of pitch to frequency result in similar functions (Zwicker, 1956). On the z-scale or along the basilar membrane, the critical band may be understood as the range over which the ear is able to integrate intensity for the formation of threshold.

The critical band represents only the first approximation of the ear's ability to separate sounds of different frequency. Since the first step should be under-

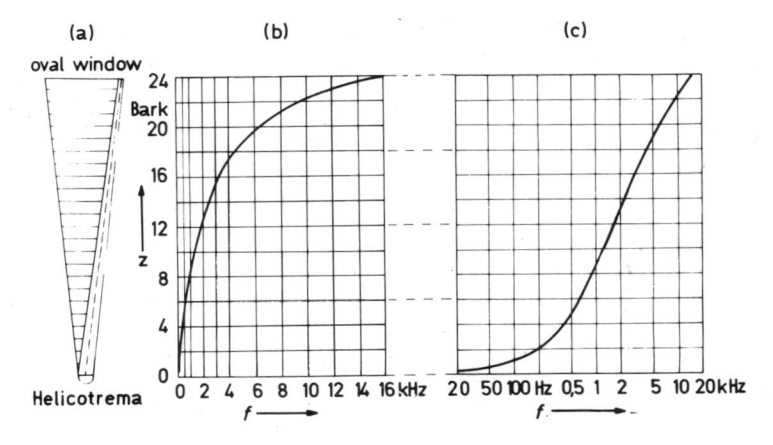

Fig. 1. Relation between critical band rate z and frequency on linear (b) and logarithmic scale (c). (a): Schematic drawing of the cochlear partition in the inner ear with bone (scratched), basilar membrane (white) and organ of Corti (dashed).

stood before the second is taken, I have to go a little more into detail. If we assume filters with rectangular shape of attenuation, continuously shiftable along the frequency scale, we can easily calculate the intensity within each critical band. This intensity will be called "psychoacoustical incitation". A broad-band noise with the same intensity in each critical band is therefore a uniform inciting noise (slightly different from uniform masking noise). Since the width of the critical band is known as function of frequency, the incitation A can be calculated (Zwicker and Feldtkeller, 1967, Ch. IX) for any sound using the intensity density dI/df in the equation

$$A(f_v) = \int_{f=f_v-\frac{1}{2}\Delta f_G(f_v)}^{f=f_v+\frac{1}{2}\Delta f_G(f_v)} \frac{dI}{df} df. \tag{1}$$

Since critical band rate z depends on frequency (Fig. 1), this equation can also be written as

$$A(z_v) = \int_{z=z_v-\frac{1}{2}\text{Bark}}^{z=z_v+\frac{1}{2}\text{Bark}} \frac{dI}{dz} dz. \tag{2}$$

As $A(z)$ is related to $A_o = I_o = 10^{-12}\,\text{W/m}^2$, we can express the incitation in the incitation factor $A(z)/A_o$. The logarithm of the incitation factor is called the incitation level

$$L_A = 10 \lg \frac{A(z)}{A_o} \text{ dB}. \tag{3}$$

In the next step or, better to say, in the second approximation of the frequency selectivity of the ear, the slopes of the "filters" have to be taken into account, since we can be sure that an infinitely steep filter slope does not exist. The frequency selectivity of the ear can be measured by masking pure tones by narrow bands of noise, not only over a large frequency range but also for a wide range of levels. Such masking curves don't have rectangular characteristics. They have the form of trapezoids with steep slopes to lower frequencies and flat slopes to higher frequencies. The threshold is raised even at frequencies well above the range in which the narrow-band noise is effective physically. To describe this, in relation to the incitation, weaker selectivity the "psychoacoustical excitation" is introduced. The term "psychoacoustical excitation" is shortened in this paper to "excitation" only, whereby always "psychoacoustical excitation" is meant, even in "excitation factor" and "excitation level". As I am, in this paper, more interested in the coherence of the whole concept than in the details, it may be allowed to neglect the attenuation factor a_o

representing the sound transformation from free field through ear canal and middle ear to oval window. In that case, the excitation factor is $E(z)/E_o$ and the excitation level becomes

$$L_E = 10 \lg \frac{E(z)}{E_o} \text{ dB}. \quad (4)$$

The excitation level at the critical band rate z_c which corresponds to the centre frequency f_c of the narrow-band noise is equal to the incitation level at z_c and will be called main excitation level. The masked threshold reaches its peak there. Furthermore, in general, the excitation level is equal to the incitation level for *that* range of critical band rate in which the masked threshold is defined by the physically existing spectral density of the sound. Outside of this range there is accessory excitation, the level of which gradually drops down to threshold in quiet. The accessory excitation is obtained by shifting the masked threshold thus far that main excitation and accessory excitation pass into each other without step. Fig. 2 shows the spectrum density level, the incitation level, the masked threshold and the excitation level of the narrow-band noise as a function of the critical band rate z. In Fig. 3 the derivation of the excitation pattern is demonstrated, for narrow-band and for white noise (a) as well as for a sound consisting out of eleven harmonics (b). The excitation pattern for tones is construed under the assumption that a one critical band wide noise and a tone of centre frequency produce the same excitation if their sound pressure levels (SPL) are equal.

For different frequencies the ear shows almost the same selectivity if frequency scale is transferred to critical band scale. In Fig. 4 the excitation for critical-band wide noises of different centre frequencies but equal SPL (60 dB) is plotted on the z-scale. The curves look very similar and may be derived from

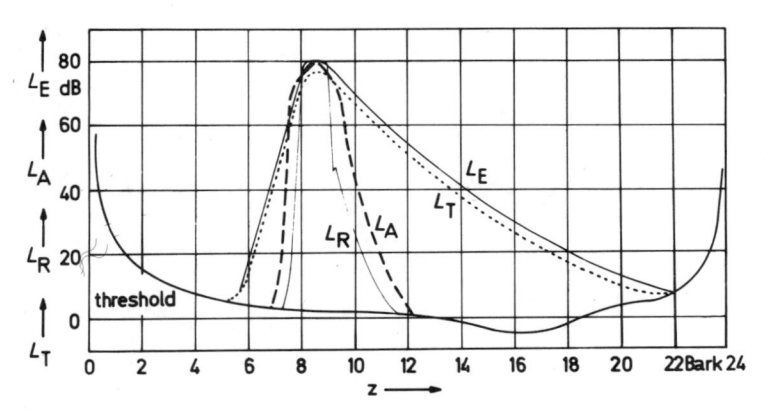

Fig. 2. Spectrum density level L_R of a one critical band wide noise, incitation level L_A of that noise, threshold level L_T of sinusoidal tones masked by that noise and excitation level L_E as a function of critical band rate z (excitation pattern).

380 E. ZWICKER

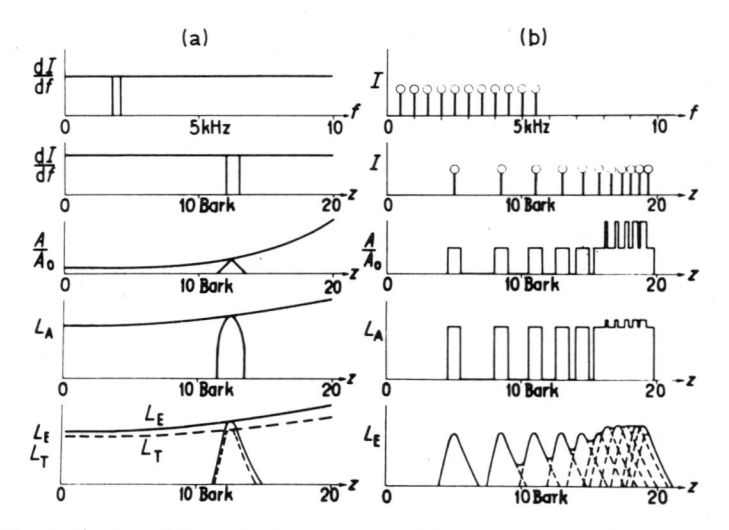

Fig. 3. The derivation of the excitation pattern; (a) for narrow band noise and white noise; (b) for 11 harmonics (L_A, L_T, L_E see Fig. 2).

each other by shifting them up and down the z-scale.

The ear's selectivity is somewhat nonlinear. Increasing the level of the masking noise, the main excitation and the lower (in frequency or critical band rate) accessory excitation grow to the same extent, whereas the slope of the upper accessory excitation decreases. This nonlinearity of the excitation pattern is illustrated in Fig. 5. Factually, this effect is the reason why the selectivity is presented in graphical form rather than by equations.

The incitation patterns behave linear. But the next approximation, the excitation, contains this nonlinear effect and describes therefore completely the

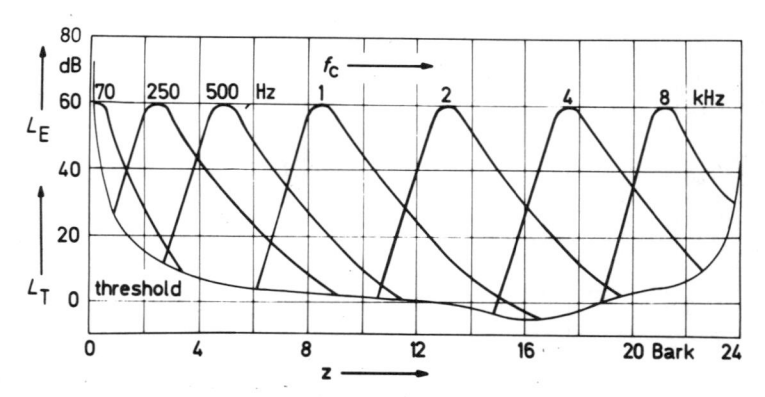

Fig. 4. Threshold level L_T in quiet and excitation level L_E for 7 narrow bands of noise, one critical band wide, with different centre frequencies f_c but equal SPL of 60 dB, as a function of critical band rate z.

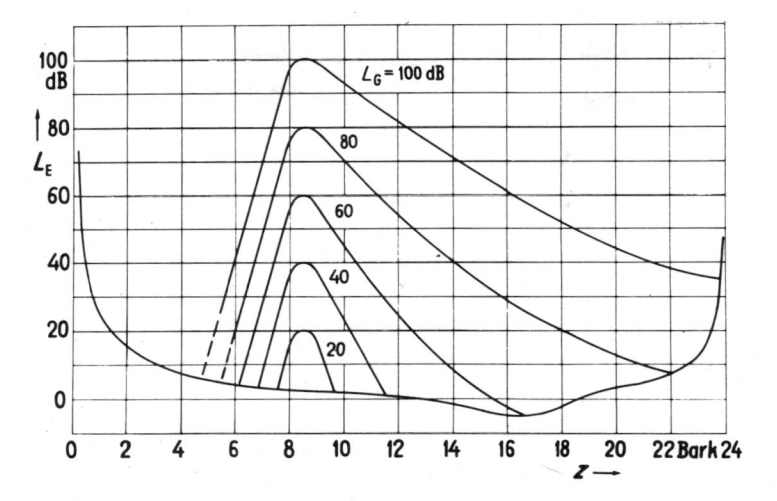

Fig. 5. Excitation level L_E for a one critical band wide noise, centred at 1 kHz, with different SPL's L_G, as a function of critical band rate z.

selectivity for the steady state condition as long as it can be taken from masking data. On the other hand, not only excitation along the critical band scale is of interest. As soon as an overlapping of different patterns occurs as, for example, in Fig. 3b, the time pattern of the relative excitation $E(t)/E_o$ may become important, too. Actually, the time-critical band rate pattern of the relative excitation $E(z,t)/E_o$ is the value we would like to know. So far we don't know whether E is correlated with intensity, sound pressure, velocity, displacement or force. Therefore, what I shall discuss about the time function $E(t)/E_o$ is not yet a fully proofed hypothesis, although it produces very interesting results.

APPLICATIONS OF THE EXCITATION MODEL

Just-noticeable differences (JND)

The assumption (Zwicker, 1956) that the ear is able to detect any change of a steady sound only if the excitation level $L_E(z)$ is changed anywhere along the critical band scale by $\Delta L_E(z_v) \geq 1$ dB, turned out to be very effective for the interpretation of JND's (see also Goldstein, 1967). Maiwald (1967a, 1967b, 1967c) extended the model even for fluctuating sounds like band-pass noise, and found good agreement between results measured on the model and results from psychoacoustical measurements. Although this extension is very interesting, let us, for didactical reasons, first deal with steady state sinusoidal tones, subjected to slow changes in amplitude or frequency.

The JND for amplitude, or better the just noticeable degree of slow (4 Hz) amplitude modulation (AM) of a sinusoidal tone, can be understood only by taking into account the nonlinear growth of the upper accessory excitation. $\Delta L = 1$ dB corresponds to a modulation depth m of 6%. This value is found psychoacoustically only over a level range from 30 dB to about 50 dB SPL. At higher levels, smaller m-values are already noticeable, at 100 dB almost 1%. In Fig. 6a the AM situation is shown schematically. The nonlinearity is responsible for the fact that the upper accessory excitation increases much more than the main excitation if the amplitude of the tone is increased. In this drawing the relation may be only a factor of 2; but the individual curves, often considerably different from each other, show points along the z-scale where the increment of the accessory excitation is 5 to 10 times larger than the increment of the main excitation. In such a case the just-noticeable degree of AM becomes 5 to 10 times smaller and drops down to about 1% at high levels. The influence of the upper accessory excitation can be checked by adding a high-pass filtered noise to the modulated tone, to mask out this part of the excitation. Doing so (Zwicker, 1956), the JND increases to almost 6%, because the main excitation has to be changed by 1 dB to produce an audible effect.

The JND for amplitude is practically independent of the level if band-pass noise is used instead of tones. Here Maiwald's (1967b) extension can be used, because the noise contains already a very well audible statistical amplitude modulation. The statistical modulation as well as the additional 4 Hz amplitude modulation are both subject to the nonlinearity of the accessory excitation. Because the relative change of the total degree of modulation is responsible for threshold of modulation, no dependence on level can be expected.

The JND for frequency (JND_f) has to be interpreted in a similar way although the configuration (see Fig. 6b) is different. During frequency modulation the excitation pattern is shifted forth and back along the z-scale. No nonlinearity is involved. The largest change of excitation level during the shift occurs at the

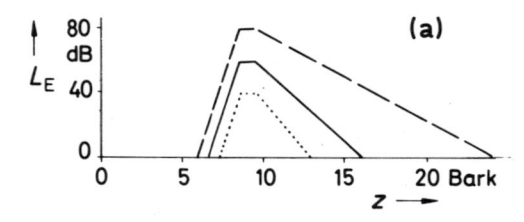

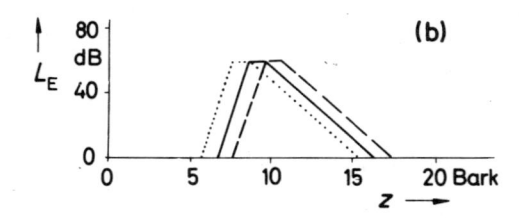

Fig. 6. Schematic drawing of the variation of the excitation pattern for JND's in amplitude (a) and for JND's in frequency (b).

lower accessory excitation with the steep slope. Since the steepness of this slope is almost independent of both the level and the frequency of the exciting tone and has, according to Maiwald (1967a), an average value of 27 dB/Bark, the JND_f should depend on frequency in the same way as the corresponding value of 1 Bark, namely Δf_G, depends on frequency. Using this relation and $\Delta L = \Delta L_E = 1$ dB, it can be calculated that

$$JND_f = 1 \text{ dB} \frac{1 \text{ Bark}}{27 \text{ dB}} = \frac{1}{27} \Delta f_G. \tag{5}$$

How well this relation fits the psychoacoustical results is shown in Fig. 7. The course of the two curves is very similar, Δf_G and JND_f grow almost parallel as a function of frequency. This means, because both scales are logarithmic, that the curves are correlated through a frequency independent factor. This factor is 30 in Fig. 7, in good agreement to the predicted value of 27.

An interesting effect can be obtained if a sound is modulated with the same modulation frequency (4 Hz) both in amplitude and in frequency. Then the phase relation between these two modulations determines the shift of excitation pattern (Zwicker, 1962a). For simplification, the main excitation may be represented by a point whose motion in the L_E-z plane is plotted in Fig. 8b for two special configurations of the mixed modulation (MM): (1) an increment in amplitude coincides with a decrement in frequency or (2) an increment in amplitude coincides with an increment in frequency. As seen in Fig. 8a, the second configuration keeps the lower accessory excitation almost stationary, whereas the upper one moves strongly. Using a high-pass noise, the latter can be masked so that for the most part the lower accessory excitation remains

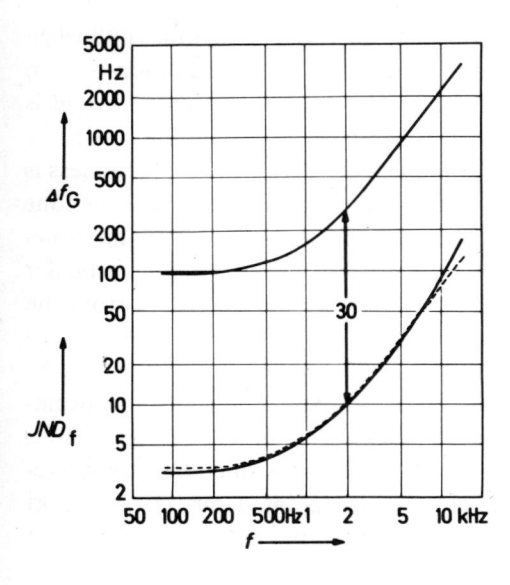

Fig. 7. The width of the critical band in relation to the JND_f (lower curve) as function of frequency. The dashed curve is constructed by shifting the upper curve down for a factor of 30.

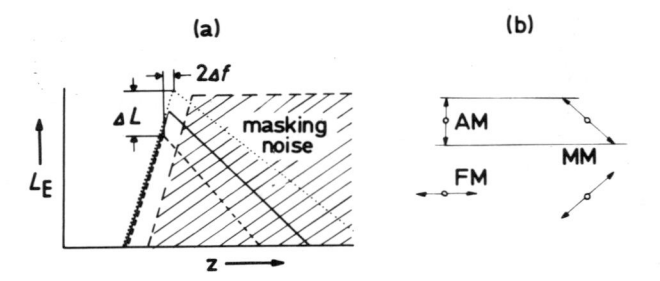

Fig. 8. The mixed modulation MM composed of simultaneous amplitude modulation (AM) and frequency modulation (FM) produces in the condition shown right down of (b) an excitation pattern (a), moving between the dashed and the dotted line. Using a high-pass filtered noise, the upper accessory excitation can be masked, resulting in a considerable decrease of the modulation sensation. Changing the phase of MM to the condition shown right up of (b) produces a considerable increase of the modulation sensation.

responsible for the audibility of modulation. The hypothesis for the JND's has been checked by Maiwald (1967c) by changing the phase of one of the modulations over 180°. Then modulation should be heard much stronger. Another possibility is to modulate an amplitude modulated sound additionally in frequency with appropriate phase, so that the audibility of modulation is weakened even by adding modulation. Both experiments show the predicted effect very clearly.

Loudness

During the last twenty years loudness became a more important measure, expecially from the technical point of view. According to this some basic studies (Zwicker, 1958) have been performed and a procedure for calculating loudness (Zwicker, 1960) was developed. In the model of loudness summation, published in co-operation with Scharf (Zwicker and Scharf, 1965), the excitation pattern (the excitation as function of the critical band rate) produced by the sound is again the fundamental measure. As shown in Fig. 9, the excitation has to be translated into the specific loudness N' (loudness density). The total loudness is found by integrating N' over z. This is demonstrated in Fig. 9a for a single tone and in Fig. 9b for white noise. In both parts of Fig. 9 the second drawing shows the displacement of the basilar membrane corresponding to the particular stimulus. It may be pointed out that, as a function of the distance x from the helicotrema, the displacement ξ is much less selective than the relative excitation E/E_0 (both linear scales). The transformation of the excitation to the specific loudness seems to decrease the selectivity of the ear. Any nonlinear transformation by power functions will give this decrease, as long as the exponent is smaller than 1. Since the exponent for specific loudness is only 0.25, the decrement of selectivity becomes effective, as was also pointed out by Zwislocki (1968). As there have been written several papers on models for loudness

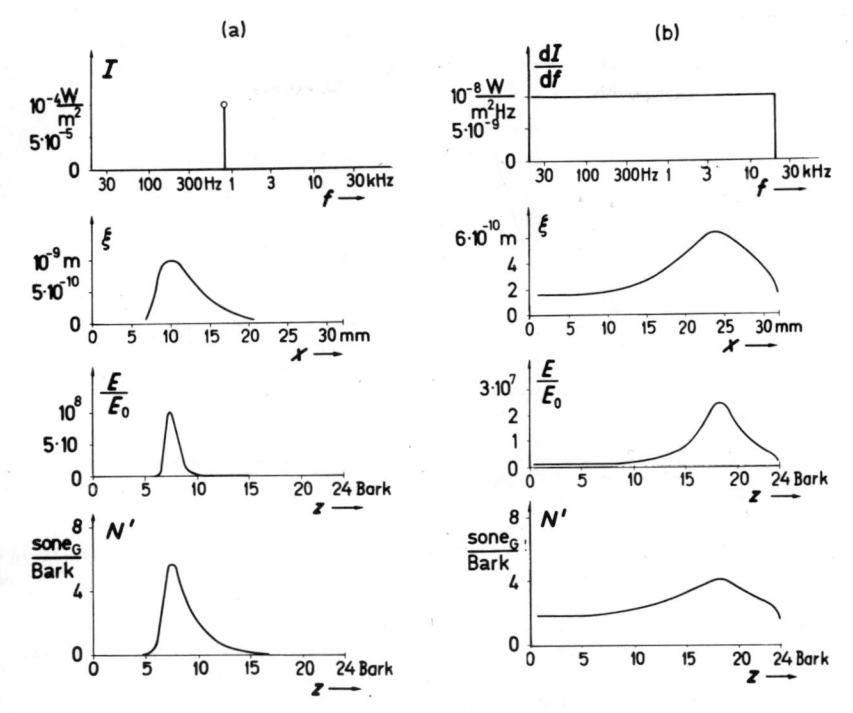

Fig. 9. Derivation of the loudness pattern for a single tone (a) and for white noise (b). Intensity I and density dI/df, respectively, as functions of frequency, corresponding displacement ξ of the basilar membrane, relative excitation E/E_0, as a function of critical band rate z, and specific loudness N' as a function of z. The attenuation factor a_0, representing the frequency dependent transformation of the middle ear, is taken into account.

formation (Zwicker, 1958; Zwicker and Scharf, 1965; Zwicker and Feldtkeller, 1967) the short summary just given may be sufficient.

The cubic difference tone

Nonlinear distortions of the ear can be measured by using two primary tones A and B with frequencies f_A and f_B ($2f_A > f_B > f_A$). The amplitude of the cubic difference tone at the frequency $2f_A - f_B$ decreases with increasing frequency distance $f_B - f_A$ in different manner, depending on the amplitude relation of the primary tones. Details of the amplitude configuration necessary to get representative results may be taken from the original publications (Zwicker, 1968; Helle, 1969/70). The threshold for sinusoidal tones masked by narrow-band noise, measured by two subjects over the same frequency range as the primary tones, allow a comparison between the inclination of the thresholds and the inclination of the accessory excitations constructed out of the levels L_{D3B} of the cubic difference tone at the frequency $2f_A - f_B$. Both results are plotted in Fig. 10 as function of the corresponding critical band rate and are in good accordance

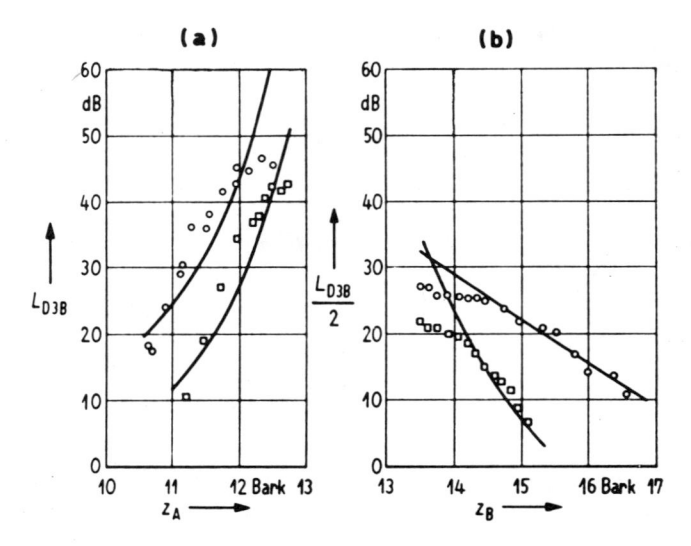

Fig. 10. Results of measurements of the level L_{D3B} of the cubic difference tone $2f_A - f_B$ as function of the critical band rate z_A and z_B, respectively, of the two primary tones. They are plotted in such a way that a comparison to the lower accessory excitation (a) and the upper accessory excitation (b) composed out of threshold measurements is possible. The solid lines correspond to the vertically shifted slopes of the thresholds of pure tones masked by narrow bands of noise with centre frequencies and levels corresponding to those of the primary tones. The threshold measurements and those of the cubic difference tone are produced by the same individual observer.

with each other. Therefore it may be concluded that excitation along the z-scale is an interesting intermediate stage along the way of sound from stimulation to sensation, important for the creation of the cubic difference tone, too. The possibility to compensate the latter leads to the conclusion that it exists already within the hydromechanical part of the inner ear, before transformation into neural processes takes place. Because the creation depends strongly on the excitation configuration of the primary tones, it is likely that excitation and cubic nonlinearity result from hydromechanics without co-operation of neural processes as inhibition.

EXCITATION-CRITICAL BAND RATE STRUCTURE: SHARP HYDROMECHANICAL FILTER OR SPATIAL INHIBITION?

Several years ago we had in mind to imitate the high selectivity of the ear, found in masking experiments, by the introduction of inhibition. The first simple model (Zwicker, 1962b), a bank of filters (realized in a recurrent ladder network with low-pass (LP) and high-pass (HP) filters) with relative flat characteristics, was built to copy the selectivity of the basilar membrane. In order to increase the selectivity, an inhibitory network was inserted, following a logarithmic transformation (LO). A strong signal in a channel influenced the first, the second, the fourth and eighth subsequent channels by reducing their gain. Thereby the difference in the amplitudes of these channels is used to adjust the servo amplifiers (SA) as indicated in Fig. 11. The expected effect was present, but only for sinusoids or narrow-band noises as input signals. If a second sinusoid at a distance of two or more critical bands was added, the additional signal was depressed completely up to a relatively high level. Rising its level even more, the original one was inhibited. Masking curves could therefore not be demonstrated. Other effects like partial masking or masking of narrow-band noise by broad-band noise could not be reproduced, either. Therefore, the system was modified and other kinds of inhibition-like forward or/and backward influence were introduced. But no inhibition system could be found which was not only relatively simple but stable too, and which, on the other hand, could imitate the effects described in the sections above. Therefore, the filter bank was changed, as published by Terhardt (1966), in such a way, that the filter characteristics (level over number of channel) corresponded to the masking curves shown in Fig. 4. Using this system of sharp filters and adding an amplitude detector, all JND effects could be reproduced on the model.

Another effect, presented in Fig. 12, may point out again the difference between the displacement along the basilar membrane in connection with inhibition on one side, and masking or excitation pattern on the other side. A narrow band of noise (NBN), one critical band wide, and a white noise (WN) having the same density level are used for both masking of pure tones and

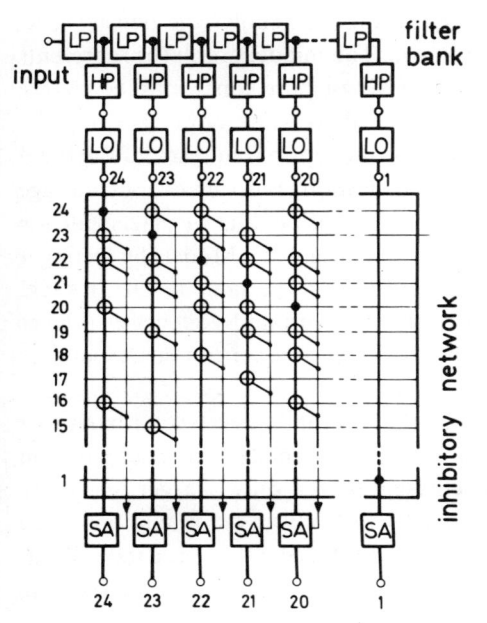

Fig. 11. Block diagram of a model of the ear containing a filter bank corresponding to the hydromechanical frequency selectivity, a nonlinear transformation corresponding to the loudness function, and an inhibitory network.

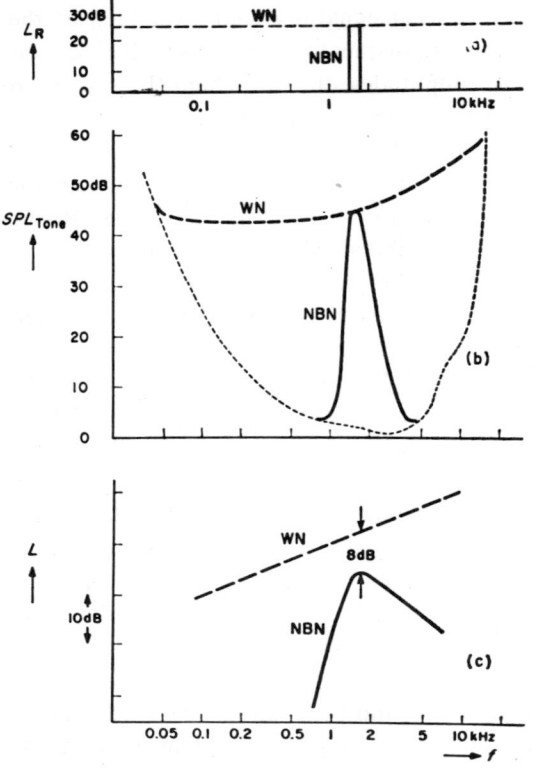

Fig. 12. (a): Density level as a function of frequency for white noise (WN) and narrow-band noise (NBN) one critical band wide.
(b): Threshold in quiet (dotted) and thresholds masked by the two noises shown in (a).
(c): Level of the displacement of basilar membrane for the two noises shown in (a) measured on models as a function of frequency.

measurements on basilar membrane models. Fig. 12a shows the density level for the narrow and for the wide band of noise. The middle graph (b) represents the psychoacoustically measured threshold level of pure tones masked by the two noises. If the narrow-band noise is widened to white noise, keeping the density level constant, the masked threshold at the centre frequency of the band remains also constant. If, under the same condition, the rms output voltage (arbitrary level) of a basilar membrane model — representing displacement — is measured, it increases 7 or 8 dB at that frequency (Fig. 12c). Models by Oetinger and Hauser (1961) and by Flanagan (1962) produced the same result. In these two models each of the many outputs is related to a certain frequency that gives maximum voltage for that particular output. These frequencies correspond to the abscissa in Fig. 12c.

Comparing the effects, it becomes clear that an inhibitory system has to be rather complicated in order to produce out of the displacement configuration a masking or excitation configuration representing not only a larger selectivity but also compensating for the difference of 7 to 8 dB at the maximum level. This behaviour of the hearing mechanism is discussed more in detail in an earlier paper (Zwicker, 1965); the conclusion was that stimulation of the hair cells may be preceded by a strong filtering process in which displacement is only one of the steps. This holds until today; all results of experiments carried out carefully can be understood in terms of this conclusion.

EXCITATION-TIME STRUCTURE

It was found that excitation along the critical band scale plays an important role in the hearing process for tones and noises with random amplitude distribution or, in other words, for steady state or quasi-steady state conditions. We do not know the physical quantity to which excitation is related. So far the sound level, which contains all information for steady state conditions, was mostly used in the excitation-critical band rate pattern $L_E(z)$. For sounds with special time structure this information is not complete. The time structure of the excitation at all places along the basilar membrane — which means along the critical band scale — may become important in this case. In order to incorporate the time structure, we have to decide whether excitation is related to intensity or pressure (velocity). Intensity correlation would require square law transformation and integration over a duration of at least several msec. By this, the time function of the quantity we like to know is changed drastically and in a way which is not observed experimentally (doubling of the frequency). Therefore, again as a very first approximation, the output voltages of a filter bank (see Fig. 11) modelling the excitation pattern $L_E(z)$ (as a pressure related value) have been studied as time functions.

As an example of a sound with specific time structure a sequence of short pulses with repetition rate $f_P = 300$ Hz was presented to a low-pass filter with a cutoff frequency of 2.5 kHz. The output of this filter was used as input of the excitation filter bank. This input is shown in the upper left oscillogram in Fig. 13 and may be understood as the ear's stimulus. The excitation as a function of time is shown for different output channels (indicated by the numbers corresponding to the critical band rate in Bark). There are 24 output channels, each of them representing the width of about one critical band. Channels with low numbers represent places near the helicotrema, those with high numbers places near the oval window (see Fig. 1). The two rows of time functions should be read as a single one. The amplitudes of the output voltages of the filter bands can be related to each other. The amplification factor of the oscilloscope was kept constant, so the ordinate is a value equivalent to a relative excitation $E(z_v = \text{const.}, t)/E_o$. The different pictures show clearly that the frequency selectivity is sufficient to separate the harmonic components of this stimulus only at places near the helicotrema (expressed in the space domain), *i.e.* at low frequencies (expressed in the frequency domain). At the channels 11 to 16 time functions with a structure related to the repetition rate are observed. This time structure may be responsible for the creation of periodicity pitch.

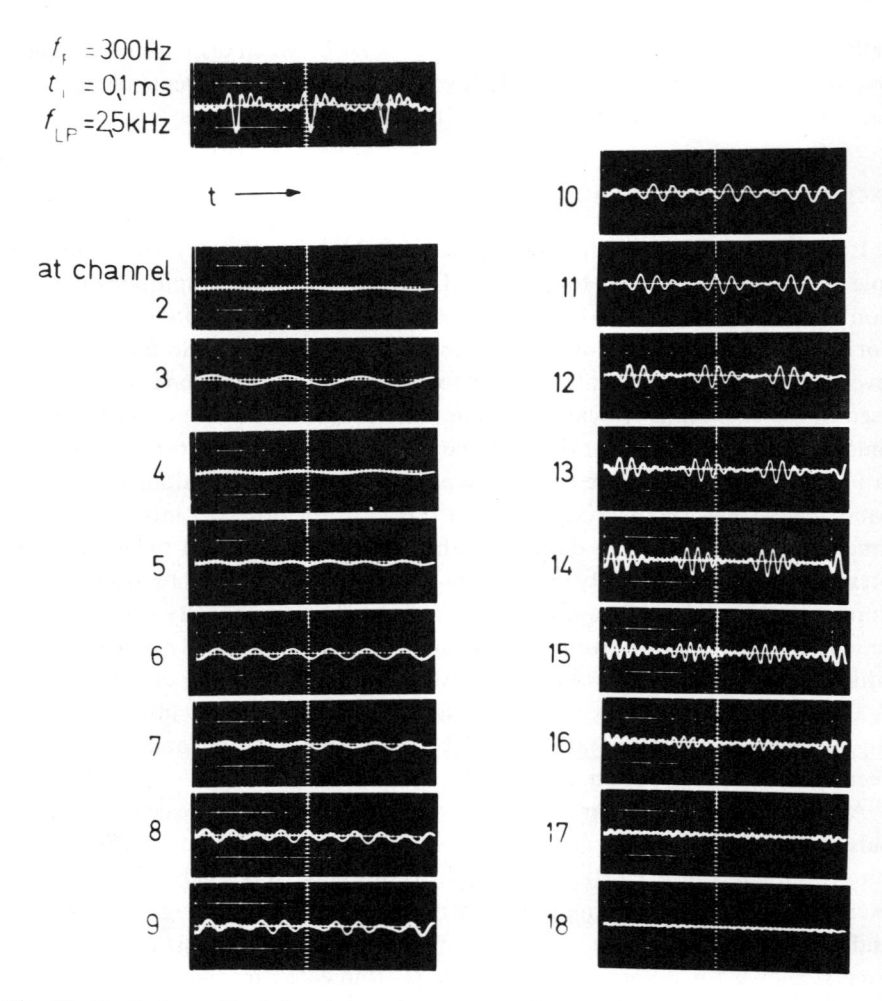

Fig. 13. Excitation-critical band rate-time pattern for a sequence of pulses low-pass filtered up to 2.5 kHz, shown in the upper left oscillogram. Excitation is measured, as a function of time, at the filter outputs of a model consisting of a bank of adjacent filters, each one critical band wide. The given numbers of the channels correspond to the critical band rate.

Two more examples of the relative excitation at different channels (places) as a function of time are added for discussion. Fig. 14 represents amplitude modulated tones. Again the input signals are shown separately in the upper oscillograms. While for $f_{mod} = 100$ Hz all channels show almost the same modulated time function, for $f_{mod} = 300$ Hz analysis of the spectrum is starting and for $f_{mod} = 525$ Hz it is so strong that almost a sinusoidal signal is found in channel 11. Such excitation-time patterns may be useful to interpret periodicity pitch and roughness.

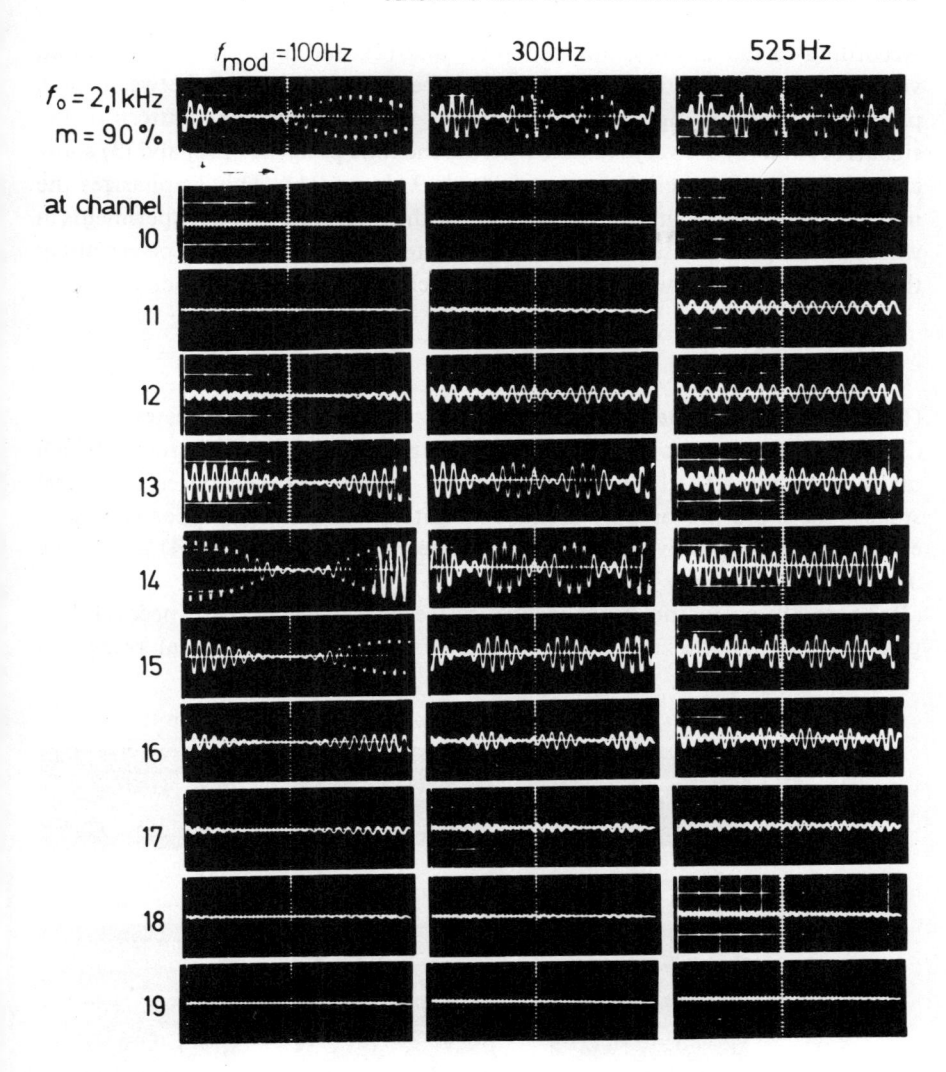

Fig. 14. Excitation-critical band rate-time pattern for a 2.1 kHz tone, modulated in amplitude with modulation frequencies of 100 Hz, 300 Hz and 525 Hz, respectively, shown in the upper 3 pictures. The excitation-time patterns show very similar configurations in the different channels for the low modulation frequency while for the high modulation frequency the configurations look different in almost each channel. At places of the lower accessory excitation the time function is almost sinusoidal, corresponding to the lower component of the stimulus.

The other example is related to measurements of Plomp and Steeneken (1969) on the influence of phase changes on timbre. They used complex sounds consisting of 10 harmonics with different successive phases: (1) cos, cos, cos, ...; (2) sin, sin, sin, ...; (3) sin, cos, sin, cos, ...; (4) cos, sin, cos, sin, ...

According to the authors, the stimuli (1) and (2), and also (3) and (4), sound very similar, whereas there is a significant perceptual difference between these pairs. In corresponding channels of Fig. 15 the excitation-time pattern of the signals (1) and (2) are very similar, whereas the time patterns of (2) and (3) show clear differences, especially at the channels 9, 11 and 13. This emphasizes the usefulness of the excitation-time pattern for the interpretation of experiments in which a change in phase produces a time pattern change at certain places along the basilar membrane which may result in a change in sensation, too.

CONCLUSIONS

(1) The excitation pattern plays an important role in hearing processes.
(2) Excitation seems to be based on a hydromechanical filtering process which takes place prior to the transformation of stimuli into sequences of neural spikes; therefore, no or only a small inhibitory effect at the upper accessory excitation (see negative after-image in hearing (Zwicker, 1964)) may be involved.
(3) There are so far no psychoacoustical results which make it necessary to give up this hypothesis. The same can be said for physiological responses

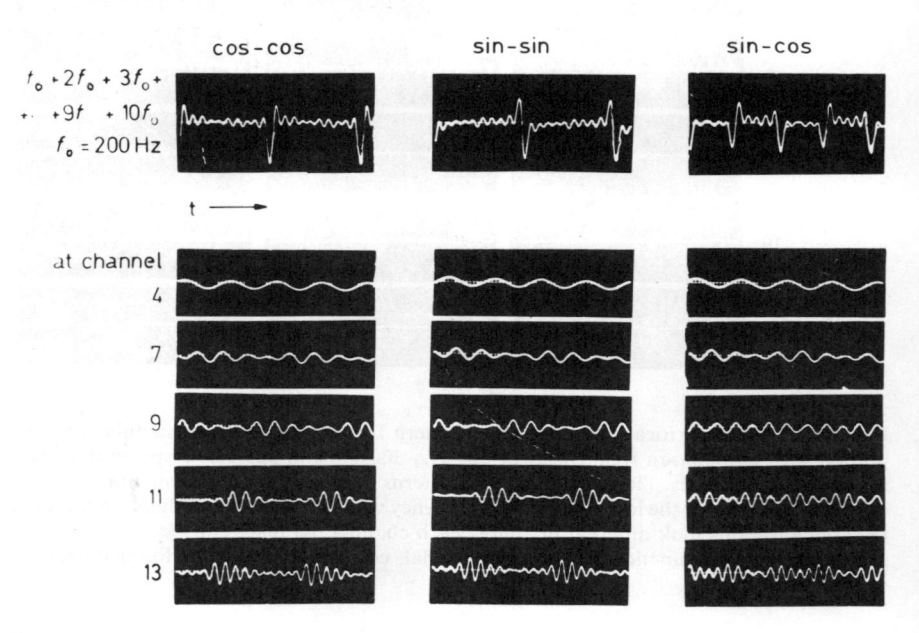

Fig. 15. Excitation-critical band rate-time pattern for signals composed of 10 harmonic components with fundamental frequency of 200 Hz but different phase relation of the components, indicated in the upper 3 oscillograms. Although the spectra are the same, the excitation-time patterns look different, especially at ranges where the excitations of the single components overlap.

picked up from single units as close as possible to the sensory cells (lowest neural level), if nonlinear transformations like the loudness function are taken into consideration.

(4) The excitation-critical band rate-time pattern is the basic pattern from which most of the hearing sensations are derived. The excitation level as a function of critical band rate seems to be especially important for JND's, loudness, pitch and timbre in steady state conditions, while excitation-time pattern may be more important for periodicity pitch, roughness, timbre, localization and binaural hearing phenomena.

(5) Instead of thinking exclusively in frequency selectivity or in periodicity selectivity hypotheses, it may be of future interest to think in more generally formulated hypotheses, including both frequency selectivity and periodicity sensitivity, like the concept of the excitation-critical band rate-time pattern from which the different sensations may be derived.

REFERENCES

Békésy, G. von (1942): Ueber die Schwingungen der Schneckentrennwand beim Präparat und Ohrenmodell, Akust. Zeits. *7*, 173-186 (J. Acoust. Soc. Amer. (1949) *21*, 233-245).
Békésy, G. von (1943): Ueber die Resonanzkurve und die Abklingzeit der verschiedenen Stellen der Schneckentrennwand, Akust. Zeits. *8*, 66-76 (J. Acoust. Soc. Amer. (1949) *21*, 245-254).
Békésy, G. von (1953): Description of some mechanical properties of the organ of Corti, J. Acoust. Soc. Amer. *23*, 770-785.
Flanagan, J. L. (1962): Models for approximating basilar membrane displacement, Part II, Bell System Tech. J. *41*, 959-1009.
Gässler, G. (1954): Ueber die Hörschwelle für Schallereignisse mit verschieden breitem Frequenzspektrum, Akust. Beih. (Acustica) H. *1*, 408-414.
Goldstein, J. L. (1967): Auditory spectral filtering and monaural phase perception, J. Acoust. Soc. Amer. *41*, 458-479.
Helle, R. (1969): Amplitude und Phase des im Gehör gebildeten Differenztones 3. Ordnung, Acustica *22*, 74-87.
Helmholtz, H. L. F. von (1863): Die Lehre von den Tonempfindungen als physiologische Grundlage für die Theorie der Musik (F. Vieweg & Sohn, Braunschweig).
Maiwald, D. (1967a): Beziehungen zwischen Schallspektrum, Mithörschwelle und der Erregung des Gehörs, Acustica *18*, 69-80.
Maiwald, D. (1967b): Ein Funktionsschema des Gehörs zur Beschreibung der Erkennbarkeit kleiner Frequenz- und Amplitudenänderungen, Acustica *18*, 81-92.
Maiwald, D. (1967c): Die Berechnung von Modulationsschwellen mit Hilfe eines Funktionsschemas, Acustica *18*, 193-207.
Oetinger, R., and Hauser, H. (1961): Ein elektrischer Kettenleiter zur Untersuchung der mechanischen Schwingungsvorgänge im Innenohr, Acustica *11*, 161-177.
Plomp, R., and Steeneken, H. J. M. (1969): Effect of phase on the timbre of complex tones, J. Acoust. Soc. Amer. *46*, 409-421.
Ranke, O. F. (1950): Hydrodynamik der Schneckenflüssigkeit, Z. Biol. *103*, 409-416.
Terhardt, E. (1966): Beitrag zur automatischen Erkennung gesprochener Ziffern, Kybernetik *3*, 136-143.
Zwicker, E. (1952): Die Grenzen der Hörbarkeit der Amplitudenmodulation und der Frequenzmodulation eines Tones, Akust. Beih. (Acustica) H. *3*, 125-133.
Zwicker, E. (1954): Die Verdeckung von Schmalbandgeräuschen durch Sinustöne, Akust.

Beih. (Acustica) H. *1*, 415-420.
Zwicker, E. (1956): Die elementaren Grundlagen zur Bestimmung der Informationskapazität des Gehörs, Acustica *6*, 356-381.
Zwicker, E. (1958): Ueber psychologische und methodische Grundlagen der Lautheit, Akust. Beih. (Acustica) H. *1*, 237-258.
Zwicker, E. (1960): Ein Verfahren zur Berechnung der Lautstärke, Akust. Beih. (Acustica) H. *1*, 304-308.
Zwicker, E. (1961): Subdivision of the audible frequency range into critical bands (Frequenzgruppen), J. Acoust. Soc. Amer. *33*, 248.
Zwicker, E. (1962a): Direct comparisons between the sensations produced by frequency modulation and amplitude modulation, J. Acoust. Soc. Amer. *34*, 1426-1430.
Zwicker, E. (1962b): Ueber ein einfaches Funktionsschema des Gehörs, Acustica *12*, 22-28.
Zwicker, E. (1964): "Negative afterimage" in hearing, J. Acoust. Soc. Amer. *36*, 1413-1415.
Zwicker, E. (1965): Temporal effects in simultaneous masking and loudness, J. Acoust. Soc. Amer. *38*, 132-141.
Zwicker, E. (1968): Der kubische Differenzton und die Erregung des Gehörs, Acustica *20*, 206-209.
Zwicker, E., and Feldtkeller, R. (1967): Das Ohr als Nachrichtenempfänger, 2nd (revised and enlarged) edition (Hirzel Verlag, Stuttgart).
Zwicker, E., Flottorp, G., and Stevens, S. S. (1957): Critical band width in loudness summation, J. Acoust. Soc. Amer. *29*, 548-557.
Zwicker, E., and Scharf, B. (1965): A model of loudness summation, Psychol. Rev. *72*, 3-26.
Zwislocki, J. J. (1948): Theorie der Schneckenmechanik, Acta Oto-Laryngol., Suppl. 72.
Zwislocki, J. J. (1968): Frequency analysis in hearing, Special report of Laboratory of Sensory Communication, Syracuse University, Syracuse, N.Y.

DISCUSSION

Hind: I think your results fit in very well with what we know about the behaviour of the primary fibres. Each fibre has a tuning curve and the adherents of the place theory tell us that the only meaning of this curve is its maximum. But, actually, the fibre responds over a range of frequencies and it has thus been customary to postulate some sort of sharpening mechanism. Physiologically, the evidence for such a sharpening mechanism is not very impressive, however. As Dr. Rose noted in discussing his paper, as one goes up the nervous system, we do not see striking improvements in the selectivity of single units, not even when we look at the auditory cortex. So, I think only the most ardent place theorists can find complete satisfaction in these tuning curves. Other difficulties face strict adherents of the periodicity theory. If we have, for example, two tones of 3000 and 4000 Hz, respectively, there is in some part of the cochlea a sort of a modulated wave with a periodicity corresponding to 1000 Hz. For a very limited range in the cochlea we will have an interspike interval histogram with a 1000 Hz synchrony but the vast majority of fibres will have interspike interval histograms corresponding, essentially, either to 3000 or 4000 Hz. Therefore, no one should be terribly dismayed by the fact that the periodicity effects are somewhat hard to find. I think that, putting it all together, the physiological results fit in quite well with the psychoacoustical results. Do you agree with that?

Zwicker: Yes, I do. I would like to summarize the most important aspects in the following way. The frequency selectivity we found and which we describe by the excitation pattern is about five times sharper than the frequency curve of the basilar membrane displacement, according to von Békésy's measurements. The tuning curves I know from the literature seem to be even sharper than the curves we found. So, there is some gap between the slopes of the displacement curve and the slopes of the excitation curve. However, what you said about the negative evidence of a sharpening mechanism seems to me very important. I think this a really unsatisfactory situation. On the other hand, in your two-tone masking experiments the dominance of one frequency changes much more quickly than I would expect on the basis of the rather wide tuning curves. So, some kind of neural process might be involved. As there seems to be no loop involved in this inhibition process, we can describe the frequency selectivity, at least at lower levels of the auditory pathway, as a linear filter. Looking at your data and what Dr. de Boer showed us, it seems to me that there is some agreement about what you found in the single cell responses and our excitation patterns.

Plomp: You explained in your paper (see p. 382) the just noticeable difference in frequency (JND_f) by the shift of the excitation level over the region with the steep slope. I wonder whether this can be checked by measuring the JND_f if one matches the pitch of a tone presented to the left ear with the pitch of a tone subsequently presented to the right ear. If your reasoning is right, one may expect that this JND_f is somewhat larger than when both tones are presented to the same ear because there is no direct comparison of peripheral excitation patterns in the case of two ears. Have you done any experiment on that?

Zwicker: No, but I agree that in the case of two ears a larger JND_f should be expected.

Plomp: I have a second point. You gave in Fig. 15 the excitation-critical band rate-time pattern of 3 complex tones consisting of 10 harmonics with equal amplitude patterns but different phase patterns. The figure shows that, in corresponding channels of your model, the excitation-time patterns differ in the same cases where I found a distinct difference in timbre (Plomp and Steeneken, 1969). During our experiments we tried this out also with 1/3-octave filters and found a similar effect. We did not include it in the paper, however, since we thought our frequency-analyzing system to be too poor a model of the ear. What I would like to emphasize is that, when we see such a waveform difference due to phase, we should be cautious to consider the envelope difference as the important thing. We had a similar situation for beats of mistuned consonances (Plomp, 1967) and found that the ear's sensitivity to phase appears to be related to the time pattern, so it is probable that this also holds for the case of complex tones involved here.

Zwicker: Well, all I can say is that we did not find any listening experiment that could not be explained by those excitation-time patterns. It would be of interest, in your experiments on complex tones, to filter out various parts of the frequency spectrum and to see what happens with the influence of phase on timbre.

Cardozo: I would like to ask Dr. Zwicker about his very nice demonstration on the effect of simultaneous amplitude modulation and frequency modulation (Fig. 8). I had the impression that also in the condition that the combined modulation was parallel to the low-frequency (steep) slope of the masking noise band a pitch shift could be heard. I think this is rather difficult to explain in terms of your model.

Zwicker: The model is formulated actually only for thresholds like masked thresholds, just noticeable differences in frequency or amplitude. In the experiment, just presented, modulations well above threshold have been used to produce a distinct demonstration. In this case the sensation produced by the frequency modulation can not be cancelled out completely and remains audible although the sensation is lowered by a large amount. The configuration of the excitation pattern shown in Fig. 8 can be used for interpretation of this effect.

Ritsma: Do you find any difference between the JND_f of, say, a high-pass noise and of a narrow-band noise with the same slope at the low-frequency side?

Zwicker: No, we do not as long as the bandwidth of the narrow-band noise is in the order of a critical band, so that the fluctuations in the lowest critical band — at the low-frequency slope — are the same for high-pass noise and narrow-band noise. For low-pass noise, however, — again in relation to narrow-band noise — there is not only a difference for JND_f, but the JND_f depends on the level, because the steepness of the high-frequency slope which is in this case responsible for the JND_f changes with level (Maiwald, 1967).

Plomp: I agree that the shift of the excitation pattern is very important for the JND_f. But I don't think that experiments with noise are so appropriate to study the origin of pitch perception.

Zwicker: Well, that was the reason why I chose sinusoidal tones in my JND_f experiments.

REFERENCES

Maiwald, D. (1967): Die Berechnung von Modulationsschwellen mit Hilfe eines Funktionsschemas, Acustica *18*, 193-207.
Plomp, R. (1967): Beats of mistuned consonances, J. Acoust. Soc. Amer. *42*, 462-474.
Plomp, R., and Steeneken, H. J. M. (1969): Effects of phase on the timbre of complex tones, J. Acoust. Soc. Amer. *46*, 409-421.

TIMBRE AS A MULTIDIMENSIONAL ATTRIBUTE OF COMPLEX TONES

REINIER PLOMP

Institute for Perception RVO-TNO
Soesterberg, The Netherlands

INTRODUCTION

Complex tones can differ in loudness, pitch and timbre. It is quite remarkable how little attention timbre, as contrasted with loudness and pitch, has received in hearing research. In modern books on audition, timbre, if mentioned at all, is mostly dispatched with only a few words. Stevens and Davis (1938) included in their well-known book, after separate chapters on pitch and loudness, a chapter on "other attributes of tones"; but only volume, density and brightness as aspects of simple (pure) tones are treated. In their extensive glossary at the end of the book, timbre is not even mentioned. The same neglect can be observed in other works (*e.g.* Fletcher, 1953; Littler, 1965). Licklider (1951) discussed also attributes of complex sounds but could only conclude: "Until careful scientific work has been done on the subject, it can hardly be possible to say more about timbre than that it is a 'multidimensional' dimension." The attention paid by Zwicker and Feldtkeller (1967) to "Klangfarbe", to be considered as synonymous to timbre, was restricted to a similar remark about its multidimensionality. In view of this situation, it is not surprising that timbre perception does not play an explicit part in hearing theory. This can be illustrated by the fact that von Békésy (1963) did not include timbre in his list of primary attributes of auditory sensation to be compared with the sensations on the skin (*cf.* also Wever, 1949).

This ignorance of timbre finds its counterpart in the vagueness of its definition. Most authors agree that it is related to the overtone structure of complex tones, but that it should be clearly distinguished from loudness and pitch. The definition approved by the American Standards Association (1960) seems to be the most acceptable one: "Timbre is that attribute of auditory sensation in terms of which a listener can judge that two sounds similarly presented and having the same loudness and pitch are dissimilar." In a note is added: "Timbre depends primarily upon the spectrum of the stimulus, but it also depends upon the waveform, the sound pressure, the frequency location of the spectrum, and the temporal characteristics of the stimulus." Apparently, the timbre concept is loaded rather negatively as the total of all aspects of sound sensation when loudness and pitch are left out of consideration. If the ambiguous restriction to "similarly presented" sounds is interpreted in a very broad sense, the five major parameters of timbre listed recently by Schouten (1968), *viz.* tonal *vs.* noiselike character, spectral envelope, time envelope, change, and acoustic prefix, may be all involved. If, however, the definition is understood in a much stricter way, more similar to the usual denotation of the word "Klangfarbe", the more dynamic aspects should be excluded and timbre should be understood as to be mainly related to frequency spectrum.

In this paper, timbre will be taken in its most narrow sense as *that attribute of sensation in terms of which a listener can judge that two steady complex tones having the same loudness, pitch and duration are dissimilar*. Since such tones are given acoustically by a periodic fluctuation of sound pressure p and can be represented by

$$p(t) = \sum_{n=1}^{\infty} a_n \sin(2\pi n f t + \phi_n),$$

the purpose of this paper is to discuss how timbre depends upon the *amplitude pattern* $a_1, a_2, a_3, \ldots, a_n$ and the *phase pattern* $\phi_1, \phi_2, \phi_3, \ldots, \phi_n$ of the successive harmonics.

The little attention paid to timbre in books on hearing reflects the situation that hardly any results of explicit experiments on timbre are available. On the other hand, there are many studies bearing on timbre perception in a more indirect way, particularly studies on the discrimination of speech sounds and on the ear's sensitivity to phase. Therefore, this review is based mainly on these investigations.

VON HELMHOLTZ'S CONCEPTION OF THE PHYSICAL CORRELATE OF TIMBRE

The neglect of timbre just discussed does not hold for von Helmholtz who carried out the first extensive experiments on its dependence on amplitude pattern and phase pattern (1859, 1863). The opinion that timbre originates from the harmonics had been advanced earlier (*e.g.* Willis, 1830; Bindseil, 1839; Seebeck, 1849; Brandt, 1861), but only von Helmholtz supplied it with a firm experimental base. Part I of his book (1863), consisting of 6 chapters and constituting in pages more than one third of the total, presents a brilliant and never surpassed treatise on timbre in its relation to the properties of sounds and of the hearing mechanism.

Von Helmholtz demonstrated that the periodic vibrations produced by musical instruments and the vocal chords consist of a series of harmonics, and that the ear is able to distinguish a number of these harmonics individually. In view of these observations, he accepted Ohm's opinion (1843) that, just as any periodic waveform can be analyzed mathematically in a series of sine waves by applying Fourier's well-known theorem, complex tones are actually analyzed in a series of simple tones by the hearing mechanism. Von Helmholtz appears to have been rather optimistic about the number of distinguishable harmonics. Recent investigations have shown that, even under most favourable conditions, this capacity is limited to the first 5 to 8 harmonics (Plomp, 1964; Plomp and Mimpen, 1968).

The extensive manner in which von Helmholtz tried to convince his readers of the audibility of the harmonics demonstrates that he was quite aware of the fact that we do not distinguish them in normal listening. He concluded that, although we do perceive the harmonics of a complex tone, we do this not always consciously (the harmonics are perceived but not "apperceived"). The harmonics fuse into the sound as a whole but their existence in our sensation is established by their influence on timbre.

Its dependence on the harmonic structure explains the great variety in timbre. By studying both the physical and perceptual differences between musical tones, von Helmholtz found the following rules:
(1) simple tones sound sweet and pleasant, without any roughness, but dull at low frequencies;
(2) complex tones with moderately loud lower harmonics up to the 6th sound more musical and rich than simple tones, but they are still sweet and pleasant if the higher harmonics are absent;
(3) complex tones consisting of only odd harmonics sound hollow and, if many harmonics are present, nasal; predomination of the fundamental gives a full tone, in the reverse case the tone sounds empty;
(4) complex tones with strong harmonics beyond the 6th or 7th sound sharp, rough and penetrating.

Having established the dependence of timbre on the amplitude pattern of the harmonics, von Helmholtz investigated its dependence on the phase pattern. He designed an ingenious apparatus consisting of 8 tuning forks, all excited electro-magnetically with 120 pulses per second. The forks were tuned to the first 8 harmonics of 120 Hz and, as a result of the excitation, their continuous vibrations were phase-locked. By means of rubber, any radiation of sound via the mountings of the forks was avoided. Each fork was provided with a resonator tuned to its frequency. By changing the distance to the tuning fork and by closing the mouth of the resonator to some extent, both the intensities and the phases of the individual harmonics could be varied over a limited range.

With this apparatus, von Helmholtz was able to simulate some vowels and musical tones and could verify the influence of phase on timbre. On the basis of many observations, not only for a fundamental of 120 Hz but also for 240 Hz, he concluded that timbre of complex tones does not depend upon the phase differences between the partials. Mostly only this conclusion by von Helmholtz is quoted, but we should not overlook his additional restriction that the timbre differences are at least not distinct enough to be detectable after a few seconds required to alter the phase pattern, implying that these changes are certainly too small to transform one vowel into another. Moreover, he did not consider it to be impossible that, since harmonics beyond the 6th to 8th give rise to dissonance and roughness, a phase effect does exist for these higher harmonics.

FURTHER DATA ON THE RELATION BETWEEN TIMBRE AND AMPLITUDE PATTERN

Searching the literature published since von Helmholtz shows that our knowledge about the relation between timbre and amplitude pattern has not increased significantly beyond the level of more than 100 years ago. We know much more now about both the production and the physical structure of complex tones, but nearly no quantitative data are available on the perceptual differences between these tones.

Some reasons for this neglect can be suggested:

(a) Von Helmholtz's emphasis on the sine wave as the basic auditory stimulus resulted in a strong preference for using simple tones in hearing research. Up to about 1950, even stimuli consisting of 2 or 3 frequency components were seldom applied.

(b) Just as amplitude pattern is a multidimensional quantity, timbre should be considered as a multidimensional attribute of tone sensation. Differences in loudness and pitch, both one-dimensional attributes, could be ordered along scales from weak to loud, and low to high, respectively, but such a scale does not exist for timbre differences. Stumpf (1890) listed, as examples of timbre connotations, no less than 20 relevant semantic categories as wide-narrow, smooth-rough, round-sharp, etc. and sighed that this wealth of adjectives is comparable only with those of wine merchants. This profusion was not very inviting to start research on timbre.

(c) As a multidimensional attribute, timbre can be investigated successfully only when adequate methods are applied. The very few attempts made in the past for studying timbre illustrates that, without good multidimensional scaling techniques, no success can be expected. These techniques were, however, not available until recently.

As a result, we owe our present knowledge of timbre perception nearly exclusively to investigations of speech perception. Within the scope of this review paper, it suffices to pay attention to the following three findings in vowel research.

(a) *Multidimensionality of the differences between vowels*. It appears that already Hellwag in 1781 represented the vowels by a triangle, implying that two dimensions are required to describe their timbre differences. A similar 2-dimensional representation was proposed by Chladni (1824) and accepted, with minor variations, by most phoneticians in the 19th century. As long as acoustical data were not available, this view was only based on the topography of the vocal tract for the various vowels in relation to the corresponding vowel sensations. Willis (1830) combined a vibrating reed [Rohrblatt] with an organ pipe provided with a piston [Kolben] and demonstrated that the vowels [i], [e], [a], [o], and [u] could be simulated in this order by increasing the distance between the reed and the piston. This was the first indication that the timbre differences between vowels are correlated with peaks in the amplitude pattern, for which since Hermann the term *formants* has been accepted. Further experiments by von Helmholtz (1859, 1863) and others confirmed this view. It has been established that at least the formant frequencies F_1 and F_2, being the two lower resonance frequencies of the coupled cavities of the vocal tract, are required to characterize each vowel. Plotting F_2 as a function of F_1 results in a vowel representation that agrees rather well with the traditional vowel triangle (*cf.* Peterson and Barney, 1952). There are clear indications, however, that two dimensions are not sufficient to describe the acoustic and perceptual differences between vowels satisfactorily, but that (at least) a third one should be added. This question is treated more extensively in the paper by Pols (see pp. 463-473 of this volume).

(b) *Absolute frequency vs. relative frequency in timbre similarity*. The finding that the differences between vowels are based on specific peaks in the amplitude pattern raised the interesting question of whether, for the same vowel, the frequencies of these peaks do or do not shift in accordance with the frequency of the fundamental. Most investigators (*e.g.* von Helmholtz, 1863; Donders, 1864; Grassmann, 1877) promoted the view that the formant frequencies are approximately constant. Observations with the aid of the phonograph invented by Edison in 1877 demonstrated clearly that the vowel quality is severely affected indeed by changing the frequencies of all harmonics by the same factor. Stumpf (1926) concluded, however, that the relative frequencies of the formants cannot be completely neglected. He observed that, in addition to a large change of the fundamental frequency, a small shift of the formant frequencies in the same direction is required to keep the vowel quality as constant as possible. This question was investigated recently much more extensively by Slawson (1968). His important study showed that when the fundamental frequency of the second one of each pair of vowel-like synthesized sounds was raised an octave above the fundamental frequency of the first sound, the difference in vowel quality estimated by a group of observers could be reduced to a minimum if the lower two formants of the sound with the higher fundamental were raised by about 10%. In order to simulate speech sounds as good as possible, the pitch of the 700 msec test stimuli was lowered in this experiment gradually during the last 450 msec of presentation. Repetition of the experiment with similar stimuli but of constant pitch, in which case the subjects were instructed to use musical timbre instead of vowel quality as a criterion, did not give a pronounced minimum but a very flat over the range -5% to $+10\%$ shift in formants. In a third experiment, also with constant pitch, both criteria gave quite similar results indicating that, when the fundamental frequency of the second one of each pair of sounds is raised by a factor of 1.5 above the fundamental frequency of the first sound, the estimated difference is minimal if the spectrum envelope is not shifted. From these observations we may conclude that, for the stimuli employed, vowel quality and timbre are highly correlated notions, and that timbre depends mainly upon the absolute position of the envelope of the amplitude pattern along the frequency scale rather than on the relative position with reference to the fundamental frequency. Slawson himself argued that the shift of about 10% in formant frequency required to obtain maximal similarity in vowel quality in the first experiment cannot be attributed to experience, but his arguments are, in the opinion of the present author, not convincing (the formants of male and female speakers differ also about 10% in frequency).

(c) *Timbre of simple tones.* Including simple tones in his classification of the timbre of tones (see p. 399), von Helmholtz recognized implicitly that also tones without harmonics are characterized by a typical frequency dependent timbre. The fact that low tones sound dull and high tones bright is nowadays completely neglected as an indication that, in addition to pitch and loudness, also timbre is an attribute of simple tones. Von Helmholtz did not emphasize this point, but it was advocated explicitly by Engel (1886) and Stumpf (1890) as a necessary implication of the fact that timbre is an attribute of complex tones. The observation by Grassmann (1877) and von Wesendonk (1909) that simple tones have some resemblance, depending upon frequency, with particular vowels supports this view. Several investigators (Köhler, 1911; Weiss, 1920; Engelhardt and Gehrcke, 1930; Fant, 1959) studied the phenomenon more carefully and demonstrated by recognition tests that, though some ranges do overlap each other to quite an extent, different vowels can be attached, in the order as found already by Willis, to a simple tone moving from high to low frequency. It is clear that this resemblance is related to the location of the most characteristic formant along the frequency scale. Morton and Carpenter (1962) showed that the rather poor result for single tones is much improved when two simultaneous simple tones are used. Since it is unlikely that this resemblance is attributable to pitch, Köhler's view that it is a proof of timbre as a third attribute of simple tones should be accepted.

In addition to these findings adopted from speech research, nearly no other contributions to timbre perception can be mentioned. Löb (1941) investigated for complex tones consisting of 8 harmonics the timbre change when the sound pressure level (SPL) of a single component is altered. Lichte (1941) compared complex tones of equal loudness and pitch and concluded that we should consider at least three further attributes: _brightness,_ being a function of the location on the frequency continuum of the midpoint of the energy distribution, _roughness,_ which is present in tones containing consecutive high partials above the 6th, and _fullness,_ being a function of the relative presence of odd- and even-numbered partials. More recently, the timbre differences between complex tones have been studied by applying the method of semantic differentials (Rahlfs, 1966). No less than 47 different categories were included in these experiments in which tones adopted from musical instruments were compared. Neither loudness nor pitch were equalized, however, and consequently corresponding factors had high loadings in the results. Generally, the application of semantic scales has the disadvantage that the subject's attitude rather than the role of the hearing mechanism itself in discriminating timbre is investigated.

EFFECT OF PHASE ON TIMBRE

It was mentioned already that von Helmholtz considered the influence of phase on timbre to be absent or very small. This conclusion was tested by many investigators in the past and only the main results shall be pointed out here (see for more details Plomp, 1967; Plomp and Steeneken, 1969). On the basis of the stimuli used, the experiments can be divided into three different groups.

(a) *Complex tones consisting of many harmonics.* König (1881) designed discs with contours corresponding to the waveform of the periodic signals to be generated and rotated these discs in front of a narrow transverse slit through which air was blown. His finding that timbre does depend upon the phase pattern of the stimuli was criticized by Hermann (1894) who considered the observed differences as to be due to the imperfections of the wave siren. Licklider (1957) applied a generator that provided control over both the amplitudes and phase angles of the first 16 harmonics. He noticed that a change in phase pattern results in almost every case in a discriminable difference in timbre. In general, changing the phase of a high-frequency component produced more effect than changing the phase of a low-frequency component; the effect was larger for low fundamental frequencies than for high ones. Similar experiments were carried out by Schroeder with complex tones containing 31 harmonics (1959). He found that timbre strongly depends upon whether or not the waveform contains large peaks. Recently, the present author (Plomp and Steeneken, 1969) used the method of triadic comparisons in comparing the timbre differences of computer-generated complex tones of equal pitch and loudness but different phase patterns. The data were analyzed with Kruskal's multidimensional scaling program (Kruskal, 1964a, 1964b). The most important results were: (1) the timbre difference between a tone consisting of only sine or cosine terms and a tone consisting of alternating sine and cosine terms represents the maximal possible effect of phase on timbre; (2) for complex tones with an amplitude pattern of the harmonics corresponding to a slope of about -6 dB/octave and a fundamental frequency of about 150 Hz, this maximal effect is quantitatively smaller than the effect of changing the slope of the amplitude pattern by 2 dB/octave and the effect is less for higher than for lower fundamental frequencies; (3) the effect of phase on timbre appears to be a specific factor independent of the effects of amplitude pattern and SPL.

(b) *Sounds consisting of three frequency components.* Already prior to von Helmholtz, Scheibler (Röber, 1834) discovered that three simultaneous simple tones of $(n-1)p$, np, and $(n+1)p$ Hz ($n=$integer), with one tone slightly mistuned, give rise to a beat sensation. Such a sound can be interpreted as a complex tone consisting of only three harmonics of which one's phase is running continuously in time, and the beat sensation should be considered as periodic changes in timbre. Thomson (1878) called attention to the fact that these beats already occur at very low SPL's, so that it is very improbable that the phenomenon is due to nonlinear distortion in the ear, as was assumed by von Helmholtz (1863) and most other investigators. The effect was studied further by ter Kuile (1902), Mathes and Miller (1947), Zwicker (1952), and Goldstein (1967), and it is now generally accepted that it is caused by the ear's sensitivity to phase.

(c) *Sounds consisting of two frequency components.* Though the beats are not so distinct as for three tones, they may also occur for two simultaneous simple tones. These so-called beats of mistuned consonances were investigated also for the first time by Scheibler (Röber, 1834). Ohm (1839) published a general formula for the rate of these beats: two tones of M and N Hz, respectively, with $M:N$ only slightly different from $m:n$ (m and n both small integers), give rise to $|mN-nM|$ beats per sec. The most promoted explanations of these beats were in terms of combination tones (Röber, 1834; von Helmholtz, 1856, 1863; Bosanquet, 1881; Stumpf, 1910; Haar, 1951) or of aural harmonics (e.g. Hermann, 1896; Wegel and Lane, 1924; Fletcher, 1930; von Békésy, 1934; Lawrence and Yantis, 1956). De Morgan (1864) proved that Ohm's formula agrees with the rate of waveform variations of the superimposed sinusoids, suggesting that the beats may be due to phase sensitivity of the ear. The latter opinion was defended by König (1876) but rejected by most investigators. Recent experiments (Plomp, 1967) have put forward new evidence that the beats of mistuned consonances cannot be explained satisfactorily by nonlinear distortion in the ear but that these beats, which are much more distinct at low than at high frequencies, have indeed their origin in the ear's phase sensitivity.

Summarizing these results, we may conclude that timbre is influenced by the phase pattern of the harmonics, but that the effect is rather small in comparison to the role of amplitude pattern. The effect is smaller for high than for low fundamental frequencies. The timbre differences are much better audible for a continuously changing phase relation than for a discontinuous move from one phase pattern to another.

RESULTS OF A CURRENT INVESTIGATION ON TIMBRE-AMPLITUDE PATTERN CORRELATION

Recently, the author started an investigation in which the dependence of timbre on amplitude pattern is studied quantitatively. A discussion of some preliminary results may illustrate how the timbre problem can be attacked. The method and apparatus used were the same as applied by Pols (pp. 463-470 of this volume), so they will be treated here only briefly.

The timbre differences between complex tones of equal loudness and pitch were compared in subsets of 3 stimuli by means of the so-called triadic comparison technique. The subject's task was only to select for each triad the most similar and most dissimilar pair of stimuli, successively. The total number of times a particular pair is judged to be more dissimilar than all other pairs with which it is compared results in a (half)matrix of dissimilarity indices. (Here, dissimilarity instead of similarity is used, to obtain numbers that are positively correlated with the physical distance measure introduced below.)

The stimuli investigated were derived from musical tones. In one experiment a single period of the same note played by 9 musical instruments was used (fundamental frequency 319 Hz). In a second one, these single periods were adopted from 10 stops of a pipe organ (fundamental frequency 263 Hz). The periods were stored in digital form in the memory of a computer and could be generated as a continuous signal. Both the signal presentation and the response processing were controlled completely by the computer. The upper triangles in Tables I and II present the cumulative dissimilarity indices for 10 subjects. Since the purpose of these experiments is to demonstrate the relation between timbre and amplitude pattern rather than to investigate by how many dimensions

Table I. Cumulative dissimilarity indices (10 subjects; upper triangle) and total differences in frequency spectrum $D_{i,j}$ ($p=1$; lower triangle) for 9 tones adopted from musical instruments.

		1	2	3	4	5	6	7	8	9
(violin)	1	—	89	25	100	19	90	88	101	96
(viola)	2	105	—	87	34	115	31	46	62	79
(cello)	3	65	72	—	92	40	40	98	72	38
(oboe)	4	124	65	97	—	107	76	74	26	67
(clarinet)	5	69	103	74	112	—	78	124	97	78
(bassoon)	6	120	47	63	73	105	—	87	36	6
(trumpet)	7	79	75	85	109	116	105	—	103	109
(French horn)	8	137	85	86	59	117	51	133	—	10
(trombone)	9	131	83	72	86	122	38	126	36	—

Table II. Cumulative dissimilarity indices (10 subjects; upper triangle) and total differences in frequency spectrum $D_{i,j}$ ($p=1$; lower triangle) for 10 tones adopted from a pipe organ.

		1	2	3	4	5	6	7	8	9	10
(reed tones)	1	—	41	46	96	119	117	39	67	134	115
	2	98	—	17	115	129	110	75	101	147	127
	3	110	107	—	84	105	80	87	79	128	101
(diapason tones)	4	144	170	134	—	17	35	90	58	66	57
	5	163	176	144	57	—	50	111	44	27	37
	6	214	227	172	111	97	—	78	93	75	75
	7	124	128	114	100	114	126	—	86	131	113
(flute tones)	8	94	148	96	101	102	162	119	—	64	22
	9	195	211	185	125	105	164	194	127	—	12
	10	133	169	126	104	91	146	143	78	72	—

the timbre differences can be described, the attention will be focussed here on the direct correlation between the dissimilarity indices and the amplitude patterns of the signals involved rather than on the results of a further analysis of the data by applying Kruskal's multidimensional scaling program.

At this point, we had to decide in which way the differences in amplitude pattern should be quantified. As it is known that the ear's frequency-analyzing power agrees rather well with 1/3-octave bands, it was decided to use these bands in the frequency analysis. It was decided also to use SPL's in dB as a measure of the contributions of the different frequency bands.

Different reasonings can be proposed concerning the way in which the difference in frequency spectrum should be expressed by one number.

(1) One approach is the one applied in earlier experiments on speech vowels (Plomp et al., 1967). We can interpret the SPL's in the m different frequency bands as the coordinates of a point in an m-dimensional Euclidean space. Different frequency spectra are represented by different points in that space and the distance between the points can be considered to be a good measure of the differences between their corresponding frequency spectra.

(2) An alternative approach, perhaps more obvious from the physiological point of view, is the following. Since the loudness of a complex tone can be considered successfully as the sum of the contributions of the different frequency bands (Zwicker and Scharf, 1965), we may suppose similarly that the total difference between two frequency spectra can be considered as the sum of the differences in each band. In other words: the area between the spectrum curves determines the difference. This means in fact a variant on the former approach: in the m-dimensional space, not the squares of the differences along the axes should be added, but the differences themselves.

Both approaches can be represented by the following equation:

$$D_{i,j} = \sqrt[p]{\sum_{n=1}^{m} |L_{i,n} - L_{j,n}|^p}$$

with $D_{i,j}$ = difference in frequency spectrum between the tones i and j; $L_{i,n}$ = SPL of tone i in band n; m = total number of frequency bands. Power $p=2$ gives the Euclidean solution (1), $p=1$ gives the area solution (2). Since we do not know *a priori* which p-value should be preferred, we may treat p as a variable and look for that value that gives the best correlation with the dissimilarity indices. As an illustration, the lower triangles of Tables I and II represent $D_{i,j}$ for $p=1$.

With the aid of a computer program in which any p-value wanted could be applied, the curves of Fig. 1 were determined, representing the correlation coefficient between the dissimilarity indices and the calculated $D_{i,j}$-values as a function of p. The third curve presents the corresponding result for 11 vowel-like sounds, of which the data were adopted from Pols' experiments.

We see that correlation coefficients between 0.81 and 0.86 were reached for the most favourable p-values. It is rather interesting that the curves are very flat, particularly for the vowels, and that their maxima do not coincide. This result should not be interpreted, however, as an indication that in timbre perception no particular p-value is involved. The invariance can be explained for the greater part by the physical properties of the stimuli used. Correlating $D_{i,j}$ for which the maximum was found ($p=0.75$ for the tones adopted from the musical instruments, $=1$ for the organ tones, $=2$ for the vowel tones) with $D_{i,j}$ for the other p-values results in the curves represented in Fig. 2, with shapes quite similar to the corresponding curves in Fig. 1. This means that the stimuli applied are not very appropriate to decide which p-value is actually involved in timbre discrimination. Experiments are going on with synthesized tones more suited to answer this question. Further calculations may also show whether a set of filters equal to the critical bandwidths and a measure more representative

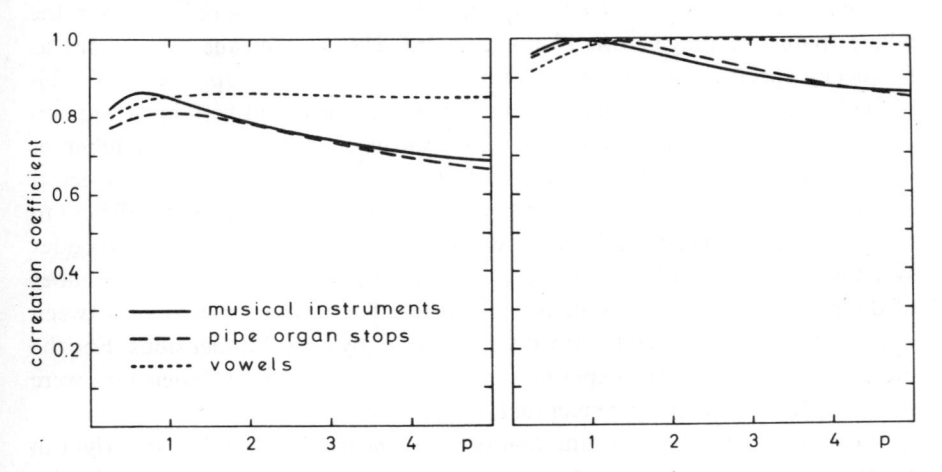

Fig. 1. Correlation between dissimilarity indices and $D_{i,j}$ as a function of p.

Fig. 2. Correlation between the best $D_{i,j}$ and the $D_{i,j}$, for variable p.

of loudness than is SPL give still better results.

Summarizing, we may conclude that the differences in frequency spectrum expressed in distance in a Euclidean space ($p=2$) or in area between the spectrum curves ($p=1$) correlate quite well with the differences in timbre observed by test subjects.

PHYSIOLOGICAL CORRELATE OF TIMBRE

We have seen that timbre as a multidimensional attribute of complex tones is the perceptual correlate of the amplitude pattern as a multidimensional physical property of the tones. Slawson's (1968) investigations on the timbre of vowel-like tones with different fundamental frequencies indicated that timbre similarity is mainly, if not completely, determined by the absolute position of the envelope of the amplitude pattern along the frequency scale rather than by the relative position with reference to the fundamental frequency. This implies that timbre is derived from frequency spectrum, measured with some specific set of band-pass filters.

In the preceding section, the use of a set of 1/3-octave filters was argued on the basis of their similarity to the critical bands. The decisive role of the critical bandwidth as a frequency dependent measure of the filtering properties of the hearing mechanism has been clearly demonstrated by numerous investigations on (masked) thresholds and loudness levels of complex sounds (*cf.* Zwicker and Feldtkeller, 1967). The upper limit of the harmonics of a complex tone that can be distinguished individually agrees also with critical bandwidth (Plomp, 1964; Plomp and Mimpen, 1968).

It appears justified to conclude that the number of dimensions in timbre perception is limited by the number of critical bands required to cover the whole frequency range from 20 up to 16,000 Hz. This means that, as far as frequency spectrum is concerned, timbre is maximally an approximately 23-dimensional attribute of sound. Excluding the very low and very high frequencies, as holds for most sounds to which we listen normally, the number of dimensions can be reduced to a theoretical value of about 15. In this respect, it is of interest that a comparable number of band filters is found to be sufficient to transmit the information bearing elements of speech in channel vocoder systems (Flanagan, 1965). For a restricted set of sounds, however, the number of dimensions involved may be much smaller. The timbre differences between vowel-like sounds can be described satisfactorily by only 3 dimensions. For the musical tones used in the experiments reported above, also 3 dimensions were sufficient to describe their perceptual differences.

Since the frequency analyzing ability of the hearing system has its origin in the frequency selective properties of the cochlea, we may state that the stimulation pattern along the basilar membrane, any possible sharpening process in the nerve cells included, should be considered as the physiological link between frequency spectrum and timbre perception. Traditionally, this frequency analyzing power is explained as the essential condition for pitch perception. In the light of recent investigations on the physical correlate of pitch (see Ritsma, pp. 250-266 of this volume) and of what is said above, it is much more likely that the function of the cochlea has to be understood primarily in relation to timbre rather than to pitch perception.

In addition the effect of phase on timbre should be mentioned. Though it is clear that this effect is only of minor importance, it cannot be neglected. We saw that the ear's sensitivity to phase cannot be explained satisfactorily by non-linear distortion. Since the influence of phase depends upon frequency separation (Plomp, 1967), it seems to be related to the extent to which the stimulation patterns of the frequency components overlap. In the author's opinion, the most likely explanation is in terms of the synchrony between nerve impulses and waveform. For overlapping stimulation patterns, phase shift gives a change in the superimposed waveform which results in a change of the time pattern of the nerve impulses. This view is supported by the fact that the influence of phase is larger at low than at high frequencies.

REFERENCES

American Standards Association (1960): Acoustical Terminology, S1.1-1960 (American Standards Association, New York).
Békésy, G. von (1934): Ueber die nichtlinearen Verzerrungen des Ohres, Ann. Phys. 20, 809-827.
Békésy, G. von (1963): Hearing theories and complex sounds, J. Acoust. Soc. Amer. 35, 588-601.
Bindseil, H. E. (1839): Akustik (Verlag der Horvath'schen Buchhandlung, Potsdam).
Bosanquet, R. H. M. (1881): On the beats of consonances of the form $h:1$, Philos. Mag. 5th Ser. 11, 420-436 and 492-506.
Brandt, S. (1861): Ueber Verschiedenheit des Klanges (Klangfarbe), Ann. Phys. Chem. 112, 324-336.
Chladni, E. F. F. (1824): Ueber die Hervorbringung der menschlichen Sprachlaute, Ann. Phys. 76, 187-216.
De Morgan, A. (1864): On the beats of imperfect consonances, Trans. Cambr. Philos. Soc. 10, 129-141.
Donders, F. C. (1864): Zur Klangfarbe der Vocale, Ann. Phys. Chem. 123, 527-528.
Engel, G. (1886): Ueber den Begriff der Klangfarbe, Philos. Vorträge, Berlin, Neue Folge, II. Ser., Heft 12, 311-355.
Engelhardt, V., and Gehrcke, E. (1930): Ueber die Vokalcharaktere einfacher Töne, Z. Psychol. 115, 16-33.
Fant, G. (1959): Acoustic analysis and synthesis of speech with applications to Swedish, Ericsson Technics No. 1, 3-108.
Flanagan, J. L. (1965): Speech Analysis, Synthesis, and Perception (Springer Verlag, Berlin).
Fletcher, H. (1930): A space-time pattern theory of hearing, J. Acoust. Soc. Amer. 1, 311-343.
Fletcher, H. (1953): Speech and Hearing in Communication (D. van Nostrand, New York).
Goldstein, J. L. (1967): Auditory spectral filtering and monaural phase perception, J. Acoust. Soc. Amer. 41, 458-479.

Grassmann, H. (1877): Ueber die physikalische Natur der Sprachlaute, Ann. Phys. Chem. *1*, 606-629.
Haar, G. (1951): Ueber den Charakter der Nichtlinearität des Ohres, Funk und Ton *5*, 248-257.
Helmholtz, H. L. F. von (1856): Ueber Combinationstöne, Ann. Phys. Chem. *99*, 497-540.
Helmholtz, H. L. F. von (1859): Ueber die Klangfarbe der Vocale, Ann. Phys. Chem. *18*, 280-290.
Helmholtz, H. L. F. von (1863): Die Lehre von den Tonempfindungen als physiologische Grundlage für die Theorie der Musik (F. Vieweg & Sohn, Braunschweig).
Hermann, L. (1894): Beiträge zur Lehre von der Klangwahrnehmung, Arch. ges. Physiol. *56*, 467-499.
Hermann, L. (1896): Zur Frage betreffend den Einfluss der Phasen auf die Klangfarbe, Ann. Phys. Chem. *58*, 391-401.
Köhler, W. (1911): Akustische Untersuchungen. II, Z. Psychol. *58*, 59-140.
König, R. (1876): Ueber den Zusammenklang zweier Töne, Ann. Phys. Chem. *157*, 177-237.
König, R. (1881): Ueber den Ursprung der Stösse und Stosstöne bei harmonischen Intervallen, Ann. Phys. Chem. *12*, 335-349.
Kruskal, J. B. (1964a): Multidimensional scaling by optimizing goodness of fit to a nonmetric hypothesis, Psychometrika *29*, 1-27.
Kruskal, J. B. (1964b): Nonmetric multidimensional scaling: a numerical method, Psychometrika *29*, 115-129.
Kuile, Th. E. ter (1902): Einfluss der Phasen auf die Klangfarbe, Arch. ges. Physiol. *89*, 333-426.
Lawrence, M., and Yantis, P. A. (1956): Onset and growth of aural harmonics in the overloaded ear, J. Acoust. Soc. Amer. *28*, 852-858.
Lichte, W. H. (1941): Attributes of complex tones, J. Exptl. Psychol. *28*, 455-480.
Licklider, J. C. R. (1951): Basic correlates of the auditory stimulus, in: Handbook of Experimental Psychology, S. S. Stevens, Ed. (Wiley, New York).
Licklider, J. C. R. (1957): Effects of changes in the phase pattern upon the sound of a 16-harmonic tone, J. Acoust. Soc. Amer. *29*, 780 (A).
Littler, T. S. (1965): The Physics of the Ear (Pergamon, Oxford).
Löb, E. (1941): Ueber die subjektive Wirkung von Klangfarbenänderungen, Akust. Z. *6*, 279-294.
Mathes, R. C., and Miller, R. L. (1947): Phase effects in monaural perception, J. Acoust. Soc. Amer. *19*, 780-797.
Morton, J., and Carpenter, A. (1962): Judgement of the vowel colour of natural and artificial sounds, Language and Speech *5*, 190-204.
Ohm, G. S. (1839): Bemerkungen über Combinationstöne und Stösse, Ann. Phys. Chem. *47*, 463-466.
Ohm, G. S. (1843): Ueber die Definition des Tones, nebst daran geknüpfter Theorie der Sirene und ähnlicher tonbildender Vorrichtungen, Ann. Phys. Chem. *59*, 513-565.
Peterson, G. E., and Barney, H. L. (1952): Control methods used in a study of the vowels, J. Acoust. Soc. Amer. *24*, 175-184.
Plomp, R. (1964): The ear as a frequency analyzer, J. Acoust. Soc. Amer. *36*, 1628-1636.
Plomp, R. (1967): Beats of mistuned consonances, J. Acoust. Soc. Amer. *42*, 462-474.
Plomp, R., and Mimpen, A. M. (1968): The ear as a frequency analyzer. II. J. Acoust. Soc. Amer. *43*, 764-767.
Plomp, R., and Steeneken, H. J. M. (1969): Effect of phase on the timbre of complex tones, J. Acoust. Soc. Amer. *46*, 409-421.
Plomp, R., Pols, L. C. W., and Geer, J. P. van de (1967): Dimensional analysis of vowel spectra, J. Acoust. Soc. Amer. *41*, 707-712.
Rahlfs, V. (1966): Psychometrische Untersuchungen zur Wahrnehmung musikalischer Klänge, Doctoral Dissertation, University of Hamburg.
Röber, A. (1834): Untersuchungen des Hrn. Scheibler in Crefeld über die sogenannten Schläge, Schwebungen oder Stösse, Ann. Phys. Chem. *32*, 333-362 and 492-520.
Schouten, J. F. (1968): The perception of timbre, in: Reports 6th International Congress

on Acoustics, Tokyo, Japan, Vol. I, GP-6-2.
Schroeder, M. R. (1959): New results concerning monaural phase sensitivity, J. Acoust. Soc. Amer. *31*, 1579(A).
Seebeck, A. (1849): Akustik, in: Repertorium der Physik, Vol. 8 (Verlag von Veit & Comp., Berlin).
Slawson, A. W. (1968): Vowel quality and musical timbre as functions of spectrum envelope and fundamental frequency, J. Acoust. Soc. Amer. *43*, 87-101.
Stevens, S. S., and Davis, H. (1938): Hearing—Its Psychology and Physiology (Wiley, New York).
Stumpf, C. (1890): Tonpsychologie, Vol. 2 (Verlag S. Hirzel, Leipzig).
Stumpf, C. (1910): Beobachtungen über Kombinationstöne, Z. Psychol. *55*, 1-142.
Stumpf, C. (1926): Die Sprachlaute (Verlag J. Springer, Berlin).
Thomson, W. (1878): On beats of imperfect harmonies, Proc. Roy. Soc. Edinburgh *9*, 602-612.
Wegel, R. L., and Lane, C. E. (1924): The auditory masking of one pure tone by another and its probable relation to the dynamics of the inner ear, Phys. Rev. *23*, 266-285.
Weiss, A. P. (1920): The vowel character of fork tones, Amer. J. Psychol. *31*, 166-193.
Wesendonk, K. von (1909): Ueber die Synthese der Vokale aus einfachen Tönen und die Theorien von Helmholtz und Grassmann, Physikal. Z. *10*, 313-316.
Wever, E. G. (1949): Theory of Hearing (Wiley, New York).
Willis, R. (1830): On the vowel sounds, and on reed organ-pipes, Trans. Cambr. Philos. Soc. *3*, 231-268.
Zwicker, E. (1952): Die Grenzen der Hörbarkeit der Amplitudenmodulation und der Frequenzmodulation eines Tones, Acustica 2, Akust. Beih. 125-133.
Zwicker, E., and Feldtkeller, R. (1967): Das Ohr als Nachrichtenempfänger, 2nd Ed. (S. Hirzel Verlag, Stuttgart).
Zwicker, E., and Scharf, B. (1965): A model of loudness summation, Psychol. Rev. *72*, 3-26.

DISCUSSION

Schouten: I am worried about the matter of terminology. In acoustics we know very well what loudness is, what pitch is, and what duration is. And then all the other aspects are put into one dustbin which we call timbre. As you said quite rightly in your paper, very little attention has been given to the sorting out of the multidimensionality of timbre. You covered only one aspect and already went into many dimensions. We should realize, however, that this is only one part of it. We can play a violin at different loudnesses, different durations and different pitches, but it is always a violin. We say, well, that is its timbre. But this includes many aspects. First of all, we can decide very well whether a sound is tonal or noiselike. What you treated is specifically a tonal sound. Secondly, we have, roughly speaking, spectral characteristics, in your case covered by the amplitudes a_n and phases ϕ_n. Thirdly, there is the time envelope; it makes quite a difference whether we listen to a piano tone or to the time inversion of this sound. Fourthly, there can be changes within the signal as we have in the spectral distribution of, for instance, diphthongs and in pitch or frequency, what we call micro-intonation. Finally, there is in many sounds in nature and in music what we call the prefix, the fact that the start is quite different from the rest of the oscillations, as we have in organ pipes, vibrating strings, etc. What we need is, I think, a better terminology for these various aspects of timbre. As you mentioned in your paper, the word "Klangfarbe" is used in a restricted sense similar to what you studied, but we need proper words for all these aspects. Do you have any idea about that?

Plomp: I agree with you that it would be very helpful to have clearly defined words and it is quite unsatisfactory to apply the same word, timbre, both in a broad sense as you put it and in a rather narrow sense as synonymous to "Klangfarbe" in German and "klankkleur" in Dutch. It cannot be denied that in literature the word timbre is used in both senses. In order to avoid misunderstandings, the best thing I could do was to define very precisely in which way the word timbre in my paper had to be understood.

Sergeant: I would like to support strongly what has been said about the need of a clear terminology for timbre. A great deal of confusion has arisen in literature about the meaning of this word. In my opinion, we should keep the word timbre to refer to the whole ongoing process of the sound and then use subdivisions for the various effects or characteristics *within* that sound. I don't think that we can present these terms, but at least we should agree to use the same terminology; it does not matter which words are adopted if we use the same ones.

Plomp: I have thought about using another word instead of timbre to describe the perceptual differences between steady complex tones; the only one I found was *tone colour*, but I don't think this is an attractive alternative.

Bosher: May I comment upon another important point of your paper? You correlated the (subjective) dissimilarity between complex tones with the difference in frequency spectrum given by the equation on p. 406 and found rather flat curves as a function of p (Fig. 1). In a way we may consider p as evidence concerning the presence of a weighting factor. A maximum at a large p-value would mean that timbre discrimination is mainly based on those frequency bands in which the SPL difference is largest. On the other hand, a maximum at a small p-value would mean that only the question whether there is a SPL difference or not, irrespective of its size, is important. The fact that the curves of Fig. 1 show no pronounced maximum is very interesting with respect to the properties of the discriminatory mechanisms in the auditory system. It suggests that no specific weighting factor is involved, that these mechanisms are neither looking exclusively for large SPL differences nor exaggerating the presence of a difference, however small it may be.

Plomp: This conclusion would be justified were it not that, as shown in Fig. 2, there is a large intrinsic insensibility of the frequency spectra investigated for differences of p. We need many more experiments with various sets of stimuli to get a better insight in the question of whether or not the ear's discriminatory mechanism is governed by a particular p-value or weighting factor. Perhaps the fact that, for speech vowels, it does not matter whether $p=1$ or 2 or even 5 (Figs. 1 and 2) is an indication that no such a particular weighting factor exists. We have in mind to investigate the extent to which the maxima of the other curves depend upon the absolute value of the spectral differences. It is not excluded that a reduction of all dB differences by the same factor will result in a shift of maximum. In any case, the results show that it is rather dangerous to conclude that an Euclidean space is appropriate in describing multidimensional data only on the basis that an excellent fit is obtained. It may be that another p-value would give the same or even better results.

Goldstein: I would like to approach the effects of phase from a somewhat different point of view. You found that phase effects are unimportant in the signals you dealt with. I have some reservations about the generality of this finding. Cancellation experiments in the perception of combination tones are in fact phase perception studies; we perceive the effects of the phase changes of the cancellation tone quite clearly. This means in my view that in the nonlinear transformation in the peripheral ear, phase change can, along with other effects, significantly alter the amplitude spectrum of the signal, as is demonstrated in Fig. 10 of my paper. Therefore, I really wonder whether the phase effects are in general unimportant perceptually. In my knowledge, all data on phase perception in tone complexes below 60-70 dB SPL indicate that frequency separation between the spectral components is a very relevant parameter in studying the role of phase. This role may be large for harmonics above the 6th, but the effects may have been masked by the lower harmonics that are present in your stimuli. Whether or not phase effects are important perceptually, I think they are valuable phenomena for investigating the physical processes in the peripheral auditory system.

Plomp: Of course, the study of phase effects is important. I am in complete agreement with you on that. I would deny the relevance of my own experiments on the beats of mistuned consonances (Plomp, 1967) if I should state that phase is of minor importance in studying the hearing processes. My only point is that, as my experiments showed, the role of phase pattern in the timbre perception of complex tones representative of speech and music is *relatively* rather small compared with the role of amplitude pattern. We should also realize that in listening to phase shifts or to small pitch shifts we have the impression that there is a strong effect because we have focussed all our attention on that part of our sensation, whereas the effect is rather small, relatively speaking.

Scharf: I would like to come back to the problem of terminology discussed earlier. We should keep in mind the distinction between subjective and physical attributes since there is no one-to-one relationship between them. This is even the case for pure tones. Stevens (1934) in the early thirties was able to demonstrate experimentally that one can distinguish four subjective attributes: loudness, pitch, density and volume, all based on changes of only two physical attributes: frequency and SPL. Now, when we turn to these complex sounds you studied, of course, the problems become much greater than for pure tones. We don't even know how to define the various subjective attributes of complex tones. I think, Dr. Plomp has begun to explore this immense problem in a very systematic way. If we would like to have names for the different aspects of timbre, we should realize that no one is able to sit down and define these aspects. It is not a question of finding names for hundreds of subjective attributes of timbre, but we should look for operational distinctions which requires further experimental work of this type.

Plomp: I'd like to comment only that we started originally with semantic scales in looking for differences in the perception of musical intervals (van de Geer *et al.*, 1962) and concluded that it would be much better to avoid these scales in the beginning. Similarly, I started with nonverbal techniques in studying the influence of different parameters on the timbre of steady complex tones. When at some time these experiments have given a clear picture of the multidimensionality of timbre perception in its dependence on the physical parameters, it would be of great interest to investigate the relationship between the dimensions found and verbal categories by which the timbre differences can be described.

REFERENCES

Geer, J. P. van de, Levelt, W. J. M., and Plomp, R. (1962): The connotation of musical consonance, Acta Psychologica *20*, 308-319.
Plomp, R. (1967): Beats of mistuned consonances, J. Acoust. Soc. Amer. *42*, 462-474.
Stevens, S. S. (1934): The attributes of tones, Proc. Nat. Acad. Sci. *20*, 457-459.

ON THE PERCEPTION OF CONCORDS

K. SCHÜGERL

Wien

1. INTRODUCTION

This paper deals with the basic element of harmony, *viz.* the simultaneous sounding of two complex sounds, firstly by theoretical considerations in the light of modern hearing theory and secondly by discussing the results of some experiments with synthesized sound. The study is, mainly for the reason of simplicity, focussed on the major and the minor third.

2. ABBREVIATIONS

The following abbreviations shall be used throughout the paper. H = harmonic; C = complex sound, composed of several Hs; CC = concord, built up of two Cs, namely the C_l = the C of the lower and C_h = the C of the higher fundamental frequency; $f_l[f_h]$ = fundamental frequency of the $C_l[C_h]$, irrespective of the presence or absence of spectral energy at $f_l[f_h]$; MT = major third, mt = minor third. The spectral composition of a C, CC, MT or mt is indicated in the following way: $C(1 \div 4)$ means that the C is composed of the Hs 1, 2, 3 and 4; $CC\frac{(5 \div 8)}{(1 \div 4)}$ means that the CC is composed of $C_l(1 \div 4)$ and of $C_h(5 \div 8)$, so that the spectrum of the CC consists of f_l, $2f_l$, $3f_l$, $4f_l$, $5f_h$, $6f_h$, $7f_h$ and $8f_h$. $f_h = 5f_l/4$ (MT) or $6f_l/5$ (mt), according to pure temperament; $f_{cc} = f_h - f_l$; $T_{cc} = 1/f_{cc}$ = period of the CC.

3. WAVE DIAGRAMS

The space-time patterns in the cochlea are represented by wave diagrams (see Figs. 1-3). Fig. 1 is the wave diagram of a $C(1 \div 8)$, all the Hs being of equal amplitude; the phase angle of the Hs 3-8 is zero at the stapes. Fig. 1 is derived from a superimposition of eight transparent wave diagrams, one for each H. Because the excursions of the basilar membrane show too broad a

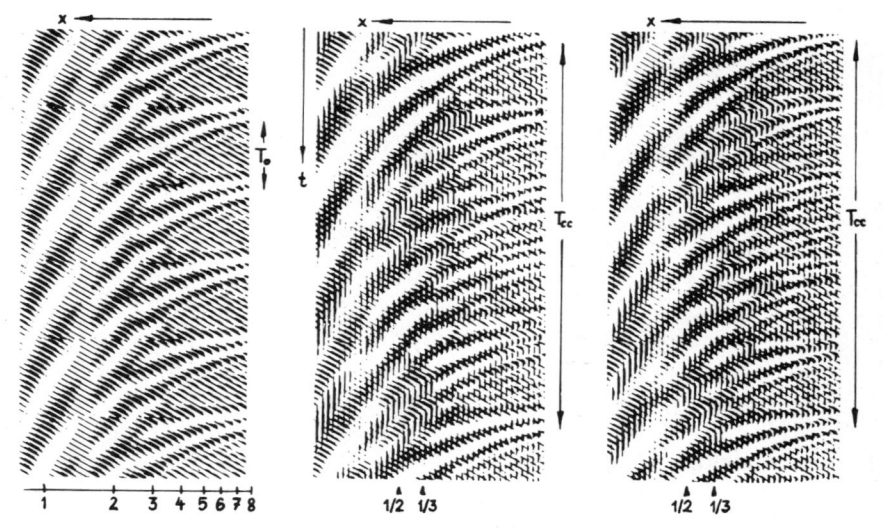

Fig. 1. Complex sound.   Fig. 2. Major third.   Fig. 3. Minor third.
Wave diagrams representing space-time patterns in the cochlea. x = spatial coordinate of the cochlea (stapes at the right); t = time; $T_o=1/f_o$ = period of the complex sound; T_{cc} = period of the concords.

distribution as to account for the resolving power of the ear, the excitation curve of each H has been "sharpened" in comparison to the curve given by von Békésy (1960) for 200 Hz. (For fuller details of how Fig. 1 was obtained see Schügerl, 1966.)

The wave diagram has to be interpreted like a map: heavily shaded areas (summits or ridges) represent excursions in one direction, open areas (depressions or valleys) excursions in the opposite direction. The deviation of the direction of the ridges and valleys from the horizontal is caused by the phase difference between stapes movement and displacement along the x-axis. The larger the phase shift, the smaller will be the local and momentaneous velocity of wave propagation. The loci of maximal displacement in the case of sounding each H separately are indicated by the scale at the bottom of Fig. 1. The number of summits (or depressions) varies in accordance with the Hs 1, 2, 3... This may be seen by covering the wave diagram by an opaque sheet with a vertical slit of approximately 1.5 mm and, then, twinkling the eyes whilst moving the sheet from left to right.

By covering the wave diagram of Fig. 1 by another one, differing from the first only in the fundamental frequency (4/5 for MT, 6/5 for mt, respectively), the wave diagrams for a MT (Fig. 2) and for a mt (Fig. 3) are obtained. The overall periodicity jumps from T_o to $T_{cc}=5T_o$. Other periodicities than T_{cc}, namely $T_{cc}/2$ and $T_{cc}/3$, can be seen too, optimally by using the slit described above (the positions of the slit for $T_{cc}/2$ and $T_{cc}/3$ are indicated by 1/2 and 1/3).

Thus, not only the subjective tone which corresponds to f_{cc}, but also the subjective tones corresponding to the multiples of f_{cc} ($2f_{cc}$, $3f_{cc}$...), which have been often observed when sounding two Cs together (*e.g.* by Riemann, 1875), find their counterpart in the time pattern of the basilar membrane and may possibly be regarded as residual phenomena. Note also the complicated variations of the fine structure during the period T_{cc}, which, for a fixed point x, may be interpreted as both an amplitude and phase modulation.

4. LOCAL DISTRIBUTION OF PERIODICITIES

In Fig. 4 the local distribution of periodicities in the cochlea is shown in another, highly schematized way for a MT (above) and a mt (below) of the composition $CC\frac{(1 \div 16)}{(1 \div 16)}$. The loci of maximal displacement in the case of sounding each H separately are plotted on a logarithmic scale, corresponding to the local distribution of frequencies in the cochlea. The excitations of any pair of Hs, whose frequency difference is small, will disturb each other, and that the more, the smaller the distance between the loci of the maxima for the two Hs. These regions of disturbance are indicated by triangles. At the dashes for the Hs $2f_l$ and $2f_h$ of the MT, the construction of the length b (basis of the triangles) is shown. The somewhat arbitrary length marked with "const" has to be chosen with regard to the ear's critical bandwidth. The repetition frequency of the pattern of the envelope equals the frequency difference of the components, namely f_{cc}, $2f_{cc}$, and so on. The triangles are arranged in lines according to those difference frequencies. In the region of lower Hs (left), Fig. 4 corresponds rather well with Figs. 2 and 3 (compare the position of the triangles for $2f_{cc}$ and $3f_{cc}$ with the marks 1/2 and 1/3). In the region of higher Hs, however, the critical bandwidth encompasses the excitation areas of more than two Hs, a fact reflected by the overlapping of the triangles, from which an enhancement of

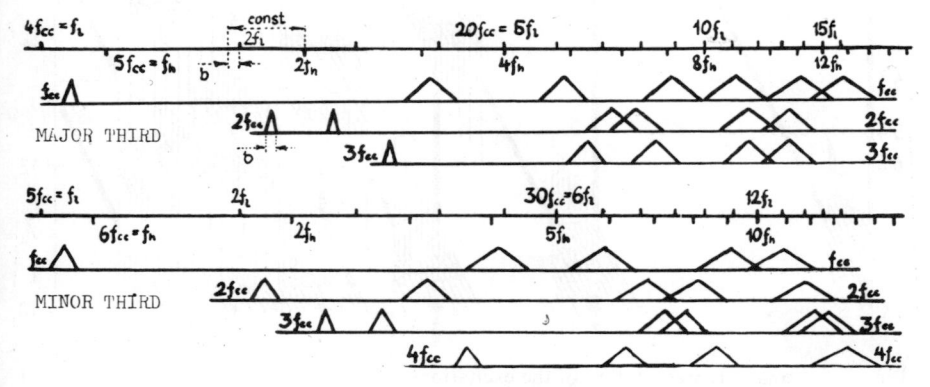

Fig. 4. Distribution of periodicities along the basilar membrane.

the period $T_{cc}=1/f_{cc}$ may be expected. (A superposition of two patterns with the repetition frequencies $2f_{cc}$ and $3f_{cc}$ yields a pattern with the repetition frequency f_{cc}.)

In any case, a certain CC (MT, mt, ...) is characterized by a specific local distribution of repetition frequencies f_{cc}, $2f_{cc}$, ... or by a kind of "formants of periodicity" (of the envelope), whereby the specificity of the distribution will be partly lost in the regions of higher Hs due to the large extent of overlapping.

5. CRITICAL BANDWIDTH, CONTINUOUSLY VARYING FREQUENCY RATIO

Another application of the concept of critical bandwidth is given in Fig. 5. In this case, the CC has the composition $CC\frac{(1 \div 6)}{(1 \div 6)}$, f_l = const and f_h varies continuously between f_l and $2f_l$. The horizontal scale is logarithmic and may also represent the spatial coordinate x of the cochlea. Hs which may (not must, of course) be heard separately, are marked by heavy full lines. Hatched areas indicate that two Hs interact in some way, as it is the case when they fall within a single critical band. The width of the hatched areas, derived in a similar manner as the basis b of the triangles in Fig. 4, is a rough measure of the extent of overlap of the excitations.

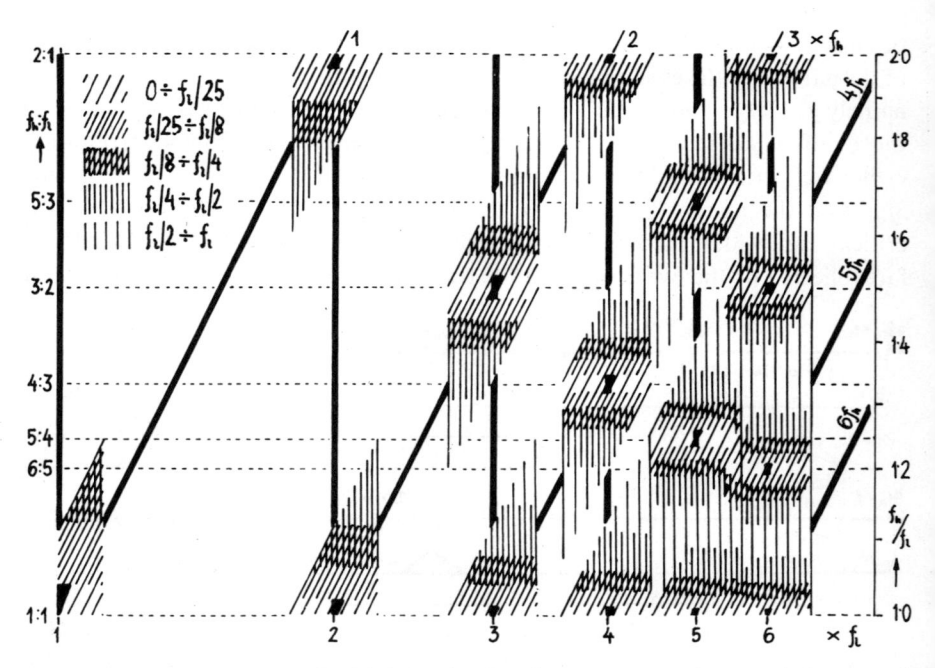

Fig. 5. Schematic representation of the excitations due to two complex sounds (composed each of the harmonics $1 \div 6$) of continuously varying frequency ratio.

In order not to overcrowd the diagram, a somewhat smaller value than in Fig. 4 is taken for the basis b and, accordingly, for the critical bandwidth; otherwise, even in the region of lower Hs, more than two Hs would fall within a single critical band. In such a case it would be difficult to determine the most prominent frequency which actually influences the sensation. Strictly speaking, there are already in Fig. 5, beyond $4f_1$, regions of three Hs per critical bandwidth; they have been neglected, however, since they are very narrow. In view of this difficulty, Fig. 5 has been restricted to CCs with the Hs $1 \div 6$. As a further consequence of the rather small value of the critical bandwidth, chosen for Fig. 5, the spectral resolution, as indicated by the smallest distance between the heavy full lines (at left) is slightly better than the spectral resolution according to Plomp (1966). Generally, the diagram should be regarded only as a qualitative illustration.

The repetition frequency of the envelope in the regions of overlap (the beat frequency) is shown by the style of hatching, as explained by the insert left hand top corner, which gives the upper and lower limits of the beat frequency in fractions of f_1. As it is well known (von Helmholtz, 1863), very slow beats are indifferent or even beneficial to the musical quality — provided they are not too regular — and very fast beats produce only negligible roughness, the maximum of disturbance occurring in a range of, say, 25-50 Hz (see also Terhardt's paper, pp. 278-290 of this volume). Assuming for f_1 a value of approximately 200 Hz, the system of hatchings illustrates this point: the heavier the hatching, the more disturbing the roughness. For higher f_1's, the zones of maximal disturbance have to be shifted toward the centre of the lozenges.

From Fig. 5, it might be inferred that a certain CC will be characterized, *inter alia*, by a system of resolved (or resolvable) "spectral lines" and of "spectral bands" (non-resolvable Hs, high degree of overlap, strongly marked periodicity) (see also Sect. 4). This specific local distribution of sources of roughness (of different quality) may serve as a basis for the recognition of a certain CC. Such distribution would supply more information than the mere sum of the individual disturbances according to von Helmholtz (1863). The latter principle, on the other hand, is useful in distinguishing a justly tuned prime or octave or fifth from a mistuned one.

6. PROCEDURE OF THE EXPERIMENTS

To check some aspects of the above considerations, experiments were carried out using Cs produced by a synthesizer of the photoelectric type. In each C, the Hs were cosine functions of equal amplitude; the phase angle was zero for all Hs. The Cs were recorded on dual tape, the C_ls on one and the C_hs on the other track. The stimuli were presented to the listeners *via* two loudspeakers, mounted in a vertical wall, one closely above the other. Each loudspeaker was connected to one of the outputs of the stereo amplifier. In this way, intermodulation products between C_l and C_h are avoided as much as possible.

Two kinds of experiments were performed. In the first one, a sequence of pairs of CCs were presented. Each pair consisted of a MT and a mt; the spectral composition of the CCs was different for different pairs. The subjects were asked to make statements about the difference in the interval character (or quality) between MT and mt, and to compare the difference in one pair with the differences in other pairs. The second test dealt with a randomized sequence of MTs and mts of different spectral composition. In this case, the subjects had to identify the CC as a MT or a mt. $(f_l+f_h)/2$ was 130 Hz, 260 Hz (best frequency for discrimination) and 520 Hz. Seven subjects, only a part of them trained musicians, participated in the experiments.

7. RESULTS WITH REGARD TO TIMBRE

In general, the characteristic differences between MT and mt were most prominent if the two Cs, which constitute the CC, were for themselves of good (e.g. $CC\frac{(1 \div 4)}{(1 \div 4)}$) or tolerable (e.g. $CC\frac{(5 \div 8)}{(5 \div 8)}$) musical quality. The latter CC is somewhat sharp, because of the Hs 7 and 8, as was already noticed by von Helmholtz (1863). The more Hs fall within the same critical band, the deeper the modulations of the vibrations in the cochlea will be, as may be seen from the right part of Fig. 1, and the sharper the sound will be. It became apparent from the tests that an increase in sharpness leads to an impairment of the individuality of MT and mt. $C(17 \div 32)$ yielded, contrary to Cs composed of lower Hs, a rather atonal residue (Schouten et al., 1962). MTs and mts of the composition $CC\frac{(17 \div 32)}{(17 \div 32)}$ showed almost no differences in quality. The sharpness or piercingness, due to the higher Hs, seems to be a nonspecific element, which distracts the attention from the specific features of a CC. This was confirmed by listening to MTs and mts of the composition $CC\frac{(1 \div 32)}{(1 \div 32)}$: the quality differences seemed to emerge mainly from the lower region of the spectrum (or from the two pitches, see Sect. 9).

8. ELIMINATION OF SELECTED HARMONICS

The specific character of MT and mt, which stood out pretty well when sounding the $CC\frac{(1 \div 8)}{(1 \div 8)}$, was almost entirely lost with $CC\frac{(3,5,7)}{(3,5,7)}$, so that identification was very difficult. As may be seen from Fig. 4, the latter CC has for a MT and a mt no adjacent Hs with frequency difference f_{cc}. Consequently, the regions of the periodicity $1/f_{cc}$ and the roughness, produced by this relatively slow periodicity, are greatly diminished (note that for $(f_1+f_h)/2=260$ Hz f_{cc} is of the order of 50–60 Hz). By adding the fundamentals, so that the $CC\frac{(1,3,5,7)}{(1,3,5,7)}$ resulted, the discriminability was improved, as is to be expected from the fact that by reintroducing the region of the periodicity $1/f_{cc}$ at the lowermost part of the CC, the usual impression of a MT or a mt is somewhat approached (see Fig. 4). On the other hand, removing the fundamentals from the $CC\frac{(1\div 8)}{(1\div 8)}$ diminished the differences in quality between a MT and a mt only by a small amount, most probably because the remaining $CC\frac{(2\div 8)}{(2\div 8)}$ produces a sufficient degree of the periodicity $1/f_{cc}$ and the corresponding roughness. The elimination of the coinciding Hs ($5f_1$, $4f_h$ or $6f_1$, $5f_h$) from the $CC\frac{(1\div 8)}{(1\div 8)}$ had only small effects on the discriminability, evidently from the reason that the regions of periodicity and the distribution of roughness remain unaffected.

9. PITCH

The results of Sect. 8 seem to be in line with the consonance theory of von Helmholtz (1863), which is based on the roughness, produced by the Hs of the frequency difference f_{cc}. But other spectral compositions with little or, perhaps, no overlapping of the excitation areas of C_1 and C_h, as $CC\frac{(5\div 8)}{(1\div 4)}$, $CC\frac{(9\div 16)}{(5\div 8)}$ and $CC\frac{(9\div 16)}{(1\div 4)}$, still could be discriminated, although the listeners judged the CCs as "dull", "not exciting" or so. A CC producing four adjacent excitation areas in the cochlea was synthesized, the first and the third areas consisting only of Hs of C_1 and the second and fourth areas only of Hs of C_h. Although the CC has a rather strange character, musically trained subjects had no difficulty in identifying MTs and mts. Evidently, one is able to hear the two pitches (corresponding to f_1 and f_h). In many cases, the two pitches are more or less fused into a complex sensation which may be judged spontaneously, without further deliberation. If, however, the evaluation of the CC is difficult, as in cases of unusual spectral composition, a "good" subject finds the right answer by scanning successively the two pitches and comparing them with the models stored in his memory.

A comparison between subjects of high and low performance demonstrated that the effect of musical training is to be able to distinguish timbre and formant tone (Meyer-Eppler et al., 1959) as derived from the frequency domain on the one side, and pitch as a result of periodicity on the other (Plomp, 1966). There is presumably a kind of competition between timbre and pitch in the process of evaluating a CC. The $\mathrm{mt}\frac{(1,2,4,8,16,32)}{(1,2,4,8,16,32)}$ was qualified as open, free, articulated and beautiful, though somewhat strange. These qualities misled often a subject to judge this mt as a MT, whereas a MT of the above composition was always correctly identified. (It should be noted, however, that this composition is well resolved by the ear and, therefore, periodicity is not pronounced.) Conversely, a MT with a high degree of unspecific roughness or piercingness (see Sect. 7) was often misinterpreted — but only by less trained subjects — as a mt, apparently because, in our musical experience, the mt figures as something dimmed or troubled in comparison with a MT.

10. ANIMATED SOUND

An animation roughly similar to the musical vibrato was produced by modulating the amplitude of C_l and C_h synchronously by one and the same function. A comparison of pairs of CCs without and with amplitude modulation showed at modulation frequencies of about 6 Hz no influence on the major-minor characteristics; at 12-15 Hz, however, an unspecific disturbance was generated, which reduced the differences in quality. Another kind of animation may be brought about by altering the frequency difference by a small amount. Although this yields amplitude modulations too, due to the slow beatings of the almost coinciding Hs, the sensations were quite different. By making the intervals wider than 5/4 or 6/5, respectively, the MT gained more brightness and the sensation of a mt showed a tendency towards the character of a MT. Narrowing the intervals makes the mt even more flat and dull, whilst the MT takes a somewhat minor-like hue. These results may be brought back, perhaps, to the predominance of equal temperament in our musical experience. They are easier to understand from the viewpoint of periodicity detection, for there was no apparent change in roughness or timbre.

11. FUSION

From Sect. 9, it may be concluded that by impairing the specific distribution of roughness or even by completely removing the roughness, the specificity of a CC does in most cases not disappear entirely. Thus, in addition to the aspect of disturbance, the aspect of fusion, introduced by Stumpf (1890), shall be discussed. Fusion ("Verschmelzung") may be regarded as the resistance against the extraction of the two pitches from a CC. This phenomenon is illustrated by interpreting the CC as a C of the frequency f_{cc}, from which certain Hs have been eliminated. Table I shows in column (A) the frequency ratio (M, N integers, $M \leq N$); in column (B) the ratio of the number of eliminated Hs to the total number of Hs before elimination, calculated under the assumption that the C

Table I. Figures for concords, interpreted as a single complex sound with selected harmonics.

(A)	(B)	(C)	(D)
N:M	$\frac{NM-N-M+1}{NM}$	$1/NM$	$M-1$
1:1	·000	1·000	0
2:1	·000	·500	0
3:2	·333	·167	1
4:3	·500	·083	2
5:3	·533	·067	2
5:4	·600	·050	3
6:5	·667	·033	4
8:5	·700	·025	4

has a large number of Hs; in column (C) the ratio of the coinciding Hs to the total number of Hs, and in column (D) the number of eliminated Hs below the lowest present $H = Mf_{cc}$. The order of the CCs in the table agrees quite well with the fusion order given by Stumpf, especially for the CCs with the lower integers. The figures of the table show schematically to what extent the spectrum and, accordingly, to what extent the spatial distribution in the cochlea for a CC depart from the spectrum or the distribution for a single C as occurring in spoken vowels or monodic music.

12. CONCLUSION

Thus, fusion may be an example for the law of "Prägnanz" as formulated by the Gestalt psychologists (Hofstätter, 1957). In the course of the neural processing of a CC a tendency prevails to fill up the "voids", the missing elements of the Gestalt, in other words, the eliminated Hs and so reconstitute the habitual Gestalt of a C. Since every change in the frequency domain has, according to the theorem of Fourier, its counterpart in a change in the time domain, fusion may also be regarded as the result of a tendency to misinterpret the more articulated time pattern (*e.g.* Fig. 2) in the manner of the customary regular time pattern of a unison (*e.g.* Fig. 1).

But how is the reverse of fusion, the extraction of the two pitches, effected? Is there in the case that the tendency stated above is inoperative, again a tendency to disentangle the excessive supply of information by subdividing the excitations into two groups of different fundamental frequency? This seems to be easier in such cases where the excitation areas of C_l and C_h do not interfere (see Sect. 9). The overlap of the excitation regions, however, complicates the problem. It seems to be very likely that separation into two groups is facilitated if, unlike the experimental situation reported in Sect. 10, each group has specific modulations, or irregularities, which distinguish the Hs of one group from the Hs of the other. In actual music, there are several factors which work along this line, onset phenomena, arpeggio-like techniques, transients, vibrato, small mistunings, cues from binaural listening and musical context.

The experiments using various spectral compositions showed a rather astonishingly great stability of the specificity of a MT and a mt. This very stability, however, is the characteristic of a highly organized pattern, an appropriate explanation of which cannot be afforded by summing up the results of the analyses of the elements. This interdependence of the elements stood out clearly from the experiments. Every change in the spectral composition involved changes in all possible dimensions of the timbre, such as roughness, brightness, fullness, similitude to vowels. This makes it nearly impossible to isolate the influence on one dimension, the more so by the quick shifting of the attention from one aspect to another as the actual experimental situation requires.

Consequently, it would not be quite correct to distinguish between a natural component corresponding to the concept of von Helmholtz and a cultural component, or the effect of learning. The information on a peripheral level of the sensory process affords several features, which may be used depending on, among others, attention, talent, education and musical context. The effect of learning is perhaps not so much the association with other data, but consists rather in directing the attention on all available features. Whereas we are rather well informed about the coarsest feature, namely the disturbance as the result of rather slow periodicities, very little is known about the other, finer and more remote features, resultant from the extraction of the two pitches and the complex quality emergent from their interaction. In addition to the introduction of the concept of critical bandwidth into the theory of von Helmholtz, the concept of periodicity detection should be extended to sounds with several periodicities. For this, a closer consideration of the space-time pattern in the ear, as a physiological counterpart of the psychological Gestalt, may be of value.

ACKNOWLEDGEMENTS

The author is very grateful to Professor W. Graf of Vienna University for many stimulating discussions. Financial aid to establish the synthesizer was granted by the Austrian "Fonds zur Förderung der wissenschaftlichen Forschung".

REFERENCES

Békésy, G. von (1960): Experiments in Hearing (McGraw-Hill, New York), p. 448.
Helmholtz, H. L. F. von (1863): Die Lehre von den Tonempfindungen als physiologische Grundlage für die Theorie der Musik, transl. by Ellis (Dover, New York, 1954), pp. 179-197.
Hofstätter, R. (1957): Psychologie (Fischer Bücherei, Frankfurt/Main), pp. 142-152.
Meyer-Eppler, W., Sendhoff, H., and Rupprath, R. (1959): Residualton und Formantton, Gravesaner Blätter 4, 70-83.

Plomp, R. (1966): Experiments on Tone Perception, Doctoral Dissertation, University of Utrecht, pp. 8-23, 48-61, 88-101.
Riemann, H. (1875): Die objektive Existenz der Untertöne (Luckhardtsche Verlagbuchhandlung, Cassel).
Schouten, J. F., Ritsma, R. J., and Cardozo, B. L. (1962): Pitch of the residue, J. Acoust. Soc. Amer. *34*, 1418-1424.
Schügerl, K. (1966): Die Abbildung von Klängen auf der Schneckentrennwand, Arch. Ohren-, Nasen- u. Kehlkopfheilk. *186*, 134-145.
Stumpf, C. (1890): Tonpsychologie (Hirzel Verlag Leipzig), Vol. 2, pp. 127-183.

DISCUSSION

Kuyper: You base your reasoning in two ways on integral numbers. First, you take complex sounds consisting of frequency components f, $2f$, $3f$, etc. I agree that this is justified because the partials of sounds produced by musical instruments and of speech vowels are mostly harmonically related, with the exception of bells. Second, you also take integral numbers in the frequency ratios of your concords. I don't think, however, that this is justified because since Bach the equally tempered scale has been accepted in music, after which we have no simple frequency ratio of, say, 4:5 for a major third, but this third is a little bit mistuned. Therefore, you should have many more of these triangular signs in your Fig. 4.

Schügerl: To answer this, look at Fig. 5. If you change from pure temperament to equal temperament, you have only to shift the horizontal dotted lines for the consonances by small amounts up or down. By these small shifts, the distribution of the roughness is not affected materially. The periodicities in the hatched areas will remain about the same. A similar result holds for the time domain. Assume you change from the values of 400 and 500 Hz to the values of 405 and 505 Hz. From a purely mathematical point of view, the exact repetition frequency of the whole period will jump from 100 Hz down to 5 Hz. But the repetition frequency of the envelope remains 100 Hz. The shape of the envelope remains also the same. The slow beats, resulting from 5 Hz, have only very small influence on the interval character of a major third. Therefore, in Fig. 4 the slight mistuning will result only in small shifts in the frequencies of the triangles, but the approach is not changed in principle.

Kuyper: Thank you. I have a second point. I understand that there is a difference between musically trained subjects and untrained subjects, although it is not very great. I suggest that another difference would be more important in performing experiments on the character of intervals. I assume that a violinist will match after the natural scale and a pianist after the equally tempered scale, more or less. Is that true?

Schügerl: There are small differences indeed, but they lie in another direction. The piano player can only influence the intensity and the moment of attack. The player of a string instrument is trained to form the onset and the timbre, to control the sound he produces in accordance to his ideas. The difference is not only the attachment to equal or to pure temperament. The keyboard player hears in a more abstract way, the string instrumentalist in a more sensual way.

Ritsma: If one takes two pure tones of, for example, 400 Hz and 500 Hz, and presents them monaurally, one gets the impression of a major third. One does not get that impression by presenting one of the tones to the left ear and the other to the right ear. Might this difference be due to the fact that the overlap is missing in the latter situation?

Schügerl: Yes, the overlap is essential. Of course, the question is of interest: what remains from a concord, if the overlap is removed? There are experiments concerning this question by Husmann (1953), who supplied one ear with the lower and the other with the higher tone. With pure tones as stimuli, the subjects could neither say whether a consonance or a dissonance was presented, nor could they name the intervals. Even with complex tones, most subjects could not name the intervals, but all subjects reported an exact and distinct difference between intervals with small integers and the other ones. Husman explains this by the coincidence of certain harmonics, in a similar manner as von Helmholtz explains successive consonances. Bone conduction may also have an influence. I am not sure whether the mechanism which is operative in conditions without overlap will be active during normal hearing with overlap.

REFERENCES

Husmann, H. (1953): Vom Wesen der Konsonanz (Verlag Müller-Thiergarten, Heidelberg).

MACH BANDS IN AUDITORY PERCEPTION

E. C. CARTERETTE, M. P. FRIEDMAN and
J. D. LOVELL

Department of Psychology
University of California
Los Angeles, Calif., U.S.A.

INTRODUCTION

Detection and enhancement of contours in the eye of the horseshoe crab *limulus* have been shown to depend on a nerve net whose elements mutually inhibit each other (Hartline *et al.*, 1961). There is strong presumptive evidence from psychophysical experiments that the human eye also possesses some type of laterally inhibiting neural network (Alpern and David, 1959). The pervasive contrast effects (Mach bands) in human visual perception have recently been reviewed in Ratliff's excellent monograph (1965).

Such laterally inhibiting neural networks almost certainly exist in the ear. It appears impossible otherwise to account for the remarkable ability of the ear in discriminating between two pure tones only a few Hz apart. Von Békésy (1928; 1960, Ch. 11) and Huggins and Licklider (1951) as well as others (*e.g.* Wever and Lawrence, 1954) have pointed out how far short of explanation are the mechanical sharpening properties of the basilar membrane.

If mutual inhibition exists among the neural elements of the auditory pathway, then tonal contours should be enhanced by analogy with the mechanism of Mach bands in vision. According to the place theory, a narrow band of noise "stimulates a certain section of the basilar membrane intensively with a drop in the excitation at both edges. These edges are emphasized by contrast phenomena like the Mach bands" (von Békésy, 1963). Von Békésy (1963) had a subject who could match pitches to the upper and lower edges, or half-power points, of an octave band of noise lying between 400 and 800 Hz.

Small and Daniloff (1967) used an easier task. Their subjects made pitch matches an octave above or below the edge of a high-pass noise, a low-pass noise, and sinusoids. The evoked pitch of the noise bands was found to be "some nearly linear" function of cutoff frequency. In general, pitches judged an octave below cutoff were systematically low, whereas pitches judged an octave above were systematically high. Small and Daniloff rejected sharpening or contrast mechanisms of the place theory, as presently conceived, to account for their results, because the effective stimulation patterns of the noise bands lacked sufficiently steep gradients.

MONAURAL MASKING WITH NOISE BANDS HAVING STEEP SKIRTS

We reasoned that by using narrow bands of noise synthesized by digital computer we could obtain very steep gradients of physical excitation, and so maximize the chance of finding edge effects, or auditory Mach bands. Of course the steepness of the gradients would be limited by the response of the recording system, the earphones, and the basilar membrane. The work discussed below has now been reported in full in Carterette, Friedman and Lovell (1969).

A digital computer was used to generate a Rayleigh (narrow-band) noise 100 Hz wide centered at about 530 Hz and having theoretically infinite attenuation rates outside the pass band. In all, 56 sinusoids were used spaced randomly by frequency, with average separations of about 1.73 Hz. After conversion from digital to analog form, the 100 Hz wide noise was recorded on an Ampex model PR-10 magnetic tape recorder at 7.5 inches/sec. When played back at double the recording speed, the result was a 200 Hz band of noise centered at about 1060 Hz. Fig. 1 shows the amplitude of the noise as a function of frequency as measured with the 3 Hz filter of a wave analyzer.

Monaural thresholds for pure tones were obtained in the quiet by means of a modified von Békésy tracking method. Tones were switched on and off smoothly (rise and decay times of 10 msec) so that tone bursts of 125 msec duration alternated with silences of 125 msec duration. Frequency of the tones ranged between about a half octave below and above the lower and upper cutoff frequencies of the noise. Twelve minutes were required to track the range. Thus the frequency of the oscillator changed at a rate of about 28 Hz/min for the low-frequency band, which was 100 Hz in width, and 56 Hz/min for the

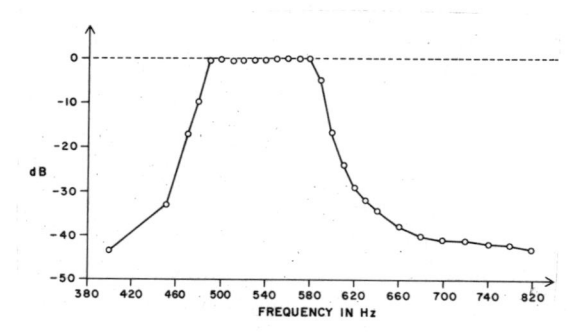

Fig. 1. Frequency spectrum in dB/Hz of the 100 Hz band-pass noise.

high-frequency band, which was 200 Hz in width. Frequency was monitored continuously by an electronic counter. Monaural masked thresholds for the pure tones in the narrow-band noises were obtained in the same way. The audiometric records were read at 10 Hz intervals, and from these values mean threshold shifts were computed.

Each of four listeners was run through four such sessions for the two bands of masking noise at five different sensation levels (SL) of the masking noise: 20, 30, 40, 50 and 60 dB SL. Frequency was swept upwards in half the sessions, downwards in the other half. Two listeners were women, two were men; their ages ranged from 20 to 40 years.

The results for the 100 Hz wide noise are shown in Fig. 2, those for the 200 Hz wide noise in Fig. 3. These functions clearly are not the smooth critical band curves to be found in the classical masking experiments. The curves show edge effects or sharpening in the neighborhood of the nominal cutoff frequencies indicated by the vertical dashed lines. Some of the curves look very much like what one might expect from the heuristic general sensory inhibitory model of von Békésy or from the Hartline-Ratliff model of the simple eye of the horseshoe crab, *limulus* (Fig. 4). With no essential exceptions it can be said that associated with the edge of the noise band there is a marked discontinuity in the masking

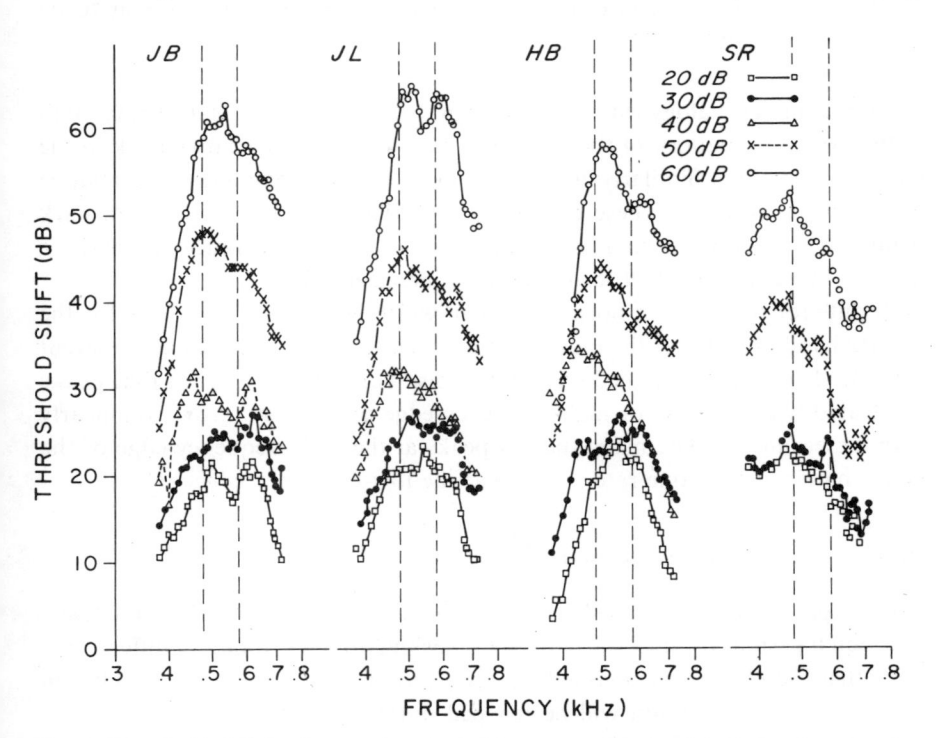

Fig. 2. Threshold shift in dB as a function of frequency, for the 100 Hz band-pass noise

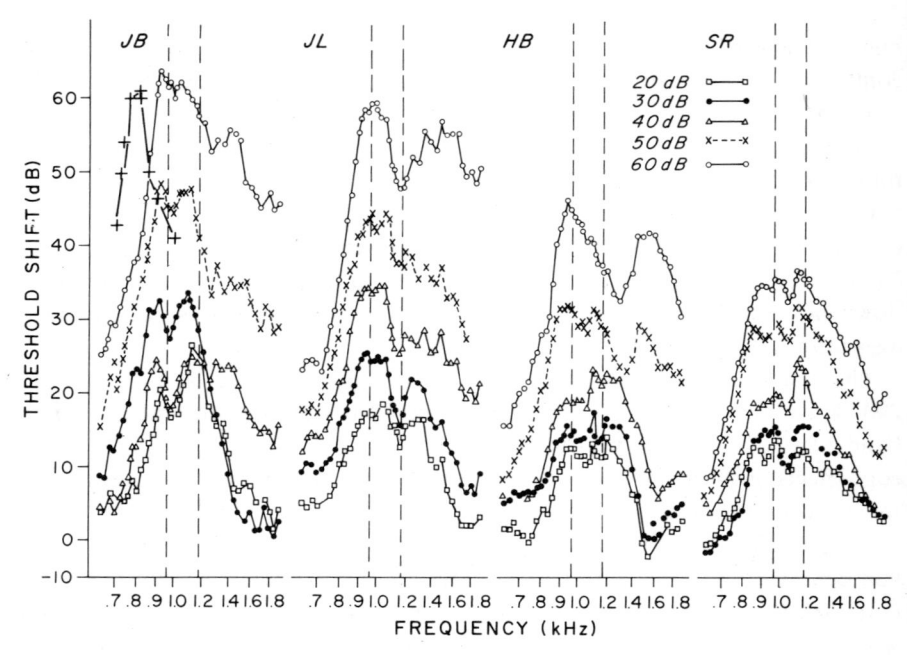

Fig. 3. Same as Fig. 2, except that above 1200 Hz measurements were taken at 20 Hz rather than 10 Hz intervals.

function. On the analogy with vision a bright band would be a relative maximum in the excitation pattern of the noise and would show as a relative peak in the masking function. Similarly a dark band would be manifest as a relative minimum in masking. These curves show evidence of both bright and dark bands. We see, too, a standard result, namely, the appearance of masking upward in frequency which increases with increasing SL.

Results for the band of noise 200 Hz in width were similar to those for the 100 Hz band except that the upward shift of masking was much more apparent (Fig. 3). The results obtained with both noise bands show (Fig. 5) that peaks associated with the low- or high-frequency edge of the band grow nonlinearly with increasing SL. The magnitude of peaks at the high-frequency edge of the band change at the same rate as those at the low-frequency edge.

EDGE EFFECTS IN THE TIME DOMAIN

We expected that if the masked tone were alternately on and off for relatively long equal periods of time, the pattern of masking found should not differ much from that found when the masked tone is on continuously, since an equilibrium state of mutual inhibition will be reached during each period of stimulation. A similar result should be observed for relatively short equal periods of alterna-

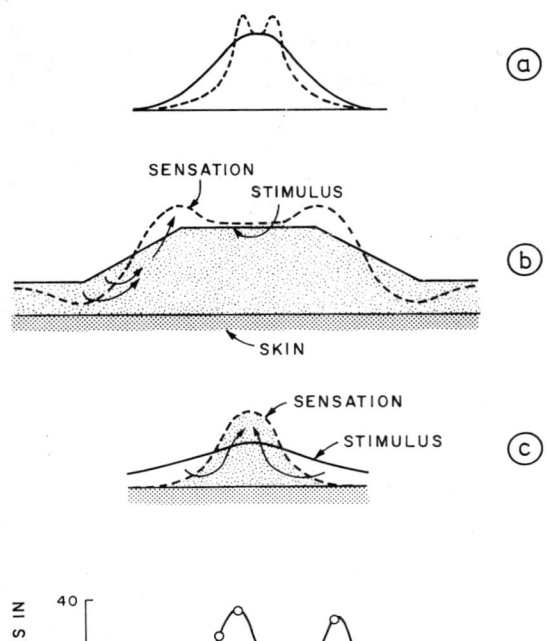

Fig. 4. (a) The amplitude distribution (solid line) on the basilar membrane of two simultaneous tones close in frequency and the expected sensory effect (dashed line). (b) The intensity distribution on the skin and the resulting sensation. (c) Sensation to be expected when stimulus has a single, gentle peak (a, b, and c modified from von Békésy, 1928). (d) Rate of discharge of impulses from the retinal ganglion cell of a cat is shown as elicited by a step function of illumination (b is bright, d is dark). The solid curve was drawn by hand through the data points. The dashed horizontal line represents the rate of discharge elicited by an extended field having the same intensity as the bright zone, b (d redrawn from Baumgartner, 1961).

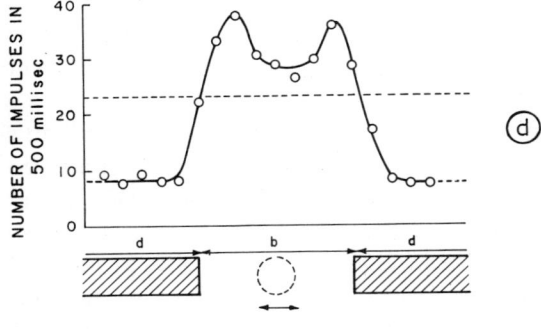

tion, since temporal summation should occur. There should be an intermediate rate at which the maximal effect occurs, one for which the duration of the masked tone is long enough to approach its maximum intensity but short enough to exert little inhibition on the masker.

As a test of this notion, subject J. B. was run for four sessions under each of four conditions of an interrupted masked tone alternately on and off for periods of 250, 125, 50, and 25 msec, respectively, and one condition of a continuous tone. The level of the continuous 100 Hz band-pass synthetic noise used as masker was always 50 dB SL.

Fig. 6 shows the results. Threshold shift was measured at 10 Hz intervals. The secondary peaks (or bright bands) near 450 and 600 Hz are most pronounced for burst durations of 125 and 250 msec, with the latter graph looking more like von Békésy's idealizations. The secondary maxima flanking the main peaks when measured from valley to peak average about 4 dB. The peaks lie

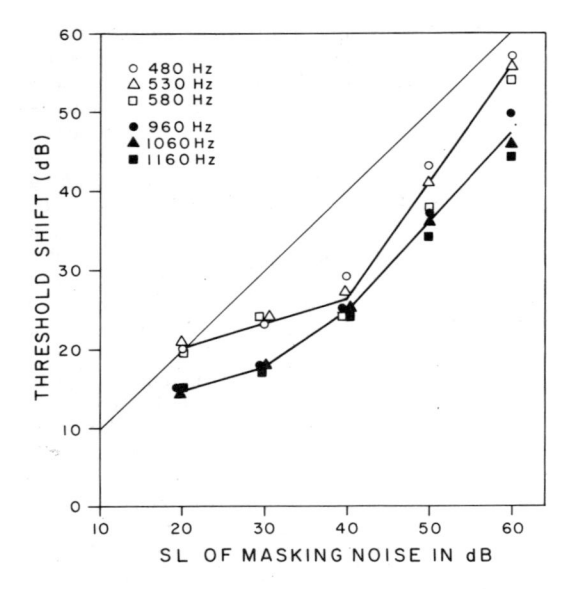

Fig. 5. Threshold shift in dB averaged over all four listeners at the nominal band edges and centers for the 100 Hz band (open points) and for the 200 Hz band (solid points). Each point is based on 16 observations. The solid lines are drawn through the over-all means.

outside the nominal cutoff frequencies of the noise, whereas the minima of the valleys lie on them. There are additional maxima, which probably represent real effects, as at 230 Hz, near 300 Hz, and perhaps near 675 Hz.

These limited data are consistent with the temporal effects seen in Elliott's (1967) experiments on backward and forward masking. Her work supports the hypothesis that the ear requires an "organization" time of about 250 msec to establish frequency contours. Evidence of the critical role of time in central masking may be seen in the study of Zwislocki et al. (1968; see also pp. 445-454 of this volume). They measured the threshold shift of 10 msec tone bursts in one ear as affected by a 1000 Hz masking tone burst of 250 msec duration in the other ear ("central masking"). The amount of threshold shift as a function of the frequency of the test tone burst of 10 msec depended upon both the SL of the masker and the time delay between the masker and test tone. Local maxima in the masking functions were seen above and below the main maximum at 1000 Hz. The local maximum below 1000 Hz shifted toward lower frequencies as the delay between masker and test tone decreased from 160 to 0 msec.

Thus it appears from the data of Elliott, from that of Zwislocki et al., and from our data that the interaction of excitation and inhibition is strongly time dependent and that transients heighten the degree of sharpening. Lateral inhibition not only produces maxima and minima in the response where none appears in the stimulus but there may be temporal and spatial amplification of the response. Amplification means that the peak-to-peak distance between maxima and minima of a response can be greater with lateral inhibition than without it. Spatial amplification has been calculated theoretically by von Békésy

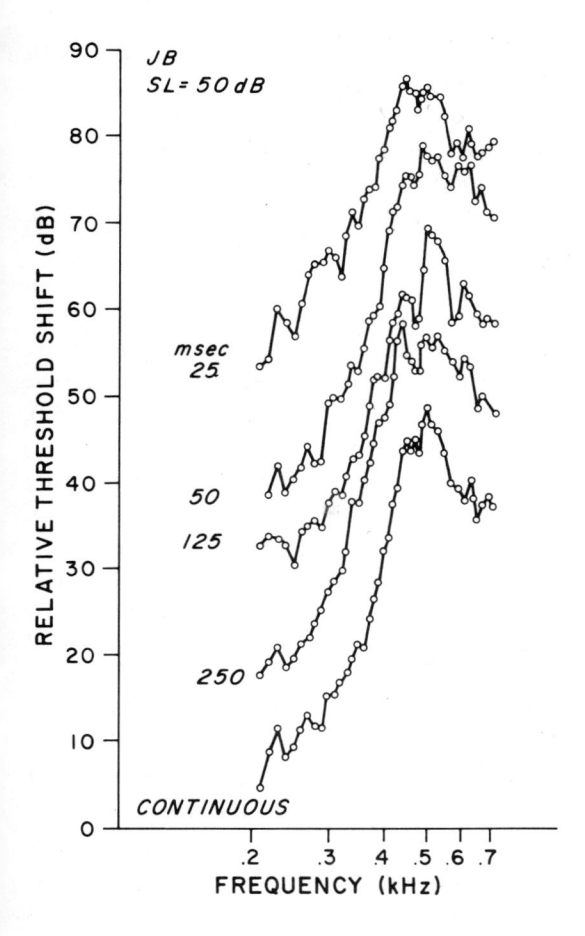

Fig. 6. Relative threshold shift as a function of frequency of the masked tone. Parameter is the on and off time. The curve labelled "continuous" shows absolute threshold shift. For the sake of legibility 10, 20, 30 and 40 dB have been added to the absolute threshold shift values for the curves labelled 250, 125, 50 and 25 msec, respectively. Hence actual values may be obtained by subtracting the appropriate constant.

(1967, Ch. 2), both spatial and temporal amplification by Ratliff, Knight and Graham (see Ratliff, 1968). Only temporal amplification has been shown experimentally (Ratliff, 1968).

SOME OTHER RELEVANT STUDIES

There are other results in the frequency domain consistent with our data. We consider a few examples. Karlin (1945) showed that very small changes near the cutoff frequency of a noise band are easily detected even with broadband noises. Rainbolt and Schubert (1968) found that listeners assigned, with some reliability, a unitary pitch to band-pass noises. Our interpretation of their data is that the assigned pitch lay below the geometric mean frequency of the band. And Greenwood (1968) suggests that pitch may be determined by the lower frequency limit of a band of noise, and that a masked tone masked in the presence of a noise band resembles the noise most when its frequency is nearest the lower limit of the noise. These last three studies, those of von Békésy (1963), Small and Daniloff (1967), and Zwislocki et al. (1968), together with our own data, point to the presence of local maxima in response functions near the cutoff frequencies of low-pass, high-pass, and band-pass filters. The maxima can be made larger by changing temporal properties of the stimuli. All these findings are consistent with lateral inhibition in auditory nerve nets.

EVIDENCE FROM ANATOMY AND PHYSIOLOGY

Two-tone inhibitory interactions between neurons in the auditory nerve are well established (Kiang et al., 1965, Ch. 9). On the basis of a very extensive study of two-tone inhibition in auditory-nerve fibers (Sachs and Kiang, 1968), it appears certain that two-tone inhibition for all auditory-nerve fibers of the cat and all fibers show generally similar properties of inhibition. Inhibitory areas (1) occur on both sides of the "characteristic" frequency (CF) of a unit, (2) are asymmetrical, and (3) overlap with the excitatory area. Also (4) sharpness of excitatory and inhibitory areas increases with increasing CF. Two-tone inhibitory interactions occur at many neural levels, for example, as shown by Suga (1964, 1965) in neurons of the cochlear nucleus, posterior colliculus, and cortex of bat.

Spoendlin (1966; pp. 10-40 of this volume) shows that the anatomical network of cochlear elements has connections suitable for complex inhibitory interaction. For example, a type of cell receiving both afferent and efferent nerves is found in the lower basal turn. Towards the apical coil this type is replaced gradually by a type of cell receiving only afferent nerves. Such an innervation pattern is consistent with the four properties of two-tone interaction stated above. The efferent nerves make extensive contact with cochlear (afferent) dendrites beneath the inner hair cells or contact with outer hair cell itself.

Clinical and experimental evidence has been detailed elsewhere (Carterette, 1969) in support of the suggestion that sharply delimited regions of loss of hair cells along the basilar membrane may be associated with gains in hearing level due to release from inhibition. Such results are expected from the properties of inhibitory neural networks. For, if an element exerting inhibition on another is removed or is itself inhibited, the response of the other element is increased or, as is often said, disinhibited. As an example, Fig. 7 shows sharpening, which may be due to disinhibition, in the audiogram of a policeman deafened by gunfire (Ward et al., 1961).

FURTHER EXPERIMENTS

Furthermore, on the assumption that the nerve nets operate over space and time by means of the interaction of excitatory and inhibitory nerve impulses, a

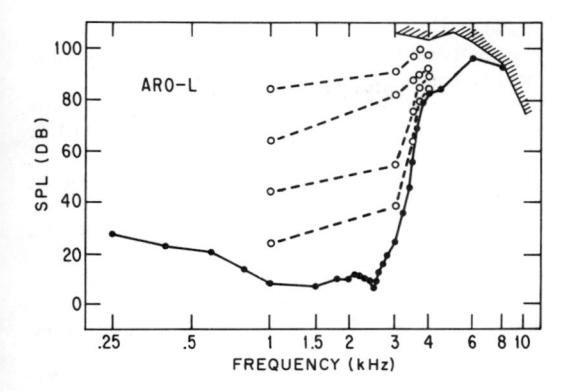

Fig. 7. Detailed audiogram (solid line) showing region of increased sensitivity often found just below a high-frequency abrupt loss or tonal gap redrawn from Ward et al., 1961).

number of experiments suggest themselves. One class we have already mentioned, aimed at understanding the temporal properties of inhibitory interaction —how long after onset or offset of stimulation before inhibition appears or disappears? If the effect of one stimulus is decreased by the presence of a second, then a third stimulus should exist whose presence removes the inhibitory effect of the second, that is, disinhibits the effect on the first stimulus. Experiments on matching the pitch of a tone in the quiet to suprathreshold tones in the noise should be particularly useful in mapping details of edge effects. It could be argued that this is a more direct way of demonstrating Mach bands in hearing.

REFERENCES

Alpern, M., and David, H. (1959): The additivity of contrast in the human eye, J. Gen. Physiol. 43, 109-127.
Baumgartner, G. (1961): Kontrastlichteffekte an retinalen Ganglienzellen: Ableitungen vom Tractus opticus der Katze, in: Neurophysiologie und Psychophysik des visuellen Systems, R. Jung and H. Kornhuber, Eds. (Springer-Verlag, Berlin), pp. 45-53.
Békésy, G. von (1928): Zur Theorie des Hörens, Die Schwingungsform der Basilarmembran, Physikal. Z. 29, 793-810.
Békésy, G. von (1960): Experiments in Hearing (translated and edited by E. G. Wever) (McGraw-Hill, New York).
Békésy, G. von (1963): Hearing theories and complex sounds, J. Acoust. Soc. Amer. 35, 588-607.
Békésy, G. von (1967): Sensory Inhibition (Princeton University Press, Princeton).
Carterette, E. C. (1969): Release from masking as a means of studying hair-cell function, J. Speech Hear. Res. 12, 497-509.
Carterette, E. C., Friedman, M. P., and Lovell, J. D. (1969): Mach bands in hearing, J. Acoust. Soc. Amer. 45, 986-998.
Elliott, L. L. (1967): Development of auditory narrow-band frequency contours, J. Acoust. Soc. Amer. 42, 143-153.
Greenwood, D. D. (1968): Comments on "The use of noise bands to establish noise pitch", J. Acoust. Soc. Amer, 44, 634-635.
Hartline, H. K., Ratliff, F., and Miller, W. H. (1961): Inhibitory interaction in the retina and its significance in vision, In: Nervous Inhibition, E. Florey, Ed. (Pergamon Press, New York), pp. 241-284.
Huggins, W. H., and Licklider, J. C. R. (1951): Place mechanisms of auditory frequency

analysis, J. Acoust. Soc. Amer. *23*, 290-299.
Karlin, J. E. (1945): Auditory tests for the ability to discriminate the pitch and the loudness of noise, U. S. Office of Research and Scientific Development Report 5294.
Kiang, N. Y.-S., Watanabe, T., Thomas, E. C., and Clark, L. F. (1965): Discharge Patterns of Single Fibers in the Cat's Auditory Nerve (Research Monograph No. 35, M.I.T. Press, Cambridge, Mass.).
Rainbolt, H. R., and Schubert, E. D. (1968): Use of noise bands to establish noise pitch, J. Acoust. Soc. Amer. *43*, 316-323.
Ratliff, F. (1965): Mach Bands: Quantitative Studies on Neural Networks in the Retina (Holden-Day, San Francisco).
Ratliff, F. (1968): On fields of inhibitory influence in a neural network, in: Neural Networks, E. R. Caianiello, Ed. (Springer-Verlag, New York).
Sachs, M. B., and Kiang, N. Y.-S. (1968): Two-tone inhibition in auditory-nerve fibers, J. Acoust. Soc. Amer. *43*, 1120-1128.
Small, A. M., and Daniloff, R. G. (1967): Pitch of noise bands, J. Acoust. Soc. Amer. *41*, 506-512.
Spoendlin, H. (1966): The Organization of the Cochlear Receptor (Karger, Basel).
Suga, N. (1964): Recovery cycles and responses to frequency modulated tone pulses in auditory neurons of echolocating bats, J. Physiol. *175*, 50-80.
Suga, N. (1965): Functional properties of auditory neurons in the cortex of echolocating bats, J. Physiol. *181*, 671-700.
Ward, W. D., Fleer, R. E., and Glorig, A. (1961): Characteristics of hearing losses produced by gunfire and by steady noise, J. Aud. Res. *1*, 325-356.
Wever, E. G., and Lawrence, M. (1954): Physiological Acoustics (Princeton University Press, Princeton).
Zwislocki, J. J., Buining, E., and Glantz, J. (1968): Frequency distribution of central masking, J. Acoust. Soc. Amer. *43*, 1267-1271.

DISCUSSION

Greenwood: I would like to comment that, if the notches you found are edge effects, you should see them also for wider bands of noise.

Carterette: Yes, quite.

Greenwood: But in my experience one does not find edge effects associated with the edge of the noise when one uses wider bands, rather than the band in your experiment (which was right on the dividing line of the average critical bandwidth). I have mapped many masked audiograms but I have never seen such notches for band-pass noise of larger bandwidths (*i.e.* greater than about critical width) nor, in a few instances, for low-pass noises whose slopes were also very steep. To illustrate this point in the case of band-pass noise, I include Fig. 1 as a single example.

Another point is that if there are Mach bands in hearing or, in words which I prefer, if there are effects attributable to inhibitory zones, then you should find lower thresholds for frequencies where the slopes of the masked audiograms intersect the quiet threshold or the raised threshold level produced by a background masking noise. Such an effect may perhaps be seen in Figs. 1 and 9 of Dr. Zwislocki's paper on central masking (pp. 446 and 452 of this volume). I found some evidence for it in my monaural experiments. This point relates

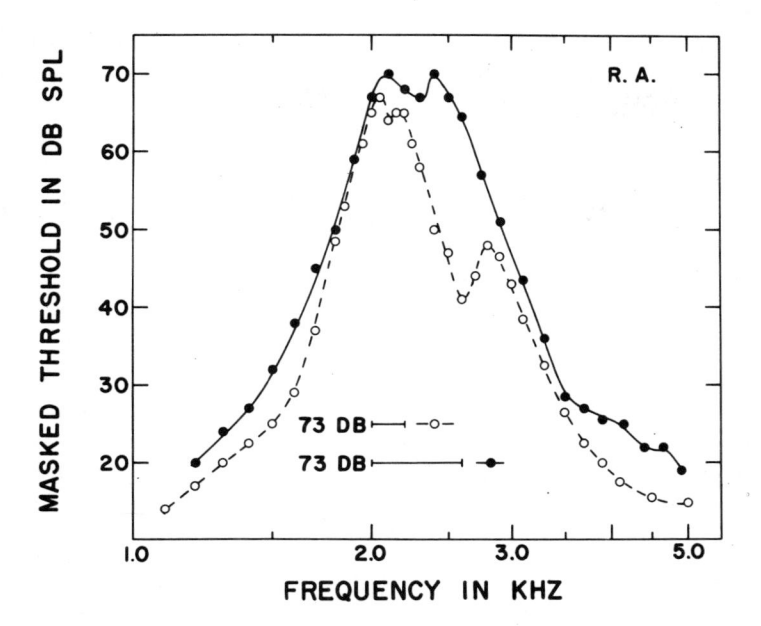

Fig. 1 (Greenwood). The figure displays two sample masked audiograms produced by bands of noise 200 Hz and 600 Hz in width, respectively, whose lower frequency limit was 2000 Hz and whose overall level was 73 dB SPL. The latter band represents a "wider" band of noise in the context of this discussion. The notch seen in the case of bands 200 Hz in width (or less) becomes smaller and disappears when bandwidth increases to about 400 Hz. (For this subject and at this level three masked audiograms were obtained at the 600 Hz width, each differing somewhat in the shape across the top, the others being flatter, but very much alike in high- and low-frequency slopes.) Wider bandwidths have not brought the notch back.

The fixed test (or signal) frequencies used to map the masked audiogram were 100 Hz apart near 2000 Hz, increasing to about 200 Hz apart near 4000 Hz. This corresponds to a small and fairly constant interval (between 1/4th to 1/3rd critical bandwidth) relative to the frequency region explored in these examples. The skirt slopes of the 200 Hz band were 240 dB/octave and 390 dB/octave on low and high sides, respectively, and the corresponding slopes of the 600 Hz band were 105 dB/octave and 160 dB/octave, respectively. The high-frequency noise slope can be decreased to at least about 125 dB/octave before it can begin to affect the masked audiogram's high-frequency side.

also to both Dr. Zwicker's conceptual excitation patterns and the importance of the movement of edges, and also to the spatial patterns of excitation that one may be able to reconstruct from single-unit data as I outlined in my paper on excitatory and inhibitory response areas of auditory neurons in the cochlear nucleus (Greenwood and Maruyama, 1965). *If* there are lower thresholds in the intersection area that are indeed related to inhibition, then these areas could correspond to areas of inhibited units outside the active population of units inferred from the tuning curve.

Finally, I would like to caution against mapping masked audiograms by means of the sweep-frequency method. There is a likelihood that there may be an overshoot near the limits of the band of noise as a partial consequence of the movement of signal frequency and also, as Dr. de Boer said yesterday in conversation, as a consequence of the fluctuations that may be more especially marked near the limits of the noise. I think you need a pulsed test tone of a fixed frequency to which the subject can listen for some time. It is my experience that finding thresholds takes longer near the edges even when one comes out with the same values from day to day.

Carterette: In reference to your first question I'd like to remark that one has to space the frequencies of the masked tone very closely in order to reveal the complete structure and your masked frequencies (Fig. 1) are far apart relative to ours, at least if you refer to your paper on auditory masking and the critical band (Greenwood, 1961). I do not know whether there are edge effects in the case of wide-band noise with steep skirts, but I do not think one should exclude the possibility. If you'd look at the literature on masking functions you would recognize that they often tend to follow the filter slope pretty closely. The steepness of the slopes suggests that somewhere sharpening might occur if measures were taken with a suitably fine grain. I think you were quite right in saying that we should not look only to maskers of width about equal to the critical bandwidth. We are in the process of investigating low-pass and high-pass noise in addition to wider bands of noise.

[In some work done subsequent to this Symposium, we dealt with Greenwood's two objections: (1) Noises wider than critical band should not show edge effects. Such a result would be consistent with Greenwood's (1961) study in which no edge effects were found for any noise bands — whether bandwidths were equal to, greater than, or less than the critical bandwidth. Incidentally, in that study, critical bandwidths were stated to be somewhat less than those used by us in the work reported at this Symposium. (2) A tracking method of the kind we had used, and used earlier by Greenwood (1961), may introduce uncertainty in location of masked frequencies due to sweeping across the frequency range.

To take account of objection (1) we used rectangular band-pass noises lasting 45 min. Such bands were generated by digital computer and contained 500 independent random sine components logarithmically spaced within the band. The sampling rate was 10 kHz. Conversion to analog form was via a low-pass filter with cutoff frequency of 4 kHz. We dealt with objection (2) by using a 2-alternative, forced-choice adaptive threshold technique based on Wetherill's (1966) UDTR (up-down, transformed response) method. The masked frequency was randomly chosen but fixed for a given threshold run. A single trial took 4 sec. A warning light 1 sec in duration was followed 500 msec later by two 125 msec bursts of noise separated in time by 250 msec. A response interval of 2 sec was followed by a feedback interval of 750 msec during which a green or red signal light signified that the response was right or wrong, respectively.

Measurements were made at minimum intervals of 50 Hz between 250 Hz and 5 kHz.

A threshold at a given frequency was an estimate of the attenuator setting for which the probability of a correct response was 0.794. A first run used a stopping rule of 2 peaks and valleys as a crude estimate of the threshold. This estimate was used as the initial setting for a series of three runs, each of which was terminated after 4 peaks and valleys. The experiment was run by a computer.

The masking noise was strictly band-limited between 1000 and 2000 Hz. The synthetic noise was presented alone as the masker at 60 dB SL in one condition and in another condition together with a wide-band noise at 30 dB SL limited between 30 Hz and 5 kHz. The resulting threshold shifts of the masked tones, computed as the dB difference: SPL(masked) − SPL(quiet), are plotted in Figs. 2 and 3. The mean of the mean deviation from the median threshold shift was less than 0.5 dB.

Fig. 2 shows the masking of tones by the 1000-2000 Hz synthetic noise, and Fig. 3 shows masking of tones by synthetic noise plus the wide-band noise. It can be seen in both figures that edges occur in the neighbourhood of the cutoff frequencies of the band-pass noise, being most marked at about 1150 Hz. Fig. 3 especially should be compared with the graphs of Fig. 4 of our paper (see p. 431). The condition in which the synthetic noise was a pedestal in a wide-band sea of noise was used in the hope of obtaining relatively silent regions ("dark bands") beyond the pedestal. Fig. 3 gives some support for the existence of silent bands — one above (3-3.5 kHz) and one below (400-800 Hz) the 1000 Hz band of synthetic noise. These effects cannot reasonably be attri-

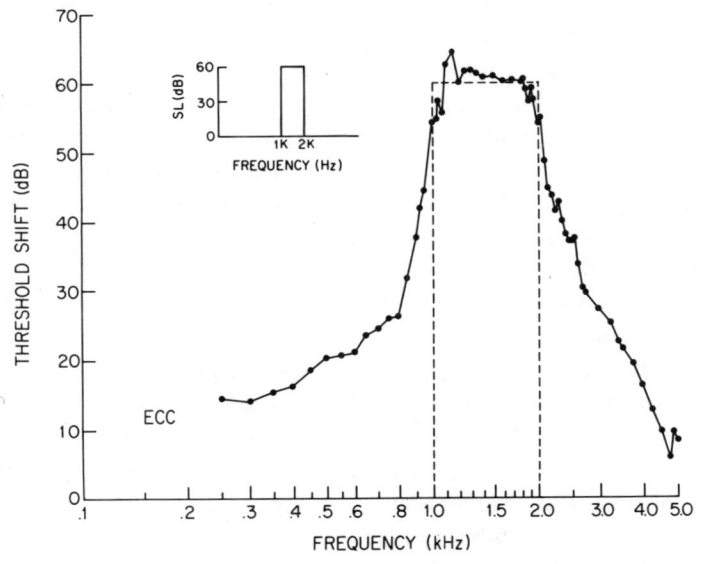

Fig. 2 (Carterette). Masking of tones by a 1000-2000 Hz band of synthetic noise.

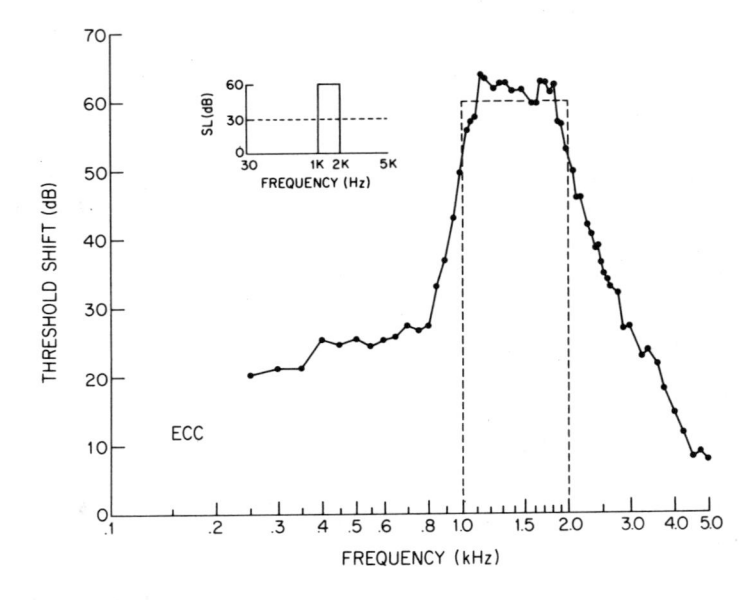

Fig. 3 (Carterette). Masking of tones by a 1000-2000 Hz band of synthetic noise plus wide-band noise.

buted to the noise being less wide than the critical band. Even the most extreme theoretical estimate has never put the width of the critical band for 1000 Hz beyond 500 Hz. All other theoretical or experimental estimates have been 200 Hz or less, mainly less (see Ahumada, 1967).

In regard to the so-called dark-band areas, and perhaps they are indeed the most critical regions to look at, our recent paper (Carterette et al., 1969) points out that we think we have some evidence for "dark bands".]

In your last remark you supposed the edge effects to be caused by temporal fluctuations. A careful analysis of the spectrum of synthetic noises reveals no temporal fluctuations that could reasonably lead to edge effects of the kind found experimentally. The physical and mathematical theory upon which the synthetic noises were constructed do not lead us to expect temporal fluctuations that would cause the effects found.

Of course, there was another class of temporal effects expected and, we believe, found. Fig. 6 of our paper (see p. 433) shows that really there seem to be temporal effects. Different rates of interruption of the masked tone do not give simple translations of the curves. The temporal effects could be associated with lateral inhibition and should be, on analogy with retinal networks. As a matter of fact, one might expect to find more powerful effects (temporal amplification) as a function of time. Ratliff (1968) has given a good discussion of temporal and spatial amplification in inhibitory nets.

De Boer: I'd like to offer an alternative explanation which was mentioned by Dr. Greenwood a moment ago. We can learn from this conference that one not only has to measure but also needs to listen. And by listening a great deal to these signals I learned a striking thing. It is rather easy to determine a masked threshold for test tones well within the noise band. But when the test tone comes to an edge of the noise band, a threshold measurement becomes increasingly difficult. The test tone at just about the masked threshold has a very fluctuating character which is much more pronounced at the edge than in the centre of the noise band. When the tone moves to the edge of a noise band, the critical band moves with it. Within the shifted critical band the noise has a smaller bandwidth and thus it fluctuates more. I'd like to offer this as an explanation because it means that when you shift the test tone from the centre of a noise band to its edge, you have to adopt a different criterion. And the general effect of fluctuations is an increased threshold.

Carterette: I understand your reasoning and there is no doubt that perhaps one's criterion might change at the edge of the band of noise. It is true that the task is somewhat more difficult at the edges. But it would be curious that the criterion changed just in such a way as to lead to the kind of effects that we see. Actually, what we would expect is an increase in the variance and no essential change in the mean of a given threshold shift in fluctuating, as compared to "stable", noise.

De Boer: I'm sorry but that is not true. Despite the increased fluctuations the slope of the psychometric function remains the same. That has been proven all over again in statistical detection theory. The whole thing behaves as a Gaussian detector. The *variance* of the data should not increase near the edge and, as such, would not be an indicator of the greater fluctuations.

Carterette: What is to be expected depends upon the theory and the experimental procedures. We set out to find edge effects on the assumption that lateral inhibition was one of the mechanisms of sharpening in the auditory system. We found both spatial (frequency) and temporal sharpening effects. They *may* be due purely to amplitude fluctuations and of course we have not yet established that they are not or could not be due to this cause.

Hind: In as much as I was involved in developing the cortical test of hearing with Schuknecht (Hind and Schuknecht, 1954), I think we really must be very cautious in interpreting that 5 dB increase of sensitivity (Fig. 7) as being significant. We found threshold shifts with the opening of the middle ear which, we thought, might explain some seemingly increased sensitivities.

Carterette: Your caution is well taken. However, there are other data which show an increase in hearing sensitivity, negative hearing losses, which are minimally 15 dB but may be much larger because the limits of the threshold were not measurable (see, *e.g.*, references cited in Carterette, 1969).

Greenwood: I agree with Dr. de Boer that the sound of the test signal, as it approaches the masked threshold, differs very much depending on the position of the test signal with respect to the noise band. It is true that the criterion, therefore, must change. But if one asks the subject to detect anything, *i.e.* any effect introduced by the test signal, then it is still possible to plot these curves. And, under those circumstances, I have in many masked audiograms, with sufficiently wide bands, not observed any peaks at the edges.

Another point is that Dr. Carterette mentioned, — as suggesting the edge of the noise to be important — the coincidence of the notch, in a number of instances, with the edge of the band of the width he used. But I would like to point out that this does not hold for my results in Fig. 4. The masked audiograms of narrow bands of noise show notches that are considerably higher in frequency than the edges of the respective bands. A pure tone as a masker also results in a notch. Of course, this notch does not coincide with the edge of anything in the acoustic spectrum.

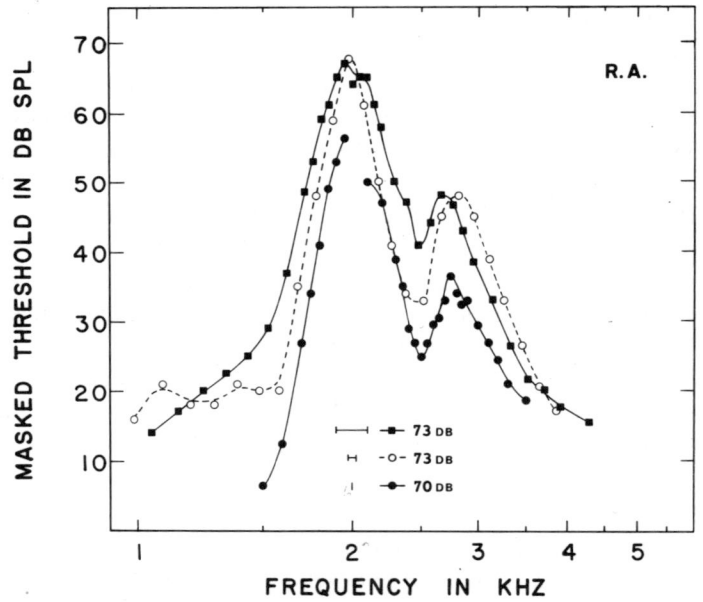

Fig. 4 (Greenwood). Masked audiograms produced by narrow bands of noise near 2000 Hz, in comparison with the masked audiogram produced by a pure tone at 2000 Hz. Data from this subject and a number of others demonstrate that sufficiently narrow bands of noise result in the same notch seen in those cases in which a pure tone masker is used, which is to say that they result in a notch that is not coincident in position with an edge in the acoustic spectrum. Secondly, the logical force of such data is that the mechanisms responsible for the notch operate equally in the two cases shown, in which maskers are either pure tones or narrow bands of noise. The 50 Hz band of noise was generated by means of a General Radio wave analyzer (skirts 30 dB down 100 Hz from centre frequency). The 200 Hz band, and data, are the same as in Fig. 1; the lower frequency limits of the 50 Hz and 200 Hz bands were at 2000 Hz, and the masked audiograms produced by the 50 Hz and 200 Hz bands were shifted downward slightly (25 Hz and 100 Hz) for better comparison with data obtained using the 2000 Hz tone.

Zwicker: Dr. Greenwood, were you perfectly careful to measure the masked threshold correctly? In the conditions where you have the notch it is probable that one does not measure the threshold of the masked test tone but the threshold of some kind of perceptible difference noise introduced by the test tone. By increasing the level of the test tone, the difference noise may be perceived

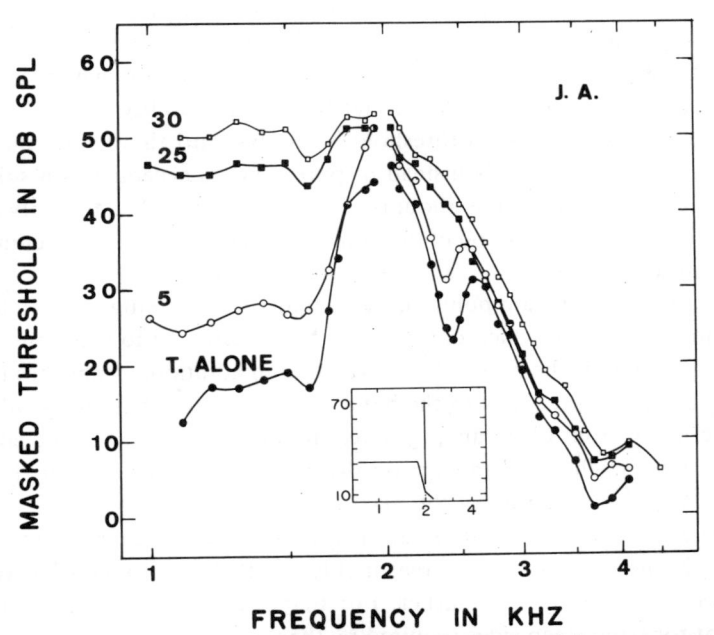

Fig. 5 (Greenwood). Portions of masked audiograms produced by pure tone maskers at 2000 Hz and at 70 dB SPL accompanied by low-pass noise presented at ascending spectrum levels; the noise cutoff is at 1800 Hz with a slope of 135 dB/octave for the first 25 dB. The inset at the bottom of the graph shows the noise at the highest spectrum level used, 30 dB SPL/1 Hz band. Curves through solid circles display masked audiograms produced by the masking tone presented alone. Low-pass noise is then added at spectrum levels of 5, 15, 25, and 30 dB SPL/Hz but some resulting masked audiograms have been omitted to permit the curves to be distinguished more easily. All are consistent in showing a progressive elimination of the notch. In addition, for three subjects (data not included here), low-pass noises were used alone at spectrum levels that, in the presence of the masking tone, were sufficient to eliminate the notch. This maneuver was followed in order to show that such noises presented alone did not produce masking in the frequency region of the notch. In accompanying sub-figures (shown in the Symposium discussion but not included here because of space considerations), masked audiograms were obtained when wide-band noise extending to 7 kHz was substituted for the low-pass noise. Although noise components were then present in the region of the notch, such wide-band noises were not any more effective in eliminating the notch than was low-pass noise. These two sets of data are taken to support the proposition that the notch is attributable to combination components. That is, the data support the idea that the *occurrence* of a signal at a frequency just above the masking tone (or the narrow band of masking noise) in the region of the notch is detected *at threshold* not because the signal per se is detected, but rather because the combination components produced by the signal and masker fall below the frequency of the masker where the events they produce are detected and where they can be masked.

before the test tone itself which introduces dips in the masked audiograms.

Greenwood: Precisely, that is exactly what I was intending to say. This notch is, I think, due to combination products and Fig. 5 gives some evidence on that score. The bottom curve is obtained with a masking tone presented alone and has a notch. By adding to the masking tone low-pass noise with a cutoff frequency near the masking tone one wipes out the notch because one masks the combination tones. The elimination of the notch is smoothly progressive which follows from more curves than have been plotted in the figure.

Carterette: We have never referred to the notches about which you are talking. As a matter of fact, I pointed out that the possibility of combination products obscuring the masked threshold is an awkward thing and you often see that the use of noise bands instead of pure tones is accepted as a solution to this problem. I have no claims about the notches you discribed but only about the notches in the neighbourhood of the nominal band edge of nearly rectangular band-pass noise.

Greenwood: But what I am pointing out is that we are talking about one and the same notch. A notch which occurs at all bandwidths up to a certain value. The fact that one finds the notch with pure tones in addition, establishes that it should not be attributed solely to the band edge, although that fact would not prevent the notch from being an edge effect in another sense; but the edge in question would somehow be an edge on the basilar membrane rather than an edge in the acoustic spectrum of the masker.

Carterette: It has been our aim in all this work to exhibit sharpening on the basilar membrane which could not reasonably be attributed to peaks or valleys in the power spectrum of the masker. For that reason we synthesized noises with the flat tops and steep sides required by theory.

REFERENCES

Ahumada, A. (1967): Detection of tones masked by noise: a comparison of human observers with digital computer simulated energy detectors of varying bandwidths, Tech. Report No. 29, Human Communication Laboratory, Department of Psychology, University of California, Los Angeles.

Carterette, E. C. (1969): Release from masking as a means of studying hair-cell function, J. Speech Hear. Res. *12*, 497-509.

Carterette, E. C., Friedman, M. P., and Lovell, J. D. (1969): Mach bands in hearing, J. Acoust. Soc. Amer. *45*, 986-998.

Greenwood, D. D. (1961): Auditory masking and the critical band, J. Acoust. Soc. Amer. *33*, 482-502.

Greenwood, D. D., and Maruyama, N. (1965): Excitatory and inhibitory response areas of auditory neurons in the cochlear nucleus, J. Neurophysiol. *28*, 863-892.

Hind, J. E., and Schuknecht, H. F. (1954): A cortical test of auditory function in experimentally deafened cats, J. Acoust. Soc. Amer. *26*, 89-97.

Ratliff, F. (1968): On fields of influence in a neural network, in: Neural Networks, E. R. Caianiello, Ed. (Springer-Verlag, New York).

Wetherill, G. B. (1966): Sequential Methods in Statistics (Wiley, New York).

CENTRAL MASKING AND AUDITORY FREQUENCY SELECTIVITY

J. J. ZWISLOCKI

Laboratory of Sensory Communication
Syracuse University
Syracuse, New York, U.S.A.

INTRODUCTION

The threshold of audibility in one ear is increased by the presence of sound in the other even when a practically complete acoustic insulation between the two ears is achieved. Under such conditions, the effect must result from a purely neural interaction and is called central masking. The phenomenon was first discovered by Wegel and Lane (1924) and was subsequently investigated on several occasions. These studies have shown the effect to be very small for a continuous masker and a pulsed test stimulus and somewhat larger when both are pulsed. They have also revealed a sharp frequency selectivity.

Our systematic studies on central masking began in 1963. They were motivated not so much by curiosity about the phenomenon itself as by the hope that its investigation may throw some light on signal processing in the lower parts of the auditory system. Psychophysiological comparisons that have become possible as a result of these studies seem to justify our expectations. A simple relationship could be established between firing rates of single neurons and the threshold shifts induced by central masking.

Typical results of central masking experiments are shown in Figs. 1 and 2. The first indicates in its upper right corner the stimulus time pattern used in most of our experiments (Zwislocki et al., 1967). The masker consists of 250 msec sound bursts repeated at a rate of about one per second; the test stimulus, of tone bursts with an approximately Gaussian envelope, which can be delayed with respect to the masker onset by a variable amount of time. The onset and decay times of both stimuli are on the order of 10 msec. Their sound frequency can be varied independently. The experimental results plotted in the same figure show the threshold shift as a function of test frequency. The masker consisted

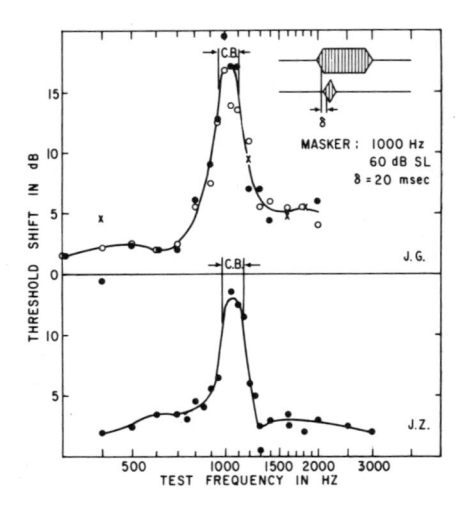

Fig. 1. Stimulus pattern and frequency distribution of central masking near the masker onset (from Zwislocki et al., 1967).

of a 1000 Hz pure tone at a sensation level of approximately 60 dB, and the test tone was delayed by 20 msec from the masker onset. Thresholds were tracked by means of an automatic Békésy attenuator. Note the similarity between the two curves which were fitted by eye to the individual data. The frequency bandwidths of both masking distributions agree very well with Zwicker's (Zwicker et al., 1957) critical band.

Fig. 2 shows the median threshold shift as a function of the time delay from the masker onset. A group of 6 listeners participated in the experiment. The thresholds were determined by means of a psychophysical method which combines Békésy's tracking with forced choice (Zwislocki et al., 1958). A descriptive name for the method is "forced-choice tracking". The stimulus and response schedules were so arranged that the thresholds were tracked at a level of approximately 79% correct responses. A critical-band noise centered at 1000 Hz and adjusted consecutively to several sensation levels served as masker. The test stimulus consisted of a 1000 Hz tone. The results show that central

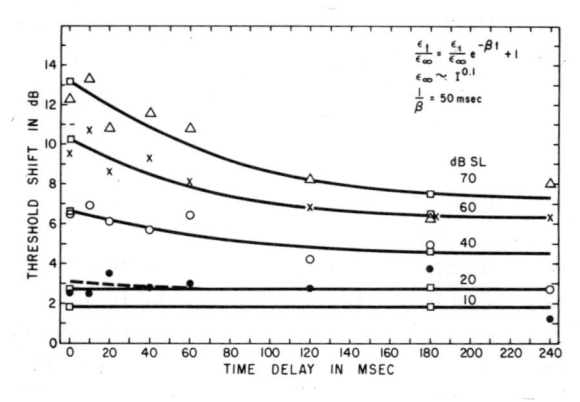

Fig. 2. Temporal decay of central masking for several masker sensation levels (median values of 6 subjects).

masking decays rapidly with the time delay and reaches an asymptote at a level dependent on the masker sensation level. Additional experiments have shown that the rapid decay does not depend on the masker spectrum or its center frequency.

I now shall endeavor to show that the data of Figs. 1 and 2 and other results of central-masking experiments are simple reflections of neural activity at low levels of the auditory system.

THEORY OF CENTRAL MASKING

The theory involved in the physiological interpretation of central masking requires only one axiom. The same axiom is accepted either explicitly or implicitly in many if not all psychoacoustical theories. It states that, for a given criterion of detection, the threshold of audibility is determined by a constant signal-to-noise ratio in the neural domain. In monaural masking by a random noise, the constancy of the signal-to-noise ratio extends to the stimulus domain. If the neural activity contributed by the signal is denoted by ε_S and that due to noise by ε'_N, the axiom may be expressed as $\varepsilon_S/\varepsilon'_N = $ const. In general, the noise has two sources, one extrinsic and one intrinsic. Accordingly, $\varepsilon'_N = \varepsilon_N + \varepsilon_I$. In the absence of extrinsic noise, the threshold is determined by $\varepsilon_S/\varepsilon_I = $ const.

Considering the relevant neural activity in the channel of the test ear when no extrinsic noise is introduced into the contralateral ear, it is possible to write

$$\varepsilon_Q = \varepsilon_{SQ} + \varepsilon_I + M\varepsilon_I. \tag{1}$$

The symbols have the following meaning: $\varepsilon_Q = $ total neural activity above the relevant place of binaural interaction; $\varepsilon_{SQ} = $ activity contributed by the signal; $\varepsilon_I = $ activity due to intrinsic noise in the test ear; $M\varepsilon_I = $ portion of the intrinsic activity in the contralateral ear, which affects the test ear channel. When extrinsic noise is introduced into the contralateral ear, Eq. (1) takes the form

$$\varepsilon_M = \varepsilon_{SM} + \varepsilon_I + M(\varepsilon_N + \varepsilon_I), \tag{2}$$

where $M\varepsilon_N$ means the contribution of the extrinsic noise. Because of the accepted axiom, the signal-to-noise ratio at the threshold of audibility must remain constant. Therefore,

$$\frac{\varepsilon_{SQ}}{\varepsilon_I + M\varepsilon_I} = \frac{\varepsilon_{SM}}{\varepsilon_I + M(\varepsilon_N + \varepsilon_I)}. \tag{3}$$

Eq. (3) may be transformed to express the ratio between the stimulus induced excitations in presence and absence of extrinsic noise

$$\frac{\varepsilon_{SM}}{\varepsilon_{SQ}} = \frac{\varepsilon_I + M(\varepsilon_N + \varepsilon_I)}{\varepsilon_I + M\varepsilon_I}. \tag{4}$$

After simplification, we obtain

$$\frac{\varepsilon_{SM}}{\varepsilon_{SQ}} = \frac{M}{1+M}\left(\frac{\varepsilon_N}{\varepsilon_I}+1\right). \tag{5}$$

Now, it is necessary to find the relationship between the excitations ε_S and the corresponding signal intensities. There is some psychophysical evidence that, for excitations $\varepsilon_S < \varepsilon_I$, the relationship approximates a direct proportionality (Zwislocki, 1960, 1965). Recently, it could be shown that the same holds for single neurons. As an example, the intensity characteristic of a typical unit of the cat's eighth nerve is shown in Fig. 3. The data points are from Wiederhold's (1967) doctoral dissertation; the curves have been fitted according to a simple equation that describes quite accurately the intensity characteristics of many sensory neurons. The upper curve corresponds to the expression $\varepsilon_S + \varepsilon_I$, and the lower to ε_S alone. Note that the lower curve indicates a direct proportionality between stimulus intensity and neural firing rate over a range of more than 10 dB. This range extends to about 10 dB above the estimated behavioral threshold marked by the letter T. In central-masking experiments, this range is rarely exceeded by the test stimulus. Accordingly, it appears reasonable to accept that $\varepsilon_S = aS$, where S denotes the stimulus intensity, and a is a proportionality constant. Eq. (5) can now be rewritten in terms of stimulus intensities:

$$\frac{S_M}{S_Q} = \frac{M}{1+M}\left(\frac{\varepsilon_N}{\varepsilon_I}+1\right). \tag{6}$$

The ratio S_M/S_Q is the linear expression of the threshold shift in the test ear. As a consequence, *this threshold shift is a simple measure of the neural activity produced by the extrinsic noise in the contralateral ear.* From Eq. (6), it is possible to obtain:

$$\frac{\varepsilon_N}{\varepsilon_I} = \frac{1+M}{M}\left(\frac{S_M}{S_Q}-1\right). \tag{7}$$

It is necessary to emphasize that not all the noise-induced neural activity can be measured by means of central masking. Only that portion which interacts with neurons excited by the test signal produces a contralateral threshold shift. This may have some disadvantages but it also has advantages. When the signal consists of a narrow-band, low-intensity sound, it excites only a small group of neurons. This group may be expected to interact with a similarly small group of neurons from the contralateral ear. As a consequence, by changing the center frequency of the signal, it should be possible to sample the excitation distribution produced by the noise. Only in this way is it possible to understand the narrow frequency distribution of the threshold shift in Fig. 1.

VALIDATION OF THE THEORY

Because of an axiom accepted without proof, the theory requires an experimental validation. The psychophysiological nature of the theory calls for psychophysiological comparisons. The neuronal characteristic of Fig. 3 can serve as a point of departure. Since the characteristic is typical of eighth-nerve units, it should be directly proportional to the summed activity of a small group of similar neurons. As a consequence, it should predict the corresponding central-masking characteristic. The neural firing rates of Fig. 3 were determined by counting the neural spikes during tone bursts of 48 msec duration. Because the firing rate decays with stimulus duration, the mean firing rate obtained from such a count is equal to the instantaneous firing rate at about 20 msec from the stimulus onset. Corresponding central-masking experiments were performed on a group of 6 listeners by measuring the threshold shift with a time delay of 20 msec from the masker onset. The masker consisted of a critical-band noise centered at 1000 Hz and the test stimulus of a 1000 Hz tone. The resulting median threshold shifts are plotted in Fig. 4 as a function of masker sensation level. The curve in the same figure has been predicted from the neuronal characteristic of Fig. 3 with the help of Eq. (6). For this purpose, the numerical values of $\varepsilon/\varepsilon_I$ were read on the graph and substituted for $\varepsilon_N/\varepsilon_I$. A value of 0.3 was accepted for the free constant M which may vary between 0 and 1. This choice resulted from the comparison of one value of the threshold shift to a corresponding value of the neural firing rate. As can be seen from Fig. 4, the predicted masking curve fits very well the experimental data up to a sensation level of 50 dB. At higher levels, the experimentally determined threshold shift increases above the predicted one and seems to indicate that a new group of neurons comes into effect.

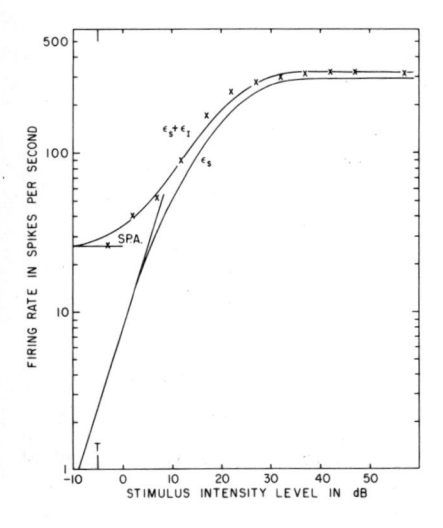

Fig. 3. Firing rate of a typical eighth-nerve unit as a function of stimulus intensity.

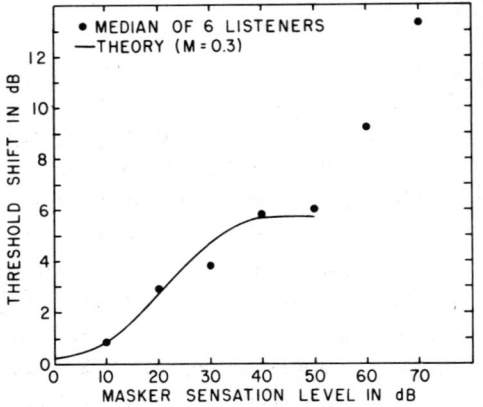

Fig. 4. Median central masking near the masker onset as a function of the masker sensation level. The points are experimental; the curve is calculated.

In order to further test the theory, predictions based on Fig. 3 and Eq. (6) were compared to individual masking results obtained on 4 listeners by means of a 1000 Hz masker. As is evident from Fig. 5, the experimental data are fitted quite well by the theoretical curves calculated for various values of the constant M. Since the variation of M accounts completely for the individual differences, they seem to be mainly due to a varying amount of binaural interaction.

In another test of the theory, steady-state threshold shifts were compared to steady-state neural firing rates. The central-masking experiment was performed on a group of 4 listeners by means of a continuous 1000 Hz masker and 1000 Hz test bursts of 250 msec duration. The thresholds were tracked with the help of a Békésy attenuator. The individual and median data are plotted in Fig. 6. The smooth curve fitted by eye to the medians was used for a calculation of a typical neuronal characteristic. The calculation was performed by means of

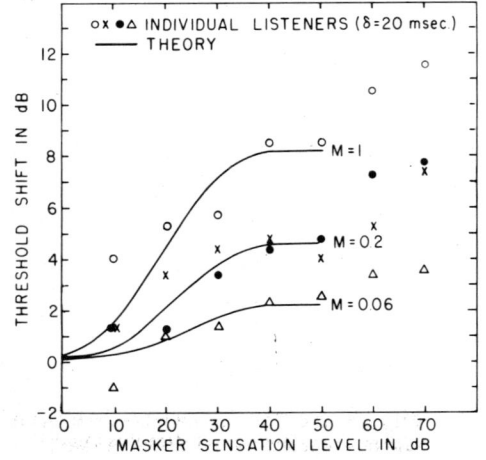

Fig. 5. Same as Fig. 4, but for individual listeners.

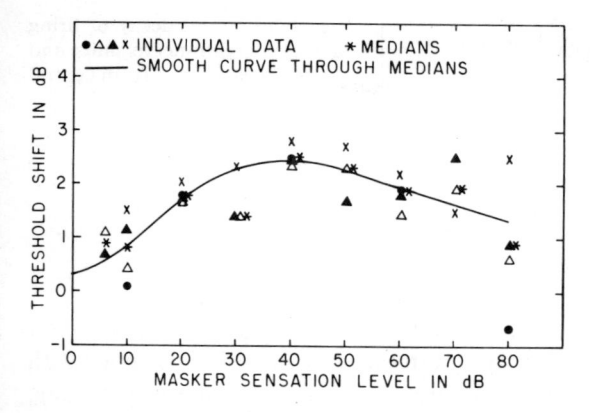

Fig. 6. Steady-state central masking as a function of masker sensation level.

Eq. (7). The constant M was estimated to be 0.6 for the group of listeners involved. The resulting curve is shown in Fig. 7 for a typical spontaneous activity ε_1 of 54 spikes per second. The crosses and closed circles in the same figure indicate firing rates of two typical eighth-nerve units (Kiang, 1965, p. 82). It is evident that the curve derived from the central-masking data follows approximately the same function as the neural firing rates. Note in particular the maximum near 40 dB and the somewhat puzzling decay at high intensities.

The last test of the theory I am able to discuss within the framework of this paper concerns the fast decay of neural firing rate with stimulus duration. The decay seems to be quite stable in that it exhibits little interneuronal variation and seems to be independent of sound frequency. Pertinent masking data are shown in Fig. 2. In order to fit available neural recordings, they were interpolated to a sensation level of 45 dB. The constant M was estimated for the group to be 0.3. The calculated neural decay function is plotted in Fig. 8. In order to eliminate

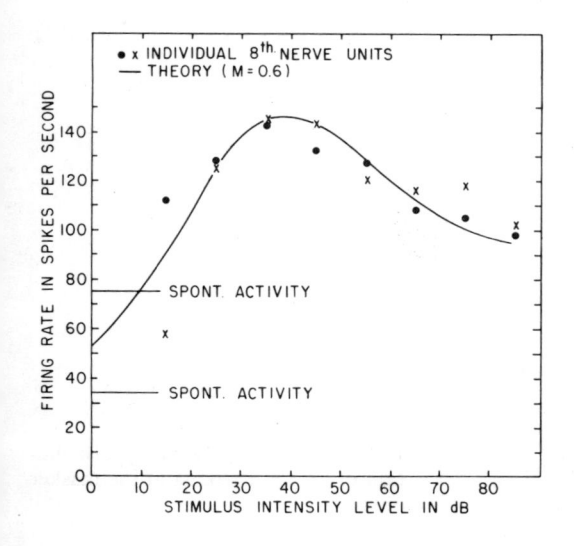

Fig. 7. Steady-state firing rates of 2 eighth-nerve units and firing rates calculated from central masking.

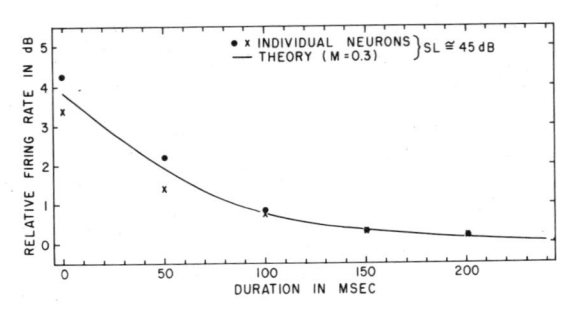

Fig. 8. Temporal decay of firing rate in 2 eighth-nerve units and the decay calculated from central masking.

the effect of the highly variable spontaneous activity, the plotted values refer to an asymptotic firing rate. Actual firing rates of two eighth-nerve units with characteristic frequencies near 1000 Hz are plotted in the same way (Kiang, 1965, p. 69). The theoretical and experimental data follow the same function.

The results illustrated in Figs. 4 through 8 should constitute a sufficient validation of the theory of central masking. They indicate that the threshold shifts induced by central masking are related to typical neuronal firing rates in the low portions of the auditory system according to Eqs. (6) and (7). Because of the form of these equations, the relationship is linear.

AN APPLICATION

It now becomes possible to use central masking for an indirect investigation of neural excitation patterns. As an example, Fig. 9 shows individual frequency distributions of threshold shift for a constant test tone of 1000 Hz and a variable masker (Zwislocki et al., 1968). This method of procedure corresponds to the recording of response areas of single neurons. The threshold shifts were

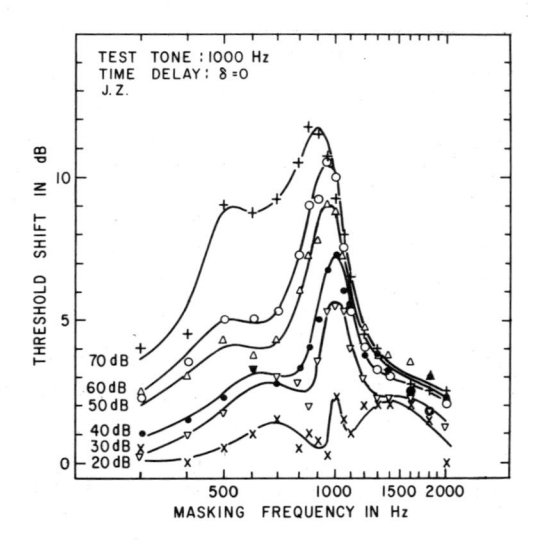

Fig. 9. Individual frequency distribution of central masking at the masker onset.

determined at the onset of the masker. In order to determine the changes in frequency distribution with sound intensity, the masker was set at several sensation levels. Several characteristic features are apparent in the figure. The distributions become broader as the sound intensity increases. At the same time, the principal maximum is slightly displaced toward low frequencies. The curves exhibit a relative minimum below the frequency of the principal maximum. The minimum drifts away from the maximum as SPL increases. At very low intensities, a second minimum appears above the frequency of the principal maximum. The same features were found in two other listeners and, consequently, appear to be typical. According to Eq. (7), they reflect linearly the corresponding distributions of neural excitation and, in this way, reveal the actual filter characteristics of the human auditory system. Such characteristics cannot be obtained by any other known method.

REFERENCES

Kiang, N. Y.-S. (1965): Discharge Patterns of Single Fibers in the Cat's Auditory Nerve (Research Monograph No. 35, M.I.T. Press, Cambridge, Mass.).
Wegel, R. L., and Lane, C. E. (1924): The auditory masking of one pure tone by another and its probable relation to the dynamics of the inner ear, Physiol. Rev. 23, 266-285.
Wiederhold, M. L. (1967): A Study of Efferent Inhibition of Auditory Nerve Activity, Doctoral Dissertation, Massachusetts Institute of Technology, Cambridge, Mass., p. 65.
Zwicker, E., Flottorp, G., and Stevens, S. S. (1957): Critical band width in loudness summation, J. Acoust. Soc. Amer. 29, 548-557.
Zwislocki, J. (1960): Theory of temporal auditory summation, J. Acoust. Soc. Amer. 32, 1046-1060.
Zwislocki, J. (1965): Analysis of some auditory characteristics, in: Handbook of Mathematical Psychology, R. D. Luce, R. R. Bush, and E. Galanter, Eds. (Wiley, New York, N. Y.), Vol. 3, pp. 1-97.
Zwislocki, J. J., Buining, E., and Glantz, J. (1968): Frequency distribution of central masking, J. Acoust. Soc. Amer. 43, 1267-1271.
Zwislocki, J. J., Damianopoulos, E. N., Buining, E., and Glantz, J. (1967): Central masking: some steady-state and transient effects, Perception and Psychophysics 2, 59-64.
Zwislocki, J., Maire, F., Feldman, A. S., and Rubin, H. (1958): On the effect of practice and motivation on the threshold of audibility, J. Acoust. Soc. Amer. 30, 254-262.

DISCUSSION

Whitfield: You mentioned the non-monotonic aspect of the masking with increasing masker SPL and referred to the non-monotonically increasing firing rate of primary fibres with increasing SPL reported by Kiang (1965). The binaural interaction you have discussed is presumably, however, somewhere beyond the cochlear nucleus. As we showed in earlier work (Hilali and Whitfield, 1953), the output of many cochlear nucleus neurons is highly non-monotonic, and at still higher levels in the central auditory pathways you can find tremendous drops in firing rate at higher intensities (*e.g.* Hind *et al.*, 1963). So, I do not think you necessarily need to refer these effects to the peculiarities of primary fibres.

Zwislocki: I agree with your view that this non-monotonic relation for primary fibres is peculiar. At higher levels of the auditory pathway such a decrease of firing rate for increasing SPL may be caused by multiple interactions. But actually I was not surprised at the physiological data; the surprise came from the psychoacoustical data. As far as I know these are the only data that show that masking decreases as the masker level increases.

Terhardt: Did you find any binaural beats in your experiments? And, if so, do you believe that these beats influenced your threshold when the test tone frequency was in the neighbourhood of the masker's frequency?

Zwislocki: Because of the theme of this symposium I thought that going into experimental details would carry us too far. It is possible to avoid these beats by using a narrow band of noise as masker. In some experiments we used a simple tone as masker. This does not affect the results if a frequency of 1000 Hz is used where one does not hear such beats. We also did some experiments at low frequencies and when the masker and the test tone are close in frequency you can abolish the masking completely because of binaural beats.

Terhardt: Does this mean that, in case of equal frequencies for test tone and masker tone, the phase relation between both tones has no influence on the masking threshold?

Zwislocki: Indeed, not at 1000 Hz.

REFERENCES

Hilali, S., and Whitfield, I. C. (1953): Responses of the trapezoid body to acoustic stimulation with pure tones, J. Physiol. *122*, 158-171.
Hind, J. E., Goldberg, J. M., Greenwood, D. D., and Rose, J. E. (1963): Some discharge characteristics of single neurons in the inferior colliculus of the cat, J. Neurophysiol. *26*, 321-341.
Kiang, N. Y.-S. (1965): Discharge Patterns of Single Fibers in the Cat's Auditory Nerve (Research Monograph No. 35, M.I.T. Press, Cambridge, Mass.). p. 82.

LOUDNESS AND FREQUENCY SELECTIVITY AT SHORT DURATIONS

B. SCHARF

Department of Psychology
Northeastern University
Boston, Mass., U.S.A.

The way in which the loudness of a complex sound changes as a function of bandwidth is closely related to the ear's frequency selectivity (Zwicker and Feldtkeller, 1967). At bandwidths less than the critical band, loudness is proportional to the total power. Beyond the critical band, loudness usually increases with bandwidth although the total power or overall sound pressure level is held constant. Several experiments (Port, 1963; Zwicker, 1965) have shown that the relation between loudness and bandwidth is independent of duration, suggesting that the critical band and frequency selectivity are intact even when stimulation lasts less than 10 msec. Various threshold measurements (Elliott, 1965; Scholl, 1962), on the other hand, suggest that the critical band broadens at short durations. The disparity between the loudness and threshold measurements leaves unresolved the question of the build-up of the critical band. The present experiment takes this question up again and measures the loudness of a pair of successive tone bursts as a function of their separation in frequency and in time.

PROCEDURE

A single 5 msec tone burst was matched in loudness to a pair of 5 msec bursts. The time interval between the components of the pair was set at a fixed value between −5 msec (making the two components simultaneous) and 800 msec. The pair and the comparison burst were repeated alternately, with an 800 msec silent interval between them, until the match was completed. The frequency separation, ΔF, between the two bursts within the pair was varied between 0 Hz and a value equivalent to 6 to 12 critical bands. The geometric mean of the two frequencies was 500, 2000, or 4000 Hz which was also the frequency of the comparison burst. Measurements were made monaurally at 40, 70, and 90 dB SPL. The subject used a sone potentiometer to adjust the SPL of one sound until it was as loud as the other.

The components of the pair were set equally loud for each subject on the basis of loudness matches previously made between each component burst and the comparison.

The duration of the tone bursts and the intervals between them were measured between the half-power points as judged from oscilloscope tracings. The rise-fall time of each burst was 1.5 msec. The earphone (Telephonics TDH-49), mounted in a MX41/AR cushion, proved remarkably good in transducing the stimuli; oscilloscope tracings of the phone's output measured via a Bruel & Kjaer microphone (4132) showed no large transients, distortion, or ringing. (Only at a center frequency of 500 Hz were there significant transients at frequencies outside the critical band.)

Six men and women, all students except the writer, served as subjects. The subject made his judgments in a soundproof room, separated by a double steel wall from the equipment and experimenter.

RESULTS

The measure of summation between the two successive tone bursts is the difference between their settings when each was matched alone to the comparison burst and when matched as a pair. A positive value indicates summation between the successive bursts because they could attain the same loudness level at a lower SPL when presented together than when presented separately. These values are given on the ordinates of Figs. 1, 2, and 3 as a function of the

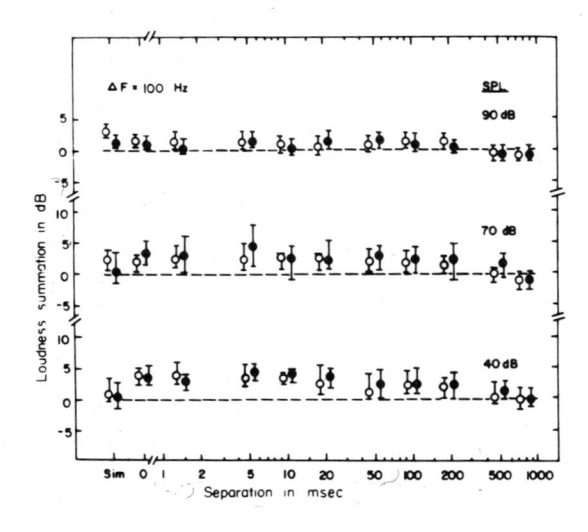

Fig. 1. The summation of loudness between two successive 5 msec tone bursts as a function of the time interval separating them. Sim means the bursts were simultaneous. The SPL of the standard is the parameter on the curves. Unfilled circles are the medians obtained when the lower frequency was presented first; filled circles when the higher frequency was first. (At simultaneity, the circles refer to replications.) A reference line is drawn at 0 dB. Vertical lines are the interquartile ranges. The value of ΔF, the difference in frequency between the two bursts, was 100 Hz.

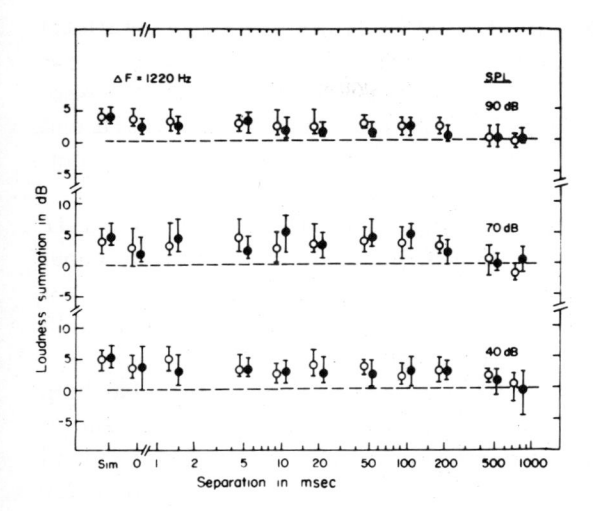

Fig. 2. Summation of two tone bursts with a frequency difference of 1220 Hz.

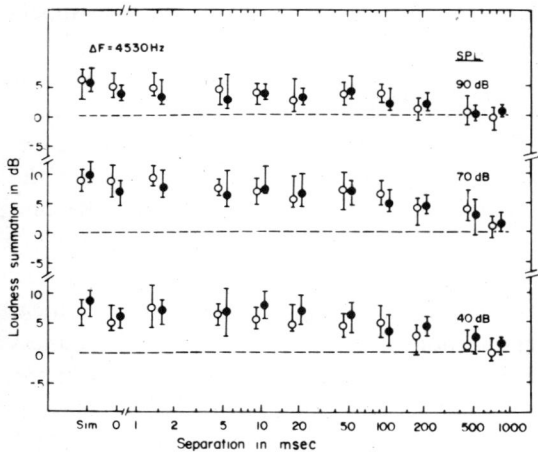

Fig. 3. Summation of two tone bursts with a frequency difference of 4530 Hz.

time interval, ΔT, between the end of the first tone burst and the beginning of the second. Results are given for 3 of the 6 ΔF's centered on 2000 Hz. Not shown are the results for ΔF's of 0, 300, and 2620 Hz, and for 6 other ΔF's centered on 500 and 4000 Hz. Each circle is the median of 24 loudness matches by the 6 subjects, each of whom made two matches by adjusting the pair and two by adjusting the single burst. Dashed lines are drawn at 0 dB where the loudness of the tone pair was no greater than either component. A second important value is at 3 dB; here the total effective energy of the successive tone bursts is equal to that of the single burst, suggesting energy summation between the successive bursts.

On the basis of measurements of long-duration tones, the bursts whose frequencies differ by a critical band or less should show 3 dB of summation when presented simultaneously. At $\Delta F = 100$ Hz, which is less than the critical band (300 Hz), summation was about 2 dB at 70 and 90 dB SPL. (It was probably less than 3 dB there, and also at 40 dB SPL, because the 5 msec samples from a simultaneous tone pair, which is separated by only 100 Hz and beats with a period of 10 msec, often contain less power than the single comparison burst or the pairs with wider ΔF's.)

At the subcritical ΔF of 100 Hz, summation stayed close to 3 dB up to a ΔT of 100 to 200 msec. However, at 90 dB SPL summation went down to about 1 dB as soon as the components were no longer simultaneous. Similar results were obtained with a ΔF of 300 Hz and 0 Hz, except that at 0 Hz and 90 dB SPL, summation was 0 dB for all time separations.

For the supercritical ΔF's, summation was greater than 3 dB, as has been noted often for long-duration complex sounds. Once the critical band is exceeded, the loudness of the components summate after allowance for mutual inhibition (Zwicker, 1958; Zwicker and Scharf, 1965). However, at $\Delta F = 1220$ Hz, the summation barely exceeded simple energy summation. At both supercritical ΔF's, 1220 and 4530 Hz, summation was less at 90 dB and 40 dB than at 70 dB. This result, especially clear at simultaneity, also duplicates those for long-duration sounds. The most striking aspect of these results is that summation between the successive bursts was measurable up to at least 200 msec. Summation did tend to decrease as ΔT was extended to 200 msec, but the decrease was more pronounced at the wider frequency separations. Beyond 100 to 200 msec summation declined rapidly to reach 0 dB at 800 msec.

The present results for simultaneous tone bursts confirm earlier results with bands of noise by Port (1963) and Zwicker (1965) that showed that the critical band for loudness is intact at brief durations. However, the amount of loudness summation was slightly less than found in some earlier measurements of the loudness of pairs of long-duration tones (Scharf, 1966). Hence, the loudness of simultaneous tones lasting 5 and 500 msec was measured as a function of ΔF. The results were very similar for the two durations; indeed, in an experimental context excluding successive tone bursts, loudness summation was somewhat greater for the brief than for the long tones, a finding contrary to what should happen if the critical band were broader at short durations. (These differences between brief and long tones were not statistically significant.)

It had been expected that summation might depend upon which tone burst came first, the higher or lower frequency. However, the results point to no clearcut advantage for one mode of presentation, i.e., the filled and unfilled circles do not differ consistently.

DISCUSSION

The results shown in the three figures are generally confirmed by the results obtained at other ΔF's and other center frequencies. On the whole, compared to earlier data with long-duration stimuli (see Zwicker and Scharf, 1965), the loudness summates at brief durations in much the same way as at long durations. This similarity suggests that the critical band is not time dependent, at least not with respect to loudness. Just as at long durations, overall loudness at brief durations increases by the same amount whether energy is increased by adding a second tone burst within the *same* critical band or by raising the level at a single frequency; moreover, the energy may be added 100 or 200 msec after the offset of the first burst. Apparently, energy is summated. However, if the added burst is in a different critical band, summation is greater than 3 dB, the value associated with energy summation. Since a critical duration of 100 to 200 msec is incompatible with the known properties of the inner ear where energy summation would have to occur, the summation is most probably neural. Moreover, Zwislocki (1960, 1969) has presented convincing arguments that temporal integration occurs at a fairly high level in the auditory nervous system.

If, however, two successive tone bursts summate neurally, their loudness ought to double, because the neural basis for loudness appears to be determined earlier in the nervous system than temporal integration (Zwislocki, 1969). Doubled loudness should appear as 10 or 11 dB of summation instead of the 3 dB measured for subcritical ΔF's. Zwislocki suggests that neural adaptation prevents loudness from doubling. According to this view, the second of two successive bursts contributes less to loudness because the system has not yet recovered from adaptation to the first. Adaptation could also explain why at 90 dB SPL there is so little summation between bursts close in frequency. Bursts far apart in frequency, and therefore less susceptible to mutual adaptational effects, show a smaller decline in summation at the high level.

A difficult problem remains. Why doesn't overall loudness increase with ΔT, at least for subcritical ΔF's, since as adaptation decreases with time, the contribution of the second burst ought to increase? Indeed, Buytendijk and Meesters (1942) mention an increase in the loudness of the second of two successive clicks as a function of ΔT. However, their results are equivocal and need to be checked. Perhaps the overall loudness of the two bursts does not increase with ΔT because the cohesiveness and summation between the two successive bursts decreases as ΔT becomes longer, as suggested by the results at the large supercritical ΔF's where loudness usually declines as a function of ΔT. At narrow ΔF's, loudness may appear to be approximately independent of ΔT because decreasing adaptation offsets decreasing loudness summation.

An unexpected aspect of these results is the lack of gross changes in either inter- or intra-subject variability as a function of ΔT. At about 50 msec, two bursts begin to be heard as separate sounds. Yet the judgments of their overall loudness did not become more difficult and, hence, more variable. Furthermore, the amount of summation was almost constant or decreased gradually with increasing ΔT up to 100 to 200 msec. There was no obvious change in summation around 50 msec. Finally, summation was greater for bursts far apart in frequency than for those close in frequency even at a ΔT of 100 msec. Apparently, subjects were able to judge, without resorting to special strategies, the overall loudness of two tone bursts whether heard as a fused sound or as separate sounds. A similar ability to judge the overall loudness of two tones localized in different ears was noted in measurements of dichotic loudness summation (Scharf, 1969). Judgments of loudness and judgments of the location of sounds either in space or time may be largely independent of each other. This independence has been suggested, on other grounds, by von Békésy (1967).

Schwarze (1964) also measured the loudness of two successive 5 msec tone bursts. He found that summation began to decrease at a separation of 10 msec and was gone by 100 msec. He observed that bursts separated by more than 50 msec are heard as two distinct sounds and cannot summate. Very likely this bias was transmitted to the subjects who then judged the loudness of individual bursts instead of the overall loudness. Niese (1956) probably biased his subjects the same way; he could measure summation up to only 25 msec. Both Schwarze and Niese presented tone bursts of the same frequency and so could hope to measure no more than 3 dB of summation. A clearer picture was obtained in the present experiments where the summation of bursts far apart in frequency was 10 dB or more. Confirmation also comes from threshold measurements; Zwislocki (1960) found that two successive clicks separated by as much as 300 msec summate and have a lower threshold than either click alone.

Research supported by a grant from the U.S. Public Health Service, National Institute of Neurological Diseases and Stroke.

REFERENCES

Békésy, G. von (1967): Sensory Inhibition (Princeton University Press, Princeton), Ch. 4.
Buytendijk, F. J., and Meesters, A. (1942): Duration and course of the auditory sensation, Commentationes Pontif. Acad. Sci. *6*, 557-576.
Elliott, L. L. (1965): Changes in the simultaneous masked threshold of brief tones, J. Acoust. Soc. Amer. *38*, 738-746.
Niese, H. (1956): Vorschlag für die Definition und Messung der Deutlichkeit nach subjektiven Grundlagen, Hochfrequenztechn. u. Elektroakust. *65*, 4-15.
Port, E. (1963): Ueber die Lautstärke einzelner kurzer Schallimpulse, Acustica *13* (Beih. 1), 212-223.
Scharf, B. (1966): Critical bands, Special Report LSC-S-3, Laboratory of Sensory Communication, Syracuse University; published in (1970): Foundations of Modern Auditory Theory, J. V. Tobias, Ed. (Academic Press, New York), Vol. 1, Chapt. 5.
Scharf, B. (1969): Dichotic summation of loudness, J. Acoust. Soc. Amer. *45*, 1193-1205.
Scholl, H. (1962): Das dynamische Verhalten des Gehörs bei der Unterteilung des Schallspektrums in Frequenzgruppen, Acustica *12*, 101-107.
Schwarze, D. (1963): Die Lautstärke von Gausstönen, Doctoral Dissertation, Technische

Universität, Berlin.
Zwicker, E. (1958): Ueber psychologische und methodische Grundlagen der Lautheit, Acustica 8 (Beih. 1), 237-258.
Zwicker, E. (1965): Temporal effects in simultaneous masking and loudness, J. Acoust. Soc. Amer. 38, 132-141.
Zwicker, E., and Feldtkeller, R. (1967): Das Ohr als Nachrichtenempfänger (S. Hirzel Verlag, Stuttgart), 2nd Edition.
Zwicker, E., and Scharf, B. (1965): A model of loudness summation, Psychol. Rev. 72, 3-26.
Zwislocki, J. (1960): Theory of temporal auditory summation, J. Acoust. Soc. Amer. 32, 1046-1060.
Zwislocki, J. (1969): Temporal summation of loudness: an analysis, J. Acoust. Soc. Amer. 46, 431-441.

DISCUSSION

Zwislocki: Recently, we made some measurements which were similar to those of Dr. Scharf, but the questions we asked the subject were somewhat different. They were presented with a pair of successive dichotic tone bursts and were asked to match the loudness of a third tone burst to the second one. Under this condition we found practically no summation between the first and the second tone burst unless there was an overlap in time. I have no idea what this means, but it is interesting that when you ask the subject a different question, his result changes very much.

Scharf: It is quite clear that instructions may influence psychophysical results drastically. In studies of dichotic loudness summation in which one tone was presented to one ear and another tone of a different frequency to the other ear, I obtained very different measures of loudness depending upon whether the subject was asked to judge the overall dichotic loudness or the loudness in just one ear (Scharf, 1969). I also measured the loudness of two tone bursts presented, with a time delay, separately to the two ears (Scharf and Weissmann, 1970). Even in this dichotic presentation, loudness summated over time separations as long as 100 to 200 msec.

Johnstone: I must say that I have not really understood all details because it is out of my experience, but in effect you are saying that you cannot find a time structure in the critical band. I always felt, I may be wrong here, that the critical band was a component that indicated lateral inhibition or sharpening. Do not your experiments, in which no time dependency was found, tend to indicate that this inhibition does not exist? I cannot imagine a sharpening effect without assuming a time structure.

Scharf: That assumption is one of the reasons why it has been interesting to study the critical band's properties. If the critical band is there immediately, so to speak, after the onset of a signal, then it probably could not be neural in nature.

Johnstone: Well, it could be neural but it should have a very short time constant. It could be electrical rather than a chemical synapse.

Scharf: I agree.

Zwicker: Let me try to make it clear. The critical band seems to be there, as you say, almost immediately. Hence, it is improbable that a large inhibition system with loops passing through higher levels is involved. I am in favour of thinking that maybe somehow there is a direct interaction without loops at the level of the hair cells. If this is the case, then we may interpret the critical band as being present already.

Johnstone: If there is lateral inhibition, it has to be some direct electrical or ephaptic inhibition. Would you agree with that?

Zwicker: I am not a physiologist but it seems to me that there is no intermediate neuron involved.

Scharf: Is it not true that the usual physiological definition of inhibition does require that a synapse is involved?

Johnstone: No, that is no condition. One can have ephaptic transmission which may be faster.

De Boer: Just think about two-tone inhibition (depression), there is no second neuron involved.

Dallos: I cannot see how a synaptic delay of 1 or 2 msec could show up in the data presented by Dr. Scharf.

Scharf: Since each tone burst lasted 5 msec, my results tell little about the first 2 or 3 msec of stimulation. We should be very careful in making any physiological interpretation.

REFERENCES

Scharf, B. (1969): Dichotic summation of loudness, J. Acoust. Soc. Amer. *45*, 1193-1205.
Scharf, B., and Weissmann, S. M. (1970): Dichotic summation of loudness over time, J. Acoust. Soc. Amer. *47*, 96-97(A).

PERCEPTUAL SPACE OF VOWEL-LIKE SOUNDS AND ITS CORRELATION WITH FREQUENCY SPECTRUM

LOUIS C. W. POLS

Institute for Perception RVO-TNO
Soesterberg, The Netherlands

INTRODUCTION

It is well-known that isolated vowel sounds, spoken or whispered, can be identified quite well. In such a presentation pitch, loudness and duration give no distinct information. The only stimulus variable in physical terms is spectral information, and in perceptual terms timbre or vowel quality. In this paper we want to investigate the relation between the spectral and perceptual information.

The usual way of describing the spectral information of vowel sounds is in terms of positions in the 2-formant plane (Peterson and Barney, 1952). The formant frequencies are those at which the frequency spectrum has pronounced maxima. These peaks correspond to the resonance frequencies of the cavities in the vocal tract during vowel pronunciation. It is, therefore, not so surprising that the formant frequencies have since long been used to describe the differences in the vowel sounds. But are those formant frequencies themselves the specific factors in vowel perception? This would imply that the hearing system works as a formant detector. It is perhaps more acceptable to suppose that the ear uses in principle all the information present in the frequency spectrum. A spectral analysis with a bandwidth smaller than the critical bands of the ear makes then no sense.

STIMULI

In order to reduce the number of physical variables that characterize the stimuli, we worked with sustained vowel-like sounds; the signals differed only in their frequency spectra. Loudness, onset, duration and pitch were equalized. This was made possible by using a digital computer (DEC PDP-7) for generating the stimuli. We started with normally spoken words of the type h(vowel)t from one male speaker. Each word was sampled 20,000 times per sec via an analog-to-digital converter. The amplitude values of the samples were successively stored in the memory of the computer. Afterwards the whole word, but also any desired part out of it, could be generated via a digital-to-analog converter. In this way, one period out of the constant vowel part of each word was taken. By repeating this period a specific number of times, a sustained vowel-like sound was produced. The 11 vowels used were œ, ɔ, a, u, y, i, a, ø, o, ɛ, and e (IPA, 1967) which are in written language the Dutch vowels u, o, a, oe, uu, ie, aa, eu, oo, e and ee. By resampling each period with a fixed number of samples, the fundamental frequency of all signals was made equal to 123.5 Hz. The loudness levels of the stimuli were equalized. By starting the signals always on or in the neighbourhood of the zero line, the onset transients were minimized.

SPECTRAL INFORMATION

The signals can be described physically in terms of their formant frequencies (F_i) and levels (L_i), which information can be extracted from the measured line spectra. We may wonder, however, whether the differences between the frequency spectra are fully described by taking into account only the positions of the characteristic peaks. With reference to the critical bandwidth of the ear's analyzing power (Plomp and Mimpen, 1968) we also made a spectral analysis by using 1/3-octave bandfilters. The sound pressure levels in dB in the 18 used frequency bands constitute an (11 × 18) data matrix. In terms of a geometrical model, we can say that the sound spectra of the 11 vowels result in a set of

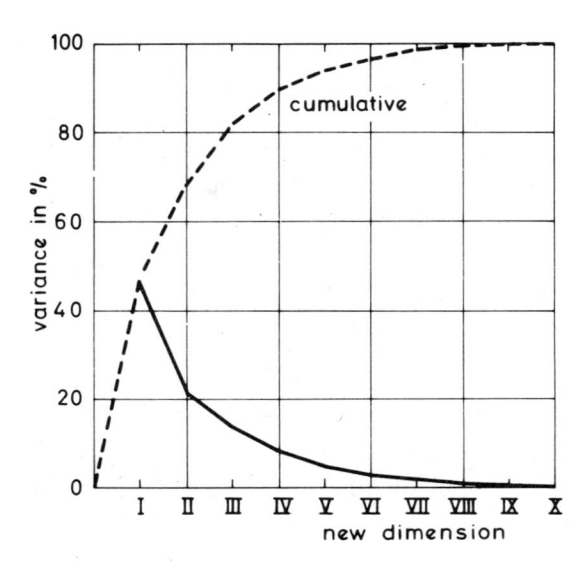

Fig. 1. Percentages of the total variance explained by the computed new dimensions.

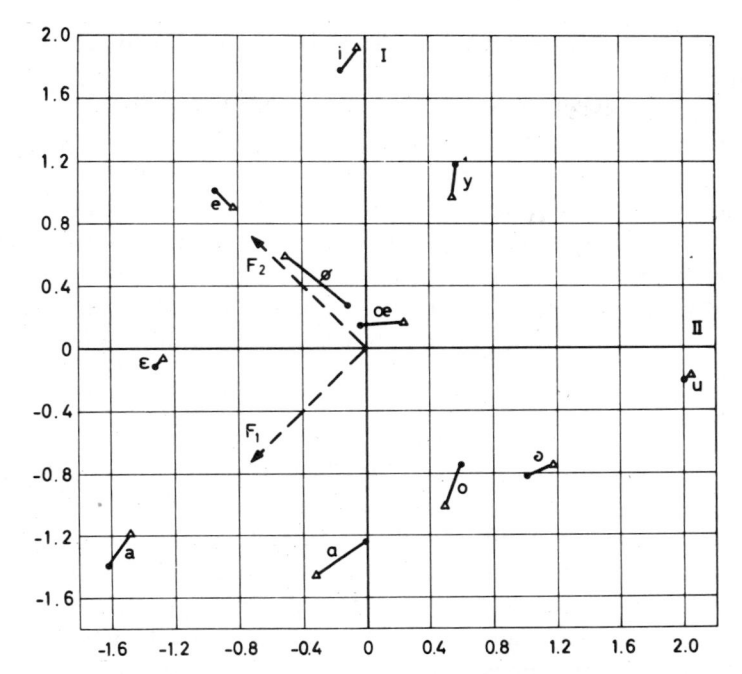

Fig. 2. Projections of the points on a plane in the 6-dimensional physical configuration (Δ) that correlates maximally with the F_1-F_2 plane (·). The original orientations of the F_1 and F_2 axes are also given.

11 points in an 18-dimensional space. It is now of interest to determine the minimal number of dimensions required to describe these data without loss of too much information. Considering the small number of stimuli and the mutual dependency between the different frequency bands, a reduction should be possible. Using a principal-components analysis it appears that just by extracting 3 dimensions, 81.7% of the total variance can already be explained. The explained variance as a function of the number of dimensions is given in Fig. 1. The 11 points can of course always be described in 10 dimensions. Also for a group of natural vowel sounds spoken by 10 (and in a later experiment by 50) male subjects, this reduction appeared to be possible (Plomp et al., 1967; Klein et al., 1970). If this physical space is a good description of the vowel sounds then also the information of formant frequencies and levels must be present in this configuration. In the 6-dimensional physical space (96.6% explained variance) a plane could be found which correlated almost perfectly (correlation coefficients 0.985 and 0.981) with the F_1-F_2 plane, see Fig. 2. This plane only makes a small angle with the one determined by the 2 first physical dimensions which explain most of the variance (68.0%). Furthermore it appears out of a multiple correlation analysis (Anderson, 1958) that the first 3 formant

frequencies and levels are not independent variables, especially for this group of 11 signals. With this method one determines directions in the physical space which correlate maximally with $F_1, F_2, F_3, L_1, L_2, L_3$, respectively. Out of the multiple correlation coefficients given in the last column of Table I, one sees that the correlation is significantly high for F_1, F_2 and L_2 in the 6-dimensional physical space. One also has, however, to determine whether these "images" of the outside variables are independent or perhaps more or less associated. The correlation coefficients given in Table I can be considered as cosines of the angles between the indicated directions. One sees that F_1 and F_2 are independent variables, and that the other variables are associated with each other.

PERCEPTUAL ANALYSIS

In the foregoing we have discussed the way in which the physical information and more precisely the frequency spectra of the signals, can be best described.

We now want to direct our attention to the way in which these signals can be judged by observers. Since it is quite clear that the stimuli only can be characterized by a complex attribute, the use of multidimensional scaling techniques is most appropriate. The aim of such an analysis is to determine a perceptual space based on observations concerning the relative similarity of the stimuli. For that, techniques like short term recall (Wickelgren, 1965), semantic scaling (Solomon, 1958), scaling based on perceptual confusion (van der Kamp and Pols, 1970), direct scaling by ratio estimation (Hanson, 1967), paired comparison (Mohr and Wang, 1968) or comparison based on triadic combinations (Levelt et al., 1966) are used in psychoacoustics. The last one has in our opinion some essential advantages over the other scaling methods. Using this method the subjects only have to decide for each possible subset of 3 stimuli, which pair is most similar and which pair is least similar, without further indicating the degree of similarity or its relation to specific categories.

Table I. Correlation coefficients between the projections of the points on the vectors corresponding maximally to F_i and L_i in the 6-dimensional physical space. The multiple correlation coefficients are given in the last column, together with the significance levels ($++ : p=.01, + : p= .05$ and n.s. : not significantly different from zero correlation).

	F_1	F_2	F_3	L_1	L_2	L_3	mult. corr. coeff.
F_1	—	−.0319	.0901	−.5201	−.0538	−.5134	.983 ++
F_2		—	.6417	.5620	−.8709	.6216	.984 ++
F_3			—	−.1296	−.4487	.5513	.679 n.s.
L_1				—	−.5478	.6135	.920 n.s.
L_2					—	−.4577	.974 +
L_3						—	.821 n.s.

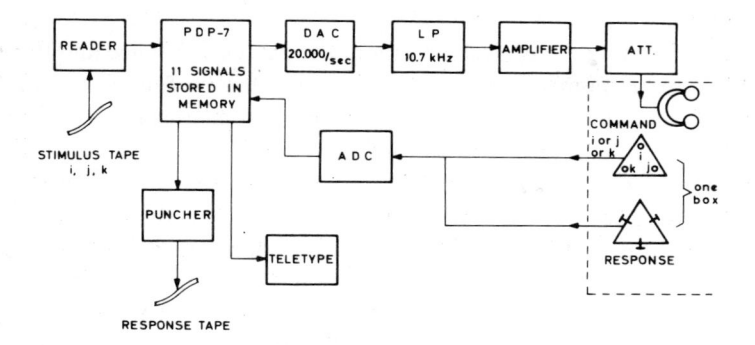

Fig. 3. Block diagram of the experimental setup.

By using the computer as a stimulus generator the triadic experiment is quite easily to carry out. Fig. 3 is a block diagram of the experimental setup. Every time the subject presses one of the buttons positioned on the vertices of an equilateral triangle he hears in his headphones one of the stimuli of the particular triad. He can listen as often as he wants and in any order to the 3 signals. The duration of each stimulus is maximally 405 msec but the subject can switch to another signal within that time. When he has selected the 2 stimuli which, in his opinion, are most similar he presses the response button positioned on the side of the triangular box between the 2 stimulus buttons; he does the same for the most dissimilar pair. Both responses are punched and typed out after which the code for the next triad is read in. The subject then at once can go on judging this new group of 3 stimuli. In fact the computer controls the whole experiment. 15 Students, all with normal hearing, participated in the experiment and executed individually the judgments of all 165 $[n(n-1)(n-2)/6, n=11]$ triads within an hour. From the single decisions of the subjects a similarity matrix (see Table II) was built up in such a way that every time a pair was judged most similar the concerning cell in the matrix was updated with 2 points, the most dissimilar pair got zero points, and the cell concerned with the

Table II. Cumulative similarity matrix of 11 vowel-like sounds (15 subjects).

	ɔ	a	u	y	i	ɑ	ø	o	ɛ	e
æ	168	140	115	145	100	108	250	180	145	201
ɔ	—	162	152	116	51	99	144	230	158	130
a		—	99	87	78	205	144	195	209	130
u			—	208	78	66	94	144	141	85
y				—	155	68	134	106	118	84
i					—	70	115	57	91	104
ɑ						—	122	141	187	127
ø							—	156	150	225
o								—	163	135
ɛ									—	160

remaining pair got one more point.

For determining the dimensionality of the perceptual space and the positions of the stimulus points in that space, a way must be found to transform the similarity judgments of the subjects to perceptual distances and next to a perceptual configuration. No uniform model exists for that; the assumptions made by Kruskal (1964a) seemed most appropriate to us. He developed a program (Kruskal, 1964b) in which, in an iterative way, that point configuration is determined of which the rank order of interpoint distances agrees as well as possible with the rank order of the dissimilarities. This means that not the absolute values of the similarity scores are used but only their rank order.

The metric of the system and the number of dimensions can still be chosen freely. As a quantitative measure of "goodness of fit" a "stress" percentage, related to the residual variance, is defined. An interpretation of the amount of stress is a matter of intuition and experience, and an objective criterion for evaluating the stress is only possible by comparing it with the distribution of stress percentages found by analyzing random data. One more problem arises, since this distribution appears to be not independent of the metric used. Keeping these points in mind, the solutions can be interpreted in a correct way.

Analysis of the cumulative similarity matrix, see Table II, results in a 3-dimensional structure with 1.6% stress, in 4 dimensions this value becomes 0.5% and in 2 dimensions 8.2%. The reduction of the stress percentage by taking 4 instead of 3 dimensions is not large enough to make a 4-dimensional solution more acceptable. Therefore, the 3-dimensional solution was chosen for further analysis. Working with non-Euclidean distance functions (variable Minkowski parameter p) did not result in more probable solutions with lower stress percentages. In fact the curve representing the minimal stress as a function of p was rather flat. Plomp has shown in his paper (pp. 405-408 of this volume) that the physical properties of the applied stimuli are not very appropriate to decide which p-value is involved in vowel perception.

RELATION BETWEEN PERCEPTUAL SPACE AND FREQUENCY SPECTRA

The last remark brings us to the relation between the perceptual and the physical space of the vowel sounds. In order to derive an optimal correspondence between two sets of points the canonical matching procedure (Cliff, 1966; Levelt et al., 1966) was applied. Both normalized configurations are rotated in such a way that the covariance between projections of the points of both configurations on corresponding axes is maximal. The amount of correspondence can be expressed in a correlation coefficient per orthogonal axis. The coefficients for a matching of the 3-dimensional physical with the 3-dimensional perceptual space are 0.992, 0.972 and 0.742. These values can still be improved by matching the 3-dimensional perceptual space with the 6-dimensional physical space

(0.999, 0.987 and 0.974), see Fig. 4. This excellent correspondence suggests that the subjects used for their judgments information comparable with that present in the multidimensional physical representation of these signals. The effect of familiarity with the Dutch vowel sounds may be considered as negligible, as was judged from the results obtained by using as subjects two foreign visitors, a Welshman and a native Japanese. Their results were comparable with the individual results of our group of 15 Dutch subjects. The perceptual differences between the stimuli, to be considered as timbre differences, appear to be qualified by their differences in frequency spectra. The results obtained by analyzing the signals with 1/3-octave filters, which are comparable in bandwidth to the critical bands of the ear, support the view that also in vowel detection the critical bandwidth plays an important role. As can also be expected, F_i and L_i are highly correlated with specific directions in the perceptual space. However, most of these directions are associated with each other; only F_1 and F_2 appear to be independent variables. This confirms the results of other investigators who found from scaling (Hanson, 1967), confusion (Pickett, 1957; Miller, 1965) and synthesizing (Cohen et al., 1967) experiments that the first two formant frequencies are important factors.

Our results show that it is not necessary to determine the spectra with narrow-band filters, but that a 1/3-octave filtering is sufficient. In order to evaluate the merits of the 1/3-octave filtering, the signals were also analyzed with other filter systems, both with constant Δf and constant $\Delta f/f$. In no case a better correlation with the results of the perceptual analysis could be achieved than was obtained with the 1/3-octave filter system. The set of stimuli was too small to determine precisely the most appropriate bandwidth.

The technique of extracting basic information out of a 1/3-octave spectral analysis is currently also successfully used by us in running speech analysis and in the development of systems for vowel and word recognition.

More details of the above experiment can be found in Pols et al. (1969).

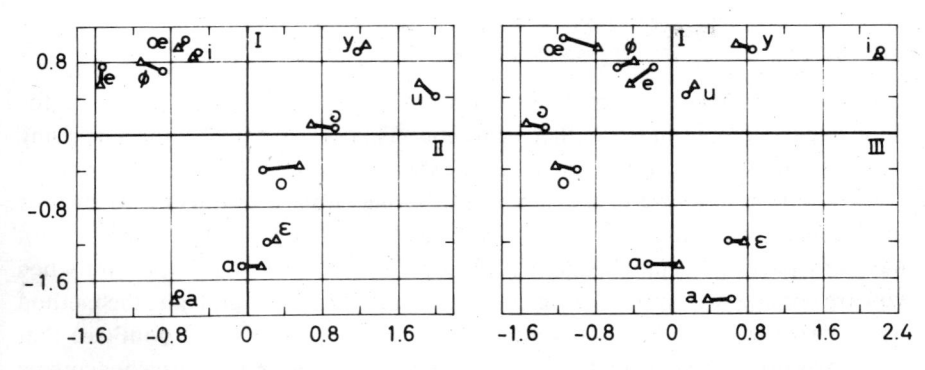

Fig. 4. Positions of the points in the optimal I-II and I-III planes when the 3-dimensional perceptual configuration (Δ) is matched with the 6-dimensional physical configuration (o).

REFERENCES

Anderson, T. W. (1958): An Introduction to Multivariate Statistical Analysis (Wiley, New York).
Cliff, N. (1966): Orthogonal rotation to congruence, Psychometrika *31*, 33-42.
Cohen, A., Slis, J. H., and 't Hart, J. (1967): On tolerance and intolerance in vowel perception, Phonetica *16*, 65-70.
Hanson, G. (1967): Dimensions in speech sound perception. An experimental study of vowel perception, Ericsson Technics *23*, 3-175.
IPA (1967): The principles of the International Phonetic Association (Dept. of Phonetics, University College, London, W.C. 1).
Kamp, L. J. Th. van der, and Pols, L. C. W. (1970): Perceptual analysis from confusions among vowels, submitted for publication in Acta Psychologica.
Klein, W., Plomp, R., and Pols. L. C. W. (1970): Vowel spectra, vowel spaces, and vowel identification, J. Acoust. Soc. Amer. *48*, 999-1009.
Kruskal, J. B. (1964a): Multidimensional scaling by optimizing goodness of fit to a nonmetric hypothesis, Psychometrika *29*, 1-27.
Kruskal, J. B. (1964b): Nonmetric multidimensional scaling: a numerical method, Psychometrika *29*, 115-129.
Levelt, W. J. M., Geer, J. P. van de, and Plomp, R. (1966): Triadic comparison of musical intervals, Brit. J. Math. Statis. Psychol. *19*, 163-179.
Miller, G. A. (1965): The perception of speech, in: R. Jakobson Essays on the Occasion of his Sixtieth Birthday, M. Halle *et al.*, Eds. (Mouton & Co., 's-Gravenhage) pp. 353-359.
Mohr, B., and Wang, W. S. I. (1968): Perceptual distances and the specification of phonological features, Phonetica *18*, 31-45.
Peterson, G. E., and Barney, H. L. (1952): Control methods used in a study of the vowels, J. Acoust. Soc. Amer. *24*, 175-184.
Pickett, J. M. (1957): Perception of vowels heard in noises of various spectra, J. Acoust. Soc. Amer. *29*, 613-620.
Plomp, R., and Mimpen, A. M. (1968): The ear as a frequency analyzer. II, J. Acoust. Soc. Amer. *43*, 764-767.
Plomp, R., Pols, L. C. W., and Geer, J. P. van de (1967): Dimensional analysis of vowel spectra, J. Acoust. Soc. Amer. *41*, 707-712.
Pols, L. C. W., Kamp, L. J. Th. van der, and Plomp, R. (1969): Perceptual and physical space of vowel sounds, J. Acoust. Soc. Amer. *46*, 458-467.
Solomon, L. N. (1958): Semantic approach to the perception of complex sounds, J. Acoust. Soc. Amer. *30*, 421-427.
Wickelgren, W. A. (1965): Distinctive features and errors in short term memory for English vowels, J. Acoust. Soc. Amer. *38*, 583-588.

DISCUSSION

Fourcin: I noticed in your demonstration that the vowel sounds used for your experiment sounded rather synthetic. This is, of course, not surprising because they are repeated single periods extracted from real vowels. One should realize that, in listening to synthetic speech, one either recognizes it as such or one hears a miscellaneous sequence of buzzes, clicks and hisses, which does not have any particular speech significance at all. I think that it is important when you are dealing with speech recognition to ensure that the stimuli and the method used strike the heart of what is involved in speech perception. I dont' see that you have done this. In fact, you are working at a peripheral psychoacoustic level, not on the higher levels which are generally employed in speech recognition. Well, this is just a general comment which, I think, is basic to the decades of relative lack of success which we now have seen go by with regard to work on speech recognition in general.

In particular, I would like to make some additional small comments. First, you don't appear to distinguish between long and short vowels but put them all together. This again is an example of, I think, a neglect of essential human dimensions. Second, you give all stimuli the same fundamental frequency. This normalization changes your formant frequencies and affects the speech-like quality of the sounds.

Pols: Since we presented the signals to the listeners as sustained sounds it is indeed true that the difference between long and short vowels, as far as duration is concerned, is neglected. That also explains why in Fig. 4 the positions of [æ] and [ø] and of [ɔ] and [o] are close to each other. The fundamental frequency of all signals was equalized by resampling the original waveform with a fixed number of samples per period. This method of normalization affects the positions of the formant frequencies in a way proportional to the ratio between the original fundamental frequency and the average fundamental frequency of 123.5 Hz. Since the original vowels were all spoken by one person, the formant shifts as a result of this normalization are small. Furthermore it was not our first claim to guarantee the speech-like quality of the sounds. We only liked to use stimuli which were *related* to sustained vowels. Sustained vowels can be recognized normally, which means that they contain about the minimal information required to know that a certain vowel is spoken. In running speech we use many more issues: all kinds of formant transitions, linguistic features, dictionary knowledge, etc. We know that the ear performs a frequency analysis; the only purpose of this research was to look for a possible correlation between the bandfilter analysis of sustained vowels and the way in which subjects judge these sounds. That is also the reason why only one speaker was involved. Single periods out of the same vowels pronounced by others would have given slightly different positions both in the physical space and in the perceptual space, but again a similar correlation would have been found. I don't pretend that the stimuli used represent the average Dutch vowels, but this is irrelevant if we would like to know whether the subjects base their judgements on frequency spectrum as analyzed. The only issue of this study is the correlation and this appeared to be quite good.

Fourcin: Your explanation just given means that you could have taken any other miscellaneous set of sounds; they don't have to be vowels. This does not imply that the criteria involved in this experiment apply for real speech and it is important to recognize that.

Pols: Remember that exactly the same technique was used by Dr. Plomp for tones derived from musical instruments (p. 405 of this volume), showing that the agreement between the physical and perceptual analyses is a general one.

We applied the same physical analysis in a speech recognition device and obtained quite promising results. This may be an indication that the correlation found in my experiments is, indeed, relevant in speech perception.

Schouten: I would just like to underline the importance of Dr. Fourcin's remarks with respect to either listening to true vowels or listening to some synthetic sound resembling a vowel. In point of fact, your stimuli were synthetic vowels. In listening to vowels we have our whole cognitive system coming into action. Our cumulated experience from babyhood onward enables us to place an actual vowel immediately in its space, and that is different from comparing synthetic vowels.

Carterette: Dr. Møller and I have done some similar work at the Speech Transmission Laboratory in Stockholm (Carterette and Møller, 1962). We used actual and synthetic Swedish vowel sounds and operated on them with high- and low-pass filters. Listening tests showed that the two formant frequencies were sufficient to describe the perceptual differences; their levels appeared to be of minor importance. We applied Kruskal's multidimensional scaling program on the confusion matrices and came out with about 5% stress for a two-dimensional solution. The results seem to conform with yours.

Pols: We also did a confusion experiment with the same signals. This is in fact a rather different approach than the one presented in my paper, but we obtained quite similar results (van der Kamp and Pols, 1970).

Carterette: One of the reasons I never published the results of multidimensional scaling of the confusion matrices (but see Carterette, 1967, for a weighting analysis based on a linear model) was that I could not meet the condition of symmetry: the probability that vowel i was confused with vowel j was not the same as the probability that j was confused with i. Can you give any comment on that problem?

Pols: This is truly a difficulty in confusion experiments. We developed in our Institute various mathematical methods to eliminate the response bias as good as possible (van der Kamp and Pols, 1970; Klein *et al.*, 1970).

Cardozo: I have a question about the very high correlation coefficients of 0.999, 0.987, etc. between the matched perceptual and physical spaces. I feel that there are more people here who wonder how we can arrive at these figures. Is the correlation coefficient really an adequate measure of similarity, or is the success partly due to the fact that you are matching multidimensional point configurations so that you have many degrees of freedom?

Pols: The problem is that, as far as I know, for the matching technique we used, there are no significance tests for the correlation coefficients. Using a Monte Carlo procedure we therefore determined cumulative probability curves for random data. It came out that matching two 3-dimensional configurations of 11 points there will be a 5% chance that the successive correlation coefficients are larger than 0.97, 0.77 and 0.43, respectively, for random data. Matching a

3- and a 6-dimensional configuration these figures are: more than 0.99, 0.94 and 0.72. You see that our experimental results are well above these 5% probability values.

Lindblom: With respect to the perceptual space, do I understand you correctly if I assume that the subjective distance between the stimuli i and j depends upon the total set of stimuli you work with? If you present the same two stimuli in a context of 9 other stimuli, would you get the same distance?

Pols: The distance is not an absolute value but related to the interpoint distances of the whole set. One can only hope that this relation is not disturbed by including other stimuli in the set. This means that if we divide the total stimulus set into 2 subsets of 6 points, then we suppose that the relative positions of the points in both subsets are the same as the relative positions of the same points in the complete set.

Plomp: In experiments on the relative effect of phase on timbre (Plomp and Steeneken, 1969) about the same results were obtained for different sets of stimuli (experiments 4 and 6). This supports the view that the context is of minor importance.

REFERENCES

Carterette, E. C. (1967): A simple linear model for vowel perception, in: Models for the Perception of Speech and Visual Form, W. Wathen-Dunn, Ed. (M.I.T. Press, Cambridge, Mass.), pp. 418-427.

Carterette, E. C., and Møller, A. (1962): The perception of real and synthetic vowels after very sharp filtering, in: Reports 4th International Congress on Acoustics, Copenhagen, Vol. I, G 54.

Kamp, L. J. Th. van der, and Pols, L. C. W. (1970) submitted for publication in Acta Psychologica.

Klein, W., Plomp, R., and Pols, L. C. W. (1970). Vowel spectra, vowel spaces, and vowel identification, J. Acoust. Soc. Amer. *48*, 999-1009.

Plomp, R., and Steeneken, H. J. M. (1969): Effect of phase on the timbre of complex tones, J. Acoust. Soc. Amer. *46*, 409-421.

NAME INDEX

Ades, H. W., 34, 37, 38, 164.
Ahumada, A., 440.
Aitkin, L. M., 153.
Allanson, J. T., 144.
Alpern, M., 427.
Anderson, D. J., 37, *116*, *193-203*, *215*.
Anderson, T. W., 465.
Arthur, 183.
Barney, H. L., 401, 463.
Baumgartner, G., 431.
Beck, C., 68.
Beickert, P., 68.
Békésy, G. von, 2, 3, 5, 8, 29, 34, 39, 43, 45, 46, 60, 61, 63, 66, 80, 81, 84, 86-89, 92, 95, 103, 104, 107, 108, 112, 113, 115, 118, 126, 127, 130, 137, 213, 227, 243, 250, 278, 294, 373, 376, 377, 395, 397, 404, 416, 427-429, 431, 432, 434, 446, 450, 460.
Beranek, L. L., 302.
Bergström, R. M., 360, 361.
Biber, K. W., 56, 57.
Bilsen, F. A., 252, 259, 266, 268, *291-302*, 312, 314, *316*.
Bindseil, H. E., 398.
Bird, C. M., 244.
Bishop, H. G., 303.
Blodgett, H. C., 319.
Boer, E. de, 29, 32, 46, 47, 51, *132*, *149*, *150*, *191*, *200*, *202*, *204-216*, 218, *245*, 246, 251, 253, 255, *264*, 265, 267, 271, 272, 281, 289, *290*, 294, 301, *326*, *360*, 395, 438, 440, 441, *462*.
Boerger, G., *147-149*, *161-167*.
Bogert, B., 118, 122.
Borghesan, E., 4.
Bosanquet, R. H. M., 404.
Bosher, S. K., 29, *37*, *58*, *79*, *213*, *317*, *361*, *412*.
Botsford, J. H., 317.
Boudreau, J. C., 144.
Boyle, A. J. F., 62, 81, 87, 107, 113, 116, 126, 130, 207.
Brandt, S., 398.
Breuing, G., 63.
Brink, G. van den, *362-374*.
Brugge, J. F., 180, 183, 185, *193-203*, 208.
Butler, R. A., 74, 104.
Buytendijk, F. J., 459.
Capps, M. J., 34, 37, 38, 164.

Cardozo, B. L., 47, 51, *151*, 271, 327, 329, *339-349*, *396*, *472*.
Carpenter, A., 402.
Carterette, E. C., *302*, *427-444*, *472*.
Chamberlain, S. C., 161.
Chistovich, L. A., 354, 355.
Chladni, E. F. F., 230, 401.
Chocholle, R., 166, 220, 355.
Christiansen, J. A., 7, 8.
Cliff, N., 468.
Cohen, A., 469.
Colburn, H. S., 337.
Comis, S. D., 140, 145.
Corliss, E. L. R., 351, 357, 358, 360.
Cortesina, G., 161.
Cramer, E. M., 327.
Crane, H. D., 243.
Creelman, C. D., 330, 352.
Crowley, D. E., 74.
Crozier, W. J., 317.
Dallos, P., 56, *117*, *132*, *133*, *218-229*, *462*.
Daniloff, R. G., 427, 428, 434.
David, E., *153-160*.
David, H., 427.
Davis, H., 29, 32, 34, 37, 63, 68, 94, 103, 104, 106, 123, 130, 131, 136, 176, 178, 181, 213, 397.
De Morgan, A., 404.
Derbyshire, A. J., 136.
Desmedt, J. E., 21, 67, 161.
Dewson, J. H., 215.
Djourno, A., 141, 142, 150.
Donders, F. C., 401.
Dunker, E., 161, 164.
Eccles, J. C., 138.
Eldredge, D. H., 22, 104, 122, 130, 229.
Elliott, D. N., 35.
Elliott, L. L., 432, 455.
Engebretson, A. M., 104, 122, 228.
Engel, F. L., 254, 267.
Engel, G., 402.
Engelhardt, V., 402.
Enger, P. S., 142.
Engström, H., 4, 7, 8, 10, 12, 67, 68.
Evans, E. F., 90.
Fant, G., 402.
Feldtkeller, R., 378, 385, 397, 408, 455.
Fernandez, C., 62, 95, 220, 221.
Fex, J., 7, 21, 22, 34, 67, 140, 161, 162, 166.
Finkenzeller, P., 63, *153-160*.

Fischler, H., 261.
Flanagan, J. L., 86, 118, 244, 251, 253, 257, 258, 388, 408.
Fletcher, H., 45, 46, 60, 79, 397, 404.
Flock, A., 8.
Fourcin, A. J., 252, 292, 293, *302*, 312, *316*, *319-328*, *470*, *471*.
Fourier, J., 230.
Friedman, M. P., *427-444*.
Frishkopf, L. S., 69, 193.
Fruhstorfer, H., 360, 361.
Gacek, R. R., 12, 21.
Galambos, R., 21, 153, 159, 161, 176, 178.
Garner, W. R., 355, 356.
Gässler, G., 377.
Geer, J. P. van de, 414.
Gehrcke, E., 402.
Geisler, C. D., 215, 336.
Goblick, T. J., Jr., 183.
Goldberg, J. M., 336.
Goldstein, J. L., *37*, 45, 49, 51, 88, *149*, *150*, *173*, *190*, *201*, *202*, 210, 222, 223, 225, 227, 227, 228, *230-247*, 261, 272, *275*, *277*, *301*, *328*, *337*, *360*, 381, 404, *413*.
Graham, 433.
Grassmann, H., 401, 402.
Green, D. M., 357, 358.
Greenwood, D. D., 2, 113, 433, *436-437*, 438, 441-444.
Grubel, G., 161.
Gruber, J., *161-167*, *300*, *301*.
La Grutta, V., 161.
Guelke, R. W., 56.
Guinan, J. J., 238, 239.
Guttman, N., 251.
Haar, G., 404.
Hall, J. L., 138, 140.
Hanson, G., 466, 469.
Harris, G. G., 327.
't Hart, J., 268.
Hartline, H. K., 67, 427, 429.
Hauser, H., 388.
Hecht, S., 71.
Held, H., 67.
Helle, R., *246*, 385.
Hellwag, 401.
Helmholtz, H. L. F. von, 44-46, 61, 222, 230, 231, 236, 250, 278, 376, 398-404, 419-421, 424, 426.
Henning, G. B., *350-361*.
Hermann, L., 44, 46, 231, 401, 403, 404.
Hilali, S., 453.
Hind, J. E., 153, 180, 181, *193-203*, 209, 210, *394*, *441*, 441, 454.
Hofstätter, R., 423.
Honrubia, V., *38*, *94-106*, *117*, *124*, 127, 151, *192*.
Hood, J. D., 317.
Hoogland, G. A., 46, 47.
Houtsma, A. J. M., 277.
Huggins, W. H., 327, 427.
Husmann, H., 426.
Huxley, A. F., 60, 61, 86.
Huygens, C., 291.
Huyssen, R. M. J., 56.
Iurato, S., 3, 12, 15, 16, 161.
Jensen, C. E., 7.
Johnstone, B. M., 32, 34, 37, *38*, 56, 61, 80, *81-93*, 107, 113, 116, *124*, 126, 130, *131*, *166*, 207, *215*, *227*, *316*, *461*, *462*.
Jones, A. T., 230.
Jones, R. C., 141.
Jongkees, L. B. W., 29, 32, 207.
Kallert, *153-160*.
Kamp, L. J. Th. van der, 466, 472.
Karlin, J. E., 433.
Katsuki, Y., 68, 70, 71, 73, 74, 177, 178, 193.
Kayser, D., 142.
Keeler, J. S., 317.
Keidel, W. D., 56, 57, *60-80*, *147*, *153-160*, *165*, *166*.
Kemp, E. H., 137.
Khanna, S. M., 103.
Kharkevich, A. A., 357, 358.
Kiang, N. Y.-S., 29, 37, 51, 52, 67, 70, 81, 84, 90, 95, 104, 136, 150, 153, 159, 164, 172, 177, 190, 191, 193, 198-202, 207, 210, 213, 225, 227, 228, 232-239, 241, 244, 246, 247, 275, 276, 434, 451-453.
Kietz, H., 61, 71.
Kimura, R., 12.
Klein, W., 465, 472.
Klinke, R., 34, 68, *149*, *161-167*, *360*, *361*.
Knight, B., 433.
Köhler, W., 402.
Kohllöffel, L. U. E., *39*, *91*, *107-117*, *124*, *228*.
Kolmer, W., 8.
König, R., 403, 404.
Kruskal, J. B., 403, 406, 468, 472.
Krutel, J., 355.
Kuile, Th. E. ter, 404.
Kupperman, R., *126-133*.
Kuyper, P., 207, *348*, *425*.
Lagrange, J. L., 230.
Lane, C. E., 404, 445.
Lawrence, M., 69, 75, 79, 218, 404, 427.
Legouix, J. P., *118-125*, 220.
Leibbrandt, C. C., 35.
Levelt, W. J. M., 159, 466, 468.
Liang, C., 354, 355.
Lichte, W. H., 403.

Lichtensteiger, W., 29.
Licklider, J. C. R., 46, 53, 253, 302, 320, 397, 403, 427.
Lindblom, B., *473*.
Littler, T. S., 397.
Löb, E., 402.
Loewenstein, R., 5.
Lorente de Nó, R., 11, 67.
Lovell, J. D., *427-444*.
Lowenstein, O., 137.
Lummis, R. C., 303.
Lüscher, E., 71.
Lynn, P. A., 213.
MacGinitie, G., 67.
Maiwald, D., 381-384.
Marsh, J. T., 74.
Maruyama, N., 437.
Mathes, R. C., 278, 404.
McClellan, M. E., 172, 252, 266, 292, 329.
McNichol, E. F., 67.
Meesters, A., 459.
Mersenne, M., 230.
Meyer-Eppler, W., 421.
Meyer zum Gottesberge, A., 68.
Michler, H., 68.
Miller, G. A., 355, 356, 469.
Miller, R. L., 278, 404.
Mills, A. W., 319.
Mimpen, A. M., 399, 408, 464.
Misrahy, G. A., 103.
Mohr, B., 466.
Møller, A., 472.
Møller, A. R., 37, *160*, *168-174*, *214*, *215*.
Morton, J., 402.
Moushegian, G., 138, 153.
Naftalin, L., 75, 86.
Neubert, K., 63, 68.
Nieder, I., 37.
Nieder, P., 37.
Niese, H., 460.
Nomoto, M., 177, 178, 193.
Nomura, Y., 12.
Nordmark, J., 252.
Oetinger, R., 388.
Ohm, G. S., 42, 44, 45, 230, 399, 404.
Peake, W. T., 161, 238, 239.
Peterson, G. E., 401, 463.
Peterson, L., 118, 122.
Pfalz, R. K. J., 67, 68, 140, 161, 165.
Pfeiffer, R. R., 153, 172, 183.
Piazza, R., *54*, *264*, *327*.
Pickett, J. M., 469.
Plattig, K. H., 68, 72, 74.
Plomp, R., 41, 45, 49, *57*, *93*, *151*, 159, *190*, 191, 222, 225, *227*, 231, 232, 255, 260, 268, 272, *277*, 285, 286, *287*, 289, *290*,
310, *327*, 391, *395*, *396*, *397-414*, 419, 421, 464, 465, 468, 471, *473*.
Pollack, I., *329-338*, 340, 345.
Pols, L. C. W., 401, 405, 407, *463-473*.
Port, E., 455, 458.
Psotka, J., 351, 355.
Rahlfs, V., 403.
Rainbolt, H. R., 433.
Ranke, O. F., 60, 63, 67, 71, 79, 124, 214, 376.
Rasmussen, G. L., 10-12, 15, 67, 161.
Ratliff, F., 67, 88, 427, 429, 433, 440.
Rayleigh, J. W. S., 231.
Reichardt, W., 67.
Reid, G., 317.
Riemann, H., 417.
Ritsma, R. J., 47, 49, 52, 144, *250-266*, 267, 268, 271, 285, 286, 289, 293-295, 297, 298, 314, 316, 327, 339-341, 348, 358, *396*, 409, *426*.
Röber, A., 404.
Rodieck, R. W., 153.
Rose, J. E., *92*, 136-138, 144, *150*, *151*, 153, *166*, *176-192*, *193-203*, 212, *336*, *337*, 394.
Rosenberg, A. E., 172, 251, 340.
Rosenblith, W. A., 303.
Ross, H. F., *92*, 104, 107, 108, 136, 145, 201, 243.
Rossi, G., 161.
Rupert, A., 153, 178, 181.
Rutherford, W., 136.
Sachs, M. B., 164, 177, 193, 198-202, 241, 275, 276, 434.
Saulnier, C., 166.
Sayers, B. McA., 213.
Scharf, B., *166*, *174*, *266*, *289*, *373*, 384, 385, 406, *413*, *455-462*.
Scholl, H., 357, 455.
Schouten, J. F., *41-58*, *151-152*, 172, *191*, 232, 251, 253, 255, 256, *263*, 267, 268, 271, 281, *289*, 294, 296, 340, *348*, 398, *411*, 420, *472*.
Schroeder, M. R., 254, 261, 403.
Schubert, E. D., 433.
Schügerl, K., *79*, *80*, *189*, *277*, *290*, *373*, *415-426*.
Schuknecht, H. F., 12, 136, 137, 441.
Schwarze, D., 460.
Schwartzkopff, J., *79*, *105*, *132*, *150*, *158*, 159, *227*, *300*.
Seebeck, A., 41-44, 46, 53, 57, 58, 398.
Sergeant, D., *412*.
Shaw, E. A. G., 359.
Siebert, W. M., 60, 62, 63, 337, 356, 360.
Simmons, F. B., 141, 150.
Six, P. D., 218.

Sjöstrand, F., 24.
Slawson, A. W., 401, 402, 408.
Small, A. M., 172, 252, 253, 266, 292, 329, 427, 428, 434.
Smith, C. A., 10, 12, 24.
Smoorenburg, G. F., 37, 144, *149*, *188*, *201*, *226*, *246*, *263*, *267-277*, *288*, *289*, 293, 296, 297, *348*.
Sohmer, H., 161.
Solomon, L. N., 466.
Somerville, T., 292.
Spoendlin, H., *2-40*, 67, 68, 434.
Spreng, M., 74.
Stange, G., 68, 72.
Steeneken, H. J. M., 287, 391, 395, 403, 473.
Stevens, S. S., 141, 279, 282, 397, 413.
Stopp, P. E., 144.
Stumpf, C., 400-402, 404, 423.
Suga, N., 183, 193, 434
Supa, M., 291.
Sweetman, R. H., 218-220.
Swets, J. A., 357.
Tartini, G., 230.
Tasaki, I., 94, 104, 107, 108, 113, 118, 121, 126, 153, 178, 220, 221.
Taylor, K., *81-93*.
Taylor, M., 330.
Teas, D. C., 104, 123, 126.
Terhardt, E., 55, 92, 213, *264*, 272, *277*, *278-290*, *338*, 387, 419, *454*.
Theissing, J., 68, 72.
Thomson, W., 404.
Thurlow, W. R., 252, 253, 292, 293.
Tonndorf, J., 8, 60, 63, 103, 104, 118, 119, 201, 218, 243.
Trahiotis, C., 35.
Tsuchitani, C., 144.
Turnbull, W. W., 355.
Vilstrup, Th., 7.
Wachsmuth, D., 161.
Walliser, K., 55, 253, 254, 265, 278, 281, 282, 284-286, 293.
Walzl, E. M., 144.
Wang, W. S. I., 466.
Ward, P. H., *94-106*, 127.
Ward, W. D., 70, 71, *159*, *189*, 244, 284, *317*, 435.
Wegel, R. L., 404, 445.
Weiss, A. P., 402.
Weiss, Th. F., 204, 205.
Weissmann, S. M., 461.
Wellek, A., 277.
Wersäll, J., 5, 12, 67, 68.
Wesendonk, K. von, 402.
Wetherill, G. B., 438.
Wever, E. G., 45, 94, 104, 120, 136, 218, 278, 397, 427.
Whitfield, I. C., 56, 92, 104, *106*, 107, 108, *131*, *136-152*, *159*, *173*, *189*, 201, *202*, *215*, 243, *265*, *453*.
Wickelgren, W. A., 466.
Wiederhold, M. L., 161, 448.
Wien, M., 61, 76.
Willis, R., 398, 401, 402.
Wilson, J. P., 39, 90, *91*, *174*, 252, *266*, *303-318*, *327*, 360.
Woolsey, C. N., 144.
Worden, F. G., 74.
Worthington, D. W., 220.
Wundt, W., 44, 46, 250.
Wüstenfeld, E., 63, 68.
Yantis, P. A., 404.
Young, T., 230.
Zwicker, E., 49, 57, 58, 92, *159*, *160*, *201*, *202*, *213*, 222, 225, 231, 232, 272, *277*, 280, 286, *288*, *289*, 303, *317*, *337*, *349*, *361*, *376-396*, 397, 404, 406, 408, 437, *443*, 446, 455, 458, 459, *462*.
Zwislocki, J. J., 37, 55, 56, 57, 60, 104, 122, 124, *149*, *151*, *166*, 172, *190*, *202*, 207, 212, *276*, *338*, *347*, *361*, 371, *373*, 376 384, 432, 434, 436, *445-454*, 459, 460, *461*.

(Italicized page numbers indicate the presence of text, rather than references to publications.)

SUBJECT INDEX

Acid mucopolysaccharides, role in mechano-electric transduction, 7, 8, 10, 38, 39.
Action potentials
  compound AP evoked by transients, 121-122.
  compound AP in adaptation, 71.
  initiation of AP, 71-74, 204-206.
Adaptation, 70, 73, 304, 311, 317-318, 459.
After-image, auditory, 303-312, 315-318.
Amplitude JND, 381-384, 396.
  effect of duration, 350-361.
Auditory pathway
  electrical stimulation, 141-142, 144, 150-151.
  preservation of periodicity, 137-138.

Basilar membrane
  amplitude at threshold, 86.
  damping, 61.
  mechanics, 60-67, 81-89.
  shift of vibration pattern (SPL), 93.
  spectral amplitude response, 61-66, 84-86, 90-92.
  spectral phase response, 86.
  structure, 3.
Beats (*see also* roughness sensation), 419, 425.
  of mistuned consonances, 189-190, 404.
  binaural, 454.

Clock mechanisms, 138, 140, 144, 157, 159.
Cochlea
  biophysics, 60-76.
  dimensions, 2.
  electrophysiology (*see also* AP, CM, SP), 67-76
  mechanics, 60-67, 81-89.
Cochlear microphonics
  in adaptation, 71.
  after-oscillations, 120.
  combination tones, 218-229.
  in different species, 100-101.
  early deflexion, 119-120.
  place of generation, 38-39, 107.
  properties, 38.
  recording techniques, 108-109, 117, 118, 124.
  shift of spatial distribution (SPL), 95-96, 151, 192.
  spatial amplitude distribution, 95-97, 109-116.
  spatial phase distribution, 97, 109-116.
  spectral amplitude response, 100-101.
  spread of potentials, 103, 108-117, 124-125.
  transients, 118-125.
Cochlear nucleus
  activity in dichotic stimulation, 161-167.
  different types of neurons, 168-174.
  firing rate (SPL), 453-454.
  responses to clicks, 168-174.
  responses to two-tone stimuli, 143-144.
Coincidence mechanisms, 140, 157.
Colouration, 297-299, 312-314, 316.
Combination tones
  difference tone, 45, 49-50, 230-233.
  of type $f_1 - n(f_2 - f_1)$, 49-51, 88, 149, 230-233, 237, 246, 247, 261, 263, 272-276, 297, 385-386, 443-444.
  as origin of beats, 404.
  physiological correlate, 187, 210-211, 218-229, 233-247.
Concords, perception of, 415-426.
Consonance
  physiological correlate, 158, 159.
  theory, 421.
Critical band, 377-384.
  in concord perception, 417-420.
  in loudness summation, 384-385, 455-462.
  in relation to pitch, 316.
  in relation to roughness, 280, 286, 288.
  in relation to timbre, 408.

Delay-line mechanisms, 139-140, 147, 150.
Detection
  of basilar membrane movements, 87-88, 91, 92.
  of pulse intervals, 138-140, 144-145, 147, 150.
Degeneration
  of afferent dendrites, 21.
  of efferent nerve fibres, 12-23, 37.
Difference tones – *see* combination tones
Diplacusis, binaural, 362-365, 368-373.
Doublicity theory, 68.

Formants, 57, 401, 463-466, 469, 472.
Frequency analysis (*see also* basilar membrane, critical band, place theory)

in central masking, 445-446, 452-453.
in concord perception, 415-417.
in masking, 376-384, 386-393.
in relation to dominance region, 257-261, 294, 312.
in relation to timbre, 399-400, 408-409, 471.
selectivity, 395.
structural basis, 2-40.
Frequency discrimination
in dichotic stimulation, 166.
role of efferent innervation, 34, 37, 164.
Frequency JND, 382-384, 395, 396.
effect of duration, 350-360.
Fundamental
fundamental tone, 44-46
generation, 187.
missing fundamental, 45, 149.
Funneling, 70.
Fusion, 422-423.

Geniculate body
frequency response curves, 155-156.
periodic spontaneous activity, 153.

Hair cells
afferent nerve supply, 21-29, 34.
difference between inner and outer hair cells, 34, 68-70.
directional sensitivity, 8-10, 39.
efferent nerve supply, 11-23, 34-35, 37-38.
function of cell body, 10.
receptor pole, 3.

Inferior colliculus, time-locked responses, 137, 153-155.
Inhibition, 67-68, 88, 91, 92.
for swept frequencies, 173.
frequency selective, 161-164, 166, 171-172, 177, 193-202.
in contralateral stimulation, 161-164, 166.
latency times, 165-166.
lateral inhibition, 177, 193-202, 386-388, 395, 427, 432, 434-437, 440, 461-462.
postsynaptic inhibition, 20-21.
presynaptic inhibition, 20-21.
two-tone inhibition, 177, 193-202, 395.
Initial segment of cochlear fibres, 29, 32, 34, 37, 38.
Innervation
afferent, 21-29, 34.
efferent, 11-23, 34-35, 37-38, 68, 161, 164.
multiple, 26, 67.
Interaural cross talk, 161-162, 325-326.
Isophones, fine structure, 370-371.

Jitter, 340-341, 347-349.
detection threshold, 329-338, 341-347.
internal, 345-348.

Kinociliar basal body, 8-10, 39.

Lateralization, auditory, 319-323, 327-328, 337.
Loudness, 384-385, 455-461.

Mach bands, 427-444.
Masking, 378-379, 428-434, 436-444.
central, 445-454.
in single-fibre responses, 181, 193-202, 395.
Mechano-electric transduction, 3, 5-10, 38-39.
role of acid mucopolysaccharides, 7, 8, 10, 38-39.
Models
autocorrelation, 261, 298, 302, 311.
explaining multi-maxima frequency responses, 157-158.
of cochlear action, 204-206.
of cochlear detection, 356-358, 360.
of cochlear mechanics, 62-63, 371.
of combination tone generation, 230-244, 246.
of hair cell function, 69-70.
of periodicity detection, 138-140, 144-145, 147, 150-152.
of pitch perception, 46-58.
of spectral filtering, 386-392, 395-396.
Mössbauer technique, 81-84.

Nerve fibres
afferent terminals, 24, 26, 68.
degeneration of efferent fibres, 12-23, 37.
diameter of afferent fibres, 28-29.
diameter of efferent fibres, 38.
efferent terminals, 11-12, 16, 20, 34, 37-38, 68.
initial segment, 29, 32, 34, 37, 38.
myelin sheath, 11, 29.
outer spiral fibres, 14, 25-26.
upper tunnel radial fibres, 14, 21-23.
Nonlinearity
in cochlear electrophysiology, 104, 218-229.
in detection, 209-211, 213-215.
in loudness, 384.
in relation to combination tones, 45, 49-50, 56, 57, 88, 149, 231, 235-239, 242-244, 246, 263, 272, 385, 404, 413.
in spectral filtering, 201-202, 380, 382.

Olivo-cochlear bundle
  electrical stimulation, 161.
  transsection, 12-23.
Organ of Corti
  afferent innervation, 21-29, 34.
  efferent innervation, 11-23, 34-35, 37-38, 68.
  supporting structures, 2-3.

Pattern recognition, 58.
  in pitch perception, 145, 152.
Phase sensitivity, 288-289, 409, 413.
  effect on pitch, 251, 252, 254, 264, 267, 293, 294-295, 312.
  effect on timbre, 391-392, 395-396, 399-400, 403-405.
Pitch
  of after-image, 311-316.
  of AM signals, 253-256, 261, 264, 267-268, 297, 373.
  ambiguity of pitch, 47.
  central pitch, 319-328.
  complex sounds, 260, 281-286.
  dominance region, 49, 257-261, 264-265, 268, 286, 290, 294-296, 312-314, 316.
  effect of added noise, 282.
  effect of SPL, 54, 151, 282-283.
  existence region, 47, 285-286, 289-290.
  first effect of pitch shift, 46, 253.
  of inharmonic complexes, 253-254, 256, 268-274, 281.
  of narrow-band signals, 253.
  of periodically gated noise, 266.
  periodicity pitch, 46, 142, 277, 281-286, 289-290, 299, 337, 340, 347-348, 389.
  physiological mechanisms, 151, 172.
  of pulse pairs, 252, 292, 294-295, 297.
  of pulse trains, 251, 257-260, 267, 340-341, 348, 372-373.
  of quasi-FM signals, 254-255, 263-264, 267.
  repetition pitch, 252, 291-302, 316.
  role of fine structure, 255, 261, 263, 265, 269-271, 294-297.
  role of subharmonics, 254, 256, 281-286.
  role of temporal envelope, 253, 281, 285, 292.
  role of time pattern, 55-57, 268, 271-272, 276.
  role of waveform, 257-258, 260-261, 271-272, 274, 294-296, 343-344.
  second effect of pitch shift, 46-47, 50, 256, 261, 263-265, 271, 296-297.
  time separation pitch, 252, 292, 299.
  of two-tone complexes, 267-277, 297.

Place-pitch theory, 46, 54-57, 144, 145, 147-148, 240-242, 316, 394, 427-428.

Refractory period, 180, 205, 208.
Residue phenomenon (see also pitch), 41-53, 56, 251, 253, 256, 264, 267, 277, 285, 289, 340, 372, 373.
Resonance theory, 60-61.
Roughness sensation, 278-281, 285-290, 399, 419-422, 425.

Sharpening mechanisms, 29, 75-76, 80, 394-395, 428, 432, 444, 461-462.
  mechanical, 66-67.
  neural, 67-68, 70, 87-88, 91, 92.
Shearing motion, 3, 7, 8, 102.
Single-fibre responses
  effect of SPL, 151, 176-177.
  firing rate (SPL), 148, 237-238, 447, 449-454.
  frequency response curves, 155-158, 176-178, 190, 192, 194-195, 207-208, 212-214.
  to noise plus delayed noise, 300-302.
  periodic discharges, 142, 153, 159, 160.
  threshold, 37, 38, 191-192, 204, 210, 213-215.
  time-dependent responses, 138, 140.
  time-locked responses, 136-138, 143-144, 153-155, 168-174, 178-192, 193-202, 204-215, 234-235, 239-242, 394.
  timing accuracy, 336-337, 339-340.
  tuning curves, 90-91, 205, 207, 394, 395.
Spatial distribution (see also CM and SP), over fibre array, 136, 144-145.
Spontaneous activity, 37, 153, 160, 161.
Stereocilia
  arrangement, 4, 10, 39.
  attachment to hair cells, 4.
  connection with tectorial membrane, 4-5.
Summating potentials
  change of polarity, 98, 99, 102, 104-106, 128.
  definition of polarity, 131.
  difference between +SP and −SP, 127-132.
  effect of stimulus duration, 98, 99, 106.
  origin and function, 104, 130-131.
  relation to decay of after-image, 317.
  remote masking of SP, 128-129, 131, 132.
  spatial distribution, 97, 98, 127-130.
  spectral response, 101-102, 127-130.

Tectorial membrane
  connection with stereocilia, 4-5.

polarization, 75, 79.
structure, 5.
Threshold
  fine structure, 366-373.
  for nerve fibres, 37, 38, 204, 210, 213-215.
  threshold shifts, 70-73.
Timbre, 264, 298, 311, 397-400, 411-414, 420-422.
  effect of amplitude pattern, 400-403, 405-409, 412-413, 463-473.
  effect of phase pattern, 391-392, 395-396, 403-405, 413.

Time structure, role of, 123, 316, 337-338, 339-340, 343, 349, 350, 360, 389-393, 394, 409, 417-419.
Travelling wave theory, 60-63.
Tuning curves – *see* single-fibre responses

Vibrato, 422.
Vigilance, influence of, 360-361.
Vowel perception, 401-402, 463-473.